AF801481

Extracellular Vesicle: Biology and Translational Application

Zongjin Li
Editor

Extracellular Vesicle: Biology and Translational Application

Springer

Editor
Zongjin Li
Department of Pathophysiology
Nankai University School of Medicine
Tianjin, China

ISBN 978-981-97-5535-6 ISBN 978-981-97-5536-3 (eBook)
https://doi.org/10.1007/978-981-97-5536-3

© The Editor(s) (if applicable) and The Author(s), under exclusive license to Springer Nature Singapore Pte Ltd. 2024
This work is subject to copyright. All rights are solely and exclusively licensed by the Publisher, whether the whole or part of the material is concerned, specifically the rights of translation, reprinting, reuse of illustrations, recitation, broadcasting, reproduction on microfilms or in any other physical way, and transmission or information storage and retrieval, electronic adaptation, computer software, or by similar or dissimilar methodology now known or hereafter developed.
The use of general descriptive names, registered names, trademarks, service marks, etc. in this publication does not imply, even in the absence of a specific statement, that such names are exempt from the relevant protective laws and regulations and therefore free for general use.
The publisher, the authors and the editors are safe to assume that the advice and information in this book are believed to be true and accurate at the date of publication. Neither the publisher nor the authors or the editors give a warranty, expressed or implied, with respect to the material contained herein or for any errors or omissions that may have been made. The publisher remains neutral with regard to jurisdictional claims in published maps and institutional affiliations.

This Springer imprint is published by the registered company Springer Nature Singapore Pte Ltd.
The registered company address is: 152 Beach Road, #21-01/04 Gateway East, Singapore 189721, Singapore

If disposing of this product, please recycle the paper.

Preface

Extracellular vesicles (EVs) are small, lipid-bound vesicles that carry cargo and can be released by virtually all cell types. Over the past decades, the field of EV research has experienced exponential growth. Once dismissed as mere cellular debris, EVs are now recognized as sophisticated messengers, facilitating the transmission of crucial information between cells and orchestrating a myriad of biological processes. Notably, stem cell-derived EVs hold particular promise in regenerative medicine. Rich in the functional components derived from stem cells, these EVs offer an alternative approach to cell therapy for tissue repair and regeneration. This transformative shift in regenerative therapy has sparked a surge in research aimed at uncovering the therapeutic potential of EVs, positioning them as promising tools to propel advancements in regenerative medicine.

The purpose of this book is to provide an overview of EVs, covering their biology, function, and potential applications. We begin by exploring the intricate biology of EVs, their cargo, and their unique ability to modulate cellular processes. Furthermore, we highlight the therapeutic potential of EVs in various diseases. Additionally, we discuss approaches for the bionanotechnological development of artificial EVs for theranostics. Large-scale EV isolation methods and their preclinical applications are also addressed.

I would like to extend my deepest appreciation and gratitude to each of the chapter authors for the time and effort spent on writing informative reviews on their respective areas. We would also like to extend our gratitude to Ms. Yin Chen and Ms. Kavitha Jayakumar, editors at Springer Nature, for their help and advice in putting together this book.

Tianjin, China Zongjin Li

Contents

Chapter 1
Extracellular Vesicles: Tiny Packages with Big Potential

Xiaoniao Chen and Zongjin Li

Abstract With their ability to self-renewal and differentiation, stem cells have been widely applied for regenerative therapy. However, ethical concerns, potential tumorigenesis, and immunogenicity of stem cells limit their clinical application. Extracellular vesicles (EVs), especially derived from stem cells, contain the functional compositions from stem cells that have been widely employed in tissue repair and regenerative medicine, offering an alternative to stem cells for regenerative therapy. Recent studies indicate that EVs exhibit similar therapeutic effects compared to their donor cells, while minimizing the drawback associated with stem cell therapy. Unlike their cell counterparts, EVs are stable enough for long-term storage and can be easily engineered and modified for drug loading and targeting therapy. In this book, we will discuss recent advances in EV secretion, cargo packages, and translational application, and emphasize the multifaceted roles of EVs in regenerative therapy.

Keywords Extracellular vesicles (EVs) · Stem cells · Regenerative therapy · Intercellular communication · Immunomodulation · Cell-free therapy

1 Introduction

With the ability of self-renewal and differentiation, stem cells hold great promise for regenerative therapy by healing and regenerating damaged tissues. Extracellular vesicles (EVs) derived from various cells have shown great promise in regenerative

X. Chen
Department of Ophthalmology, The Third Medical Center of Chinese PLA General Hospital, Beijing, China

National Key Laboratory of Kidney Diseases, Chinese PLA General Hospital, Beijing, China

Z. Li (✉)
Department of Pathophysiology, Nankai University School of Medicine, Tianjin, China
e-mail: zongjinli@nankai.edu.cn

© The Author(s), under exclusive license to Springer Nature Singapore Pte Ltd. 2024
Z. Li (ed.), *Extracellular Vesicle: Biology and Translational Application*,
https://doi.org/10.1007/978-981-97-5536-3_1

medicine with similar therapeutic potential compared to their donor cells [1, 2]. EVs, derived from stem cells, including embryonic stem (ES) cells, induced pluripotent stem (iPS) cells, and mesenchymal stem cells (MSCs), have been investigated in clinical for a variety of diseases, including COVID-19, lung diseases [3]. This paradigm shift in regenerative therapy has fueled a surge in research dedicated to unraveling the therapeutic applications of EVs, positioning them as promising tools to advance regenerative therapies.

2 Definition and Classification of EVs

EVs encompass a heterogeneous group of membranous structures that can be released by virtually all kinds of cell types. EVs can be classified into exosomes (30–150 nm), microvesicles (100–1000 nm), and apoptotic bodies (1000–5000 nm), which are distinguished by biogenesis and size [4]. EVs, which range from exosomes to microvesicles and apoptotic bodies, constitute a diverse group of membranous structures released by cells into the extracellular environment and interact with neighboring or remote cells. Exosomes, also called small EVs, originate from endosomal pathways, and microvesicles bud directly from the cell membrane and play a distinctive role in cellular communication and regenerative processes [5].

The complex contents of EVs are a critical factor in their functional diversity, which is varied upon cell type and the microenvironment of the cell. EVs encapsulate a myriad of bioactive molecules, including proteins, peptides, lipids, miRNA, and mRNA [6]. The specific cargo composition of EVs varies depending on the cell of origin and physiological state, influencing the therapeutic application and functional impact of EVs on recipient cells.

3 Intercellular Communication of EVs

The fundamental role of EVs lies in their ability to facilitate intercellular communication. These intricate processes play a pivotal role in modulating various physiological and pathological responses within organisms [7]. By shuttling bioactive molecules from donor to recipient cells, EVs modulate a spectrum of cellular processes, including proliferation, differentiation, and immune responses. This communication is not limited to local interactions but extends systemically, emphasizing the potential of EVs to act as messengers in the bloodstream that cross various biological barriers, including the blood–brain barrier [8].

Intercellular communication through EVs begins with their biogenesis within parent cells. EVs are generated through different mechanisms, mainly involving the endosomal pathway for exosomes and direct growth from the cell membrane for microvesicles [9]. Once formed, these vesicles encapsulate various contents, including proteins, lipids, and nucleic acids, which are subsequently transported to

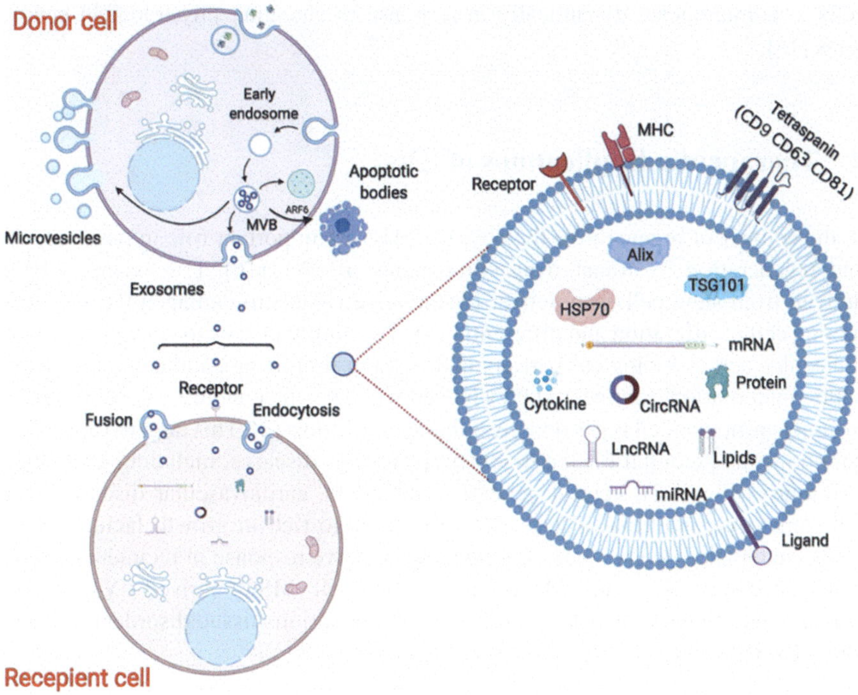

Fig. 1.1 Extracellular vesicles (EVs) serve as mediators of cell communication. Target cells internalize EVs through endocytosis and fusion, while membrane-bound receptors facilitate EV uptake and regulate cellular signaling pathways. (Reproduced under a Creative Commons Attribution 4.0 International License from Chen K, Li Y, Xu L, Qian Y, Liu N, Zhou C, et al. Comprehensive insight into endothelial progenitor cell-derived extracellular vesicles as a promising candidate for disease treatment. Stem cell research & therapy. 2022;13(1):238)

recipient cells. EVs can be uptaken by target cells through various mechanisms, including membrane fusion, endocytosis, micropinocytosis, and phagocytosis (Fig. 1.1). The specific uptake mechanism often depends on the characteristics of both EVs and the recipient cells [6]. Cargo carried by EVs plays a central role in dictating their functional impact on recipient cells. This cargo includes proteins/peptides, which can alter cellular signaling pathways; lipids, which contribute to membrane remodeling; and nucleic acids/mRNA/microRNA, which can modulate gene expression [10].

EVs participate in immune regulation by modulating the activity of immune cells [11–13]. For example, EVs released by immune cells can carry signaling molecules that promote or suppress immune responses, influencing the behavior of neighboring immune cells. This immunomodulatory function is crucial for maintaining immune homeostasis and responding appropriately to various challenges. The microenvironment profoundly influences EV-mediated communication. Cellular stress, inflammation, and exposure to specific stimuli can modulate the content of EVs and alter their functional impact on recipient cells [14]. This adaptability allows cells to communicate dynamically in response to changing physiological conditions [15].

4 Therapeutic Implications of EVs

In the context of regenerative therapy, EV plays an important role in tissue repair and regeneration by transferring components of EVs [16]. For example, EVs derived from stem cells can deliver regenerative signals to damaged tissues, promoting cell proliferation and differentiation. The ability of EVs to convey bioactive molecules across biological barriers makes them promising candidates for therapeutic intervention. Engineered EVs loaded with specific contents can be designed to target particular cell types or deliver therapeutic drugs [2]. This targeted approach has shown its potential for the treatment of various diseases, including acute kidney injury (Fig. 1.2), neurodegenerative disorders, cardiovascular diseases, and inflammatory conditions. These EVs carry a cargo rich in growth factors, cytokines, and nucleic acids, orchestrating a regenerative response in recipient cells of damaged tissues or organs. Mesenchymal stem cell (MSC)-derived EVs, in particular, have shown potent regenerative effects in various tissue disorders or damages [16–18].

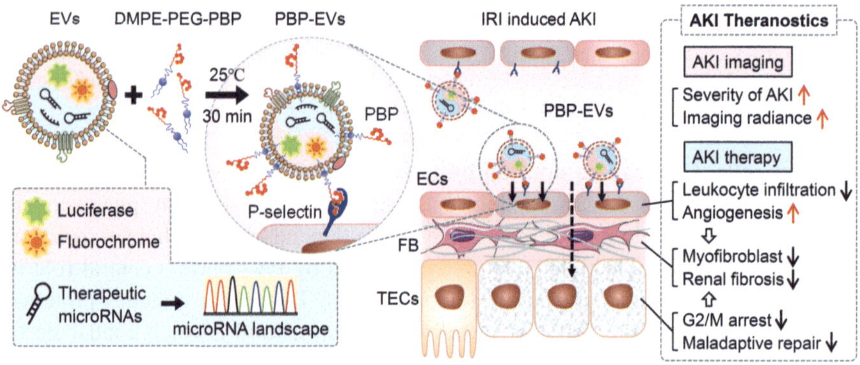

Fig. 1.2 Targeted approach for the treatment of acute kidney injury. P-selectin binding peptide (PBP)-engineered EVs (PBP-EVs) with imaging and therapeutic functions are developed for targeting injured endothelial cells. (Reproduced under a Creative Commons Attribution 4.0 International License from Zhang K, Li R, Chen X, Yan H, Li H, Zhao X, et al. Renal Endothelial Cell-Targeted Extracellular Vesicles Protect the Kidney from Ischemic Injury. Adv Sci. 2023;10(3):e2204626)

5 Therapeutic Effects of EVs Through Immunomodulation

The immunomodulatory properties of EVs add another layer of complexity to their regenerative potential [19]. EVs released by various cell types, including MSCs and regulatory T cells (Tregs), exhibit potent anti-inflammatory effects. These EVs carry a cargo of anti-inflammatory cytokines, such as interleukin-10 (IL-10) and transforming growth factor-β (TGF-β), as well as other immunomodulatory molecules [20–22]. When in contact with immune cells, these EVs can suppress the production of pro-inflammatory cytokines, dampening excessive inflammatory responses. Furthermore, EVs can carry antigens, promoting antigen-specific T cell responses and influencing the adaptive immune system. For example, Treg-derived EVs can induce a regulatory phenotype in target immune cells, which contributes to immune tolerance. This property has implications for the treatment of autoimmune diseases and conditions in which immune hyperactivity must be controlled [1, 23].

EVs interact with various types of immune cells, including macrophages, dendritic cells (DCs), and B cells, influencing their functions. EVs can modulate the maturation and activation of DCs, affecting antigen presentation and subsequent T cell responses. Additionally, EVs can alter macrophage polarization, shifting the balance between the pro-inflammatory M1 phenotype and the anti-inflammatory M2 phenotype. The immunomodulatory properties of EVs make them potential candidates for the treatment of autoimmune diseases. By suppressing aberrant immune responses and promoting regulatory phenotypes, EVs offer a targeted approach to modulate the dysregulated immune system observed in conditions such as rheumatoid arthritis, multiple sclerosis, and systemic lupus erythematosus (SLE) [24].

The immunomodulatory properties of EVs have implications for organ transplantation. For example, MSC-derived EVs can suppress immune rejection responses and promote graft tolerance. This finding suggests a potential therapeutic role for these cells in improving organ transplantation and reducing the need for immunosuppressive drugs [25].

6 Challenges and Future Directions

One notable advantage of EV-based therapies lies in their cell-free nature. Unlike traditional cell-based therapies, which can pose challenges related to cell viability, immunogenicity, and tumorigenicity, EVs offer a safer and more scalable alternative [8]. Eliminating live cells from the therapeutic equation addresses ethical concerns and improves the translational potential of EV-based treatments. The inherent properties of EVs contribute to their efficient delivery to target tissues. EVs exhibit stability in various biological fluids and resist degradation by enzymes. Furthermore, their nanoscale size facilitates their ability to cross biological barriers, including the

Table 1.1 The advantage of EVs versus stem cells in regenerative therapy

Feature	EVs	Stem cells
Production and purification	Easier	More complex
Stability	High	Low
Storage	Longer	Shorter
Immunogenicity	Lower	Higher
Delivery	Targeted	Less targeted
Drug delivery	Efficient	Less efficient
Therapeutic potential	Wide range	More limited
Ethical issues	No	Yes

blood–brain barrier, opening avenues for targeted delivery in neurological disorders (Table 1.1).

Despite promising strides in EV research, challenges related to standardization and characterization persist. Standardized isolation methods and clear definitions of EV subtypes are crucial to advance the field [26]. The heterogeneity of EV populations and variations in isolation techniques pose challenges in comparing and replicating results in multiple studies. A deeper understanding of the underlying mechanisms by which EVs exert their regenerative effects is essential to optimize therapeutic strategies. Elucidating the specific interactions between EV cargo and recipient cells, as well as the signaling pathways involved, will provide critical insight into improving the efficacy of EV-based therapies.

The scalability of EV production and the associated production costs represent practical challenges in translating EV-based therapies from the laboratory to the clinic [27, 28]. Developing cost-effective and scalable production methods is essential to realize the full therapeutic potential of EVs and make them available for widespread clinical use. Regulatory guidelines that address aspects such as safety, efficacy, and quality control are essential for navigating the path from experimental studies to approved clinical interventions.

7 Conclusions

In conclusion, EVs are at the forefront of regenerative therapy, offering a versatile and promising approach to tissue repair and regeneration. Their ability to mediate drug delivery, modulate immune responses, and serve as a cell-free alternative position in the landscape of regenerative medicine. Although challenges remain in clininal application, ongoing research is poised to reveal the full therapeutic potential of EVs, driving the evolution of regenerative therapies and opening new avenues for treating a variety of diseases, such as degenerative diseases, injury, and damage. EVs are a promising new therapeutic modality with the potential to revolutionize regenerative medicine. However, more research is needed to overcome the challenges associated with EV purification, characterization, targeting, and regulatory approval.

Acknowledgments This study was financially supported by open funding from the Nankai University Eye Institute (NKYKD202203), the Research Project on Skin Injury & Repair (BKJ21J016, 2021XB008), and the Tianjin Natural Science Foundation (22JCZXJC00170, 21JCZDJC00070).

References

1. Kalluri R, LeBleu VS. The biology, function, and biomedical applications of exosomes. Science. 2020;367(6478):eaau6977. https://doi.org/10.1126/science.aau6977.
2. Zhang K, Li R, Chen X, Yan H, Li H, Zhao X, et al. Renal endothelial cell-targeted extracellular vesicles protect the kidney from ischemic injury. Adv Sci (Weinh). 2023;10(3):e2204626. https://doi.org/10.1002/advs.202204626.
3. Liu L, Ye Y, Lin R, Liu T, Wang S, Feng Z, et al. Ferroptosis: a promising candidate for exosome-mediated regulation in different diseases. Cell Commun Signal. 2024;22(1):6. https://doi.org/10.1186/s12964-023-01369-w.
4. Thery C, Witwer KW, Aikawa E, Alcaraz MJ, Anderson JD, Andriantsitohaina R, et al. Minimal information for studies of extracellular vesicles 2018 (MISEV2018): a position statement of the International Society for Extracellular Vesicles and update of the MISEV2014 guidelines. J Extracell Vesicle. 2018;7(1):1535750. https://doi.org/10.1080/20013078.2018.1535750.
5. Hendrix A, Lippens L, Pinheiro C, Théry C, Martin-Jaular L, Lötvall J, et al. Extracellular vesicle analysis. Nat Rev Methods Prime. 2023;3(1):56. https://doi.org/10.1038/s43586-023-00240-z.
6. Chen K, Li Y, Xu L, Qian Y, Liu N, Zhou C, et al. Comprehensive insight into endothelial progenitor cell-derived extracellular vesicles as a promising candidate for disease treatment. Stem Cell Res Ther. 2022;13(1):238. https://doi.org/10.1186/s13287-022-02921-0.
7. van Niel G, Carter DRF, Clayton A, Lambert DW, Raposo G, Vader P. Challenges and directions in studying cell–cell communication by extracellular vesicles. Nat Rev Mol Cell Biol. 2022;23(5):369–82. https://doi.org/10.1038/s41580-022-00460-3.
8. van Niel G, Carter DRF, Clayton A, Lambert DW, Raposo G, Vader P. Challenges and directions in studying cell-cell communication by extracellular vesicles. Nat Rev Mol Cell Biol. 2022;23(5):369–82. https://doi.org/10.1038/s41580-022-00460-3.
9. Phinney DG, Pittenger MF. Concise review: MSC-derived exosomes for cell-free therapy. Stem Cells. 2017;35(4):851–8. https://doi.org/10.1002/stem.2575.
10. Dixson AC, Dawson TR, Di Vizio D, Weaver AM. Context-specific regulation of extracellular vesicle biogenesis and cargo selection. Nat Rev Mol Cell Biol. 2023;24(7):454–76. https://doi.org/10.1038/s41580-023-00576-0.
11. Hezam K, Mo R, Wang C, Liu Y, Li Z. Anti-inflammatory effects of mesenchymal stem cells and their secretomes in pneumonia. Curr Pharm Biotechnol. 2022;23(9):1153–67. https://doi.org/10.2174/1389201022666210907115126.
12. Ren J, Liu Y, Yao Y, Feng L, Zhao X, Li Z, et al. Intranasal delivery of MSC-derived exosomes attenuates allergic asthma via expanding IL-10 producing lung interstitial macrophages in mice. Int Immunopharmacol. 2021;91:107288. https://doi.org/10.1016/j.intimp.2020.107288.
13. Zhang K, Chen S, Sun H, Wang L, Li H, Zhao J, et al. In vivo two-photon microscopy reveals the contribution of Sox9+ cell to kidney regeneration in a mouse model with extracellular vesicle treatment. J Biol Chem. 2020;295(34):12203–13. https://doi.org/10.1074/jbc.RA120.012732.
14. Yu L, Wen H, Liu C, Wang C, Yu H, Zhang K, et al. Embryonic stem cell-derived extracellular vesicles rejuvenate senescent cells and antagonize aging in mice. Bioactive Mater. 2023;29:85–97. https://doi.org/10.1016/j.bioactmat.2023.06.011.
15. Maas SL, Breakefield XO, Weaver AM. Extracellular vesicles: unique intercellular delivery vehicles. Trends Cell Biol. 2017;27(3):172–88. https://doi.org/10.1016/j.tcb.2016.11.003.

16. Zhang K, Zhao X, Chen X, Wei Y, Du W, Wang Y, et al. Enhanced therapeutic effects of mesenchymal stem cell-derived exosomes with an injectable hydrogel for Hindlimb ischemia treatment. ACS Appl Mater Interfaces. 2018;10(36):30081–91. https://doi.org/10.1021/acsami.8b08449.
17. Zhang C, Shang Y, Chen X, Midgley AC, Wang Z, Zhu D, et al. Supramolecular nanofibers containing arginine-glycine-aspartate (RGD) peptides boost therapeutic efficacy of extracellular vesicles in kidney repair. ACS Nano. 2020;14(9):12133–47. https://doi.org/10.1021/acsnano.0c05681.
18. Pham PV. Concise review: extracellular vesicles from mesenchymal stem cells as cellular therapy. Biomed Res Ther. 2017;4(08):98. https://doi.org/10.15419/bmrat.v4i08.287.
19. Robbins PD, Dorronsoro A, Booker CN. Regulation of chronic inflammatory and immune processes by extracellular vesicles. J Clin Invest. 2016;126(4):1173–80. https://doi.org/10.1172/jci81131.
20. Li R, Wang C, Zhou M, Liu Y, Chen S, Chai Z, et al. Heparan sulfate proteoglycan-mediated internalization of extracellular vesicles ameliorates liver fibrosis by targeting hepatic stellate cells. Extracell Vesicle. 2022;1:100018. https://doi.org/10.1016/j.vesic.2022.100018.
21. Tang Y, Li J, Wang W, Chen B, Chen J, Shen Z, et al. Platelet extracellular vesicles enhance the proangiogenic potential of adipose-derived stem cells in vivo and in vitro. Stem Cell Res Ther. 2021;12(1):497. https://doi.org/10.1186/s13287-021-02561-w.
22. Lei X, He N, Zhu L, Zhou M, Zhang K, Wang C, et al. Mesenchymal stem cell-derived extracellular vesicles attenuate radiation-induced lung injury via miRNA-214-3p. Antioxid Redox Signal. 2021;35(11):849–62. https://doi.org/10.1089/ars.2019.7965.
23. Bobrie A, Colombo M, Raposo G, Thery C. Exosome secretion: molecular mechanisms and roles in immune responses. Traffic. 2011;12(12):1659–68. https://doi.org/10.1111/j.1600-0854.2011.01225.x.
24. Zhao Y, Wei W, Liu M-L. Extracellular vesicles and lupus nephritis - new insights into pathophysiology and clinical implications. J Autoimmun. 2020;115:102540. https://doi.org/10.1016/j.jaut.2020.102540.
25. Sailliet N, Ullah M, Dupuy A, Silva AKA, Gazeau F, Le Mai H, et al. Extracellular vesicles in transplantation. Front Immunol. 2022;13:800018. https://doi.org/10.3389/fimmu.2022.800018.
26. Ng KS, Smith JA, McAteer MP, Mead BE, Ware J, Jackson FO, et al. Bioprocess decision support tool for scalable manufacture of extracellular vesicles. Biotechnol Bioeng. 2019;116(2):307–19. https://doi.org/10.1002/bit.26809.
27. Paganini C, Capasso Palmiero U, Pocsfalvi G, Touzet N, Bongiovanni A, Arosio P. Scalable production and isolation of extracellular vesicles: available sources and lessons from current industrial bioprocesses. Biotechnol J. 2019;14(10):e1800528. https://doi.org/10.1002/biot.201800528.
28. Ter-Ovanesyan D, Norman M, Lazarovits R, Trieu W, Lee JH, Church GM, et al. Framework for rapid comparison of extracellular vesicle isolation methods. elife. 2021;10:10. https://doi.org/10.7554/eLife.70725.

Chapter 2
Regulation and Mechanisms of Exosome Cargo Selection and Secretion

Rui Li, Benjamin Hinnant, and Zongjin Li

Abstract Exosomes, which are double-layered nanoscale vesicles, belong to a category of extracellular vesicles and are released by diverse cells following the fusion of multivesicular bodies (MVBs) with the plasma membrane. They play a crucial role in intercellular communication by carrying specific cargo, including proteins, lipids, RNAs, and metabolites. Exosomes are involved in various processes, such as angiogenesis, apoptosis, antigen presentation, inflammation, cell homeostasis, proliferation, differentiation, and signaling. However, the molecular mechanisms underlying the sorting of cargo into exosomes and their secretion have not been fully elucidated. This review aimed to investigate the molecular mechanisms involved in exosome cargo sorting and secretion, along with techniques for labeling exosomes, which are valuable for studying exosome secretion and delivery. These investigations lay a solid foundation for understanding the physiological and pathological functions of exosomes and advancing their clinical applications.

Keywords Exosomes · Cargo · Sorting · Secretion · Labeling

1 Introduction

There are three main subcategories of extracellular vesicles (EVs): microvesicles, apoptotic bodies, and exosomes [1, 2]. Among these vesicles, exosomes are the smallest in size, with a diameter of approximately 30–150 nm, while microvesicles and apoptotic bodies typically have diameters greater than 100 nm [3]. Exosomes

R. Li
School of Medicine, Nankai University, Tianjin, China

B. Hinnant
Division of Biological Sciences, University of California San Diego, La Jolla, CA, USA

Z. Li (✉)
Department of Pathophysiology, Nankai University School of Medicine, Tianjin, China
e-mail: zongjinli@nankai.edu.cn

© The Author(s), under exclusive license to Springer Nature Singapore Pte Ltd. 2024

Z. Li (ed.), *Extracellular Vesicle: Biology and Translational Application*,
https://doi.org/10.1007/978-981-97-5536-3_2

were initially defined as "membrane-enclosed vesicles" responsible for the elimination of cell waste by releasing unwanted proteins, lipids, or nucleic acids. Subsequent extensive studies have aimed to characterize exosomes and investigate their impact on both physiological and pathological conditions. Exosomes allow the exchange of cargo molecules with neighboring and distant cells through the extracellular environment, thus contributing to diverse biological processes in both physiological and pathological states, including immune responses, wound healing, and tumorigenesis [4].

The role of exosomes in intercellular communication is mediated mainly by the membrane of the exosome [5]. Membrane-bound molecules enable exosomes to target neighboring or remote cells by binding to the plasma membrane. Upon docking with their targets, exosomes can initiate receptor–ligand interactions, leading to subsequent signaling activation. Alternatively, they can be internalized by recipient cells through endocytosis, phagocytosis, or fusion with target cells to facilitate cargo transfer [6]. Exosomes induce autophagy and defend against infection by viruses (such as human cytomegalovirus, herpes simplex virus type 1, and poliovirus) by delivering chromosome 19 miRNA clusters from placental trophoblasts to non-placental cells [7]; mesenchymal stem cell-derived miR-21a-5p aids in preventing placental infection and reduces myocardial apoptosis, facilitating repair by down-regulating the protein levels of PTEN-related genes in the apoptosis signaling pathway (including Peli1, PDPD4, FasL) and providing cardiac protection [8, 9]. Additionally, exosomes derived from cancer cells, such as pancreatic cancer, breast cancer, and colon cancer, are enriched with glypican 1 (GPC1)-positive exosomes, which can be utilized for cancer detection and potentially reducing the number of GPC1-positive exosomes in patients to alleviate tumor burden [10–12]. Therefore, thorough investigation into the secretion and packaging mechanisms of exosomes can elucidate their specific roles in these diseases and establish a theoretical foundation for the development of pertinent treatment strategies. Additionally, exosomes have emerged as a prominent subject in biomedical research owing to their innate nanoscale dimensions and biocompatibility. Investigating the mechanisms of exosome secretion and packaging can optimize the production and functionality of exosomes, expand their potential as drug delivery vehicles, and offer new avenues for treating diverse diseases.

With the gradual development of various types of omics, the composition of exosomes can be accurately analyzed through proteomics, transcriptomics, lipidomics, and metabolomics [13, 14]. These studies also establish a solid foundation for comprehending exosome biogenesis, cargo sorting, and release processes. To further investigate exosome secretion and delivery, it is crucial to select suitable labeling technology to monitor the dynamic changes of exosomes in different environments, thus revealing the regulatory mechanism of their release. Furthermore, exosome labeling offers an effective approach to tracking their destination, targets, and interactions in vivo, leading to a more comprehensive understanding of their functions in physiological and pathological conditions. A comprehensive exploration of exosome secretion and packaging mechanisms, along with the use of appropriate labeling technologies, will uncover their potential implications in cell biology, disease etiology, and treatment, providing solid support for future biomedical research and clinical applications. This chapter provides a comprehensive summary of exosome cargo sorting, secretion, and labeling.

2 Cargo Sorting into Exosomes

The cargo loaded into exosomes mainly includes proteins, nucleic acids, lipids, and metabolites (Fig. 2.1). Essentially, exosomes mediate for autocrine, paracrine and endocrine intracellular or intercellular communication, depending on what is loaded in the exosomes [15]. The proteins loaded into exosomes consisted primarily of tetraspanins (CD9, CD63, CD81, and CD82), ESCRT proteins involved in transport (ALIX, Tsg101), integrins, heat shock proteins (Hsp90, Hsp70), actin, and flotillins. Although some proteins (such as Hsp, CD63, the endosomal sorting complex required for transport (ESCRT), and cytoskeletal components) are relatively common in all exosomes, others (such as MHC class I and class II) are present in some

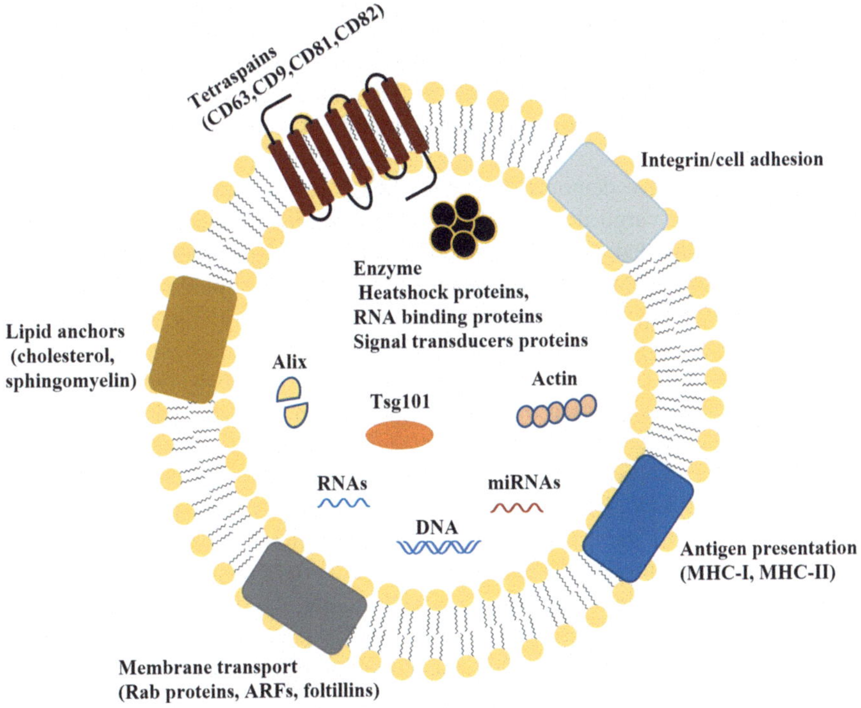

Fig. 2.1 Composition of exosomes. A schematic diagram illustrating the loading of cargo into exosomes is shown. The surface proteins of exosomes mainly consist of immunomodulatory proteins, integrins, and tetraspanins. In addition, the lumen of exosomes contains various cargo molecules, including mRNAs, DNAs, miRNAs, ESCRT proteins, and cytosolic proteins

unique cell types [16]. The exosome composition also includes nucleic acid complexes such as DNA, mRNA, and noncoding RNA, with miRNA being one of the most abundant noncoding RNA species [17]. Furthermore, the bilayer membrane of exosomes contains lipid components such as sphingomyelin, cholesterol, and ceramides, which can affect cargo sorting, exosome secretion, and signal transduction [18]. The content of exosomes plays a crucial role in intercellular communication, and the use of exosomes for accurate diagnosis and treatment of diseases has become a prominent research area. Regarding the sorting mechanism of exosomal cargo, exosomes carry cargo mainly through ESCRT-dependent and ESCRT-independent mechanisms [19]. Furthermore, certain conditioned stimuli, such as hypoxia, can induce alterations in the content of exosomes.

The ESCRT complex plays a crucial role in cargo sorting, as well as in membrane budding, ILV formation, and MVB formation, which have been extensively documented. The ESCRT complex comprises the subcomplexes ESCRT-0, -I, -II, and -III, along with the ATPase VPS4 and the auxiliary protein ALIX from the ESCRT-III complex [20]. Proteins are usually loaded into exosomes with the help of ESCRT cargo. The ESCRT-0 complex, composed of hepatocyte growth factor-regulated tyrosine kinase substrates (HRS) and STAM, contains ubiquitin-binding domains (UBDs) capable of recognizing and binding monoubiquitinated or polyubiquitinated cargoes [21]. Both ESCRT-I and ESCRT-II contain UBD, whereas the ESCRT-III complex does not contain the UBD [22]. In summary, the ESCRT complex plays a pivotal role in the recognition of ubiquitinated cargoes, the regulation of membrane shape, the sorting of cargoes in and out, and the loading of cargoes into ILVs [23].

Recent studies have demonstrated that ESCRT complexes are not exclusively involved in sorting cargo through ubiquitination-dependent mechanisms. Certain membrane proteins that lack ubiquitination can also be incorporated into ILVs. Interleukin-2 receptor beta (IL-2Rβ) interacts with the C-terminus of HRS within ILVs through an ubiquitination-independent mechanism. Similarly, ALIX can recognize cargo through three distinct ubiquitin-independent mechanisms, with the Syndecan-Syntenin-ALIX pathway being the most extensively investigated. Syndecan, a widely expressed transmembrane protein, interacts with the adapter protein syntenin. Syndecan and Syntenin can recognize the fibroblast growth factor receptor (FGFR) and lysyl-tRNA synthetase (KRS) in a ubiquitination-independent manner [24, 25]; ALIX binds to tetraspanins, such as CD9, CD63, and CD81, facilitating the sorting of these proteins into exosomes; and ALIX also binds to the YPX(3) L motif of protease-activated receptor-1 (PAR1) and purinergic receptor (P2Y1) through its V domain [26, 27]. Additionally, the Tsg101 subunit of the ESCRT-I complex specifically recognizes P(S/T)AP sequences via its amino-terminal ubiquitin E2 variant (UEV) domain, facilitating the sorting of proteins that harbor these sequences into exosomes [28].

In addition, exosomes are capable of cargo sorting in an ESCRT-independent manner, and these exosomes are commonly enriched in cholesterol, sphingolipids, phosphatidylserine, and ceramides, which are similar in composition to membrane lipid rafts. Furthermore, the presence of leopterin and caveolin in exosomes further highlights their importance as components of lipid rafts [18, 29]. Lipid rafts play a role in protein sorting, membrane curvature, and vesicle budding. The neutral sphingomyelinase 2 (nSMase2) ceramide pathway is an extensively studied mechanism for ESCRT-independent cargo sorting. nSMase2 serves as the key enzyme responsible for the conversion of SM to ceramide. The nSMase2-ceramide pathway regulates the sorting of ILVs or exosomes containing various cargos, including prion proteins and RNAs [30]. Similarly, stimulation with epidermal growth factor (EGF) triggers the activation of the epidermal growth factor receptor (EGFR), which in turn activates Rab31. Rab31 interacts directly with the membrane scaffolding protein flotillin, facilitating lipid raft-dependent sorting of exosomes via receptor tyrosine kinases. This process requires ceramide and cholesterol and does not require the involvement of the ESCRT complex [31]. Additionally, tetraspanins play a role in the ESCRT-independent sorting of various cargoes into ILVs. For example, the coupling of CD63 with apolipoprotein E promotes the formation of ILV and facilitates the sorting of the melanocyte-specific protein PMEL into ILVs [32]. Exosomes bind cargo mainly through the two pathways, but the detailed molecular mechanisms of sorting different cargoes have not been well characterized. Consequently, we will provide a brief overview of the mechanisms governing the sorting of proteins, nucleic acids, and lipids into exosomes.

2.1 *Protein*

Currently, exosome-associated proteins are the most characterized exosomal cargo, and revealing the mechanism of protein loading into exosomes is particularly urgent. While proteomic analysis methods can identify proteins enclosed within exosomes, they offer limited insight into the specific mechanisms governing protein packaging. Proteins can be loaded into exosomes through various mechanisms, which will be discussed in the following sections.

2.1.1 Post-Translational Modifications

After translation, proteins undergo enzymatic catalysis, resulting in chemical modifications known as post-translational modifications (PTMs). PTMs include precursor cleavage, disulfide bond formation, covalent addition, or removal of functional groups, etc. Certain PTMs are reversible, enabling rapid alterations in protein function by covalently adding or removing functional groups. Furthermore, PTMs can also impact diverse physicochemical properties of proteins, including polarity, solubility, stability, interprotein interactions, and subcellular localization [33]. During exosome loading, several post-translational protein modifications are involved in the sorting of proteins in exosomes, the most common being ubiquitination [34], phosphorylation [35], glycosylation [36], and sumoylation [37].

Ubiquitination

Ubiquitination is achieved by the covalent binding of regulatory protein ubiquitin molecules to lysine residues in proteins. Ubiquitinated proteins are guided into multivesicular bodies by ESCRT, and when cargo is loaded into multivesicular bodies, ubiquitin is cleaved by specific deubiquitinases. More than 600 ubiquitinated proteins have been identified in urine-derived exosomes using immunoaffinity purification and LC–MS/MS quantification. Components of ESCRT-0, such as HRS and STAM1, recognize ubiquitinated proteins and facilitate their transport into luminal vesicles of multivesicular bodies [38]. Furthermore, certain proteins also require ubiquitination to be loaded into exosomes. The Fas ligand (FasL), an apoptosis-inducing type II transmembrane protein, undergoes ubiquitination before being loaded into exosomes [39]. Specifically, FasL mutations that lack lysine residues are crucial for ubiquitination, thereby affecting MVB transport and exosome secretion [40]. In summary, ubiquitination plays a role in the packaging of proteins into exosomes.

Phosphorylation

Protein phosphorylation regulates diverse cellular and signaling processes, exerting a significant influence on protein activity, structure, and subcellular localization [41]. Researchers have isolated urine-derived exosomes through differential centrifugation and conducted pioneering phosphoproteomic studies of exosomes. They identified 19 phosphorylation sites on 14 proteins. Furthermore, proteomic studies analyzed phosphorylated proteins in exosomes derived from colon cancer cells and revealed more than 300 phosphorylated proteins and 1000 phosphorylated sites [42].

Glycosylation

Protein glycosylation is the addition of sugar moieties to proteins by glycosyltransferases and glycosidases in the endoplasmic reticulum and Golgi apparatus. Protein glycosylation plays a crucial role in various processes, including cell communication, cell adhesion, and immune response [43]. Protein glycosylation can occur in two primary forms: N- and O-chain glycosylation. Furthermore, glycosylated proteins were detected in exosomes, and analysis of exosomes derived from milk using LC–MS/MS revealed the presence of 44 O-glycans and 94 N-glycans. Furthermore, bovine milk exosomes were found to contain 86 glycoproteins, 767 complete glycopeptides, 113 glycosylation sites, and 80 glycans [44].

2.1.2 Protein–Protein Interactions

Some cargoes, especially cytoplasmic proteins, can be selectively loaded into luminal vesicles or exosomes through protein–protein interactions. For example, Hsp90 can be sorted into exosomes through interaction with Rab coupling proteins [45]; the fusion protein Rab22a-NeoF1 and its binding partner PYK2 can be sorted into MVBs through interaction with Hsp90, either through an ESCRT-dependent pathway or in exosomes [46]. GPR143 interacts with the HRS subunit of ESCRT-0, promoting the recruitment of cargo to endosomes and the recognition of protein cargo. This interaction regulates the protein composition of exosomes during MVB loading of intraluminal vesicles (ILVs) and plays a role in cancer metastasis [47]. Cytoplasmic Ago2 can also be loaded into ILVs or exosomes through the association with ALIX [48]. Furthermore, LAMP2A coordinates with HSC70 and other molecules to load cytoplasmic proteins containing KFERQ motifs into luminal vesicles and subsequently into exosomes, where HSC70 can interact with related cytoplasmic proteins containing KFERQ motifs [49].

2.1.3 Molecular Clustering

The clustering of proteins with other molecules has been poorly characterized, but these proteins can facilitate the loading of associated proteins into exosomes. During the maturation of the reticulocyte, the transferrin receptor (TfR) is detected in exosomes. Artificial clustering of TfRs with lectins disrupts the TfR recycling process, leading to increased secretion of TfRs in exosomes [50]. Furthermore, artificial targeting of oligomeric cytoplasmic proteins to the plasma membrane facilitates the budding of these proteins into exosomes [51]. In summary, the clustering of proteins with other molecules can facilitate the sorting of certain proteins into exosomes and their subsequent release into the extracellular environment.

2.2 *Nucleic Acids*

Exosomes contain DNA, mRNA, miRNA, and other noncoding RNAs that can be delivered to recipient cells through exosomes and ultimately affect various biological functions, such as the expression and signaling of target genes in recipient cells [15, 52]. Several studies have shown that gDNA and nucleoproteins are sorted into exosomes [53, 54]. In cancer cells, such as ovarian cancer cells, gDNA, nuclear protein micronuclei, and CD63 interact and are sorted into exosomes [55]. Furthermore, mitochondrial DNA has been loaded into exosomes [56]. However, the sorting mechanism of RNA and DNA into exosomes is still a relatively new field, and its complexity makes it challenging to determine whether sorting proteins are independently or cooperatively involved in the larger sorting process. Therefore, the sorting mechanism underlying this phenomenon has not been fully described, as only the sorting of miRNAs into exosomes has been extensively studied.

Several studies have shown that short sequence motifs play a crucial role in controlling the sorting of miRNAs into exosomes. Heterogeneous ribonucleoprotein A2B1 (hnRNPA2B1) specifically binds 30 exosomal miRNAs by recognizing GGAG motifs and regulating their loading into exosomes. The exosomal hnRNPA2B1 was first sumoylated, leading to the binding of the modified hnRNPA2B1 to miRNA [57]. hnRNPA2B1 also binds to two short RNA recognition motifs (AGG and UAG) to regulate RNA sorting into exosomes [58]. Similarly, the RNA-interacting protein human heterogeneous ribonucleoprotein Q (HNRPQ) can also sort exosomal miRNAs by recognizing a specific sequence (GCUG) at the 3′ end of miRNAs [59]. In addition, the RNA-binding protein Y-box protein I (YBX1) can specifically recognize and bind mRNA motifs containing ACCAGCCU, CAGUGAGC, and UAAUCCCA [60, 61]. YBX1 also controls the loading of specific long-chain lncRNAs, as well as miR-144 and miR-223, into exosomes [62, 63].

In addition to directly binding miRNA sequences, certain proteins regulate miRNA sorting in other ways. Membrane proteins, such as Vps4A and nSMase2, play crucial roles in the sorting of miRNAs into exosomes. Overexpression of nSMase2 can increase the levels of exosomal miRNA-16 and miRNA-146a [64]. Similarly, overexpression of Vps4A enhances the levels of miRNA-27b-3p and miRNA-92a-3p in exosomes derived from hepatocellular carcinoma cells (HCC) [65], whereas inhibition of Vps4A expression decreases the exosomal loading of miR-92a and miR-150 [66]. In addition, Ago2, associated with the RNA silencing complex (RISC), is thought to regulate the loading of specific miRNAs (such as let-7a, miR-100, and miR-320a) into exosomes. Knockdown of Ago2 reduces the usual abundance of miRNAs, such as miR-150 and miR-142-3p, which are enriched in exosomes [67].

2.3 *Lipids*

Exosomes are loaded with certain lipids (such as cholesterol, phospholipids, sphingomyelin, triglycerides, ceramides, and saturated fatty acid chains), which play an important role in maintaining biological activity [30, 68]. On the one hand, loaded lipids can cause exosomes to acquire rigidity and maintain stability [69]; on the other hand, exosomes contain lipids that can increase the intrinsic role of exosomes in intercellular communication and disease progression. Furthermore, studies have shown that exosomal lipids can potentially serve as biomarkers for noninvasive diagnosis and as therapeutic targets to control disease progression and pathogenesis [70]. The composition of lipids and metabolites in exosomes can be modulated by different cellular conditions, influencing their biological functions. For example, co-incubation in vitro of cells with the ether lipid precursor hexadecylglycerol also revealed an increase in ether lipids and the presence of other proteins in exosomes released from cells containing high levels of lipids [71]. However, there are still some limitations in research on how lipids are loaded into exosomes and on the types and quantities of lipids contained in exosomes. There are only a few limited reports that lipids may be packaged in exosomes through lipid rafts [72]. However, more research is needed to elucidate the in vivo mechanisms of lipid sorting and provide potential therapeutic solutions for diseases lacking effective treatments.

3 Exosome Secretion Mechanism

ESCRT is a family of proteins responsible for sorting cargo and generating intraluminal vesicles (ILVs) within multivesicular bodies (MVBs). Subsequently, these ILVs are released as exosomes through the fusion of MVBs with the plasma membrane or via exocytosis [73, 74]. Components of the ESCRT complex selectively target MVBs and ILVs, facilitating exosome secretion [75]. The formation of ILVs within MVBs requires the recruitment of ESCRT complexes to sites of ILV formation [76, 77]. This recruitment process depends on the presence of phosphatidylinositol 3-phosphate (PIP3) on the MVB-limiting membrane. And HRS binds to the ubiquitinated cytoplasmic tail of endocytic proteins, facilitating their recruitment to the MVB [78].

In addition to the pivotal role of ESCRT in the process of exosome secretion, the release of exosomes is regulated by various other factors. Among them, Rabs and SNARE (soluble NSF attachment protein receptor, where NSF stands for N-ethylmaleimide-sensitive fusion protein) proteins are involved in exosome release. Rabs are important regulators of exosome trafficking between different compartments, with Rab5 and Rab7 facilitating the transport of cargo to early endosomes [79]. Rab11 and Rab35 are associated with the recycling and early sorting of endosomes [80], while Rab27a and Rab27b are involved in later stages, including MVB translocation, docking with the plasma membrane, and promoting fusion [81]. This fusion process is facilitated by the interaction between vesicular SNAREs (v-SNAREs) on the MVB and target SNAREs (t-SNAREs) on the target membrane [82]. SNARE proteins, such as vesicle-associated membrane protein 7 (VAMP7) and t-SNARE SNAP23, play crucial roles in the fusion of MVBs with the plasma membrane and subsequent release of exosomes [83, 84]. Additionally, other Rabs, such as Ral-1, Rab11, and Rab35, are directly or indirectly involved in the regulation of MVB secretion [81] (Fig. 2.2).

Some membrane lipids are also involved in the regulation of exosome secretion. Phosphatidic acid has been reported to affect exosome secretion [85, 86]. Diacylglycerol kinase α, an enzyme that consumes diacylglycerol and acts as a negative regulator of MVB formation, has been shown to inhibit the release of LAMP1/CD63-positive exosomes [85]. The final step in exosome secretion involves membrane cleavage and subsequent release, which are based on the interaction of actin and myosin and subsequent ATP-dependent contraction [87]. Stimuli such as stress also induce exosome release, leading to an increase in intracellular calcium levels and cytoskeletal remodeling [88]. In particular, some exosomes may be released

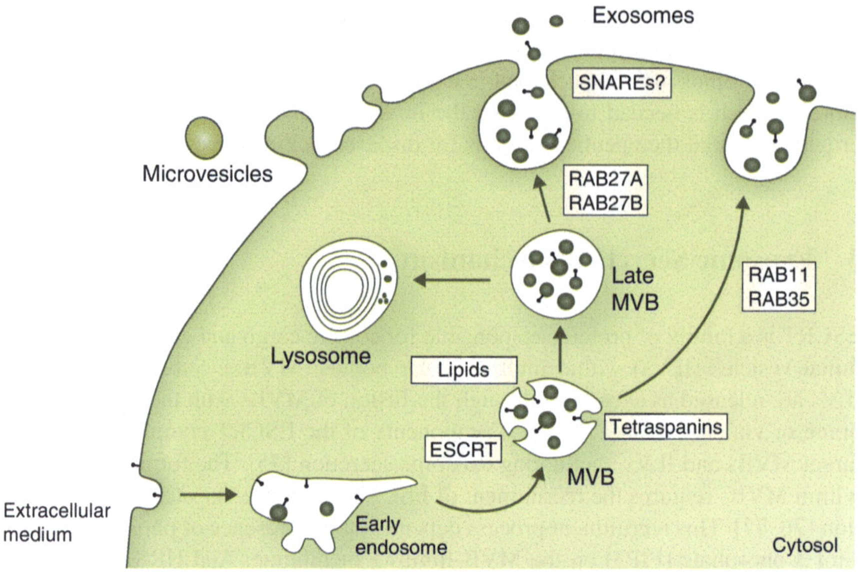

Fig. 2.2 The secretion mechanism of exosomes. Schematic representation of the biogenesis and release of exosomes. Exosomes form intraluminal vesicles in multivesicular bodies with the participation of the ESCRT mechanism and lipids such as ceramides or tetraspanins. MVBs can follow two pathways: fusion with lysosomes for degradation or fusion with the plasma membrane to release their contents (such as exosomes) from the cell. Several Rab proteins (Rab11, Rab27, and Rab35) are involved in exosome secretion. In addition, SNAREs are involved in the secretion of exosomes. In contrast, microvesicles, another type of extracellular vesicle, are released directly through budding from the plasma membrane. (Reproduced with permission from Curr Opin Cell Biol. 2014; 29:116–125)

through direct outward budding and membrane fission, similar to the process of apoptosis.

4 Labeling Methods for Tracking Exosome Delivery and Release

Proper labeling of exosomes is essential for a comprehensive understanding of their release processes and delivery. Labeling exosomes also enables researchers to monitor their changes during disease development, assess the efficacy of exosome-loaded drug delivery, track disease progression, and utilize them as biomarkers for disease diagnosis and prediction. To enable in vivo tracking of exosomes, different labeling techniques have been developed. Most are labeled outside the body and then returned to the body. Two primary approaches are used for exosome labeling: direct labeling and indirect labeling. Direct labeling is based on labeling exosomes with reagents after exosome isolation, whereas indirect labeling involves labeling parental cells by introducing exogenous reagents and subsequently loading the labeled exogenous reagents into secreted exosomes. Both approaches are suitable and reliable for real-time in vivo imaging of exosomes.

4.1 Direct Labeling for Exosome Delivery

Biodistribution studies commonly employ small-molecule fluorophores, such as lipophilic dyes, to track exosomes. Various dyes, such as DiD, DiI, DiO, DiR, PKH26, PKH67, and Cy7 dyes, have been developed for exosomal membrane labeling [89–92]. Since exosome membranes have lipid bilayers similar to those of cell membranes, lipophilic dyes bind to exosomes by embedding in the lipid bilayer, which is the most commonly used method for labeling exosomes (Fig. 2.3a). Lipophilic dyes offer stable fluorescent signals, enabling their use in both in vitro and in vivo visualization. This straightforward labeling approach allows real-time monitoring of single cells and facilitates the study of exosome–cell interactions. However, the low fluorescence intensity and limited tissue penetration depth of

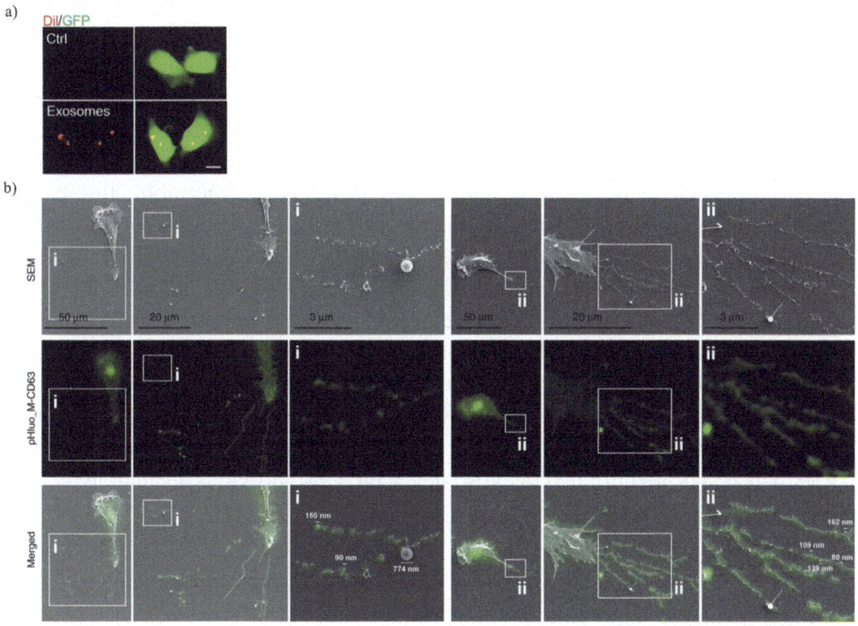

Fig. 2.3 Fluorescent labeling of exosomes. (**a**) Endothelial cells expressing green fluorescent protein (GFP, green) could take up CM-DiI labeled exosomes (red). (Reproduced with permission from Biomaterials. 2017; 133:70–81) (**b**) HT1080 cells stably expressing pHluo_M153R-CD63 (green) were subjected to correlative light-electron microscopy. Increasing magnifications of the images are presented on the right, labeled as i or ii. (Reproduced under a Creative Commons Attribution 4.0 International License from Nat Commun. 2020;11(1):2092)

lipophilic dyes restrict their utility in studying the spatiotemporal dynamics of exosome secretion and cell migration.

4.2 Indirect Labeling for Exosome Secretion

In addition, genetic engineering techniques are widely employed to label exosomes using fluorescent protein reporters fused to exosomal membrane proteins. Genetic engineering involves manipulating genes at the molecular level by introducing foreign genes into recipient cells through in vitro recombination, after which the genes are expressed in recipient cells [93]. By fusing a fluorescent protein reporter gene with an exosomal membrane protein, such as CD63, exosomes can be effectively labeled. For example, the CD63-EGFP plasmid can be transfected into HEK293T cells to obtain exosomes labeled with enhanced green fluorescent protein (EGFP). However, it should be noted that this method yields fewer EGFP-positive exosomes than the other methods [94].

To investigate the dynamic role of exosomes in intercellular communication and cell migration, researchers developed a pH sensitive fluorescent reporter gene called CD63-pHluorin. This reporter gene was applied in zebrafish to study the release, transfer, and uptake of endogenous exosomes derived from the yolk syncytial layer (YSL) [95]. Since this reporter gene cannot be stably expressed in cells and is not highly expressed, the researchers developed the stable and bright pH sensitive reporter gene pHlo_M153R-CD63 using a single amino acid mutation (Fig. 2.3b). This modified reporter gene enabled real-time imaging of multivesicular body (MVB) fusion and cell–exosome interactions [96, 97]. Furthermore, fluorescent proteins can be expressed not only by fusion with exosome proteins but also within the exosome lumen. By expressing PalmGFP in donor cells and allowing it to attach to the exosome membrane via lipid tails, the loading of GFP into exosomes was enhanced [98, 99]. In addition to labeling exosomes with fluorescent proteins, nucleic acids can also be used as tracer elements. By engineering donor cells to introduce miR-155 into the lumen of exosomes, mouse cells lack miR-155, so exosome biodistribution in mice can be analyzed by quantitative PCR [100]. In addition to the above labeling methods, mCherry recombinant proteins containing KFERQ-like sequence tags are enriched in exosomes, which also provides a valuable method to track exosome secretion [49].

5 Conclusions and Perspectives

The exploration and development of exosomes have contributed significantly to the diagnosis and treatment of diseases, and the loading, secretion, and tracing of exosome cargoes have allowed us to further understand exosomes. However, the current understanding of exosomes, especially the loading of specific cargo, the mechanism by which exosomes deliver their contents to recipient cells, and the tracking of exosomes, is still incomplete. Exosome cargo loading is regulated by multiple mechanisms, including post-translational modification pathways for cargo proteins, lipid raft-mediated pathways, and cargo–cargo interaction-dependent pathways. A deeper understanding of these cargo-loading mechanisms will enable better application of exosomes. In particular, exosomes exhibit prolonged retention in the bloodstream and can cross the blood–brain barrier [101, 102]. Furthermore, exosomes can be modified through genetic engineering and other methods to enhance their targeting and therapeutic effects [103].

Although exosomes are capable of loading cargo, the limited amount of exosomes secreted will also hinder the widespread use of exosomes. Future research should aim to explore methods to regulate the secretion of a larger number of exosomes. Additionally, labeling exosomes is an essential tool for gaining in-depth insights into their release mechanism, uncovering their biological functions, and broadening their clinical applications. This technology enables researchers to comprehensively and accurately investigate and harness the role of exosomes in physiological and pathological processes. In summary, understanding exosome loading, secretion, and labeling methods will facilitate their application in disease diagnosis, drug delivery, and other areas, thereby promoting their further clinical use.

Acknowledgments This study was financially supported by open funding from the Nankai University Eye Institute (NKYKD202203) and the Tianjin Natural Science Foundation (22JCZXJC00170, 21JCZDJC00070).

References

1. van der Pol E, Böing AN, Harrison P, Sturk A, Nieuwland R. Classification, functions, and clinical relevance of extracellular vesicles. Pharmacol Rev. 2012;64(3):676–705. https://doi.org/10.1124/pr.112.005983.
2. Raposo G, Stoorvogel W. Extracellular vesicles: exosomes, microvesicles, and friends. J Cell Biol. 2013;200(4):373–83. https://doi.org/10.1083/jcb.201211138.
3. Hessvik NP, Llorente A. Current knowledge on exosome biogenesis and release. Cell Mol Life Sci. 2018;75(2):193–208. https://doi.org/10.1007/s00018-017-2595-9.
4. Tkach M, Théry C. Communication by extracellular vesicles: where we are and where we need to go. Cell. 2016;164(6):1226–32. https://doi.org/10.1016/j.cell.2016.01.043.
5. Mulcahy LA, Pink RC, Carter DR. Routes and mechanisms of extracellular vesicle uptake. J Extracell Vesicles. 2014;2014:3. https://doi.org/10.3402/jev.v3.24641.
6. Mathieu M, Martin-Jaular L, Lavieu G, Théry C. Specificities of secretion and uptake of exosomes and other extracellular vesicles for cell-to-cell communication. Nat Cell Biol. 2019;21(1):9–17. https://doi.org/10.1038/s41556-018-0250-9.
7. Delorme-Axford E, Donker RB, Mouillet JF, Chu T, Bayer A, Ouyang Y, et al. Human placental trophoblasts confer viral resistance to recipient cells. Proc Natl Acad Sci USA. 2013;110(29):12048–53. https://doi.org/10.1073/pnas.1304718110.
8. Luther KM, Haar L, McGuinness M, Wang Y, Lynch Iv TL, Phan A, et al. Exosomal miR-21a-5p mediates cardioprotection by mesenchymal stem cells. J Mol Cell Cardiol. 2018;119:125–37. https://doi.org/10.1016/j.yjmcc.2018.04.012.

9. Mayourian J, Ceholski DK, Gorski PA, Mathiyalagan P, Murphy JF, Salazar SI, et al. Exosomal microRNA-21-5p mediates mesenchymal stem cell paracrine effects on human cardiac tissue contractility. Circ Res. 2018;122(7):933–44. https://doi.org/10.1161/circresaha.118.312420.
10. Etayash H, McGee AR, Kaur K, Thundat T. Nanomechanical sandwich assay for multiple cancer biomarkers in breast cancer cell-derived exosomes. Nanoscale. 2016;8(33):15137–41. https://doi.org/10.1039/c6nr03478k.
11. Frampton AE, Prado MM, López-Jiménez E, Fajardo-Puerta AB, Jawad ZAR, Lawton P, et al. Glypican-1 is enriched in circulating-exosomes in pancreatic cancer and correlates with tumor burden. Oncotarget. 2018;9(27):19006–13. https://doi.org/10.18632/oncotarget.24873.
12. Melo SA, Luecke LB, Kahlert C, Fernandez AF, Gammon ST, Kaye J, et al. Glypican-1 identifies cancer exosomes and detects early pancreatic cancer. Nature. 2015;523(7559):177–82. https://doi.org/10.1038/nature14581.
13. Greening DW, Xu R, Gopal SK, Rai A, Simpson RJ. Proteomic insights into extracellular vesicle biology—defining exosomes and shed microvesicles. Expert Rev Proteomics. 2017;14(1):69–95. https://doi.org/10.1080/14789450.2017.1260450.
14. Pocsfalvi G, Stanly C, Vilasi A, Fiume I, Capasso G, Turiák L, et al. Mass spectrometry of extracellular vesicles. Mass Spectrom Rev. 2016;35(1):3–21. https://doi.org/10.1002/mas.21457.
15. Boriachek K, Islam MN, Möller A, Salomon C, Nguyen NT, Hossain MSA, et al. Biological functions and current advances in isolation and detection strategies for exosome Nanovesicles. Small. 2018;14(6):1. https://doi.org/10.1002/smll.201702153.
16. Kimiz-Gebologlu I, Oncel SS. Exosomes: large-scale production, isolation, drug loading efficiency, and biodistribution and uptake. J Control Release. 2022;347:533–43. https://doi.org/10.1016/j.jconrel.2022.05.027.
17. Ibrahim A, Marbán E. Exosomes: fundamental biology and roles in cardiovascular physiology. Annu Rev Physiol. 2016;78:67–83. https://doi.org/10.1146/annurev-physiol-021115-104929.
18. Skotland T, Hessvik NP, Sandvig K, Llorente A. Exosomal lipid composition and the role of ether lipids and phosphoinositides in exosome biology. J Lipid Res. 2019;60(1):9–18. https://doi.org/10.1194/jlr.R084343.
19. Villarroya-Beltri C, Baixauli F, Gutiérrez-Vázquez C, Sánchez-Madrid F, Mittelbrunn M. Sorting it out: regulation of exosome loading. Semin Cancer Biol. 2014;28:3–13. https://doi.org/10.1016/j.semcancer.2014.04.009.
20. Babst M, Sato TK, Banta LM, Emr SD. Endosomal transport function in yeast requires a novel AAA-type ATPase, Vps4p. EMBO J. 1997;16(8):1820–31. https://doi.org/10.1093/emboj/16.8.1820.
21. Ren X, Hurley JH. VHS domains of ESCRT-0 cooperate in high-avidity binding to polyubiquitinated cargo. EMBO J. 2010;29(6):1045–54. https://doi.org/10.1038/emboj.2010.6.
22. Henne WM, Buchkovich NJ, Emr SD. The ESCRT pathway. Dev Cell. 2011;21(1):77–91. https://doi.org/10.1016/j.devcel.2011.05.015.
23. Raiborg C, Stenmark H. The ESCRT machinery in endosomal sorting of ubiquitylated membrane proteins. Nature. 2009;458(7237):445–52. https://doi.org/10.1038/nature07961.
24. Kim SB, Kim HR, Park MC, Cho S, Goughnour PC, Han D, et al. Caspase-8 controls thefig secretion of inflammatory lysyl-tRNA synthetase in exosomes from cancer cells. J Cell Biol. 2017;216(7):2201–16. https://doi.org/10.1083/jcb.201605118.
25. Baietti MF, Zhang Z, Mortier E, Melchior A, Degeest G, Geeraerts A, et al. Syndecan-syntenin-ALIX regulates the biogenesis of exosomes. Nat Cell Biol. 2012;14(7):677–85. https://doi.org/10.1038/ncb2502.
26. Dores MR, Chen B, Lin H, Soh UJ, Paing MM, Montagne WA, et al. ALIX binds a YPX(3)L motif of the GPCR PAR1 and mediates ubiquitin-independent ESCRT-III/MVB sorting. J Cell Biol. 2012;197(3):407–19. https://doi.org/10.1083/jcb.201110031.

27. Dores MR, Grimsey NJ, Mendez F, Trejo J. ALIX regulates the ubiquitin-independent lysosomal sorting of the P2Y1 purinergic receptor via a YPX3L motif. PLoS One. 2016;11(6):e0157587. https://doi.org/10.1371/journal.pone.0157587.
28. Pornillos O, Alam SL, Rich RL, Myszka DG, Davis DR, Sundquist WI. Structure and functional interactions of the Tsg101 UEV domain. EMBO J. 2002;21(10):2397–406. https://doi.org/10.1093/emboj/21.10.2397.
29. Dawson G. Isolation of lipid rafts (detergent-resistant microdomains) and comparison to extracellular vesicles (Exosomes). Methods Mol Biol. 2021;2187:99–112. https://doi.org/10.1007/978-1-0716-0814-2_6.
30. Trajkovic K, Hsu C, Chiantia S, Rajendran L, Wenzel D, Wieland F, et al. Ceramide triggers budding of exosome vesicles into multivesicular endosomes. Science. 2008;319(5867):1244–7. https://doi.org/10.1126/science.1153124.
31. Wei D, Zhan W, Gao Y, Huang L, Gong R, Wang W, et al. RAB31 marks and controls an ESCRT-independent exosome pathway. Cell Res. 2021;31(2):157–77. https://doi.org/10.1038/s41422-020-00409-1.
32. van Niel G, Charrin S, Simoes S, Romao M, Rochin L, Saftig P, et al. The tetraspanin CD63 regulates ESCRT-independent and -dependent endosomal sorting during melanogenesis. Dev Cell. 2011;21(4):708–21. https://doi.org/10.1016/j.devcel.2011.08.019.
33. Mann M, Jensen ON. Proteomic analysis of post-translational modifications. Nat Biotechnol. 2003;21(3):255–61. https://doi.org/10.1038/nbt0303-255.
34. Huebner AR, Cheng L, Somparn P, Knepper MA, Fenton RA, Pisitkun T. Deubiquitylation of protein cargo is not an essential step in exosome formation. Mol Cell Proteomics. 2016;15(5):1556–71. https://doi.org/10.1074/mcp.M115.054965.
35. Gonzales PA, Pisitkun T, Hoffert JD, Tchapyjnikov D, Star RA, Kleta R, et al. Large-scale proteomics and phosphoproteomics of urinary exosomes. J Am Soc Nephrol. 2009;20(2):363–79. https://doi.org/10.1681/asn.2008040406.
36. Lee H, Li C, Zhang Y, Zhang D, Otterbein LE, Jin Y. Caveolin-1 selectively regulates microRNA sorting into microvesicles after noxious stimuli. J Exp Med. 2019;216(9):2202–20. https://doi.org/10.1084/jem.20182313.
37. Kunadt M, Eckermann K, Stuendl A, Gong J, Russo B, Strauss K, et al. Extracellular vesicle sorting of α-Synuclein is regulated by sumoylation. Acta Neuropathol. 2015;129(5):695–713. https://doi.org/10.1007/s00401-015-1408-1.
38. Ageta H, Tsuchida K. Post-translational modification and protein sorting to small extracellular vesicles including exosomes by ubiquitin and UBLs. Cell Mol Life Sci. 2019;76(24):4829–48. https://doi.org/10.1007/s00018-019-03246-7.
39. Frängsmyr L, Baranov V, Nagaeva O, Stendahl U, Kjellberg L, Mincheva-Nilsson L. Cytoplasmic microvesicular form of Fas ligand in human early placenta: switching the tissue immune privilege hypothesis from cellular to vesicular level. Mol Hum Reprod. 2005;11(1):35–41. https://doi.org/10.1093/molehr/gah129.
40. Zuccato E, Blott EJ, Holt O, Sigismund S, Shaw M, Bossi G, et al. Sorting of Fas ligand to secretory lysosomes is regulated by mono-ubiquitylation and phosphorylation. J Cell Sci. 2007;120(Pt 1):191–9. https://doi.org/10.1242/jcs.03315.
41. Huttlin EL, Jedrychowski MP, Elias JE, Goswami T, Rad R, Beausoleil SA, et al. A tissue-specific atlas of mouse protein phosphorylation and expression. Cell. 2010;143(7):1174–89. https://doi.org/10.1016/j.cell.2010.12.001.
42. Zhang Y, Wu X, Andy TW. Characterization and applications of extracellular vesicle proteome with post-translational modifications. Trends Analyt Chem. 2018;107:21–30. https://doi.org/10.1016/j.trac.2018.07.014.
43. Macedo-da-Silva J, Santiago VF, Rosa-Fernandes L, Marinho CRF, Palmisano G. Protein glycosylation in extracellular vesicles: structural characterization and biological functions. Mol Immunol. 2021;135:226–46. https://doi.org/10.1016/j.molimm.2021.04.017.

44. Chen W, Wang R, Li D, Zuo C, Wen P, Liu H, et al. Comprehensive analysis of the glycome and glycoproteome of bovine milk-derived exosomes. J Agric Food Chem. 2020;68(45):12692–701. https://doi.org/10.1021/acs.jafc.0c04605.
45. Zhang S, Wang C, Ma B, Xu M, Xu S, Liu J, et al. Mutant p53 drives cancer metastasis via RCP-mediated Hsp90α secretion. Cell Rep. 2020;32(1):107879. https://doi.org/10.1016/j.celrep.2020.107879.
46. Zhong L, Liao D, Li J, Liu W, Wang J, Zeng C, et al. Rab22a-NeoF1 fusion protein promotes osteosarcoma lung metastasis through its secretion into exosomes. Signal Transduct Target Ther. 2021;6(1):59. https://doi.org/10.1038/s41392-020-00414-1.
47. Lee YJ, Shin KJ, Jang HJ, Ryu JS, Lee CY, Yoon JH, et al. GPR143 controls ESCRT-dependent exosome biogenesis and promotes cancer metastasis. Dev Cell. 2023;58(4):320–34.e8. https://doi.org/10.1016/j.devcel.2023.01.006.
48. Iavello A, Frech VS, Gai C, Deregibus MC, Quesenberry PJ, Camussi G. Role of Alix in miRNA packaging during extracellular vesicle biogenesis. Int J Mol Med. 2016;37(4):958–66. https://doi.org/10.3892/ijmm.2016.2488.
49. Ferreira JV, da Rosa SA, Ramalho J, Máximo Carvalho C, Cardoso MH, Pintado P, et al. LAMP2A regulates the loading of proteins into exosomes. Sci Adv. 2022;8(12):eabm1140. https://doi.org/10.1126/sciadv.abm1140.
50. Vidal M, Mangeat P, Hoekstra D. Aggregation reroutes molecules from a recycling to a vesicle-mediated secretion pathway during reticulocyte maturation. J Cell Sci. 1997;110(Pt 16):1867–77. https://doi.org/10.1242/jcs.110.16.1867.
51. Fang Y, Wu N, Gan X, Yan W, Morrell JC, Gould SJ. Higher-order oligomerization targets plasma membrane proteins and HIV gag to exosomes. PLoS Biol. 2007;5(6):e158. https://doi.org/10.1371/journal.pbio.0050158.
52. Zhang J, Li S, Li L, Li M, Guo C, Yao J, et al. Exosome and exosomal microRNA: trafficking, sorting, and function. Genomics Proteomics Bioinformatics. 2015;13(1):17–24. https://doi.org/10.1016/j.gpb.2015.02.001.
53. Takahashi A, Okada R, Nagao K, Kawamata Y, Hanyu A, Yoshimoto S, et al. Exosomes maintain cellular homeostasis by excreting harmful DNA from cells. Nat Commun. 2017;8:15287. https://doi.org/10.1038/ncomms15287.
54. Torralba D, Baixauli F, Villarroya-Beltri C, Fernández-Delgado I, Latorre-Pellicer A, Acín-Pérez R, et al. Priming of dendritic cells by DNA-containing extracellular vesicles from activated T cells through antigen-driven contacts. Nat Commun. 2018;9(1):2658. https://doi.org/10.1038/s41467-018-05077-9.
55. Yokoi A, Villar-Prados A, Oliphint PA, Zhang J, Song X, De Hoff P, et al. Mechanisms of nuclear content loading to exosomes. Sci Adv. 2019;5(11):eaax8849. https://doi.org/10.1126/sciadv.aax8849.
56. Rabas N, Palmer S, Mitchell L, Ismail S, Gohlke A, Riley JS, et al. PINK1 drives production of mtDNA-containing extracellular vesicles to promote invasiveness. J Cell Biol. 2021;220(12):e202006049. https://doi.org/10.1083/jcb.202006049.
57. Villarroya-Beltri C, Gutiérrez-Vázquez C, Sánchez-Cabo F, Pérez-Hernández D, Vázquez J, Martin-Cofreces N, et al. Sumoylated hnRNPA2B1 controls the sorting of miRNAs into exosomes through binding to specific motifs. Nat Commun. 2013;4:2980. https://doi.org/10.1038/ncomms3980.
58. Wu B, Su S, Patil DP, Liu H, Gan J, Jaffrey SR, et al. Molecular basis for the specific and multivariant recognitions of RNA substrates by human hnRNP A2/B1. Nat Commun. 2018;9(1):420. https://doi.org/10.1038/s41467-017-02770-z.
59. Santangelo L, Giurato G, Cicchini C, Montaldo C, Mancone C, Tarallo R, et al. The RNA-binding protein SYNCRIP is a component of the hepatocyte exosomal machinery controlling MicroRNA sorting. Cell Rep. 2016;17(3):799–808. https://doi.org/10.1016/j.celrep.2016.09.031.
60. Yanshina DD, Kossinova OA, Gopanenko AV, Krasheninina OA, Malygin AA, Venyaminova AG, et al. Structural features of the interaction of the 3′-untranslated region of mRNA con-

taining exosomal RNA-specific motifs with YB-1, a potential mediator of mRNA sorting. Biochimie. 2018;144:134–43. https://doi.org/10.1016/j.biochi.2017.11.007.
61. Batagov AO, Kuznetsov VA, Kurochkin IV. Identification of nucleotide patterns enriched in secreted RNAs as putative cis-acting elements targeting them to exosome nano-vesicles. BMC Genomics. 2011;12(Suppl 3(Suppl 3)):S18. https://doi.org/10.1186/1471-2164-12-s3-s18.
62. Kossinova OA, Gopanenko AV, Tamkovich SN, Krasheninina OA, Tupikin AE, Kiseleva E, et al. Cytosolic YB-1 and NSUN2 are the only proteins recognizing specific motifs present in mRNAs enriched in exosomes. Biochim Biophys Acta Proteins Proteom. 2017;1865(6):664–73. https://doi.org/10.1016/j.bbapap.2017.03.010.
63. Shurtleff MJ, Temoche-Diaz MM, Karfilis KV, Ri S, Schekman R. Y-box protein 1 is required to sort microRNAs into exosomes in cells and in a cell-free reaction. elife. 2016;5:e19276. https://doi.org/10.7554/eLife.19276.
64. Kosaka N, Iguchi H, Yoshioka Y, Takeshita F, Matsuki Y, Ochiya T. Secretory mechanisms and intercellular transfer of microRNAs in living cells. J Biol Chem. 2010;285(23):17442–52. https://doi.org/10.1074/jbc.M110.107821.
65. Wei JX, Lv LH, Wan YL, Cao Y, Li GL, Lin HM, et al. Vps4A functions as a tumor suppressor by regulating the secretion and uptake of exosomal microRNAs in human hepatoma cells. Hepatology. 2015;61(4):1284–94. https://doi.org/10.1002/hep.27660.
66. Jackson CE, Scruggs BS, Schaffer JE, Hanson PI. Effects of inhibiting VPS4 support a general role for ESCRTs in extracellular vesicle biogenesis. Biophys J. 2017;113(6):1342–52. https://doi.org/10.1016/j.bpj.2017.05.032.
67. Guduric-Fuchs J, O'Connor A, Camp B, O'Neill CL, Medina RJ, Simpson DA. Selective extracellular vesicle-mediated export of an overlapping set of microRNAs from multiple cell types. BMC Genomics. 2012;13:357. https://doi.org/10.1186/1471-2164-13-357.
68. Wubbolts R, Leckie RS, Veenhuizen PT, Schwarzmann G, Möbius W, Hoernschemeyer J, et al. Proteomic and biochemical analyses of human B cell-derived exosomes. Potential implications for their function and multivesicular body formation. J Biol Chem. 2003;278(13):10963–72. https://doi.org/10.1074/jbc.M207550200.
69. Record M, Carayon K, Poirot M, Silvente-Poirot S. Exosomes as new vesicular lipid transporters involved in cell-cell communication and various pathophysiologies. Biochim Biophys Acta. 2014;1841(1):108–20. https://doi.org/10.1016/j.bbalip.2013.10.004.
70. Donoso-Quezada J, Ayala-Mar S, González-Valdez J. The role of lipids in exosome biology and intercellular communication: function, analytics and applications. Traffic. 2021;22(7):204–20. https://doi.org/10.1111/tra.12803.
71. Phuyal S, Skotland T, Hessvik NP, Simolin H, Øverbye A, Brech A, et al. The ether lipid precursor hexadecylglycerol stimulates the release and changes the composition of exosomes derived from PC-3 cells. J Biol Chem. 2015;290(7):4225–37. https://doi.org/10.1074/jbc.M114.593962.
72. Tan SS, Yin Y, Lee T, Lai RC, Yeo RW, Zhang B, et al. Therapeutic MSC exosomes are derived from lipid raft microdomains in the plasma membrane. J Extracell Vesicles. 2013;2:1. https://doi.org/10.3402/jev.v2i0.22614.
73. Kalluri R, LeBleu VS. The biology, function, and biomedical applications of exosomes. Science. 2020;367(6478):eaau6977. https://doi.org/10.1126/science.aau6977.
74. Hurley JH. ESCRTs are everywhere. EMBO J. 2015;34(19):2398–407. https://doi.org/10.15252/embj.201592484.
75. Colombo M, Moita C, van Niel G, Kowal J, Vigneron J, Benaroch P, et al. Analysis of ESCRT functions in exosome biogenesis, composition and secretion highlights the heterogeneity of extracellular vesicles. J Cell Sci. 2013;126(Pt 24):5553–65. https://doi.org/10.1242/jcs.128868.
76. Juan T, Fürthauer M. Biogenesis and function of ESCRT-dependent extracellular vesicles. Semin Cell Dev Biol. 2018;74:66–77. https://doi.org/10.1016/j.semcdb.2017.08.022.

77. Christ L, Raiborg C, Wenzel EM, Campsteijn C, Stenmark H. Cellular functions and molecular mechanisms of the ESCRT membrane-scission machinery. Trends Biochem Sci. 2017;42(1):42–56. https://doi.org/10.1016/j.tibs.2016.08.016.
78. Hanson PI, Cashikar A. Multivesicular body morphogenesis. Annu Rev Cell Dev Biol. 2012;28:337–62. https://doi.org/10.1146/annurev-cellbio-092910-154152.
79. Hutagalung AH, Novick PJ. Role of Rab GTPases in membrane traffic and cell physiology. Physiol Rev. 2011;91(1):119–49. https://doi.org/10.1152/physrev.00059.2009.
80. Stenmark H. Rab GTPases as coordinators of vesicle traffic. Nat Rev Mol Cell Biol. 2009;10(8):513–25. https://doi.org/10.1038/nrm2728.
81. van Niel G, D'Angelo G, Raposo G. Shedding light on the cell biology of extracellular vesicles. Nat Rev Mol Cell Biol. 2018;19(4):213–28. https://doi.org/10.1038/nrm.2017.125.
82. Parolini I, Federici C, Raggi C, Lugini L, Palleschi S, De Milito A, et al. Microenvironmental pH is a key factor for exosome traffic in tumor cells. J Biol Chem. 2009;284(49):34211–22. https://doi.org/10.1074/jbc.M109.041152.
83. Fader CM, Sánchez DG, Mestre MB, Colombo MI. TI-VAMP/VAMP7 and VAMP3/cellubrevin: two v-SNARE proteins involved in specific steps of the autophagy/multivesicular body pathways. Biochim Biophys Acta. 2009;1793(12):1901–16. https://doi.org/10.1016/j.bbamcr.2009.09.011.
84. Verweij FJ, Bebelman MP, Jimenez CR, Garcia-Vallejo JJ, Janssen H, Neefjes J, et al. Quantifying exosome secretion from single cells reveals a modulatory role for GPCR signaling. J Cell Biol. 2018;217(3):1129–42. https://doi.org/10.1083/jcb.201703206.
85. Alonso R, Rodríguez MC, Pindado J, Merino E, Mérida I, Izquierdo M. Diacylglycerol kinase alpha regulates the secretion of lethal exosomes bearing Fas ligand during activation-induced cell death of T lymphocytes. J Biol Chem. 2005;280(31):28439–50. https://doi.org/10.1074/jbc.M501112200.
86. Laulagnier K, Grand D, Dujardin A, Hamdi S, Vincent-Schneider H, Lankar D, et al. PLD2 is enriched on exosomes and its activity is correlated to the release of exosomes. FEBS Lett. 2004;572(1–3):11–4. https://doi.org/10.1016/j.febslet.2004.06.082.
87. D'Souza-Schorey C, Clancy JW. Tumor-derived microvesicles: shedding light on novel microenvironment modulators and prospective cancer biomarkers. Genes Dev. 2012;26(12):1287–99. https://doi.org/10.1101/gad.192351.112.
88. Pasquet JM, Dachary-Prigent J, Nurden AT. Calcium influx is a determining factor of calpain activation and microparticle formation in platelets. Eur J Biochem. 1996;239(3):647–54. https://doi.org/10.1111/j.1432-1033.1996.0647u.x.
89. Dabrowska S, Del Fattore A, Karnas E, Frontczak-Baniewicz M, Kozlowska H, Muraca M, et al. Imaging of extracellular vesicles derived from human bone marrow mesenchymal stem cells using fluorescent and magnetic labels. Int J Nanomedicine. 2018;13:1653–64. https://doi.org/10.2147/ijn.S159404.
90. Macklin R, Wang H, Loo D, Martin S, Cumming A, Cai N, et al. Extracellular vesicles secreted by highly metastatic clonal variants of osteosarcoma preferentially localize to the lungs and induce metastatic behaviour in poorly metastatic clones. Oncotarget. 2016;7(28):43570–87. https://doi.org/10.18632/oncotarget.9781.
91. Nicola AM, Frases S, Casadevall A. Lipophilic dye staining of Cryptococcus neoformans extracellular vesicles and capsule. Eukaryot Cell. 2009;8(9):1373–80. https://doi.org/10.1128/ec.00044-09.
92. Simonsen JB. Pitfalls associated with lipophilic fluorophore staining of extracellular vesicles for uptake studies. J Extracell Vesicles. 2019;8(1):1582237. https://doi.org/10.1080/20013078.2019.1582237.
93. Khan FA, Pandupuspitasari NS, Chun-Jie H, Ao Z, Jamal M, Zohaib A, et al. CRISPR/Cas9 therapeutics: a cure for cancer and other genetic diseases. Oncotarget. 2016;7(32):52541–52. https://doi.org/10.18632/oncotarget.9646.

94. Wiklander OP, Nordin JZ, O'Loughlin A, Gustafsson Y, Corso G, Mäger I, et al. Extracellular vesicle in vivo biodistribution is determined by cell source, route of administration and targeting. J Extracell Vesicles. 2015;4:26316. https://doi.org/10.3402/jev.v4.26316.
95. Verweij FJ, Revenu C, Arras G, Dingli F, Loew D, Pegtel DM, et al. Live tracking of inter-organ communication by endogenous exosomes in vivo. Dev Cell. 2019;48(4):573–89.e4. https://doi.org/10.1016/j.devcel.2019.01.004.
96. Sung BH, Ketova T, Hoshino D, Zijlstra A, Weaver AM. Directional cell movement through tissues is controlled by exosome secretion. Nat Commun. 2015;6:7164. https://doi.org/10.1038/ncomms8164.
97. Sung BH, von Lersner A, Guerrero J, Krystofiak ES, Inman D, Pelletier R, et al. A live cell reporter of exosome secretion and uptake reveals pathfinding behavior of migrating cells. Nat Commun. 2020;11(1):2092. https://doi.org/10.1038/s41467-020-15747-2.
98. Lai CP, Kim EY, Badr CE, Weissleder R, Mempel TR, Tannous BA, et al. Visualization and tracking of tumour extracellular vesicle delivery and RNA translation using multiplexed reporters. Nat Commun. 2015;6:7029. https://doi.org/10.1038/ncomms8029.
99. Lu Y, Eguchi T, Sogawa C, Taha EA, Tran MT, Nara T, et al. Exosome-based molecular transfer activity of macrophage-like cells involves viability of Oral carcinoma cells: size exclusion chromatography and concentration filter method. Cells. 2021;10(6):1328. https://doi.org/10.3390/cells10061328.
100. Bala S, Csak T, Momen-Heravi F, Lippai D, Kodys K, Catalano D, et al. Biodistribution and function of extracellular miRNA-155 in mice. Sci Rep. 2015;5:10721. https://doi.org/10.1038/srep10721.
101. Kamerkar S, LeBleu VS, Sugimoto H, Yang S, Ruivo CF, Melo SA, et al. Exosomes facilitate therapeutic targeting of oncogenic KRAS in pancreatic cancer. Nature. 2017;546(7659):498–503. https://doi.org/10.1038/nature22341.
102. Das CK, Jena BC, Banerjee I, Das S, Parekh A, Bhutia SK, et al. Exosome as a novel shuttle for delivery of therapeutics across biological barriers. Mol Pharm. 2019;16(1):24–40. https://doi.org/10.1021/acs.molpharmaceut.8b00901.
103. Xitong D, Xiaorong Z. Targeted therapeutic delivery using engineered exosomes and its applications in cardiovascular diseases. Gene. 2016;575(2 Pt 2):377–84. https://doi.org/10.1016/j.gene.2015.08.067.

Chapter 3
Extracellular Vesicles: A Significant Mediator Through Interorgan Communication in Mesenchymal Stem Cell Therapy

Anan Huang

Abstract Intravascular infusion of mesenchymal stem cells (MSCs) is a maneuverable, noninvasive, and safe treatment in clinical practice. After transplantation, MSCs are redistributed to the liver and lung, but how these intravenously transplanted MSCs promote injury repair via other organs is unclear. In this chapter, we aim to elucidate how intravenously transplanted MSCs indirectly interact with other organs, including the heart. These intravenously transplanted MSCs may play a significant role in tissue repair through MSC-derived exosomes. These MSC-derived exosomes always deliver or transfer key factors to the injured area to promote tissue repair. A significant immune response may also participate in tissue repair mediated by intravenously transplanted MSC-derived exosomes.

Keywords Mesenchymal stem cell · Exosome · Intravenous infusion · Immunomodulation · Interorgan communication · Circulation

Transplantation of mesenchymal stem cells (MSCs) is a prominent treatment for all types of diseases. Normally, MSCs can be transplanted in situ or intravascularly. However, intravascular infusion is a convenient, noninvasive, and safe method that is widely used in clinical practice. Previous studies have demonstrated that intravenously infused MSCs are distributed into lung and liver tissue but not into other organs based on the current tracking techniques [1–3]. Therefore, for diseases involving other organs, such as the heart and kidney, how can intravenous MSC treatment repair these injured organs and play a therapeutic role? Previous animal experiments and clinical trials confirmed the protective effects of these compounds on these types of injuries in the kidney and heart [4–7]. Recently, several studies elucidated that the therapeutic effects of intravenous MSC therapy on cardiovascular or renal diseases may be the result of indirect factors (exosomes) [8–11]. These studies led to their own hypotheses, including the immune response and transorgan

A. Huang (✉)
Nankai University School of Medicine, Tianjin, China

© The Author(s), under exclusive license to Springer Nature Singapore Pte Ltd. 2024

Z. Li (ed.), *Extracellular Vesicle: Biology and Translational Application*,
https://doi.org/10.1007/978-981-97-5536-3_3

communication, regarding the therapeutic effect of intravenous MSCs, but exosomes play a significant role in the therapeutic process. This chapter aims to elucidate the significant role of exosomes in modulating the immune response or communicating with the injured myocardium between different organs.

1 Intravenously Injected MSCs and MSC-Secreted Exosomes in Vivo

To investigate the therapeutic mechanisms of cell or exosome treatment, accurate and proper tracking of the biodistribution visually in vivo can provide solid support regarding the therapeutic effects of these agents on different diseases. In recent years, a series of imaging techniques have been used for tracking MSC transplantation [12, 13]. Previous literature showed that once MSCs were transplanted into mice, intravenously transplanted MSCs were detected in lung and liver tissue within the first 2 h= through bioluminescence imaging (BLI) [3]. The signal density changes dynamically, and the cells can be tracked for no more than 7 days, which suggests that transplanted MSCs can reside continuously in vivo. In addition, similar results were also observed in intravenously injected MSCs tracked by near-infrared (NIR) imaging [1]. In this study, mice were sacrificed, and lung and liver tissues were isolated and imaged after MSCs were transplanted intravenously. NIR imaging revealed that the transplanted MSCs had left their lung tissue for approximately 3 days but left their liver tissue for nearly 10 days. In our study, bioluminescence imaging (BLI) revealed that intravenous-transplanted MSCs were present in the lung tissue after approximately 3 days [2]. However, in the case of several specific diseases, intravenously injected MSCs can move into ischemic tissue, such as ischemic hindlimb muscles. A study from Japan demonstrated that intravenously injected MSCs were tracked in ischemic hindlimbs on day 1 after intravenous MSC infusion [14]. In their work, they modified MSCs through coexpressing GFP, and these gene-modified MSCs were conducive to observing the fate of intravenously injected MSCs in vivo. The expression of GFP in ischemic hindlimb muscles indicated that transplanted MSCs can be retained in the ischemic area from the circulation. Overall, current studies support that intravenously transplanted MSCs survive in vivo for 3–7 days in the liver, lung tissue or other ischemic areas.

After MSCs are transplanted intravenously, whether these MSCs perform their biological functions needs to be further elucidated. As exosomes or extracellular vesicles (EVs) are recognized as critical carriers that are encapsulated within a series of effectors for the treatment of several diseases, previous work has also verified that intravenously injected MSCs perform their biological functions through the secretome [2, 15]. In myocardial infarction (MI), intravenous infusion of MSCs promotes cardiac repair and improves cardiac function, and in animal experiments, MSCs were infused into mice through the tail vein, and a specific protein called high-level TNF-α-induced protein 6 (TSG-6) was detected in the peripheral blood [16]. Our study revealed that intravenously infused MSCs were localized in the lung and secreted specific proteins into the peripheral blood to promote cardiac repair from the lung to the heart. Unfortunately, this work did not investigate whether TSG-6 was released into circulation from MSCs directly or was loaded in exosomes. Exosomes are widely recognized for their crucial biological effects in combination with MSCs. Several studies have gradually hypothesized that intravenously injected MSCs might release exosomes to promote tissue repair via other organs but not the lung or liver [2, 15]. Nakamura et al. also demonstrated that intravenously injected MSCs always reside in lung tissue, and they verified that exosomes from intravenously transplanted MSCs could be detected in the peripheral blood [15]. This study suggested that intravenously injected MSCs, like released exosomes, still have biological activity. The concentration of exosomes derived from MSCs changed dynamically after the MSCs were intravenously infused. Likewise, our group also found that intravenously injected MSCs gradually released exosomes into peripheral blood, and we detected the expression of MSC-derived exosomes in peripheral blood at different time points through FACS analysis [2]. Furthermore, Nakazaki et al. confirmed that exosomes derived from MSCs can be observed in peripheral blood after infusion of MSCs and that these exosomes can subsequently repair severe spinal cord injury [17]. In the present study, DiR-labeled MSCs were generated to track MSC-derived exosomes in vivo. Through confocal imaging, DiR and CD63 were both detected among the macrophages within the spinal cord, which suggested that the exosomes secreted by intravenously injected MSCs were uptaken by macrophages within the spinal cord. These studies revealed that intravenously injected MSCs could secrete exosomes into the peripheral blood and that exosomes in the peripheral blood could also be delivered into the injured area. In addition, previous literature has demonstrated that intravenous infusion of specific gene-modified MSCs could improve the release of exosomes, and this increase ultimately increased the concentration of MSC-derived exosomes in peripheral blood [15]. Overall, these results suggested that intravenously injected MSCs could release exosomes into the circulation.

2 MSC-Derived Exosomes and Tissue Repair

The infusion of MSC-derived exosomes is a promising cell-free treatment alternative to MSC transplantation. Current studies support that this cell-free treatment efficiently promotes tissue repair [9]. The outstanding therapeutic effect of MSC-derived exosomes always depends on their intrinsic components, which include cytokines, growth factors, mRNAs, and miRNAs [9]. As MSC-derived exosomes are released in vivo, these types of exosomes need to be transported or delivered and then transferred within certain cell types in other organs from the peripheral blood. This kind of biological behavior may be called interorgan communication and is carried out by the transport of peripheral blood. This interorgan communication of exosomes has been previously found in endogenous exosomes [18]. A previous study verified this interorgan communication through living tracking. In their work, CD63-pHluorin was expressed in zebrafish embryos, and they combined imaging techniques and proteomic analysis to demonstrate that endogenous exosomes are released into the blood in a syntenin-dependent manner and are ultimately captured, endocytosed, and degraded by immune cells and endothelial cells [18]. Similarly, a concise review also described that endogenous exosomes can be secreted into the interstitial space and ultimately into the peripheral blood [19]. After being transported through the circulation, these endogenous exosomes perform intercellular communication or interorgan communication [19]. This review highlights the critical role of exosomal miRNAs as cargo in intercellular communication and interorgan communication. However, the combination of imaging techniques and proteomic analysis is difficult to replicate in murine models. Due to the limited data, the current study can partly focus on miRNAs, which are significant components of intravenous MSC-derived exosomes, and these miRNAs always downregulate related signaling pathways to carry out their biological functions [2, 15]. In our previous study, we found that miR-181a-5p was highly expressed in both MSCs and MSC-derived exosomes. Once MSCs were intravenously infused into mice, MSC-derived exosomes were released into the peripheral blood. These MSC-derived exosomes with high levels of miR-181a-5p were subsequently delivered into the injured myocardium to alleviate endoplasmic reticulum (ER) stress. High levels of miR-181a-5p in MSC-derived exosomes affect cardiomyocytes and subsequently inhibit the signaling pathway involved in ER stress to alleviate injury to cardiomyocytes [2]. This biological process contributes to alleviating cardiac dysfunction. In addition, Nakamura et al. reported that intravenously injected MSCs released exosomes, which were delivered from the lung to the heart. In their study, they confirmed the cardioprotective effect of doxorubicin-induced cardiomyopathy; they also considered miRNA from intravenously injected MSC-derived exosomes as an effector molecule, but they could not confirm the specific miRNA of intravenously injected MSC-derived exosomes [15]. In addition to miRNAs, no additional studies have investigated whether proteins, including cytokines or growth factors, and mRNAs play important roles in tissue repair. In a murine model of MI, TSG-6 was found to be a significant factor after it was released from intravenously infused MSCs into peripheral blood and subsequently promoted cardiac repair [16]. Although this study did not reveal whether TSG-6 was secreted directly or through the encapsulation of exosomes, further studies are needed to verify whether TSG-6 may be encapsulated in exosomes and subsequently delivered to the injured myocardium. Exosomes are carriers of different types of activated molecules, and whether factors other than miRNAs can be observed to promote injury repair also needs to be elucidated in the future.

3 MSC-Derived Exosomes and the Immune Response

The outstanding immunomodulatory effect of MSCs has been well described for the treatment of diseases [20]. In addition, the promising immunomodulatory effects of exosomes have also been well described [21, 22]. Therefore, previous studies have demonstrated that the biological effect of MSCs can be attributed to MSC-derived exosomes [11]. Afterward, studies focused on the direct infusion of MSC-derived exosomes also obtained a few positive results [23, 24]. MSC-derived exosomes play important roles in both innate and adaptive immunity with incredible anti-inflammatory effects [25]. In innate immunity, multiple types of cells, including macrophages, natural killer (NK) cells, and dendritic cells (DCs), are affected by MSC-derived exosomes. Previous studies have shown that the anti-inflammatory potential of these materials is partly attributed to macrophage polarization from the M1 type to the M2 type [26]. For host pathogen defense, NK cells are overactivated and subsequently regulate the activity of other immune cells, while MSC-derived exosomes have an immunosuppressive effect on NK cells, which affects their proliferation, activation, and cytotoxicity [27]. As antigen-presenting cells, DCs can present antigens to B and T lymphocytes. When MSC-derived exosomes are administered, the maturation and activation of DCs can be inhibited, which eventually decreases the expression of proinflammatory cytokines [28]. In addition, MSC-derived exosomes can also modulate DCs and subsequently suppress T-cell activation and proliferation. For adaptive immunity, B cells and T cells are both crucial adaptive immune cells. When MSC-derived exosomes were injected, both B cells and T cells were inhibited in terms of cell proliferation and activation to complete their immunomodulatory functions. Currently, MSC-derived exosomes are being used in treating different types of autoimmune diseases, including multiple sclerosis, systemic lupus erythematosus (SLE), type 1 diabetes, uveitis, rheumatoid arthritis (RA), and inflammatory bowel disease (IBD) [20].

In addition to their role in inflammatory diseases, MSC-derived exosomes also play a significant role in cardiovascular diseases through immunomodulatory effects. This cell-free treatment is being applied to several cardiovascular diseases, such as MI, dilated cardiomyopathy (DCM), and atherosclerosis [25, 29, 30]. Multiple anti-inflammatory mechanisms participate in this process. In the MI model, MSC-derived exosomes downregulated the expression of CD68 and/or suppressed the immune response by attenuating the release of several proinflammatory cytokines, such as IL-6, TNF-α, and IL-1β. Similarly, MSC-derived exosomes alleviated the progression of atherosclerosis by promoting macrophage polarization from the M1 to the M2 phenotype. This mechanism could also be replicated in DCM and chronic heart failure (HF) models [31]. Therefore, MSC-derived exosomes have been recognized as promising treatments for cardiovascular diseases.

When exogenous MSCs are delivered intravenously, the transplanted MSCs are trapped in the lung and liver [2]. However, the current literature suggests that intravenously injected MSCs ameliorate cardiac repair by modulating the immune response. In our previous work, intravenously injected MSCs were shown to home to the lung and might be recognized as an endocrine reservoir [2]. Like other kinds of endogenous exosomes secreted by other organs but not the heart, MSC-derived exosomes are released into the peripheral blood and might have similar biological functions to modulate the immune response in the heart, similar to other types of endogenous exosomes. However, few works support this topic. In the future, additional work may need to be carried out to verify how circulating MSC-derived exosomes activate the immune response and eventually promote cardiac repair in CVD patients.

4 MSC-Derived Exosomes and Interorgan Communication

Due to the specific homing of intravenously injected MSCs in vivo, for those diseases originating from other organs, including the heart and kidney, intravenously injected MSCs cannot be delivered to those organs directly, which may rely on direct communication. Thus, whether indirect communication occurs among intravenously injected MSCs, which act as endocrine reservoirs that affect recipient cells from other organs, needs to be elucidated. Recently, interorgan communication has been identified as an important way to work among organs [17, 18]. Limited by imaging techniques and other relevant methods, tracking the dynamic movement of exosomes in vivo among organs has been difficult. Recently, Verweij et al. verified that this interorgan communication among organs can occur via endogenous exosomes [18]. In their work, they established a new imaging strategy, CD63-pHluorin, with specific advantages and suitability for in vivo imaging. Although the quantification of fluorescence signals is limited, at least this method can confirm that exosomes secreted by donor cells are delivered and internalized by recipient cells. In addition to the fact that this hypothesis was also verified in a zebrafish model, whether this conclusion can be replicated in rodents needs further investigation. As imaging techniques have improved recently, in vivo tracking of exosomes can be performed appropriately [17]. Nakazaki et al. demonstrated that intravenously injected MSCs released exosomes into peripheral blood, after which these exosomes were even delivered into macrophages in the spinal cord [17]. The authors detected the fluorescence signals of exosomes released by intravenously injected MSCs, which directly optimized the whole route through which the intravenously injected MSCs secreted exosomes into peripheral blood and were subsequently delivered into recipient cells from other organs. In the present study, transplanted MSCs were labeled with DiR, after which the fluorescence signals of DiR and CD63 were detected as specific surface markers of exosomes in macrophages. Their findings implied that intravenously injected MSCs could secrete exosomes to communicate with recipient cells from at least the spinal cord. Afterward, the induction of interorgan communication by intravenously injected MSCs through secreted exosomes was also verified in our recent study. We found that in doxorubicin-induced cardiotoxicity, intravenous infusion of MSCs could alleviate cardiac injury and provide cardioprotective effects in a rodent model. In our work, multiple molecular imaging methods and FACS were used, and we further found that miR-181a-5p in exosomes affects cardiomyocytes by downregulating overactivated ER stress [2]. Moreover, Nakamura et al. reported this interorgan communication in a rodent model of heart disease. The authors verified the therapeutic effect of intravenously injected MSCs on heart failure. Exosomes secreted by transplanted MSCs were detected in the peripheral blood, and regrettably, this study could not confirm that intravenously injected MSC-derived exosomes were delivered and swallowed into specific recipient cells in the heart, which may be limited by imaging techniques. However, they investigated the factor called adiponectin, which can promote the secretion of exosomes from intravenously injected MSCs. Compared to those in MSCs cultured without adiponectin, the quantity of exosomes was increased significantly in MSCs cultured with adiponectin. In addition, the transplantation of MSCs cultured with adiponectin had a better therapeutic effect, which suggested that intravenously injected MSC-derived exosomes participated in cardiac repair. Therefore, the interorgan communication of intravenously injected MSC-derived exosomes is an important mechanism for promoting tissue repair and regeneration. However, more advanced imaging techniques are urgently needed to track these exosomes after they are secreted by intravenously injected MSCs in vivo.

5 Conclusion

Overall, intravenous infusion of MSCs has been recognized as a promising treatment for tissue repair. The potential therapeutic effects of these agents have been well characterized in both animal experiments and clinical trials. Although the potential mechanisms underlying the therapeutic efficacy of these materials have not been fully elucidated until recently, multiple possible mechanisms have been identified to explain their promising therapeutic efficacy in treating CVD. Of these mechanisms, exosomes play a crucial role in delivering or transferring key factors to the injured area to promote tissue repair. Due to the secretion of exosomes by intravenously injected MSCs, these MSCs could work in the ischemic area, including the myocardium, through interorgan delivery.

References

1. Pan Y, Li J, Wang J, Jiang Q, Yang J, Dou H, et al. Ferroptotic MSCs protect mice against sepsis via promoting macrophage efferocytosis. Cell Death Dis. 2022;13(9):825. https://doi.org/10.1038/s41419-022-05264-z.
2. Huang A, Liu Y, Qi X, Chen S, Huang H, Zhang J, et al. Intravenously transplanted mesenchymal stromal cells: a new endocrine reservoir for cardioprotection. Stem Cell Res Ther. 2022;13(1):253. https://doi.org/10.1186/s13287-022-02922-z.
3. Takamiya S, Kawabori M, Yamazaki K, Yamaguchi S, Tanimori A, Yamamoto K, et al. Intravenous transplantation of amnion-derived mesenchymal stem cells promotes functional recovery and alleviates intestinal dysfunction after spinal cord injury. PLoS One. 2022;17(7):e0270606. https://doi.org/10.1371/journal.pone.0270606.
4. Caiati C, Jirillo E. Transplantation of mesenchymal stem cells as a new approach for cardiovascular diseases: from bench to bedside: a perspective. Endocr Metab Immune Disord Drug Targets. 2023;23:1359. https://doi.org/10.2174/1871530323666230411142308.
5. Glassberg MK, Minkiewicz J, Toonkel RL, Simonet ES, Rubio GA, DiFede D, et al. Allogeneic human mesenchymal stem cells in patients with idiopathic pulmonary fibrosis via intravenous

delivery (AETHER): a phase I safety clinical trial. Chest. 2017;151(5):971–81. https://doi.org/10.1016/j.chest.2016.10.061.
6. Fu Z, Zhang Y, Geng X, Chi K, Liu C, Song C, et al. Optimization strategies of mesenchymal stem cell-based therapy for acute kidney injury. Stem Cell Res Ther. 2023;14(1):116. https://doi.org/10.1186/s13287-023-03351-2.
7. Perico N, Remuzzi G, Griffin MD, Cockwell P, Maxwell AP, Casiraghi F, et al. Safety and preliminary efficacy of mesenchymal stromal cell (ORBCEL-M) therapy in diabetic kidney disease: a randomized clinical trial (NEPHSTROM). J Am Soc Nephrol. 2023;34(10):1733–51. https://doi.org/10.1681/ASN.0000000000000189.
8. Eirin A, Lerman LO. Mesenchymal stem/stromal cell-derived extracellular vesicles for chronic kidney disease: are we there yet? Hypertension. 2021;78(2):261–9. https://doi.org/10.1161/HYPERTENSIONAHA.121.14596.
9. Phinney DG, Pittenger MF. Concise review: MSC-derived exosomes for cell-free therapy. Stem Cells. 2017;35(4):851–8. https://doi.org/10.1002/stem.2575.
10. Rezaie J, Feghhi M, Etemadi T. A review on exosomes application in clinical trials: perspective, questions, and challenges. Cell Commun Signal. 2022;20(1):145. https://doi.org/10.1186/s12964-022-00959-4.
11. Ohayon L, Zhang X, Dutta P. The role of extracellular vesicles in regulating local and systemic inflammation in cardiovascular disease. Pharmacol Res. 2021;170:105692. https://doi.org/10.1016/j.phrs.2021.105692.
12. Nose N, Nogami S, Koshino K, Chen X, Werner RA, Kashima S, et al. [18F]FDG-labelled stem cell PET imaging in different route of administrations and multiple animal species. Sci Rep. 2021;11(1):10896. https://doi.org/10.1038/s41598-021-90383-4.
13. Dhada KS, Hernandez DS, Suggs LJ. In vivo photoacoustic tracking of mesenchymal stem cell viability. ACS Nano. 2019;13(7):7791–9. https://doi.org/10.1021/acsnano.9b01802.
14. Maeda S, Kawamura T, Sasaki M, Shimamura K, Shibuya T, Harada A, et al. Intravenous infusion of bone marrow-derived mesenchymal stem cells improves tissue perfusion in a rat hindlimb ischemia model. Sci Rep. 2022;12(1):16986. https://doi.org/10.1038/s41598-022-18485-1.
15. Nakamura Y, Kita S, Tanaka Y, Fukuda S, Obata Y, Okita T, et al. Adiponectin stimulates exosome release to enhance mesenchymal stem-cell-driven therapy of heart failure in mice. Mol Ther. 2020;28(10):2203–19. https://doi.org/10.1016/j.ymthe.2020.06.026.
16. Lee RH, Pulin AA, Seo MJ, Kota DJ, Ylostalo J, Larson BL, et al. Intravenous hMSCs improve myocardial infarction in mice because cells embolized in lung are activated to secrete the anti-inflammatory protein TSG-6. Cell Stem Cell. 2009;5(1):54–63. https://doi.org/10.1016/j.stem.2009.05.003.
17. Nakazaki M, Morita T, Lankford KL, Askenase PW, Kocsis JD. Small extracellular vesicles released by infused mesenchymal stromal cells target M2 macrophages and promote TGF-beta upregulation, microvascular stabilization and functional recovery in a rodent model of severe spinal cord injury. J Extracell Vesicles. 2021;10(11):e12137. https://doi.org/10.1002/jev2.12137.
18. Verweij FJ, Revenu C, Arras G, Dingli F, Loew D, Pegtel DM, et al. Live tracking of inter-organ communication by endogenous exosomes in vivo. Dev Cell. 2019;48(4):573–89 e4. https://doi.org/10.1016/j.devcel.2019.01.004.
19. Isaac R, Reis FCG, Ying W, Olefsky JM. Exosomes as mediators of intercellular crosstalk in metabolism. Cell Metab. 2021;33(9):1744–62. https://doi.org/10.1016/j.cmet.2021.08.006.
20. Shen Z, Huang W, Liu J, Tian J, Wang S, Rui K. Effects of mesenchymal stem cell-derived exosomes on autoimmune diseases. Front Immunol. 2021;12:749192. https://doi.org/10.3389/fimmu.2021.749192.
21. Riazifar M, Mohammadi MR, Pone EJ, Yeri A, Lasser C, Segaliny AI, et al. Stem cell-derived exosomes as Nanotherapeutics for autoimmune and neurodegenerative disorders. ACS Nano. 2019;13(6):6670–88. https://doi.org/10.1021/acsnano.9b01004.

22. Losurdo M, Pedrazzoli M, D'Agostino C, Elia CA, Massenzio F, Lonati E, et al. Intranasal delivery of mesenchymal stem cell-derived extracellular vesicles exerts immunomodulatory and neuroprotective effects in a 3xTg model of Alzheimer's disease. Stem Cells Transl Med. 2020;9(9):1068–84. https://doi.org/10.1002/sctm.19-0327.
23. Yang G, Fan X, Liu Y, Jie P, Mazhar M, Liu Y, et al. Immunomodulatory mechanisms and therapeutic potential of mesenchymal stem cells. Stem Cell Rev Rep. 2023;19:1214. https://doi.org/10.1007/s12015-023-10539-9.
24. Yang C, Sun J, Tian Y, Li H, Zhang L, Yang J, et al. Immunomodulatory effect of MSCs and MSCs-derived extracellular vesicles in systemic lupus erythematosus. Front Immunol. 2021;12:714832. https://doi.org/10.3389/fimmu.2021.714832.
25. Dabrowska S, Andrzejewska A, Janowski M, Lukomska B. Immunomodulatory and regenerative effects of mesenchymal stem cells and extracellular vesicles: therapeutic outlook for inflammatory and degenerative diseases. Front Immunol. 2020;11:591065. https://doi.org/10.3389/fimmu.2020.591065.
26. Cao L, Xu H, Wang G, Liu M, Tian D, Yuan Z. Extracellular vesicles derived from bone marrow mesenchymal stem cells attenuate dextran sodium sulfate-induced ulcerative colitis by promoting M2 macrophage polarization. Int Immunopharmacol. 2019;72:264–74. https://doi.org/10.1016/j.intimp.2019.04.020.
27. Bai L, Shao H, Wang H, Zhang Z, Su C, Dong L, et al. Effects of mesenchymal stem cell-derived exosomes on experimental autoimmune uveitis. Sci Rep. 2017;7(1):4323. https://doi.org/10.1038/s41598-017-04559-y.
28. Reis M, Mavin E, Nicholson L, Green K, Dickinson AM, Wang XN. Mesenchymal stromal cell-derived extracellular vesicles attenuate dendritic cell maturation and function. Front Immunol. 2018;9:2538. https://doi.org/10.3389/fimmu.2018.02538.
29. Zhu D, Liu S, Huang K, Wang Z, Hu S, Li J, et al. Intrapericardial exosome therapy dampens cardiac injury via activating Foxo3. Circ Res. 2022;131(10):e135–e50. https://doi.org/10.1161/CIRCRESAHA.122.321384.
30. Wang C, Borger V, Mohamud Yusuf A, Tertel T, Stambouli O, Murke F, et al. Postischemic neuroprotection associated with anti-inflammatory effects by mesenchymal stromal cell-derived small extracellular vesicles in aged mice. Stroke. 2022;53(1):e14–e8. https://doi.org/10.1161/STROKEAHA.121.035821.
31. Sun X, Shan A, Wei Z, Xu B. Intravenous mesenchymal stem cell-derived exosomes ameliorate myocardial inflammation in the dilated cardiomyopathy. Biochem Biophys Res Commun. 2018;503(4):2611–8. https://doi.org/10.1016/j.bbrc.2018.08.012.

Chapter 4
Extracellular Vesicle Isolation Methods

Jingjing Kang and Dingbin Liu

Abstract Efficient isolation of extracellular vesicles (EVs) holds the premises of exosome (normally small EVs) research, and a method that can extract intact and pure exosomes from real samples is the first step to conduct exosome-based liquid biopsies and therapeutics. The past years have witnessed various methods for exosome isolation, which show different effects on the morphology and biological activity of EVs. Therefore, it is particularly important to obtain high-purity EVs for subsequent studies by choosing a suitable exosome isolation method. By means of the physical, chemical, and biochemical properties of EVs, a variety of new techniques have been developed for exosome isolation, such as ultracentrifugation (UC), size-exclusion chromatography (SEC), ultrafiltration (UF), precipitation, and affinity-based isolation and microfluidic-based techniques.

Keywords Extracellular vesicles (EVs) · Ultracentrifugation · Size-exclusion chromatography · Microfluidics · Ultrafiltration · Immunoaffinity

Extracellular vesicles (EVs) are tiny sacs released by cells and have exciting potential for research and therapeutic applications. However, isolating them from biological samples like blood or cell culture media can be tricky. The choice of method depends on several factors, including the source of the EVs, desired purity, and downstream applications. In this chapter, we will discuss the main EV isolation methods.

J. Kang · D. Liu (✉)
State Key Laboratory of Medicinal Chemical Biology, Tianjin Key Laboratory of Molecular Recognition and Biosensing, Frontiers Science Centers for Cell Responses and New Organic Matter, College of Chemistry, Nankai University, Tianjin, China
e-mail: liudb@nankai.edu.cn

© The Author(s), under exclusive license to Springer Nature Singapore Pte Ltd. 2024

Z. Li (ed.), *Extracellular Vesicle: Biology and Translational Application*,
https://doi.org/10.1007/978-981-97-5536-3_4

1 Ultracentrifugation

Ultracentrifugation is one of the most commonly employed EV isolation approaches, which involves a high centrifugal force to cause the deposition of particles according to their sizes [1, 2]. It is estimated that, thus far, 56% of exosome studies chose centrifugation as their isolation technique [3]. The EVs are suspended in a fixed position in the centrifuge tube to achieve the purpose of separation, concentration, and purification by increasing the centrifugal speed or using a specific medium. The ultracentrifugation methods can be divided into two methods: differential ultracentrifugation and density gradient centrifugation. We will next discuss the application and main features of standard ultracentrifugation methods to display the details of this "gold standard" exosome isolation strategy.

1.1 Differential Ultracentrifugation

Differential ultracentrifugation (dUC) is the most common exosome separation strategy, which has contributed to most pioneering exosome studies. The principle of dUC is very simple—under a specific centrifugal force, different extracellular components of fluid samples can be separated according to density, size, and shape [4, 5]. As demonstrated in Fig. 4.1, depending on the nature of the tested samples, a cleaning step may be first conducted to eliminate large bio-particles by

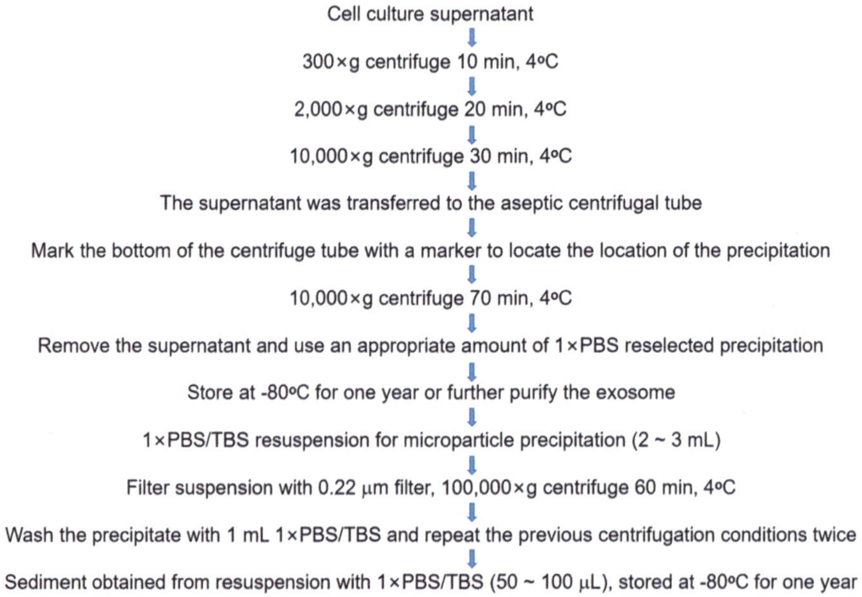

Fig. 4.1 Schematic of ultracentrifugation of EVs extracted from cell culture supernatant

centrifugation at low speed, namely, centrifugation at 300–400 × g for 10 min to sediment a main portion of the cells, at 2000 × g to remove cell debris, and at 10,000 × g to remove the aggregates of biopolymers, apoptotic bodies, and protein aggregates. Afterward, to isolate the smallest category of EVs, exosome (30–140 nm), the remaining supernatant is centrifuged at 100,000 to 120,000 × g for 70 min at 4°C. Each EV pellet is resuspended in phosphate-buffered saline (PBS) for further analysis or application.

Significantly, this method can be easily scaled up for large-scale exosome preparation [6]. dUC has been widely employed for the separation of EVs from a variety of sources over the past 30 years due to its specific advantages, such as ease of operation, low technical expertise, and the ability to perform high-volume preparations without complex sample pretreatment [4, 7, 8]. Although ultracentrifugation has been the "gold standard" for exosome separation due to its high processing capacity, high levels of protein aggregate, and lipoprotein contamination in exosome samples prepared through this method significantly compromises their quantification and functional analysis [9, 10]. Due to the need for overspeed centrifuges, the number of samples to be processed is usually limited to the capacity of six rotors. It cannot be provided by small- and medium-sized laboratories. To overcome the above barrier, liquid concentration devices can facilitate the process of volumes of up to 200 mL with a 5-mL loading capacity ultracentrifugation tube by Wei and colleagues [11].

To note, the exosomal yield can be increased using longer centrifugation during the 100,000 × g. However, it has been demonstrated that if >4 h is used, there is significant mechanical damage to the EVs and higher levels of soluble protein contamination in the final preparation [12, 13]. Even when less than 4 h is used during the 100,000 × g, dUC only results in an enrichment of EVs, not a complete separation of EVs from other components in the extracellular space [13, 14]. In addition, dUC is time-consuming and requires large starting volumes (hundreds of mL) of sample, making it challenging to process several biological samples quickly [14]. Moreover, ultracentrifugation disrupts the integrity of EVs and has a high rate of protein contamination [15]. Although it has been shown that UC is effective in extracting EVs from less viscous fluids, recovery of EVs from viscous fluids is less efficient [16].

1.2 Density Gradient Centrifugation

Density gradient centrifugation is an exosome isolation method based on ultracentrifugation, which is an improvement and optimization of the ultracentrifugation method. Likewise, in the principle of density gradient centrifugation-under a specific centrifugal force, the particles are distributed to some specific positions in the gradient to form different zones. The difference is that this is done with a preconstructed density gradient in a centrifuge tube, which is usually made of sucrose or iodixanol [14, 17]. The EVs can then be readily enriched in sucrose solutions,

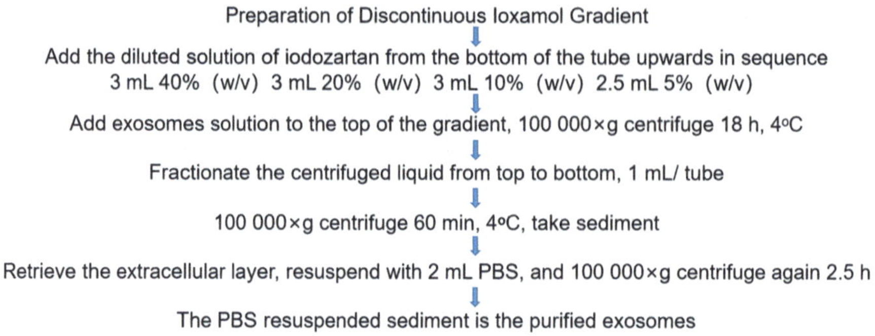

Fig. 4.2 Schematic diagram of exosome extraction from cell culture supernatant by density gradient centrifugation

typically in the density range of 1.13–1.19 g/mL [18]. In this method, the sample was added from the top to a sucrose solution of varying concentration gradients placed at the bottom of the centrifuge tube. Then, the above solutions were ultracentrifuged at 100,000 × g at 4°C, as shown in Fig. 4.2. The Tasso team describes a protocol on a valuable method for dissecting EVs heterogeneity based on density gradient ultracentrifugation [19]. Compared to Lo Sicco et al. previously published paper, where a mixed population of vesicles was considered, they proposed an advanced isolation protocol useful for isolating microvesicles from EVs [20].

Despite being widely used, sucrose density gradient ultracentrifugation has several drawbacks. It takes a great deal of time to complete multiple centrifugations. It also requires relatively high-volume sampling, rendering it unusable for low-volume clinical sample situations [21]. Finally, high-velocity ultracentrifugation increases the risk of vesicle rupture and loss [18, 22]. As a result, fewer EVs can be successfully isolated. It has also been suggested that this method is not suitable for clinical biomarker studies because of the low purity of EVs extracted from serum and urine. It is only ideal for the extraction of EVs from cell cultures.

The ultracentrifugation method risks contaminating other non-vesicular vesicles, including aggregates and lipoproteins, which are also granulated by high-speed rotation [23]. To effectively address this technical issue, this method combines ultracentrifugation with a sucrose density gradient to separate exocrine and non-vesicular particles, such as protein and protein / RNA aggregates [24]. Studies have shown that the coupling of differential centrifugation and density gradient ultracentrifugation effectively improves the purity and quality of isolated particles. Still, the only disadvantage is that the preparation and additional maintenance of gradients require extra time to prevent gradient damage during centrifugation [25]. However, appropriate centrifugation time must be used for the combined method. Otherwise, contaminated particles can still be found in EVs because contaminated particles have a similar density. In 2018, Li et al. improved this protocol with higher recoveries, purity, and better structural and functional retention [26].

2 Size-Exclusion Chromatography

Size-exclusion chromatography (SEC) separates macromolecules based on size rather than molecular weight. The solid phase media in size-exclusion chromatography columns are usually porous gel particles, and commonly used gel materials include Sephadex, Sepharose, and Sephacryl or biogel [27]. Agarose beads are highly scalable and flexible and can, therefore, be adjusted to satisfy the specific sample requirements [28]. Various studies have also utilized qEV SEC columns available in different bed volumes and size-exclusion limits (intra-pore sizes) [29].

SEC uses a column filled with beads with pores smaller than EVs. Specifically, particles smaller than the pore size enter the pore and move slowly in the presence of the buffer solution, while particles larger than the pore size bypass the beads and leave the column quickly. Based on the above principle, EVs of different sizes in the sample will pass through the size-exclusion chromatography column at different rates [30]. SEC for the separation and preparation of EVs has significant advantages, such as simplicity, economy, stability, and high recovery. Notably, SEC is usually performed at low pressures. It can even be driven by gravity, preserving to a large extent the intact structure as well as the biological activity of the EVs. The natural state of EVs can be further enhanced by selecting elution buffers with physiological osmolarity and viscosity. Apart from maintaining exosome function, SEC has additional advantages. Note that the SEC requires deficient volume. With commercially available SEC columns, volumes as small as 15 μL can be processed to achieve high-resolution, standardized, and reproducible exosome isolation suitable to exosome-based fingertip analysis [31].

Currently, SEC has been used for the isolation and preparation of EVs from a variety of samples from prokaryotes and eukaryotes, including cell culture samples, blood samples (plasma and serum) [32–34], urine [35], milk [2], saliva [36], nasal rinses [37], synovial fluids [38], cerebrospinal fluids [39], ascites [40], and tears [36]. Of course, SEC has some shortcomings. Component size-based SEC separation is difficult to distinguish between particles of similar size but different characteristics, including soluble proteins, high-density lipoproteins, low-density lipoproteins, as well as most of the coeliac particles and a small number of smaller-sized EVs [41]. In addition, the method is time-consuming and unsuitable for processing a large number of samples. The SEC is usually used with a strict requirement on the volume of uploaded samples, usually within 0.5 mL, which significantly affects the efficiency of exosome isolation. Therefore, SEC needs to be combined with other techniques to improve the purity and yield of EVs.

Interestingly, by comparing UC and SEC, Takov found that UC could address the limitations of SEC and that the combined SEC-UC approach could cleverly address the shortcomings of SEC and optimize the quality and purity of the EVs in the blood much better than using one method alone [42]. Typically, low-speed centrifugation is performed to remove cells, as well as larger cellular debris, organelles, etc., from the sample, followed by an ultrafiltration step to obtain pre-concentrated EVs. The concentrated sample is then separated by size-exclusion chromatography to collect the effluent fractions. Finally, the fractions emerging from the EVs are subjected to ultracentrifugation to isolate the vesicles [43].

In addition to ultracentrifugation, there are also ultrafiltration methods that can be considered for application in conjunction with SEC. Stassen et al. have reported the EV-yield and the EV-to-protein rate of purity were similar for UF-SEC and UC, but tended to be superior for UF-SEC [44]. Nordin et al. reported a five-fold higher EV-yield for UF-SEC than for UC [45]. Moreover, EV separation by size-exclusion chromatography, combined with a polyethylene glycol-based EV enrichment method, is highly reproducible [46].

3 Ultrafiltration

Ultrafiltration (UF) is an isolation technique based on the size of EVs. Ultrafine nanofiltration membranes with different MWCO (molecular weight cut-off values) can be used to separate extracellular vesicles from clinical samples or cell culture media and separate exocrine and co-vesicles through large cells [47]. Usually, the permeation carrier material, which is smaller than the selected membrane pore size passes through the membrane, leaving larger particles. The most common filter membranes have pore sizes of 0.8 μm, 0.45 μm, or 0.22 μm, which can be used to collect EVs larger than 800 nm, 400 nm, or 200 nm. They can also be designed as microcolumn porous silicon cilia to separate 40–100 nm EVs [48]. If the pore size of the filtration membrane is adjusted, it is possible to separate particles with a well-defined particle size. Kim et al. proposed a circulating tangential flow filtration (TFF) system (Fig. 4.3) [49], which consists of two membranes with pore sizes of 200 nm and 30 nm connected to a peristaltic pump.

Compared with the UC method, the exosome separation based on UF greatly shortens the treatment time. It does not need special equipment, which provides an ideal substitute for the classical UC strategy [50]. Importantly, by easily resizing the

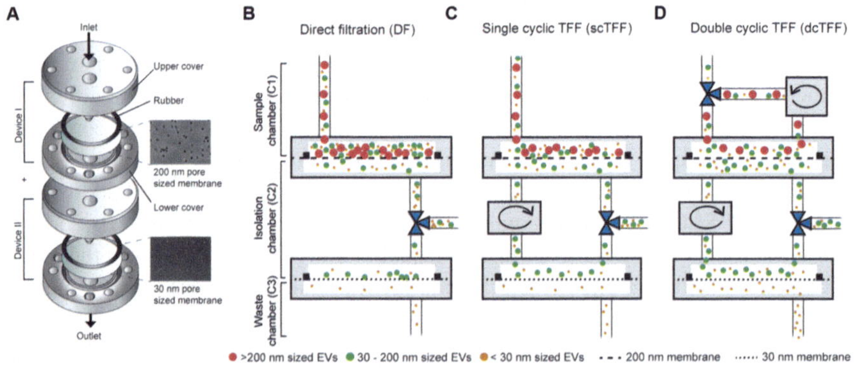

Fig. 4.3 Schematics of filtration systems for isolation of EVs

filter, UF allows researchers to classify specific subsets of small extracellular vesicles (including exocrine bodies) with a defined particle size [50]. It is reported that UF has the highest recovery rates for particles smaller than 100 nm, including EVs。 .

Besides, not much sample volume and reagents are required to support the whole process, and the on-site analytical capabilities make it a user-friendly, clinically reliable, and cost-effective choice for EVs isolation. Furthermore, UF is simple to handle with no additional need for special equipment and can separate EVs with well-defined particle sizes by adjusting the pore size of the filtration membrane.

4 Precipitation

Polymer precipitation is now a commonly used method for EVs isolation. As early as half a century ago, polymer precipitation has been used for the enrichment and purification of virus particles [51]. Viruses and EVs have similar biophysical and chemical properties. Therefore, polymer precipitation in recent years has also been considered for the enrichment and purification of EVs according to their solubility. As research continues, the use of polymer precipitation for the enrichment and purification of EVs is becoming more and more effective due to the emergence of methods that are constantly being improved. With the development of research, the effect of enrichment and purification of EVs by polymer precipitation method is becoming increasingly influential. Depending on the principle, polymer precipitation methods for the enrichment and purification of EVs include the classification into charge methods and hydrophilic polymer precipitation.

4.1 Charge-Based Method

Principle of protamine precipitation method: EVs are negatively charged, and positively charged protamine is used to aggregate and separate EVs. Firstly, centrifuge the sample at a low speed (1500 ~ 3000 × g) to remove larger impurity particles in the sample. Then, mix the sample with the precipitation buffer (0.25 mg/mL protamine solution) in a ratio of 4: 1 and leave it overnight. Finally, centrifuge the sample at room temperature at 1500 × g for 30 min, and the EVs were enriched in the precipitate. Using this method, it should be noted that the resulting EVs should preferably be purified by Sephadex G-100 columns to remove lipoproteins and protamine from the co-precipitated downstream before downstream analyses [52]. In addition, it has been shown that the addition of an extra 0.2 g/mL of PEG 35000 to the precipitation buffer helps to improve the efficiency of EVs enrichment. The advantages of this method are simple operation, simultaneous enrichment of large-volume samples, intact EVs membrane structure, and relative cheapness. The disadvantages are that the separation process is time-consuming, and the downstream resuspension and purification are troublesome.

Principle of sodium acetate precipitation method: EVs have negatively charged phosphatidylserine (PS) on their surface, and sodium acetate can neutralize the negative charge carried by PS, thus precipitating the aggregation of EVs through a hydrophobic effect [53]. The first step was two times vulgar centrifugation to remove cells, cell debris, and large impurities in the samples. After that, mix the samples with 0.1 times the volume of sodium acetate solution, ice bath for 30–60 min, and leave it at 37°C for 5 min. Then, centrifuge at 5000 × g for 10 min to precipitate EVs. Finally, wash the samples with the above sodium acetate solution once and resuspend the EVs in the HEPES salt solution. The point of this method is that it can handle large volumes of samples and, at the same time, it does not require expensive equipment and reagents or removal of the chemicals used. The disadvantage is that in using this method, abdominal muscle from serum or plasma EVs, the contamination of proteins of non-EVs origin will limit its use.

Organic solvent precipitation of proteins is a rapid method for precipitating EVs using organic solvents (acetone, chloroform, trichloroacetic acid). Biological fluids contain a large number of proteins, soluble factors, and lipoproteins. Removal of proteins from biological fluids is an important requirement for the separation of EVs. Proteins with hydrophilic and hydrophobic regions have a specific dielectric strength in aqueous solution. Oppositely charged amino acid residues attract organic solvents and promote protein aggregation. The ion-pair effect contributes to the precipitation efficiency of PROSPR [54]. In addition, organic solvents containing salts increase the removal of proteins. PROSPR has less protein contamination relative to the ultrafast separation method and avoids exosome loss due to high rotational speeds.

4.2 Hydrophilic Polymer Precipitation

Among various hydrophilic polymers, it is well known that polyethylene glycol (PEG) due to its non-toxicity, solubility in water and various organic solvents, and good thermal stability [55].

In the preparation of biomacromolecules, PEG with molecular weights of 6000–20,000 is used more often. PEG molecules with polar groups have a significant affinity for water and can attract water molecules or dissolve in water. Adding hydrophilic PEG to the solution under certain salt concentration conditions leads to exosome precipitation by interacting with water molecules around the EVs to form a hydrophobic microenvironment. The precipitation effect of PEG is mainly related to its concentration and molecular weight and is also affected by factors such as ionic strength, solution pH, and temperature [56]. At a certain pH, the higher the salt concentration, the lower the concentration of PEG required, and the closer the pH

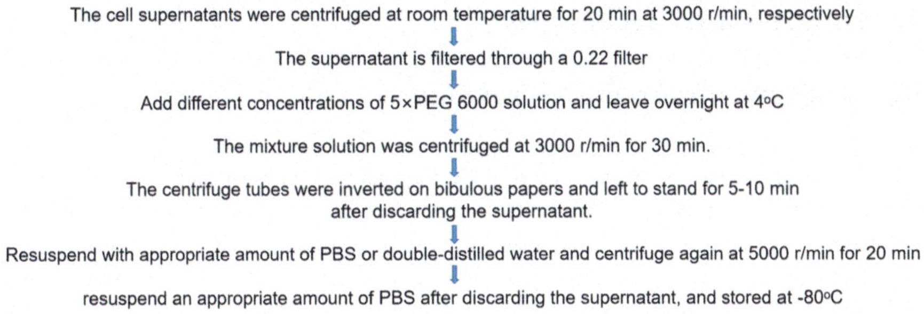

Fig. 4.4 Schematic representation of the isolation of EVs from cell cultures by the PEG precipitation method

of the solution is to the isoelectric point of the target, the lower the concentration of PEG required for precipitation. Within a specific range, high molecular weight and high concentration of PEG precipitate with high efficiency.

Usually, the extraction of EVs from cell supernatants by polymer precipitation is operated as shown in Fig. 4.4. First, pretreatment is required to remove large contaminant particles, such as cellular debris and apoptotic vesicles. Then, the pretreated samples are incubated with PEG solution at 4°C overnight. Next, precipitated EVs were collected by low-speed centrifugation (1500 × g) [6, 57]. With such a simple and straightforward protocol, this method has been widely used to isolate EVs from various types of samples (e.g., blood, cell culture media, cerebrospinal fluid, urine, and ascites) [58, 59]. Since the polymer precipitation method does not require expensive equipment, the method can be easily scaled up to high-volume preparations with high yields. This method also allows for rapid disease diagnosis by integrating various assay platforms for EVs (or protein/genetic material content) analysis [60].

The excellent adaptability and enrichment function of PEG have made it a hotspot for the development of commercial EVs enrichment reagents at present. One of them, Exoquick, can be used to separate EVs from various types of samples [61]. However, polymer-based precipitation methods may simultaneously separate non-vesicular contaminants (including lipoproteins). Moreover, the mixing of polymeric materials may affect downstream analyses. Based on the mentioned deficiencies of the PEG precipitation method, researchers have also made improvements to give a better life to the technique. Rider reported that the 8% PEG (8% PEG precipitation followed by ultracentrifugation for purification) significantly improved the purity of EVs isolation without affecting the sensitivity of their isolation [62]. Chang et al. also reported that the use of PEG-coated Fe_3O_4 nanoparticles for the enrichment and purification of EVs can effectively reduce the interference of non-EVs proteins [63]. Re-enrichment by immunomagnetic beads after PEG precipitation and density gradient ultracentrifugation after PEG precipitation are also being explored by scholars in different application fields.

5 Affinity-Based Techniques

5.1 Immunoaffinity-Based Isolation

The observation that the abundance of proteins and receptors on the surface of EVs provides a basis for the development of a method for immunoaffinity-based EVs isolation via specific interactions between exosome surface proteins and the corresponding receptors [64]. Over the past few years, various EVs proteins have been recorded, including tetraspanin (such as CD8, CD9, CD61, CD63, CD81, and CD82), cytoplasmic proteins (such as tubulin, actin, actin-binding proteins, annexins, and Rab proteins), and heat shock proteins (such as Hsp70 and Hsp90) [65, 66], which can be used as specific markers for isolating EVs. To develop immunoaffinity methods, the ligands are immobilized or conjugated onto/into various solid materials, such as magnetic beads or polymeric materials. Among these materials, immunomagnetic beads are one of the most representative supports, as they can enrich and separate EVs by a simple external magnetic field [6].

The combination of immunoaffinity and magnetic beads to enrich EVs has been reported in numerous works of literature. Immunomagnetic bead separation technology has been widely applied to cell separation and immunological and microbiological assays due to the ability to combine the high specificity of the immune response with the unique magnetic responsiveness of magnetic beads.

When an external magnetic field is present, the magnetic microspheres are magnetized and become magnetically based on the superparamagnetism of the beads, and conversely, the magnetic microsphere magnetism disappears. This property enables magnetic beads to aggregate in a magnetic field environment and is re-dispersed when the magnetic field is removed. At the same time, it also exerts the flexibility of magnetic beads in the application of exosome enrichment analysis. A result that the magnetic bead method enriches EVs with higher purity while being able to enrich different target subpopulations of EVs than ultracentrifugation of EVs products has been found by several researchers.

An increasing number of researchers believe that the immunomagnetic bead method holds promise as a new standard for EVs enrichment because of its many advantages, such as high specificity, easy operation, and high purity. Zhao et al. [67] used a microfluidic chip based on immunomagnetic beads with the detection of CA-125, EpCAM, and CD24 exosomal proteins. They were able to sense ovarian cancer at an early period with similar accuracy to the standard Bradford method. Immunoaffinity superparamagnetic has applied nanoparticles (ISPNs) to bind EVs in a recent study, where the researchers generated ISPNs by linking an anti-CD63 antibody to the nanoparticles and used them to isolate EVs from body fluids. In addition, it is possible to achieve efficient isolation of EVs by modifying magnetic particles with other antibodies, such as EpCAM, anti-A33, PSMA, anti-CD105, anti-CD171, anti-CD34, anti-CD61, anti-CD41, and anti-CD235a [68–71]. Halvorsen reported the immunomagnetic bead method for the isolation of urinary EVs from healthy individuals [68]. They coupled a cocktail of antibodies against

CD8, CD61, and CD81 to magnetic beads. By digesting the beads in a detergent-free buffer, they studied the outer membrane-associated proteins of the EVs and found 49 proteins that were membrane-associated or membrane-bound.

The immunoaffinity magnetic bead method is often used as a "new standard" to compare with the "traditional gold standard" of separated products. It was demonstrated by Jeppesen's team that the products enriched by the immunoaffinity magnetic bead method have more distinctive characteristics of classical EVs-associated proteins and nucleic acids [72] as well as fewer impurities or contaminants relative to UC. Enrichment of EVs by the immunoaffinity magnetic bead method is gradually being regarded as the standard for verifying the purity of EVs, and it is the most frequently chosen method for the clinical application of EVs.

In addition to magnetic beads, other solid media for EVs separation have also been reported, such as polymer monolithic disc posts, magnetic nanowires, and paper (cellulose). Compared to nanoparticles, elongated nanowires can envelop a large number of magnetic nanoparticles and provide a high specific surface area, which allows modification with more EVs-specific antibodies, thereby improving the recovery and purity of EVs separations. Compared to nanoparticles, elongated nanowires can envelop a large number of magnetic nanoparticles and provide a high specific surface area, which allows modification with more EVs-specific antibodies, thereby improving the recovery and purity of EVs separations. Based on these characteristics, anti-CD9, anti-CD63, and anti-CD81 antibodies have been functionalized on magnetic nanowires through streptavidin-biotin interactions. M. Grant Norton's group [73] has demonstrated the selective capture of PSMA+ tumor-derived EVs by high-affinity ligand-activated silica nanoparticles. This was achieved using biotinylated PSMA ligands to capture tumor-derived EVs from prostate cancer cells via avidin-functionalized nanoparticles. Immunoaffinity chromatographic methods using polymeric monolithic disc columns for the separation of CD9+ and CD61+ platelet-derived EVs from human plasma were developed in the Riekkola laboratory and were also investigated [74]. These methods allow for the easy one-step separation of intact CD9+ and CD61+ EVs from plasma in less than 30 min, and the resulting EVs are readily available for further analysis. By modifying the paper surface through chemical coupling, Chen developed a paper-based immunoaffinity platform [75]. Cheng describes that a new diagnostic method using cellulose-based devices and microfluidic chip techniques for isolating EVs from aqueous humor is less cumbersome and saves time.

Immunoaffinity methods can obtain high-purity EVs by specific antibodies or affinity ligands. Still, it is challenging to ensure that the surface protein receptor of EVs is not affected and the yield is relatively low. On the other hand, the heterogeneity of EVs may introduce analytical bias to the clinical application of such methods. The immunoaffinity capture method ensures the specificity of capture. Still, the firmness of antibody-antigen binding increases the difficulty of eluting the stationary phase with the enriched product, and the corresponding elution method is also prone to cause damage to the structure and function of EVs, which affects the downstream analyses. To adopt a gentler means of elution to obtain intact EVs products, people have focused on the lipoprotein receptor and ligand binding methods.

5.2 Non-immune Affinity-Based Isolation

The phospholipids that form the basis of EVs membranes are complementary to the characteristic surface proteins of EVs and can be considered as "universal" markers for EVs isolation. Therefore, targeting surface phospholipids has emerged as an alternative method for EVs isolation from various biological samples. In addition, this lipid-based isolation method is more likely to enrich purified EVs independent of CD63 expression. Wataru et al. used T cell immunoglobulin mucin 4 (Tim4) to isolate extracellular vesicles [76]. Tim4 is a type I transmembrane protein expressed on macrophages, which strongly binds phosphatidylserine, displayed not only on apoptotic cells but also on EVs and microvesicles. They immobilized Tim4 on conventional magnetic beads and applied it to cell culture supernatants of hematopoietic and cancer cells. They performed extensive downstream analyses of the isolated EVs and showed that PS-based isolation of EVs is feasible. Similar to Wataru's study, Huiying et al. used Tim4 beads to purify EVs before quantifying CD63-positive EVs [77]. They used ten serum samples from patients with hepatocellular carcinoma and showed that PS-based EVs purification could distinguish cancer patients from healthy donors. However, the number of PS on phospholipid membranes led to some limitations in the capture of EVs.

To surmount the challenges inherent in EVs isolation and purification efficiency, the pivotal step involves judicious selection of markers exhibiting elevated abundance within EVs for deployment as capture targets. In this vein, Liu et al. have ingeniously pioneered the development of choline phosphate (CP)-modified magnetic hydrogel particles (MB@CP) (Fig. 4.5), effectuated through the direct affinity design of phosphatidylcholine (PC) in the membrane architecture of EVs. This innovation facilitates the proficient extraction of EVs from intricate samples under an applied magnetic field [78]. The PC and CP charges in this context exhibit opposing directions, allowing for the formation of zwitterionic coordination via electrostatic interactions. This coordination surpasses capture strategies reliant solely on positively charged materials [79, 80]. Notably, MB@CP excludes other impurity components, cells, and their debris, typically eliminated during the preliminary processing of clinical samples due to the absence of PC structures. Of paramount significance, the CP-PC coordination bond can be disrupted at 42°C, liberating the EV from MB@CP without necessitating the introduction of additional reagents, thereby preserving the bioactivity of the EV. The outcomes revealed that the yield, purity, and recovery in this investigation exceeded 90% and were achievable within a concise timeframe of 30 min, outclassing the majority of reported methodologies. To further harness the potential of this strategy, the exploration of newly discovered biomarkers should be undertaken through extensive cohort studies. Additionally, the team advocates for the incorporation of a more diverse array of bodily fluids, including tear fluid, cerebrospinal fluid, bile, breast milk, etc., to underscore the universality of MB@CPs in EV isolation, catering to the diverse demands of EV studies in various contexts.

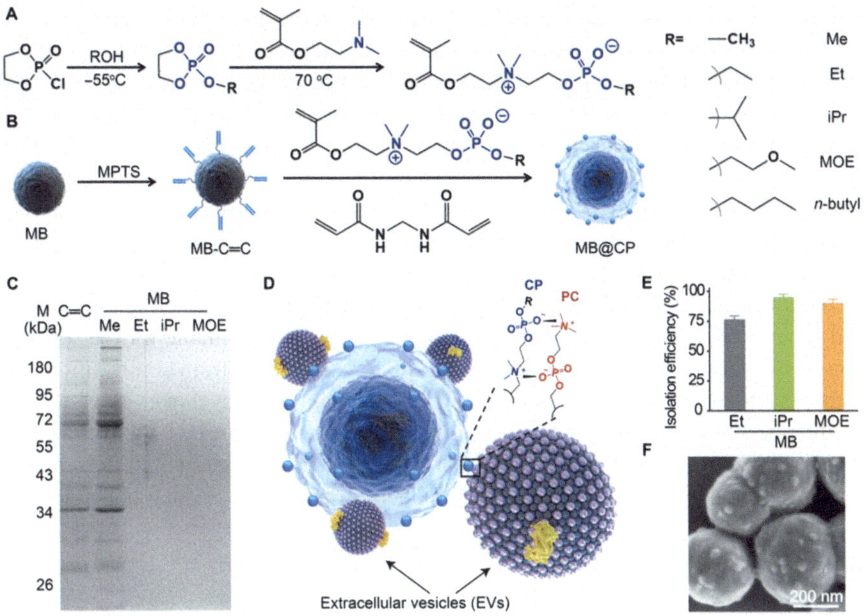

Fig. 4.5 Preparation of MB@CPs and optimization of CP monomers

6 Microfluidic-Based Isolation Techniques

With the escalating biomedical EV research, there is a need for reliable, rapid, acceptable, automated, high-yield, and high-purity separation methods with the ability to collect specific subpopulations of extracellular vesicles. Microfluidics is a kind of science and technology featuring precise manipulation of fluids in micro- and nanoscale space, which can shrink the basic functions of biology and chemistry laboratories, such as sample preparation, reaction, separation, and detection, to a few square centimeters on a chip [81, 82]. It is also known as "lab-on-a-chip" technology. Microfluidics-based separation and detection have the advantages of small sample size, fast detection speed, and low detection cost; therefore, some researchers have successfully developed a variety of novel exosomes separation methods based on microfluidics technology by combining acoustic waves, dielectric electrophoresis, microfluidic viscoelasticity, and other physical properties [83].

6.1 Separation of EVs by Acoustic Fluid Technology

Researchers have cleverly combined microfluidics and acoustics to form acoustofluidics, which can effectively sort and manipulate particles such as cells [84]. The principle of this method is that particles of different sizes are subjected to differentiated acoustic radiation and viscous forces in the microfluidic acoustic field [85, 86]. The viscous force is proportional to the radius of the particles, while the acoustic radiation force is proportional to the volume of the particles. As shown in Fig. 4.6a, for the size of the larger particles, the acoustic radiation force plays a dominant role, the particles to the acoustic node movement, while for the size of the smaller particles, viscous forces offset most of the acoustic radiation force, particles sideways movement is weak. In the combined effect of acoustic radiation and viscous forces, the size of the particles will move to different exports and thus achieve the separation of particles.

Although acoustofluidic technology has been widely used in cell manipulation and separation, the initial acoustofluidic devices were only suitable for the separation of two targets, thus making it challenging to isolate EVs from compositionally complex blood. To address this deficiency, Wu et al. [87] developed an acoustic-fluidic microfluidic chip, which can rapidly and efficiently separate EVs from whole blood samples in a label-free and contact-free manner. The device and the separation process are shown in Fig. 4.6b, which is divided into a microcell removal

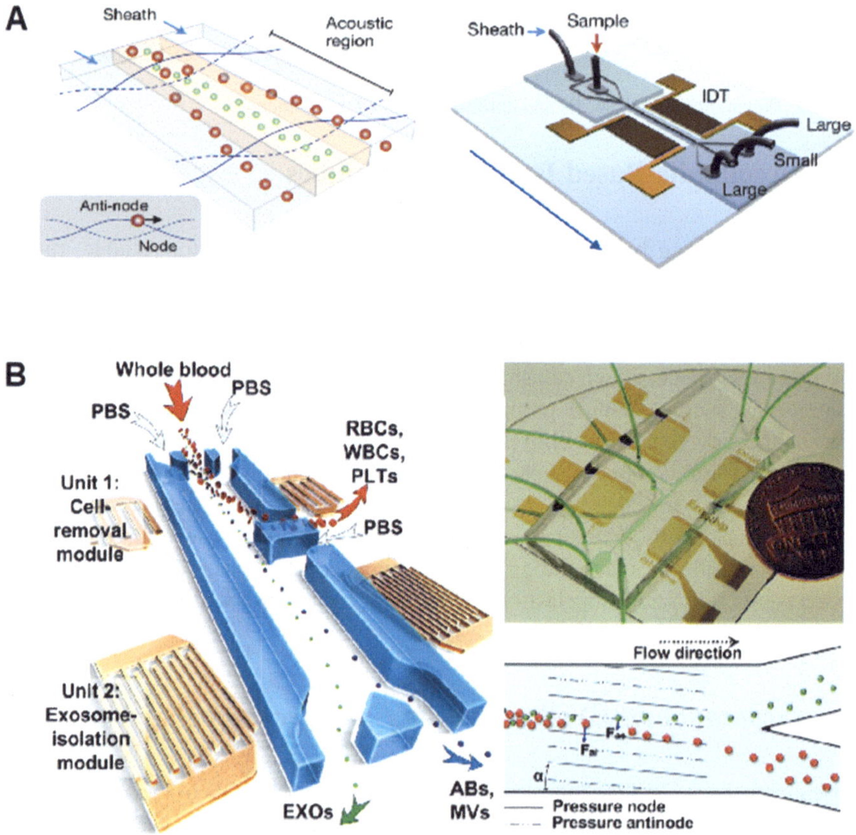

Fig. 4.6 Schematic representation of the separation of EVs using hydrodynamic properties

module and an exosome separation module. Firstly, in the microcell removal module, the lower frequency (19.6 MHz) sound wave is used to remove the larger-sized red blood cells, white blood cells, and platelets in the blood; then the higher frequency (39.4 MHz) sound wave is used in the exosome separation module to separate the EVs from the sub-component of extracellular vesicles (apoptotic vesicles, larger extracellular). Similarly, Lee et al. [88] demonstrated a microfluidic chip based on an acoustic nanofiltration system for the separation of EVs. The microchip approach is based on the fact that there are size and density differences between extracellular vesicles and other components, which in turn allows for the use of ultrasound standing waves to separate nanoscale vesicles (< 200 nm) from cell culture fluid and erythrocyte products.

Acoustic fluid-based exosome separation is a novel method with the advantages of obtaining EVs with good biological properties, ideal purity (~98%) and recovery (~82%), and a small sample size. However, the separation principle is based on the size and acoustic impedance characteristics of the object, so it will inevitably be interfered with by the plasma components with similar size and acoustic impedance characteristics as those of EVs. The separated EVs are also mixed with protein aggregates of comparable size to the EVs, which may need to be optimized by adjusting the parameters or adding a separation module.

6.2 Separation of EVs Based on Dielectrophoretic Techniques

EVs possess dielectric properties similar to those of cells. The dielectric properties depend on the composition of the vesicles, including membrane potential, solute conductivity, and diameter. Therefore, based on the difference in dielectric properties of different EVs, the force in the dielectrophoretic field is different to achieve the purpose of separation. Dielectrophoretic (DEP) force is a force generated by the polarization of dielectric particles in a non-uniform electric field and their interaction with the electric field, which causes the polarized particles to move in a directional manner. Since the magnitude of the dielectrophoretic force is related to the dielectric properties of the manipulated particles and the solution medium, and is proportional to the size of the manipulated particles, unique electrode structures need to be designed to achieve the separation and purification of EVs from blood using dielectrophoresis.

Ibsen et al. [89] designed an AC electrokinetic microarray chip based on dielectrophoresis to rapidly separate and recover EVs from undiluted plasma samples of glioblastoma. The separation principle of this microarray chip is based on the fact that EVs have different dielectric properties from other plasma components and therefore are subjected to different dielectrophoretic forces, so that nanoscale EVs are attracted to the edge region of the microelectrodes, i.e., the strong-field region of the dielectrophoresis. In contrast, the cells and macromolecules of proteins are pulled into the weak-field region of the dielectrophoresis between the electrodes, thus realizing the separation of EVs. To prevent direct contact between the electrode and the aggregated particles from damaging the integrity of the particles, the electrode is coated with a hydrogel layer. When it is necessary to release the sample of particles, the sample is disengaged from the electrode by reverse. Heineck et al. [90] demonstrated that the planar array electrode model can be unaffected by electrothermal flow under high-conductivity conditions by comparing a parallel wire arrangement with a planar array, which provides a new way of thinking about the isolation of EVs from high-conductivity solutions such as cell cultures or blood. Sonnenberg et al. [91] used a similar method of microarray electrodes to isolate from blood DNA and other nanoscale substances successfully.

Separation of EVs by dielectrophoresis has high recovery and purity and requires a small plasma sample (30–50 μL), which can be enriched in a very short time (15 min) [89]. However, a potential disadvantage of this method is that EVs may be damaged by electrochemical phenomena resulting from direct contact with the electrodes. It can be considered to lay a porous hydrogel layer on the array of electrodes to avoid direct contact between EVs and electrodes and electrochemical effects.

6.3 Separation of EVs Using Hydrodynamic Properties

6.3.1 Separation of EVs Based on Viscoelastic Microfluidic Properties

In viscoelastic microfluidics, the viscoelastic force acting on a particle can cause it to move [92], and the magnitude of the viscoelastic force and its resulting displacement are directly proportional to the size of the particle, so the viscoelastic force of microfluidics can be used as a simple, label-free technique for the manipulation and separation of EVs. Viscoelastic forces of microfluidics have been widely used for the manipulation and separation of micro- and nano-objects such as tumor cells, blood cells, bacteria, droplets, and microspheres [92–94]. In contrast to other technological methods such as acoustic fluid, dielectrophoresis, and magnetism, separation methods using microfluidic viscoelasticity allow for precise manipulation of EVs without applying additional field forces.

Despite the advantages of microfluidic viscoelastic separation, it does not generate large enough viscoelastic forces on nanoparticles. To address the limitation that general viscoelastic microfluidics cannot directly separate EVs, Liu et al. [95] proposed the use of a biocompatible polymer, poly-oxyethylene polymer (PEO), as a media additive to modify the viscoelasticity of the fluid. The polymer PEO can increase the size of microfluidic viscoelasticity, thus increasing the viscoelastic force acting on the exosome. As shown in Fig. 4.7a, during the separation process, the sheath fluid containing PEO and the sample were injected into the two injection ports, respectively. Accompanied by the sheath fluid, the sample was firstly on the two sides of the channel and then moved to the middle with the viscoelastic force of the larger-sized EVs in the sample gradually increasing, while the EVs were kept in the downstream microchannels in the microchannels at the exits of the two sides of the wall of the microchannels, thus separating both of them. The flow rates of the inlet and sheath streams are controlled so that the sample starts on either side of the

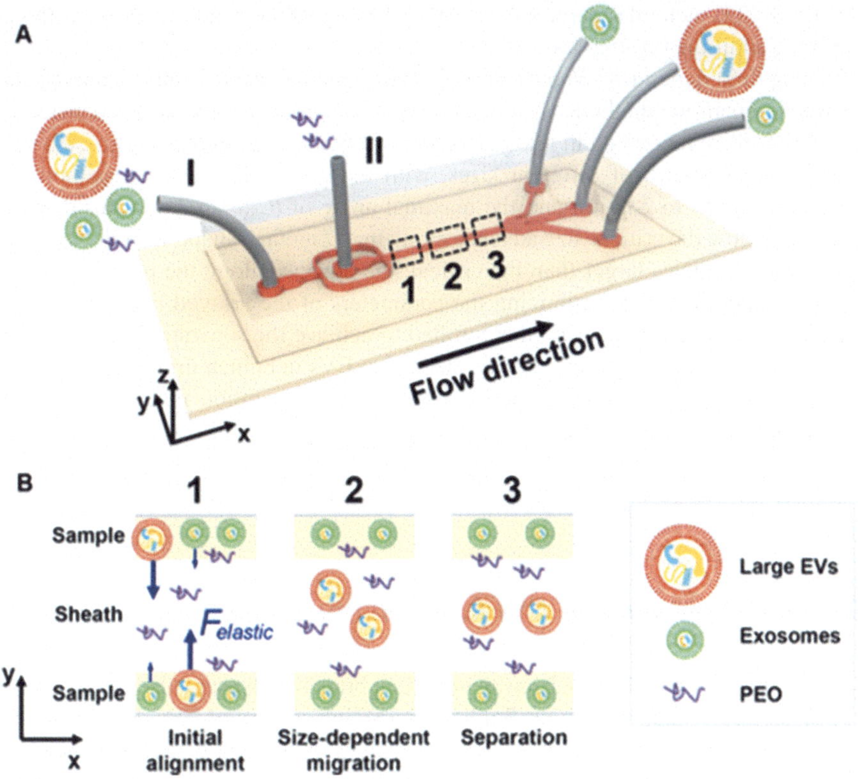

Fig. 4.7 Schematic representation of the separation of EVs using hydrodynamic properties

microchannel, the larger extracellular vesicles are subjected to greater viscoelastic forces to move toward the center of the channel, and the smaller EVs are subjected to limited viscoelastic forces and smaller displacements so that the EVs and the larger extracellular vesicles are moved to different exits.

6.3.2 Separation of EVs Based on the Principle of Deterministic Lateral Displacement

Deterministic lateral displacement (DLD) is a size-based cell sorting technique that consists of a set of ordered micropillar arrays. Laminar fluid interacts with the arrays, forcing the particles or cells into a specific trajectory predetermined by the device [96]. In other words, depending on the size of the particles to be separated, the geometry of the gradient column array is designed to determine the cut-off diameter of the particles; particles smaller than the cut-off direct "zig-zag" and flow out in a streamline [97], while particles larger than the cut-off diameter undergo a lateral displacement, so that particles of different sizes are collected at different locations. This method separates the particles by microfluid dynamics alone without applying additional source forces.

Using fluid dynamics, Wunsch et al. [97] designed a lateral displacement-based exosome separation method, as shown in Fig. 4.7b. In this method, arrayed nanopillars DLDs were fabricated in microchannels by photolithography to control the trajectories of particles of different sizes, with the gap of the arrayed nanopillars ranging from 25 to 235 nm and the maximal angle of $\theta_{max} = 5.7°$. In this design, particles below the critical size advance along the "Zig-zag" shape with the laminar flow, while particles larger than the critical size advance along the path set by the arrayed nanopillars. By optimizing the parameters of the arrayed nanopillars, the trajectory of the particles can be controlled to achieve the separation of EVs. The group also used this method to isolate 60–70 nm EVs in human urine successfully.

Compared with other microfluidic methods, the separation method using the flow characteristics of microfluidics does not need to apply additional field force, is convenient for separation, and has little damage to EVs, providing a new idea for exosome separation and quantitative analysis of microsamples.

6.4 EVs Separation of EVs Based on Immunoaffinity

Immuno-based methods have been an important branch of exosome isolation, which can also be combined with microfluidics for specific capture of EVs. Microfluidics provides a richer and more effective way to achieve immunophilic isolation of EVs. Exosome-specific proteins such as CD9, CD63, and CD81 are commonly used as markers. In addition, EVs also carry specific proteins, such as EpCAM and other proteins, which can also be captured with antibodies to the corresponding proteins.

Im et al. [98] reported an exosome assay using transmission SPR, a nanopore array functionalized with antibodies specific to exosome surface proteins. Using this method, they identified EVs purified from ovarian cancer cell cultures and ascites from ovarian cancer patients. In contrast, Hansford et al. [99] constructed a herringbone groove microfluidic device functionalized with antibodies, and in their experiments, the researchers used it for the analysis of EVs in the serum of patients with high-grade plasma ovarian cancer and used fluorescence microscopy to characterize EVs captured and released by the microfluidic chip, thus enabling dynamic monitoring of EVs in the elution process of EVs. A technique that allows quantitative analysis of EVs in real time was designed and fabricated by researchers [100], which achieves exosome capture by modifying antibodies to exosome-specific proteins on the bottom of the planar surface, followed by analyzing the signal by surface plasmon resonance imaging (SPRi). In addition, modifying graphene oxide and polydopamine with nanostructures on the planar surfaces and then modifying them with antibodies to capture EVs can significantly improve the efficiency of exosome capture [101].

A simple microfluidic method (ExoSearch) was developed that provides an enriched plasma exosome preparation for in situ multiplexed assays using immunomagnetic beads. The ExoSearch chip provides a robust continuous flow design for quantitative separation and release of plasma EVs in various preparation volumes (10 μL to 10 mL).

In addition to the above exosome separation methods, some separation methods are developed for specific applications and special needs. For example, Chiu et al. [102] proposed a single-cell exosome assay to achieve real-time detection of EVs in single cells. In this method, a PDMS substrate with mesh holes was designed and fabricated. Centrifugal force was applied to make each mesh hole capable of loading single cells and culturing them. Then the EVs secreted by the cells in each mesh hole were collected by using coverslips immobilized with the CD63 antibody corresponding to the EVs secreted by the cells in each mesh hole. This method can be used to quantify the number of EVs secreted by single cells in an in vitro environment, and to study the effects of drugs and acidic environment on exosome secretion and protein expression.

In recent years, microfluidic technology has been widely used in exosome separation. The microfluidic-based separation method has the advantages of faster separation rate, higher separation yield and purity, and less sample requirement, which is more suitable for the analysis of micro samples compared with the above traditional methods, and it is expected to be widely used in the future of personal healthcare and precision therapy. However, the current microfluidic-based exosome separation method is still in the experimental research stage, and further development is needed.

7 Conclusion and Perspectives

Enhancing the yield and purity of EVs stands as the foremost imperative task, perennially impeding the transformative application of these entities. Recent investigations underscore that the judicious amalgamation of diverse techniques for extracting and purifying EVs can redress these challenges appreciably. Yet, the optimization of their synergistic utilization warrants exploration. Furthermore, the requisite purity standards for EVs are contingent upon the specific objectives of distinct research inquiries. Exploring the functionalities of different biomolecules typically necessitates elevated EVs purity. However, in instances where known EVs biomarkers like miRNA and surface molecules are the focal points of analysis within the EVs population, stringent purity prerequisites might be discretionary. Hence, the inclination toward expeditious separation methods gains precedence. Currently, a paramount focus involves the incessant refinement and formulation of straightforward, efficient, cost-effective, and highly reproducible isolation techniques. The standardization of isolation strategies emerges as pivotal, ensuring the precision of downstream experimental outcomes and augmenting the comparability of research findings. Encouragingly, the isolated EVs have been employed in preclinical and clinical studies for disease diagnosis, monitoring, and treatment results. The anticipation is that the resolution of these pivotal issues will pave the way for the utilization of EVs as an innovative strategy in clinical applications, with ongoing advancements in isolation techniques poised to unfold in the imminent future.

References

1. Martins TS, et al. A review on comparative studies addressing exosome isolation methods from body fluids. Anal Bioanal Chem. 2022;415(7):1239–63. https://doi.org/10.1007/s00216-022-04174-5.
2. Vaswani K, et al. A method for the isolation of exosomes from human and bovine Milk. J Nutr Metab. 2019;2019:5764740. https://doi.org/10.1155/2019/5764740.
3. Zarovni N, et al. Integrated isolation and quantitative analysis of exosome shuttled proteins and nucleic acids using immunocapture approaches. Methods. 2015;87:46–58. https://doi.org/10.1016/j.ymeth.2015.05.028.
4. Konoshenko MY, et al. Isolation of extracellular vesicles: general methodologies and latest trends. Biomed Res Int. 2018;2018:1–27. https://doi.org/10.1155/2018/8545347.
5. Su W, et al. Microfluidic strategies for label-free exosomes isolation and analysis. TrAC Trends Anal Chem. 2019;118:686–98. https://doi.org/10.1016/j.trac.2019.06.037.
6. Yang D, et al. Progress, opportunity, and perspective on exosome isolation—efforts for efficient exosome-based theranostics. Theranostics. 2020;10(8):3684–707. https://doi.org/10.7150/thno.41580.
7. Hu HT, et al. Ultracentrifugal separation, characterization, and functional study of extracellular vesicles derived from serum-free cell culture. STAR Protocols. 2021;2(3):100625. https://doi.org/10.1016/j.xpro.2021.100625.
8. Hiemstra TF, et al. Human urinary exosomes as innate immune effectors. J Am Soc Nephrol. 2014;25(9):2017–27. https://doi.org/10.1681/ASN.2013101066.
9. Langevin SM, et al. Balancing yield, purity and practicality: a modified differential ultracentrifugation protocol for efficient isolation of small extracellular vesicles from human serum. RNA Biol. 2019;16(1):5–12. https://doi.org/10.1080/15476286.2018.1564465.
10. Paolini L, et al. Residual matrix from different separation techniques impacts exosome biological activity. Sci Rep. 2016;6:23550. https://doi.org/10.1038/srep23550.
11. Tran PHL, et al. Development of a nanoamorphous exosomal delivery system as an effective biological platform for improved encapsulation of hydrophobic drugs. Int J Pharm. 2019;566:697–707. https://doi.org/10.1016/j.ijpharm.2019.06.028.
12. Cvjetkovic A, et al. The influence of rotor type and centrifugation time on the yield and purity of extracellular vesicles. JExtracell Vesicles. 2014;3(1):23111. https://doi.org/10.3402/jev.v3.23111.
13. Doyle L, et al. Overview of extracellular vesicles, their origin, composition, purpose, and methods for exosome isolation and analysis. Cells. 2019;8(7):727. https://doi.org/10.3390/cells8070727.
14. Zaborowski MP, et al. Extracellular vesicles: composition, biological relevance, and methods of study. Bioscience. 2015;65(8):783–97. https://doi.org/10.1093/biosci/biv084.
15. Zhang M, et al. Methods and Technologies for Exosome Isolation and Characterization. Small Methods. 2018;2(9):1800021. https://doi.org/10.1002/smtd.201800021.
16. Rito-Palomares M, et al. Isolation of exosomes from blood plasma: qualitative and quantitative comparison of ultracentrifugation and size exclusion chromatography methods. PLoS One. 2015;10(12):e0145686. https://doi.org/10.1371/journal.pone.0145686.
17. Szatanek R, et al. Isolation of extracellular vesicles: determining the correct approach (review). Int J Mol Med. 2015;36(1):11–7. https://doi.org/10.3892/ijmm.2015.2194.
18. Wang W, et al. Recent Progress in isolation and detection of extracellular vesicles for cancer diagnostics. Adv Healthc Mater. 2018;7(20):1800484. https://doi.org/10.1002/adhm.201800484.

19. Gorgun C, et al. Isolation and flow cytometry characterization of extracellular-vesicle subpopulations derived from human mesenchymal stromal cells. Curr Protoc Stem Cell Biol. 2019;48(1):e76. https://doi.org/10.1002/cpsc.76.
20. Lo Sicco C, et al. A method for isolating and characterizing mesenchymal stromal cell-derived extracellular vesicles. Curr Protoc Stem Cell Biol. 2018;46(1):e55. https://doi.org/10.1002/cpsc.55.
21. Chen B-Y, et al. Advances in exosomes technology. Clin Chim Acta. 2019;493:14–9. https://doi.org/10.1016/j.cca.2019.02.021.
22. Caradec J, et al. Reproducibility and efficiency of serum-derived exosome extraction methods. Clin Biochem. 2014;47(13–14):1286–92. https://doi.org/10.1016/j.clinbiochem.2014.06.011.
23. Wang J, et al. Toward microfluidic-based exosome isolation and detection for tumor therapy. Nano Today. 2021;37:101066. https://doi.org/10.1016/j.nantod.2020.101066.
24. Brennan K, et al. A comparison of methods for the isolation and separation of extracellular vesicles from protein and lipid particles in human serum. Sci Rep. 2020;10(1):1039. https://doi.org/10.1038/s41598-020-57497-7.
25. Anderson NG. An introduction to particle separations in zonal centrifuges. Natl Cancer Inst Monogr. 1966;21:9–39.
26. Raffai RL, et al. Cushioned-density gradient ultracentrifugation (C-DGUC) improves the isolation efficiency of extracellular vesicles. PLoS One. 2019;14(4):e0215324. https://doi.org/10.1371/journal.pone.0215324.
27. Ruysschaert T, et al. Liposome retention in size exclusion chromatography. BMC Biotechnol. 2005;5(1):11. https://doi.org/10.1186/1472-6750-5-11.
28. Liangsupree T, et al. Modern isolation and separation techniques for extracellular vesicles. J Chromatogr A. 2021;1636:461773. https://doi.org/10.1016/j.chroma.2020.461773.
29. Hicks DA, et al. Extracellular vesicles isolated from human induced pluripotent stem cell-derived neurons contain a transcriptional network. Neurochem Res. 2020;45(7):1711–28. https://doi.org/10.1007/s11064-020-03019-w.
30. Burgess RR. A brief practical review of size exclusion chromatography: rules of thumb, limitations, and troubleshooting. Protein Expr Purif. 2018;150:81–5. https://doi.org/10.1016/j.pep.2018.05.007.
31. Stranska R, et al. Comparison of membrane affinity-based method with size-exclusion chromatography for isolation of exosome-like vesicles from human plasma. J Transl Med. 2018;16(1):1. https://doi.org/10.1186/s12967-017-1374-6.
32. Askeland A, et al. Mass-spectrometry based proteome comparison of extracellular vesicle isolation methods: comparison of ME-kit, size-exclusion chromatography, and high-speed centrifugation. Biomedicines. 2020;8(8):246. https://doi.org/10.3390/biomedicines8080246.
33. Povero D, et al. Characterization and proteome of circulating extracellular vesicles as potential biomarkers for NASH. Hepatol Commun. 2020;4(9):1263–78. https://doi.org/10.1002/hep4.1556.
34. O'Neil EV, et al. Characterization and regulation of extracellular vesicles in the lumen of the ovine uterus. Biol Reprod. 2020;102(5):1020–32. https://doi.org/10.1093/biolre/ioaa019.
35. Zheng H, et al. Deconstruction of heterogeneity of size-dependent exosome subpopulations from human urine by profiling N-Glycoproteomics and Phosphoproteomics simultaneously. Anal Chem. 2020;92(13):9239–46. https://doi.org/10.1021/acs.analchem.0c01572.
36. Aqrawi LA, et al. Proteomic and histopathological characterization of sicca subjects and primary Sjögren's syndrome patients reveals promising tear, saliva and extracellular vesicle disease biomarkers. Arthritis Res Ther. 2019;21:181. https://doi.org/10.1186/s13075-019-1961-4.
37. Bartel S, et al. Human airway epithelial extracellular vesicle miRNA signature is altered upon asthma development. Allergy. 2019;75(2):346–56. https://doi.org/10.1111/all.14008.
38. Foers AD, et al. Enrichment of extracellular vesicles from human synovial fluid using size exclusion chromatography. J Extracell Vesicles. 2018;7(1):1490145. https://doi.org/10.1080/20013078.2018.1490145.

39. Prieto-Fernández E, et al. A comprehensive study of vesicular and non-vesicular miRNAs from a volume of cerebrospinal fluid compatible with clinical practice. Theranostics. 2019;9(16):4567–79. https://doi.org/10.7150/thno.31502.
40. Czystowska-Kuzmicz M, et al. Small extracellular vesicles containing arginase-1 suppress T-cell responses and promote tumor growth in ovarian carcinoma. Nat Commun. 2019;10(1):3000. https://doi.org/10.1038/s41467-019-10979-3.
41. Gámez-Valero A, et al. Size-exclusion chromatography-based isolation minimally alters extracellular vesicles' characteristics compared to precipitating agents. Sci Rep. 2016;6(1):33641. https://doi.org/10.1038/srep33641.
42. Takov K, et al. Comparison of small extracellular vesicles isolated from plasma by ultracentrifugation or size-exclusion chromatography: yield, purity and functional potential. J Extracell Vesicle. 2018;8:1. https://doi.org/10.1080/20013078.2018.1560809.
43. Wei R, et al. Combination of size-exclusion chromatography and ultracentrifugation improves the proteomic profiling of plasma-derived small extracellular vesicles. Biol Proced. 2020;22:12. https://doi.org/10.1186/s12575-020-00125-5.
44. Benedikter BJ, et al. Ultrafiltration combined with size exclusion chromatography efficiently isolates extracellular vesicles from cell culture media for compositional and functional studies. Sci Rep. 2017;7:15297. https://doi.org/10.1038/s41598-017-15717-7.
45. Nordin JZ, et al. Ultrafiltration with size-exclusion liquid chromatography for high yield isolation of extracellular vesicles preserving intact biophysical and functional properties. Nanomedicine. 2015;11(4):879–83. https://doi.org/10.1016/j.nano.2015.01.003.
46. Stam J, et al. Isolation of extracellular vesicles with combined enrichment methods. J Chromatogr B. 2021;1169:122604. https://doi.org/10.1016/j.jchromb.2021.122604.
47. He L, et al. A highly efficient method for isolating urinary exosomes. Int J Mol Med. 2018;43(1):83–90. https://doi.org/10.3892/ijmm.2018.3944.
48. Taylor N, et al. Retention characteristics of sterile filters—effect of pore size and structure. J Membr Sci. 2021;635:119436. https://doi.org/10.1016/j.memsci.2021.119436.
49. Kim K, et al. Cyclic tangential flow filtration system for isolation of extracellular vesicles. APL Bioeng. 2021;5:016103. https://doi.org/10.1063/5.0037768.
50. Zeringer E, et al. Strategies for isolation of exosomes. Cold Spring Harb Protoc. 2015;2015(4):319–23. https://doi.org/10.1101/pdb.top074476.
51. Albertsson P-Å, et al. Partition of virus particles in a liquid two-phase system. Biochim Biophys Acta. 1960;37(2):230–7. https://doi.org/10.1016/0006-3002(60)90228-6.
52. Gunter K, et al. A method of resuspending small vesicles separated from suspension by protamine aggregation and centrifugation. Anal Biochem. 1982;120(1):113–24. https://doi.org/10.1016/0003-2697(82)90326-8.
53. Brownlee Z, et al. A novel "salting-out" procedure for the isolation of tumor-derived exosomes. J Immunol Methods. 2014;407:120–6. https://doi.org/10.1016/j.jim.2014.04.003.
54. Gallart-Palau X, et al. Enrichment of extracellular vesicles from tissues of the central nervous system by PROSPR. Mol Neurodegener. 2016;11:41. https://doi.org/10.1186/s13024-016-0108-1.
55. García-Romero N, et al. Polyethylene glycol improves current methods for circulating extracellular vesicle-derived DNA isolation. J Transl Med. 2019;17:75. https://doi.org/10.1186/s12967-019-1825-3.
56. Weng Y, et al. Effective isolation of exosomes with polyethylene glycol from cell culture supernatant for in-depth proteome profiling. Analyst. 2016;141(15):4640–6. https://doi.org/10.1039/c6an00892e.
57. Dou YQ, et al. Smooth muscle SIRT1 reprograms endothelial cells to suppress angiogenesis after ischemia. Theranostics. 2020;10(3):1197–212. https://doi.org/10.7150/thno.39320.
58. Kanchi Ravi R, et al. A modified precipitation method to isolate urinary exosomes. J Vis Exp. 2015;95:51158. https://doi.org/10.3791/51158.
59. Wan Z, et al. Mononuclear phagocyte system blockade improves therapeutic exosome delivery to the myocardium. Theranostics. 2020;10(1):218–30. https://doi.org/10.7150/thno.38198.

60. Merdalimova A, et al. Identification and analysis of exosomes by surface-enhanced Raman spectroscopy. Appl Sci. 2019;9(6):1135. https://doi.org/10.3390/app9061135.
61. Amarnath S, et al. Bone marrow-derived mesenchymal stromal cells harness purinergenic signaling to tolerize human Th1 cells in vivo. Stem Cells. 2015;33(4):1200–12. https://doi.org/10.1002/stem.1934.
62. Rider MA, et al. ExtraPEG: a polyethylene glycol-based method for enrichment of extracellular vesicles. Sci Rep. 2016;6:23978. https://doi.org/10.1038/srep23978.
63. Chang M, et al. Exosome purification based on PEG-coated Fe_3O_4 nanoparticles. PLoS One. 2018;13(6):e0199438. https://doi.org/10.1371/journal.pone.0199438.
64. Kastelowitz N, et al. Exosomes and microvesicles: identification and targeting by particle size and lipid chemical probes. Chembiochem. 2014;15(7):923–8. https://doi.org/10.1002/cbic.201400043.
65. Andreu Z, et al. Tetraspanins in extracellular vesicle formation and function. Front Immunol. 2014;5:442. https://doi.org/10.3389/fimmu.2014.00442.
66. Smolarz M, et al. Proteome profiling of exosomes purified from a small amount of human serum: the problem of co-purified serum components. Proteomes. 2019;7(2):18. https://doi.org/10.3390/proteomes7020018.
67. Zhao Z, et al. A microfluidic ExoSearch chip for multiplexed exosome detection toward blood-based ovarian cancer diagnosis. Lab Chip. 2016;16(3):489–96. https://doi.org/10.1039/c5lc01117e.
68. Hildonen S, et al. Isolation and mass spectrometry analysis of urinary extraexosomal proteins. Sci Rep. 2016;6:36331. https://doi.org/10.1038/srep36331.
69. Mathivanan S, et al. Proteomics analysis of A33 immunoaffinity-purified exosomes released from the human colon tumor cell line LIM1215 reveals a tissue-specific protein signature. Mol Cell Proteom. 2010;9(2):197–208. https://doi.org/10.1074/mcp.m900152-mcp200.
70. Taylor DD, et al. MicroRNA signatures of tumor-derived exosomes as diagnostic biomarkers of ovarian cancer. Gynecologic Oncology. 2008;110(1):13–21. https://doi.org/10.1016/j.ygyno.2008.04.033.
71. Ueda K, et al. Antibody-coupled monolithic silica microtips for highthroughput molecular profiling of circulating exosomes. Sci Rep. 2014;4:6232. https://doi.org/10.1038/srep06232.
72. Jeppesen DK, et al. Reassessment of exosome composition. Cell. 2019;177(2):428–45. https://doi.org/10.1016/j.cell.2019.02.029.
73. Ziaei P, et al. Silica nanostructured platform for affinity capture of tumor-derived exosomes. J Mater Sci. 2017;52(12):6907–16. https://doi.org/10.1007/s10853-017-0905-0.
74. Multia E, et al. Fast isolation of highly specific population of platelet-derived extracellular vesicles from blood plasma by affinity monolithic column, immobilized with anti-human CD61 antibody. Anal Chim Acta. 2019;1091:160–8. https://doi.org/10.1016/j.aca.2019.09.022.
75. Nishat S, et al. Paper-based microfluidics: simplified fabrication and assay methods. Sensors Actuators B Chem. 2021;336:129681. https://doi.org/10.1016/j.snb.2021.129681.
76. Nakai W, et al. A novel affinity-based method for the isolation of highly purified extracellular vesicles. Sci Rep. 2016;6:33935. https://doi.org/10.1038/srep33935.
77. Xu H, et al. Magnetic-based microfluidic device for on-chip isolation and detection of tumor-derived exosomes. Anal Chem. 2018;90(22):13451–8. https://doi.org/10.1021/acs.analchem.8b03272.
78. Li Q, et al. Reversible zwitterionic coordination enables rapid, high-yield, and high-purity isolation of extracellular vesicles from biofluids. Science. Advances. 2023;9(15):eadf4568. https://doi.org/10.1126/sciadv.adf4568.
79. Kanao E, et al. Rapid and highly efficient purification of extracellular vesicles enabled by a TiO2 hybridized spongy-like polymer. Anal Chem. 2023;95(38):14502–10. https://doi.org/10.1021/acs.analchem.3c03411.
80. Chen M, et al. Ti3C2 and Ti2C MXene materials for high-performance isolation of extracellular vesicles via coprecipitation. Anal Chim Acta. 2023;1269:341426. https://doi.org/10.1016/j.aca.2023.341426.

81. Beebe DJ, et al. Physics and applications of microfluidics in biology. Annual Rev Biomed Eng. 2002;4:261–86. https://doi.org/10.1146/annurev.bioeng.4.112601.125916.
82. Whitesides GM. The origins and the future of microfluidics. Nature. 2006;442(7101):368–73. https://doi.org/10.1038/nature05058.
83. Li P, et al. Progress in exosome isolation techniques. Theranostics. 2017;7(3):789–804. https://doi.org/10.7150/thno.18133.
84. Wu M, et al. Acoustofluidic separation of cells and particles. Microsyst Nanoeng. 2019;5:32. https://doi.org/10.1038/s41378-019-0064-3.
85. Ahmed D, et al. Rotational manipulation of single cells and organisms using acoustic waves. Nat Commun. 2016;7:11085. https://doi.org/10.1038/ncomms11085.
86. Augustsson P, et al. Iso-acoustic focusing of cells for size-insensitive acousto-mechanical phenotyping. Nat Commun. 2016;7:11556. https://doi.org/10.1038/ncomms11556.
87. Wu M, et al. Isolation of exosomes from whole blood by integrating acoustics and microfluidics. Proc Natl Acad Sci. 2017;114(40):10584–9. https://doi.org/10.1073/pnas.1709210114.
88. Lee K, et al. Acoustic purification of extracellular microvesicles. ACS Nano. 2015;9(3):2321–7. https://doi.org/10.1021/nn506538f.
89. Ibsen SD, et al. Rapid isolation and detection of exosomes and associated biomarkers from plasma. ACS Nano. 2017;11(7):6641–51. https://doi.org/10.1021/acsnano.7b00549.
90. Heineck DP, et al. Electrokinetic device design and constraints for use in high conductance solutions. Electrophoresis. 2017;38(11):1475–82. https://doi.org/10.1002/elps.201600563.
91. Sonnenberg A, et al. Dielectrophoretic isolation of DNA and nanoparticles from blood. Electrophoresis. 2012;33(16):2482–90. https://doi.org/10.1002/elps.201100700.
92. Lu X, et al. Continuous microfluidic particle separation via Elasto-inertial pinched flow fractionation. Anal Chem. 2015;87(12):6389–96. https://doi.org/10.1021/acs.analchem.5b01432.
93. Liu C, et al. Size-based separation of particles and cells utilizing viscoelastic effects in straight microchannels. Anal Chem. 2015;87(12):6041–8. https://doi.org/10.1021/acs.analchem.5b00516.
94. Zhang J, et al. A novel viscoelastic-based ferrofluid for continuous sheathless microfluidic separation of nonmagnetic microparticles. Lab Chip. 2016;16(20):3947–56. https://doi.org/10.1039/c6lc01007e.
95. Liu C, et al. Field-free isolation of exosomes from extracellular vesicles by microfluidic viscoelastic flows. ACS Nano. 2017;11(7):6968–76. https://doi.org/10.1021/acsnano.7b02277.
96. Salafi T, et al. A review on deterministic lateral displacement for particle separation and detection. Nano-Micro Lett. 2019;11(1):77. https://doi.org/10.1007/s40820-019-0308-7.
97. Wunsch BH, et al. Nanoscale lateral displacement arrays for the separation of exosomes and colloids down to 20 nm. Nat Nanotechnol. 2016;11(11):936–40. https://doi.org/10.1038/nnano.2016.134.
98. Im H, et al. Label-free detection and molecular profiling of exosomes with a nano-plasmonic sensor. Nat Biotechnol. 2014;32(5):490–5. https://doi.org/10.1038/nbt.2886.
99. Hisey CL, et al. Microfluidic affinity separation chip for selective capture and release of label-free ovarian cancer exosomes. Lab Chip. 2018;18(20):3144–53. https://doi.org/10.1039/c8lc00834e.
100. He M, et al. Integrated immunoisolation and protein analysis of circulating exosomes using microfluidic technology. Lab Chip. 2014;14(19):3773–80. https://doi.org/10.1039/c4lc00662c.
101. Zhu L, et al. Label-free quantitative detection of tumor-derived exosomes through surface plasmon resonance imaging. Anal Chem. 2014;86(17):8857–64. https://doi.org/10.1021/ac5023056.
102. Chiu YJ, et al. A single-cell assay for time lapse studies of exosome secretion and cell behaviors. Small. 2016;12(27):3658–66. https://doi.org/10.1002/smll.201600725.

Chapter 5
Technologies for EV Surface Modification and its Application in Targeted Therapy

Mengyuan Geng, Ying Chang, and Zongjin Li

Abstract In the new era of precision medicine, a personalized drug delivery system with better biocompatibility and higher drug loading efficiency needs to be established to overcome a series of obstacles in drug delivery to target tissues. As natural nanoscale biological vesicles, extracellular vesicles (EVs) are attractive novel drug carriers due to their inherent low immunogenicity, long circulation time, and an enhanced capacity to cross biological barriers. However, efficient and targetable drug delivery is still the bottleneck for further application of EVs. To solve this problem, various surface modification technologies for EVs have emerged in recent years and can be divided into two main categories: genetic manipulation and direct surface modification. Through design and manipulation, functionalized EVs exhibit more specific biological distributions and enhanced pharmacokinetics, which improve the application potential of EVs in drug loading. In this chapter, we summarize the specific methods of EV surface modification and typical schemes for using functionalized EVs for targeted therapy and discuss the advantages, current limitations, and future development directions of EVs as carriers for targeted drug delivery.

Keywords EVs · Surface modification · Genetic manipulation · Direct surface modification · Targeted therapy

Abbreviations

aHSC	Activated hepatic stellate cells
BBB	Blood–brain barrier
CNS	Central nervous system

M. Geng · Y. Chang
Department of Clinical Laboratory, Tianjin Central Hospital of Obstetrics and Gynecology, Affiliated Maternity Hospital of Nankai University, Tianjin, China

Z. Li (✉)
Department of Pathophysiology, Nankai University School of Medicine, Tianjin, China
e-mail: zongjinli@nankai.edu.cn

© The Author(s), under exclusive license to Springer Nature Singapore Pte Ltd. 2024

Z. Li (ed.), *Extracellular Vesicle: Biology and Translational Application*,
https://doi.org/10.1007/978-981-97-5536-3_5

CVD	Cardiovascular disease
DSPE	Distearoyl Phosphoethanolamine
EGFR	Epidermal growth factor receptor
EVs	Extracellular vesicles
FRET	Förster resonant energy transfer
HA	Hyaluronic acid
HRP	Horseradish peroxidase
lamp-2b	Lysosome-associated membrane protein-2b
mAb	Monoclonal antibody
MSCs	Mesenchymal stem cells
NHS	N-hydroxysuccinimidyl
OVA	Ovalbumin
PDGFR	Platelet-derived growth factor receptor
PEG	Polyethylene glycol
PS	Phosphatidylserine
RVG	Rabies virus glycoprotein
scFv	Single-chain variable fragment
siRNA	Small interfering RNA
ssDNA	Single-stranded DNA

1 Introduction

Extracellular vesicles (EVs) are membrane-bound bionanoparticles secreted or released by most cells into the extracellular space; these vesicles harbor biomolecular payloads such as proteins, lipids, RNA, and DNA [1], and there is no doubt that they play a significant role in intercellular communication [2]. EVs can be classified into three main populations: exosomes, microvesicles (MVs), and apoptotic bodies (APOs) [3]. However, the terminology used to describe these secretory vesicles is confusing because it is difficult to distinguish between the different types of vesicles after secretion. To avoid confusion in the present article, we used "extracellular vesicles (EVs)" as a hypernym for all types of vesicles that are present in the extracellular space. In the physiological environment, EVs exhibit very low immunogenicity, can penetrate physiological barriers [4], and have high potential in the treatment of various diseases. For example, EVs derived from mesenchymal stem cells (MSCs) have great regenerative and protective potential in animal models of myocardial infarction [5, 6], lung injury [7], skin injury [8], diabetes [9, 10], and stroke and neurological injury [4, 11]. Prabhu et al. reported that stem cell ($CD34^+$ cell)-derived EVs can repair limb ischemia [12]. The innate therapeutic potential of cell-derived EVs has been described in detail in another review [13].

With the strict requirements of precision medicine, natural EVs have problems such as low targeting ability, low drug loading efficiency, unstable circulation and insufficient raw materials, which can no longer meet the needs of multiple treatments. Therefore, many EV surface modification technologies have been developed so that proteins, lipids, nucleic acids and other types of therapeutic agents or targeted molecules can be effectively loaded on the surface of EVs. Its purpose is to increase the loading dose of therapeutic agents, improve the local drug concentration of lesions, and maximize the therapeutic effect of drugs. With the help of a stable phospholipid bilayer, the loaded cargo can resist the action of immune cells and digestive enzymes and be delivered to cells or tissues to play a therapeutic role. EVs release therapeutic agents after being taken up by cells to avoid degradation by lysosomes, improve drug targeting and reduce side effects, and become natural and effective drug carriers. By now, researchers worldwide have registered a total of 121 clinical trials of EV therapy involving different types of disease. Of the 121 trials, 21 clinical trials have been completed and demonstrated the positive therapeutic effects of EV-based therapy in various cancers, neurodegenerative diseases, and blood diseases [14]. With the strict requirements of precision medicine, many EV surface modification technologies have been developed to enable the effective loading of various types of therapeutic agents or targeted molecules, such as proteins, lipids, and nucleic acids, on the surface of EVs. The purpose of this approach is to increase the loading dose of therapeutic agents, improve the local drug concentration of lesions, and maximize the therapeutic effect of drugs. In this review, we introduce in detail the effective technologies available for surface modification of EVs and their specific application in different diseases and reveal the potential of surface-modified EVs for disease treatment.

2 Techniques for Surface Modification of EVs

Surface modification is the process of treating EV surface proteins as anchoring devices or affinity tags and modifying the protein or peptide components that meet the requirements of the particle surface by certain means [15]. Surface modification of EVs via gene manipulation and direct modification techniques is considered to be an effective strategy for improving their targeted delivery capability, minimizing untargeted toxicity, reducing drug dosage, increasing circulation time in vivo, and enhancing disease therapeutic efficacy [16]. To engineer EVs via parent cells, researchers primarily genetically modified donor cells and subsequently collected EVs with functional proteins or peptides. In contrast, direct modification seems simpler and more convenient. EVs derived from body fluids or cells are often used as raw materials for direct modification with the aid of their surface characteristics, such as membrane-associated proteins, lipids, and charge. Genetic manipulation and direct modification technologies have their advantages and disadvantages and need to be appropriately selected for application (Table 5.1). In the next two subsections, we discuss these two major technologies separately.

Table 5.1 Comparison of advantages and disadvantages between gene manipulation and direct modification

Methods	Principles	Advantages	Disadvantages
Genetic manipulation	Based on transmembrane proteins and membrane-associated proteins	Maintenance of membrane integrity Sustained secretion of engineered EVs	Uncontrollability of modifying efficiency Features losing with cell passages increase A time-consuming and labor-intensive process Potential gene editing risks Limitations of the modified molecular species Cytotoxicity to the donor cells
Direct modification	Protein–protein interaction	Simple operation A rapid and effective method	Dependent on the expression levels of specific proteins
	Click chemistry	Maintenance of membrane integrity Good stability of covalent bonding Without apparent toxicity or physiological perturbation	Safety needs to be verified
	Hydrophobic insertion	A rapid and simple method High efficiency of cargo incorporation Diversity of loaded cargo	Have a strong tendency to self-assemble Possible alteration of membrane proteins Lack of site specificity May cause side effects
	Membrane fluidity	Allow more complex membrane modifications	Risk of cargo leakage during fusion Development of efficient fusion methods
	Membrane curvature	Low synthesis cost strong modifiability	The possibility of combining with other types of vesicles

2.1 Surface Modification of EVs Via Genetic Manipulation of Cells

As promising drug delivery vehicles, modifying targeting molecules and drugs could increase the applicability of EVs and expand the available approaches for precision therapy. Targeting molecules, proteins or peptide drugs are typically displayed on the surface of EVs by fusing with transmembrane proteins or membrane-associated proteins [17]. Frequently used transmembrane proteins for EV surface

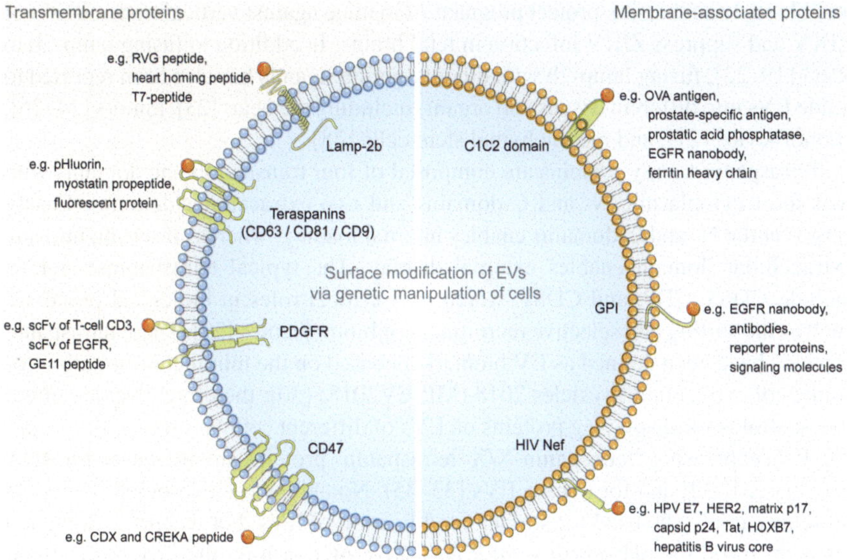

Fig. 5.1 Schematic of EV surface modifications via genetic manipulation

modification include lysosome-associated membrane protein-2b (lamp-2b), tetraspanin family proteins (CD63/CD81/CD9), and receptor membrane proteins, of which tetraspanins are the most commonly used. By comparison, membrane-associated proteins (the C1C2 domain, GPI domain, and HIV Nef protein) are effective and useful for displaying cargo on the EV surface. In this section, we summarize specific approaches for identifying EV surface modifications via genetic manipulation of different proteins (Fig. 5.1).

2.1.1 Manipulation by Fusion with Transmembrane Proteins

Lamp-2b is a protein on the EV surface with a large N-terminal extramembrane domain and a C-terminal transmembrane region containing a very short cytoplasmic tail region. Functional molecules could be displayed on the EV surface by fusion to the N-terminal extracellular domain of lamp-2b. Currently, lamp-2b is the most widely used exosomal surface protein to display a targeting motif, especially when it is fused to rabies virus glycoprotein (RVG) peptide [18] to achieve central nervous system (CNS) targeting. In a pioneering work, Alvarez-Erviti and colleagues engineered immature dendritic cell-derived EVs to express RVG-Lamp2b to target acetylcholine receptors expressed on neuronal cells [19]. The secreted EVs were endowed with neuron-specific targeting ability for efficient gene silencing and treatment of Alzheimer's disease. The same strategy of EV-engineering was adopted for developed to suppress fetal brain Zika virus (ZIKV) infection [20]. RVG-Lamp2b-labeled small EVs (encapsulated antiviral siRNA) could cross placental and blood–brain barrier, protect pregnant AG6 mice against vertical transmission of ZIKV and suppress ZIKV infection in fetal brains. In addition to fusing lamp-2b to RVG [19–22], fusing lamp-2b with other targeting ligands has also been reported to guide EVs into different tissues and organs, including the heart [23], tumors [24–26], chondrocytes [27], and mesenchymal stem cells [28].

Tetraspanin family proteins are composed of four transmembrane domains with two short cytoplasmic N- and C-domains and two extracellular domains. Fusing cargos at the N- and C-domains enables internal loading, whereas inserting into any extracellular domain enables external display. The typical tetraspanins on EVs include CD63, CD9, and CD81, which play critical roles in biological processes such as the sorting and selective recruitment of biomolecules [29]. These tetraspanin proteins have been defined as EV biomarkers based on the minimal information for studies of extracellular vesicles 2018 (MISEV2018) [30]; therefore, they are effective scaffolds for displaying proteins on EVs of different cellular origins. For example, EVs expressing "tetraspanin-X/X- tetraspanin" proteins are often used for RNA delivery [31, 32] and for tracing EVs [33–35]. Notably, these exosomal membrane proteins can also be used for surface modification of EVs. For example, Bong and his team fused the pH-sensitive GFP derivative pHluorin on the first extracellular loop of the CD63 protein (the surface of EVs) to observe dynamic vesicle fusion events [36]. Ran et al. demonstrated that the serum stability, delivery efficiency and efficacy of propeptide can be significantly enhanced by fusing the inhibitory domain of the myostatin propeptide into the second extracellular loop of CD63 (the surface of EVs). Recently, Natalie et al. described new and topologically distinct CD63 scaffolds. They generated sequential deletions of the transmembrane helix of CD63 to truncate a series of CD63 proteins, each of which was genetically fused to a fluorescent protein. Molecular and cellular characterization studies have shown that truncated CD63 proteins harboring transmembrane helix 3 (TM3) are correctly targeted and anchored to the EV membrane [37]. These authors indicated that TM3 alone was sufficient to bind to target proteins on the surface of EVs and provided a new idea for improved application of the CD63 protein for surface modification of EVs. In addition to modifying cargo on the surface outside of EVs, tetraspanins have also been applied to load cargo on the inner surface of EVs. Yim et al. described the use of EVs for protein loading via the use of optically reversible protein–protein interactions (EXPLORs). They fused CIBN, which can transiently and reversibly bind to cryptochrome 2 (CRY2) under blue light, to the N-terminus of CD9 and co-transfected with another construct consisting of CRY2 and cargo proteins into HEK293 cells, enabling a reversible protein–protein interaction controlled by blue light and loading cargo proteins into newly generated EVs [38, 39].

In addition to tetraspanins, there are abundant receptor membrane proteins on the surface of EVs, especially PDGFR and CD47. The transmembrane domain of the human platelet-derived growth factor receptor (PDGFR) was exploited as a fusion partner to display proteins on the surface of EVs; this domain has been widely used for expressing proteins on mammalian cell surfaces, and even a commercial vector named the pDisplay vector has been developed. In a beneficial application, Cheng et al. linked single-chain variable fragment (scFv) antibodies against T-cell CD3 and cancer cell-associated epidermal growth factor receptor (EGFR) to the transmembrane domain of PDGFR genetically, resulting in cross-linking of T cells and EGFR-expressing breast cancer cells and potent antitumor immunity both in vitro and in vivo [40]. This strategy was also used to anchor the epidermal growth factor receptor (EGFR)-binding peptide GE11 on the surface of HEK293 cells. The produced GE11-EVs could directly and specifically deliver let-7a miRNA to EGFR-expressing xenograft breast cancer tissues in immunodeficient mice [41].

CD47 is also a functional transmembrane protein on EVs, and its N-terminus is free of CD47 on the EV outer surface, where it mediates the protection of EVs from phagocytosis by monocytes and macrophages [42]. Yang et al. fused CDX and CREKA peptides to the N-terminus of CD47 to target PTEN-deficient human U87 cells and a murine GL261 glioma model, respectively. This design was successful in achieving increased uptake of targeted ligand-modified EVs in U87 and GL261 cells as well as active translation of the PTEN protein [43], illustrating the potential of CD47 as a bridge to display targeting peptides on the surface of EVs. Nevertheless, the fact that the N-terminus of CD47 is an important functional region for its physiological function should not be ignored [44]. The addition of other amino acid sequences at the N-terminus may lead to changes in CD47 crystal structures, affect the function of the CD47-SIPRα axis, and engulf engineered EVs by immune cells, decreasing therapeutic efficacy.

2.1.2 Manipulation by Fusion with Membrane-Associated Proteins

C1C2 is the general name for the C1 and C2 domains and is a phosphatidylserine (PS)-binding domain of lactadherin (also named milk fat globule-EGF factor 8 protein, MFGE8). C1C2 binds to EV lipids; hence, it is frequently used as an anchoring partner when engineering EVs. Christine and his colleagues associated the ovalbumin (OVA) antigen with EVs by fusing OVA to C1C2 cells. The authors demonstrated that the EV-OVA-C1C2 vaccine favored antigen-specific $CD4^+$ T lymphocyte induction and reduced tumor progression when administered to tumor-bearing mice [45]. Rountree et al. fused prostate-specific antigen (PSA) and prostatic acid phosphatase (PAP) on the surface of EVs via C1C2 for prostate cancer immunotherapy [46]. In addition to tumor antigen delivery [47, 48], Liu et al. designed a fusion protein of ferritin heavy chain (FTH1) and C1C2 to visualize mesenchymal stem cell (MSC)-derived EVs in vivo and in vitro by MRI [49]. Alternatively, Kooijmans et al. constructed recombinant C1C2-anti-EGFR nanobody fusion proteins to confer tumor cell targeting to isolated EVs in a "plug-and-play" manner [50]. This approach combines C1C2 with PS to develop a completely new method to directly modify EVs. Similarly, Maria et al. prepared lactadherin-fluorophore conjugates and inserted them onto the surface of milk EVs to study the uptake of EVs by intestinal epithelial cells [51].

Glycosylphosphatidylinositol (GPI)-anchored proteins strongly target the lipid bilayer membrane [52] and GPI-anchored protein decay-accelerating factor (DAF) was demonstrated to selectively sort into EVs during reticulocyte maturation [53].

Kooijmans et al manipulated EVs without altering their general characteristics by modifying GPI-anchored anti-EGFR nanobodies. Based on these observations, the capacity of EVs to bind to $EGFR^+$ tumor cells was improved under static and flow conditions. This study highlights GPI anchoring as a new tool for identifying a variety of functional ligands, such as antibodies, reporter proteins, and signaling molecules, in EVs [54].

The HIV Nef protein lacks enzymatic activity and can localize to the surfaces of EVs by anchoring its N-terminal myristoylation to membrane lipid raft microdomains [55]. Laura et al. identified a functionally defective G3C V153L E177G Nef mutant that could be incorporated into EVs at quite high levels. The authors developed an EV-based $CD8^+$ T-cell vaccine platform by fusing a foreign antigen to the C-terminus of the Nef mutant protein [56]. Paola et al. also used this method to display HPV E7 and HER2 on the surface of muscle EVs, which elicited a continuous and efficient antigen-specific cytotoxic T lymphocyte immune response [57, 58]. In addition, by fusing Nef mutant proteins to matrix p17, capsid p24, Tat, the hepatitis B virus core and HOXB7, multifunctional EVs could be prepared to treat infectious viruses (AIDS, hepatitis B) and tumors [59].

Notably, engineered cells can secrete EVs with corresponding characteristics continuously without compromising the membrane integrity of the EVs. Although genetic approaches have been widely applied to modify EVs of cell lines such as MSCs, there are still many disadvantages that deserve attention. In terms of genetic engineering, the following criteria can be used: (1) the low transfection efficiency and the low posttransfection survival of stem cells decrease the surface modification efficiency of the EVs; (2) some cells for which it is difficult to establish cell lines (e.g., T cells) are unable to adopt this method to modify their EVs; (3) the engineered features may be progressively reduced or even missing as engineered cell culture passages increase, which requires real-time monitoring of the presence of EV properties and repeatedly boosting the engineered features of parental cells; (4) whenever a new target molecule is desired to be modified, it is time- and labor-intensive to establish a new cell line; and (5) potential gene editing risks exist. In terms of cargo properties, (1) only polypeptides or proteinaceous molecules can be modified, with more restricted molecular species; (2) the modification efficiency of different proteins is inconsistent and highly dependent on the nature of the cargo and the parental cells; and (3) cytotoxic molecules or proteins that may cause changes in the characteristics of the donor cell (e.g., functional molecules that affect cell growth and proliferation) cannot be surface modified by this method. Therefore, there is an urgent need for other modification modalities that are simple, rapid, and applicable to compensate for the genetic engineering approach for surface functionalization of EVs.

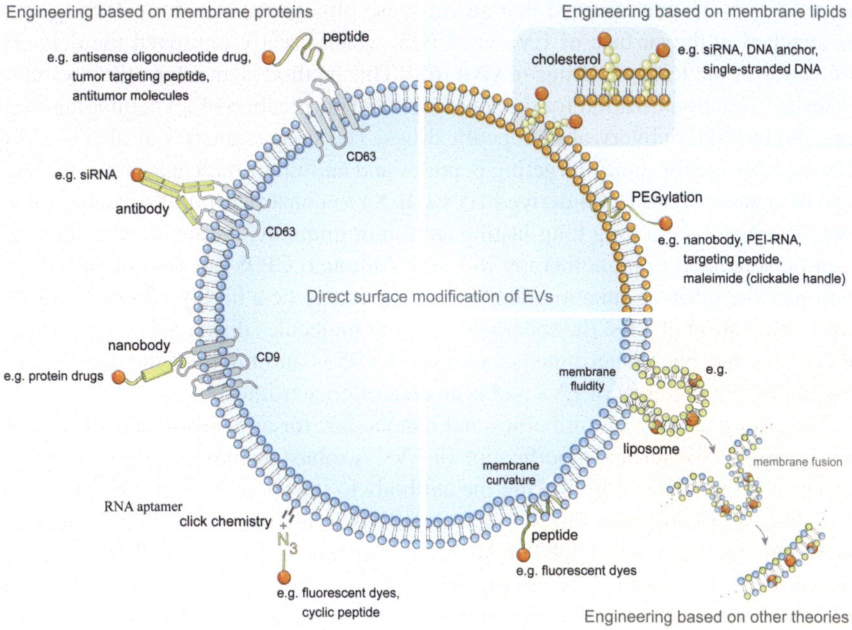

Fig. 5.2 Schematic for direct surface modification of EVs

2.2 *Direct Surface Modification of EVs*

As the surface composition, structure, and properties of EVs continue to be resolved, researchers have developed many methods to utilize their membrane proteins, membrane lipids, and membrane charges for EV surface functionalization. These methods have simple operation and high modification efficiency as well as better applicability and lower cost. Moreover, direct modification methods can be used to modify the surface of EVs derived from human body fluids in vivo [60] and in vitro [61] and assist personalized therapy based on EVs toward clinical application, which is unattainable for genetic approaches to modify EVs. We will review three representative methods of direct modification in the next subsections based on the principles of modification (Fig. 5.2).

2.2.1 Engineering Based on Membrane Proteins

An effective way to modify EVs is by using surface proteins (e.g., tetraspanins) as anchors. A quintessential example is the EV-anchored short peptide reported by Gao et al. They targeted the second extracellular loop of the EV surface membrane protein CD63 and used phage display technology for ligand screening, eventually resulting in a short peptide that could bind to it via a noncovalent bond, which was named CP05. They confirmed that an antisense oligonucleotide drug (PMO_{25}) can be attached to the surface of EVs via CP05, which greatly improved the delivery efficiency of nucleic acid drugs in vivo [61]. This method is an interesting example of surface functionalization for the separation and purification of EVs, imaging tracing, and targeted delivery of therapeutic drugs. The same strategy was also used by Zuo et al. to anchor tumor-targeting peptides and antitumor molecules onto the surface of dendritic cell (DC)-derived EVs (DEX) to construct multifunctional DEX vaccines, thereby eliciting long-lasting antitumor immunity, tumor suppression and even personalized immunotherapy [62, 63]. Although CP05 has been reported not to impair the intrinsic function of EVs and appears to be a highly efficient tool for functionalization of EVs, the specific number of molecules displayed on the surface of EVs has not been determined; moreover, CP05 is an important indicator of the drug delivery efficiency of EVs and is an area of greater interest.

The strong affinity of antibodies and nanobodies for antigens makes them efficient scaffolds for surface modification of EVs. A robust display of cargo on the EV surface can be achieved by linking the antibody to the cargo to be modified. Soma et al. used a polyarginine linker to attach siRNA to a CD63 monoclonal antibody (mAb) to construct anti-CD63 mAb drug conjugated complexes. They demonstrated that EVs modified by conjugates could be taken up by multiple myeloma cells, achieving siRNA endocytosis and mRNA degradation inside the cells [64]. This approach successfully circumvents the problems of inefficient delivery of siRNA into EVs and membrane disruption by conventional electroporation, providing a new strategy for siRNA delivery via EVs. However, the stability of siRNA exposed on the surface of EVs in the circulatory system in vivo awaits further validation. Correspondingly, Milica et al. screened a CD9 nanobody by phage display, which could capture EVs from human plasma and cell supernatant [65]. Compared with antibodies, mAb may be less expensive and equally effective surface-functionalized scaffolds.

Click chemistry is a method for conjugating ligands to the EV surface though the primary amine groups of lysine residues on membrane proteins. The N-hydroxysuccinimidyl (NHS) group of DBCO reacts with the primary amine, and the alkyne group of DBCO is subsequently coupled to azide-functionalized molecules by a copper-free click reaction [66]. Tyson et al. first used click chemistry to modify fluorescent dyes on the surface of EVs, and Xu's group subsequently used the same method to track the uptake of EVs by pancreatic cancer cells in vitro. Through click chemistry, Tian et al. modified mesenchymal stromal cell (MSC)-derived EVs with c(RGDyK) peptides and found that peptide-EV conjugates can be specifically distributed in cerebral vascular endothelial cells after cerebral ischemia, which shows that click chemistry has the potential to modify functional molecules on the surface of EVs [67]. Copper-free click chemistry is biocompatible and bioorthogonal and has been applied in vivo without apparent toxicity or physiological perturbation [68, 69]. Therefore, this process may not affect the membrane integrity of EVs or the function of conjugates. More research is needed to expand the application of this method and evaluate its safety.

2.2.2 Engineering Based on Membrane Lipids

Cholesterol can be inserted into the lipid membrane of EVs spontaneously and with high affinity through hydrophobic interactions to achieve effective modification of molecules (especially nucleic acid anchors) on EVs [70, 71]. O'Loughlin et al. optimized the method to modify EVs with cholesterol-conjugated small interfering RNA (siRNA), resulting in the efficient loading of siRNA on the surface of EVs and effective silencing of human antigen R (HuR), a tumor therapeutic target [70]. An ingenious design is that Yereni et al. developed a rapid, versatile and scalable method using cholesterol-conjugated single-stranded DNA (ssDNA) tethers to enable diverse surface functionalization of both human and murine EVs. In the tether, cholesterol binds to the lipid membrane of EVs, and ssDNA itself or complementary DNA can be linked to small molecules (such as dyes and biotin) or macromolecules (antibodies or other proteins), thus completing the modification of a variety of cargoes on the surface of EVs [72]. Furthermore, Xia and his colleagues developed a method for qualitative and quantitative detection of EVs using cholesterol anchors. They used an anti-CD63 protein antibody to capture EVs and confirmed that "cholesterol-DNA-biotin," which is enriched on the EV membrane, combined with streptavidin-tagged horseradish peroxidase (HRP) to emit an amplified fluorescence signal under the action of the substrate, which can indirectly reflect the number of captured EVs [73]. This method eliminates the interference of miscellaneous proteins and other vesicles with lipid membranes during EV quantification and is worthy of further reference. Notably, the high lipid content [74] in the EV membrane and relatively high membrane rigidity [75] decrease the vulnerability of cholesterol conjugates to release from the EV membrane, maintaining the robustness of the modification. Nevertheless, cholesterol is amphiphilic and prone to self-assembly [76]. The selection of appropriate solvents and appropriate pH are the pre-requisites for effective modification.

PEGylation, a simple method for modifying EVs directly, has been explored for its ability to prolong the circulation time and tissue accumulation of EVs [77]. Distearoyl Phosphoethanolamine (DSPE) and N-Hydroxysuccinimide (NHS) are common modifiers of polyethylene glycol (PEG). DSPE-PEG and NHS-PEG are used as bridges for binding to the hydrophobic EV outer membrane and proteins, respectively, to complete the surface modification of EVs. Kooijmans et al. engineered EVs by modifying conjugates of epidermal growth factor receptor nanobodies and phospholipid-PEG, resulting in specific uptake of functionalized EVs on $EGFR^+$ cells and prolonging the circulation time of EVs in plasma by nearly 60 min [78]. Li and his colleagues constructed an EV-based RNA delivery system using PEGylation. They constructed PEG/PEI and loaded mTOR siRNA on the surface of dendritic cell-derived EVs through stepwise self-assembly. The modified EVs successfully inhibited the mTOR pathway and effectively induced immune tolerance during transplantation, indicating the potential of PEGylated EVs for loading nucleic acid drugs [79]. Recent studies indicate that PEG couples with RGD [80], biotin [81], maleimide [82], and other ligands to increase the targetability of EVs. In particular, maleimide is a highly functional clickable handle that can attach fluorescent dyes, magnetic nanoparticles, and gold nanoparticles to the surfaces of EVs [82]. Apparently, the combination of PEGylation and click chemistry provides additional options for displaying different molecules on EVs. It is generally believed that PEG insertion does not affect the membrane integrity or protein function of EVs [83, 84]. However, it has been reported that PEGylated EVs are recognized by lysosomes after entering cells, resulting in degradation instead of functional delivery of the cargo (e.g., therapeutic nucleic acids) [85]. However, whether the decreased cellular uptake ability of pegylated EVs hinders their re-uptake in targeted tissues and cells remains to be elucidated and may be a subject of further studies.

2.2.3 Engineering Based on Other Theories

EVs are liquid combinations of phospholipids, cholesterol, and proteins, and their components can diffuse laterally to the plane of the membrane due to the fluidity of the membrane. This property has been developed for the surface functionalization of EVs, usually with the participation of synthetic functionalized liposomes. Bo et al. confirmed EV-liposome fusion events though the Förster resonant energy transfer (FRET) dequenching assay. They found that FRET activity decreased with the addition of liposomes without fluorescence to EVs dual labeled with DiI and DiD dyes, emphasizing the effective membrane fusion of EVs and liposomes [86]. Max achieved EV-liposome fusion with PEG8000 and improved the cellular delivery efficiency of a chemotherapeutic compound by a factor of 3–4 [87]. Hong's latest research has even successfully integrated a CRISPR/Cas13a-based microRNA detection system into EVs using EV-liposome fusion, enabling real-time monitoring of miRNAs in EVs without the need for traditional extraction and detection methods [88]. Real-time monitoring of proteins and RNAs in single EVs is expected to become an effective new method. In summary, EV surface functionalization by EV-liposome fusion has multiple advantages. Compared with EVs of biological origin, synthetic liposomes are more easily modified by a variety of drugs, especially RNA, which is not easily modified on the surface of EVs. On the other hand, EVs allow for the bio-camouflage of liposomes by enriching the lipid bilayer and inner compartment with biogenic molecules. It is feasible to use these properties to fuse EVs with functionalized synthetic liposomes to achieve surface modification of multiple functional molecules on EVs.

EVs are nanoscale biological vesicles with extremely small radii and high membrane curvatures. Professor Yin Hang's team designed the cyclic polypeptide C2BL3L [89], which can bind to high-curvature membranes and can detect EVs with a diameter < 100 nm. Subsequently, BK [90], a bradykinin-based peptide probe, and MARCKS-fragment peptide (MARCKS-ED) [91], a peptide probe derived from Myristoylated alanine-rich C kinase substrate (MARCKS), were developed for binding to the high-curvature membrane of EVs and phosphatidylserine (PS) on their surface. MARCKS-ED, a short peptide composed of 25 amino acid residues, has the advantages of low synthesis cost and strong modifiability and is expected to become a systematic research tool for the characterization and surface modification of EVs. However, MARCKS-ED also binds to non-EV membrane structures due to the high content of lysine, especially bind to apoptotic bodies (PS overexpression) [92], resulting in a decrease in the specificity of the membrane for binding to EV membranes.

3 Modified EVs for Targeted Therapy

EVs have a strong ability to home to target tissues or cells and can penetrate biological barriers (such as the blood–brain barrier); thus, they have natural drug loading advantages and are promising targeted drug carriers [93, 94]. However, the targeting potential of natural EVs is weak, and the effectiveness of targeted therapy is poor. Therefore, they are usually modified to form engineered EVs. The modified targeting molecules can be antibodies, ligands, short peptides, RNAs, etc., and the corresponding targets can be specific proteins in organs, tissues, lesion areas, or cells. It is worth noting that in the process of EVs targeted therapy, several details are easy to be ignored that we need to consider. First, the targeted molecules cannot hinder the cellular uptake of engineered EVs, preventing them from being effectively absorbed by cells. Second, molecules that are expressed on the target tissue as much as possible should be chosen, and molecules with important physiological functions should not be chosen (unless they are conducive to disease treatment under certain conditions, such as cancer) to avoid affecting the normal function of tissues or cells. If some protein targets are not only important functional molecules but also candidate targets with good tissue specificity, they can be anchored in nonfunctional regions by avoiding functional regions to achieve targeting while preserving protein function. Most importantly, when targeting cells, the target region must be located on the surface of the target cell. If the target region is located inside the cell, the targeting molecule cannot bind to it in the systemic circulation and loses its targeting function. In the next section, we summarize some application examples of engineered EV-targeted therapy, showing the potential of EV-targeted application in different tissues and organs (Fig. 5.3).

3.1 Targeting the Central Nervous System

The blood–brain barrier (BBB) is an important biological barrier of the central nervous system that can decrease or even prevent damage to brain tissue caused by harmful substances in the blood circulation to maintain the basic stability of the internal environment in brain tissue and has an important biological significance for maintaining the normal physiological state of the central nervous system [95]. However, the presence of the BBB also hinders the delivery of therapeutic drugs to the brain under disease conditions, which is a major challenge for drug-related brain

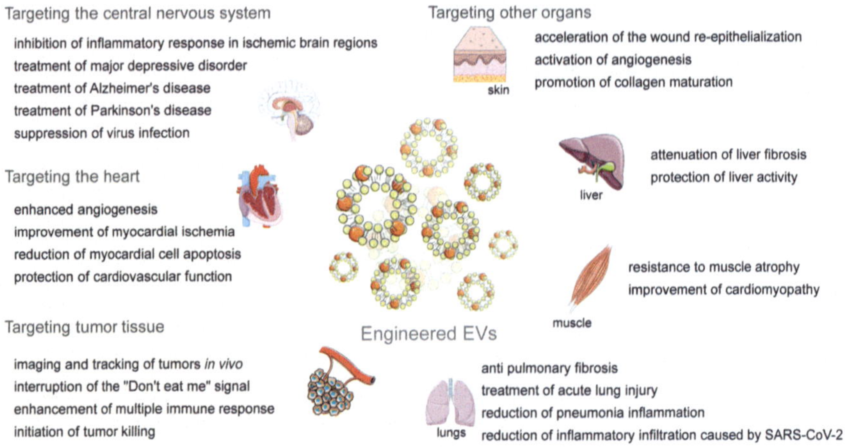

Fig. 5.3 Application of surface-modified EVs in targeted therapy

transport. As natural nanoscale vesicles, EVs are reliable carriers for drug loading to penetrate the BBB. Several studies have reported that EVs derived from specific cells can cross the BBB [96, 97], but their penetration efficacy is unsatisfactory. Therefore, targeted modification of EVs is needed to enable them to carry drugs to penetrate the barrier efficiently and achieve therapeutic purposes. Tian et al. took advantage of the characteristic that short RGD peptides can bind to integrin αvβ3, which is highly expressed in the ischemic region, and used integrin αvβ3 as a target to target the ischemic region of the brain. They modified RGD on the surface of EVs to deliver curcumin-coated EVs specifically to the ischemic area of the brain, alleviating the inflammatory response and apoptosis in the lesion area [67]. Additionally, as a common and recognized effective brain-targeting short peptide, RVG often modifies EVs to help EVs cross the BBB, thus facilitating the treatment of a variety of neurological diseases. Yu et al. recently reported a study on the treatment of major depressive disorder (MDD) with RVG-modified EVs. They purified and recovered RVG-circDYM-EVs from 293 cells by gene manipulation and observed that systemic administration of RVG-circDYM-EVs effectively delivered circDYM to the brain and alleviated depression-like behavior in mice [98]. In a typical study of the use of RVG-EVs in the treatment of Alzheimer's disease, Cui et al. reported that MSC-RVG-EVs could significantly improve learning and memory ability, reduce plaque deposition and αβ levels, and normalize inflammatory cytokines [99]. Cooper et al. realized the brain-targeted delivery of siRNA under systemic injection conditions using RVG-modified EVs; the level of alpha synuclein (α-SYN) decreased after 7 days of treatment, and the pathological conditions reversed [100]. Zhang et al. even confirmed that RVG-EVs can directly reach fetal brain tissue through the maternal placenta in the maternal fetus. They injected RVG-EVs loaded with siRNA (ZIKV-specific siRNA) into pregnant AG6 mice and found that the cells could resist the vertical invasion of the maternal ZIKV virus into the fetal mouse brain [20]. Although RVG-mediated brain targeting of EVs was significant, RVG-EVs injected systemically were found to be enriched in the liver to a high degree [61, 98]. Therefore, more powerful brain anchoring tools need to be identified, and the detection of drug-related liver toxicity, which is easy to ignore, needs to be considered.

3.2 *Targeting the Heart*

CVD is the disease with the highest morbidity and mortality worldwide [101]. Due to the tight junctions between the myocardium, the effective delivery of therapeutic drugs has always been a problem. The emergence of EV-targeted drug delivery systems has provided hope for cardiac targeting. EVs have achieved good efficacy in cardiac targeted therapy due to their high histocompatibility and easy modification. It is most common to modify cardiac-targeting peptides on the surface of EVs for cardiac targeting [102–105]. For example, Wang and his colleagues coupled cardiac stem cell-derived EVs with cardiac homing peptide (CHP) through the DOPE-NHS linker and observed its targeted therapeutic effect in an ischemia/reperfusion rat model. Twenty one days after systemic administration, cardiac fibrosis in model mice was alleviated, and cell proliferation and angiogenesis were increased [103]. Jiang et al. constructed a biomimetic drug delivery system with a platelet membrane fused with circulating human plasma-derived extracellular vesicles (P-hEVs) and reported that intravenously infused P-hEVs could accumulate at the site of cardiac injury and have a protective effect on myocardial ischemia/reperfusion injury (I/RI) [106]. The same method has also been used to treat myocardial infarction (MI). The hybridization of platelets and EVs can induce macrophage pinocytosis, significantly enhance the cellular uptake of EVs by endothelial cells and cardiomyocytes, and enhance the therapeutic effect on myocardial infarction [107]. Although many studies have demonstrated the ability of EVs to target CVD, the concentration of drugs delivered by EVs is still low after systemic injection, and most of these drugs still fail to target the site; why this topic is worthy of further research.

3.3 *Targeting Tumor Tissue*

Cancer is the second leading cause of death worldwide. The success of tumor-targeted therapy has completely changed our understanding of tumor biology and fundamentally changed the therapeutic prospects of cancer. EVs have dual functions in terms of drug delivery and targeted modification and have multiple advantages, such as low immunogenicity and long circulation time in vivo. It has become a popular new drug carrier in the process of tumor-targeted therapy. In addition, tumor tracking has also become a research hotspot in tumor therapy. Accurate and real-time tumor tracking can aid in effective surgical navigation for clinicians and help surgeons proceed quickly and smoothly or change treatment strategies in a timely manner [66].

The strategies by which EVs target tumor tissues can be divided into two main categories. One category relies on the preferential recruitment of immune cell-derived EVs by tumor cells. Compelling evidence has shown that EVs derived from immune cells carry functional molecules similar to those of parental cells, increasing the likelihood of infiltrating the tumor microenvironment and binding to tumor cells [108]. However, this natural targeting ability is insufficient and targeted modified engineered EVs are generated accordingly. Engineered EVs often target molecules that are specifically expressed on the tumor surface, and anchored target molecules are firmly modified on the EV surface, thus achieving ideal tumor targeting and drug delivery. For tumor tracking, Mendt et al. confirmed that DiR dye-labeled EVs derived from bone marrow MSCs exhibited greater fluorescence in pancreatic tumors than in the liver and spleen, revealing the specific retention of EVs in tumors [109]. Molavipordanjani et al. used ^{99m}Tc-radiolabel HER2-targeted EVs (^{99m}Tc-EVs) for tumor imaging [110]. They observed that ^{99m}Tc-EVs could accumulate in the tumors of nude mice bearing SKOV-3 tumors, facilitating tumor visualization. For tumor-targeted therapy, many studies have preloaded EVs with antitumor drugs and modified tumor-targeting molecules on the surface of EVs by surface modification technology to target specific markers of different tumors, such as EGFR [41] and HER2 [111], to achieve targeted therapy. An interesting study on the treatment of tumors with engineered EVs involved CD47-blocking EVs (EV-SIRPα), which were developed by Koh and his team [112]. EV-SIRPα can bind with CD47 specifically on the surface of tumor cells to target the tumor tissue while preventing CD47 on the tumor from binding to SIRPα on the surface of macrophages, interrupting the "do not eat me" signal and initiating the immune response mediated by macrophages.

Although considerable evidence has confirmed that engineered EVs greatly improve the bioavailability of drugs in the process of tumor-targeted therapy, reduce the toxicity and side effects of drugs, and maximize the antitumor efficacy of drugs [113], there are still many challenges in the clinical treatment of tumors caused by EVs, such as the development of purified engineered EV technology and extensive verification of the safety and effectiveness of EVs in large-scale clinical trials. In any case, we can see the potential and advantages of EVs in tumor-targeted therapy, and we also expect them to become a new tool for precision tumor therapy.

3.4 Targeting Other Organs

3.4.1 Skin

In recent years, the application of stem cell-derived EVs in the field of medical cosmetology has undergone rapid development. Due to the strong barrier effect on the skin surface, the targeting effect of EVs in dermatology and medical esthetics mainly depends on specific intake methods, such as local injection at the wound site and the use of biological materials (such as hydrogels) as dressings. Zhao et al. used local multipoint subcutaneous injection to subcutaneously inject chitosan hydrogel-loaded EVs derived from MSCs into naturally aged mice and confirmed that MSC-derived EVs can promote skin rejuvenation by improving the aging of dermal fibroblasts [114]. Tao et al. overexpressed microRNA-126-3p in synovial mesenchymal stem cells (SMSCs) and combined it with chitosan to form a wound dressing for wound healing treatment in a diabetic rat model. The authors found that the dressing could controllably release EVs derived from miR-126-3p-overexpressing SMSCs, accelerate re-epithelialization, activate angiogenesis, and promote collagen maturation in vivo [115]. It is believed that there will be additional effective ways for EVs to penetrate the skin barrier, supporting the wide application of EVs in the field of cosmetology.

3.4.2 Liver

Many studies have shown that EVs are mostly distributed in the liver and spleen (especially the liver) after systemic injection and metabolism [61, 116], which undoubtedly reflects the passive targeting characteristics of EVs in the liver. Therefore, targeting EVs to the liver for the treatment of related diseases has additional advantages. For example, EVs enriched with miRNA-155 were successfully used to target CCl4-induced liver fibrosis through the passive targeting of systemically injected EVs to the liver [31]. In addition, anchors targeting the liver or specific cells in the liver are often modified on the EV surface for targeting. Gong et al. modified the CD44-specific ligand hyaluronic acid (HA) on the surface of EVs and used HA to interact with CD44 (which is highly expressed in the fibrotic liver) to achieve liver targeting for the treatment of liver fibrosis [117]. They were also loaded with forsythiaside A (FA), a traditional Chinese medicine component with significant liver-protective activity, on the surface of milk-derived EVs (HA-MEV-FA), which can be used to construct a compound drug. The results showed that HA-MEV-FA could improve liver fibrosis symptoms in zebrafish larvae while protecting liver function in vivo. The study of engineered EVs targeting activated hepatic stellate cells (aHSCs) has led to the realization of precise treatment of aHSCs in complex liver tissue, providing a new strategy for the clinical treatment of liver fibrosis [118]. They linked the targeting peptide of aHSCs (named HSTP1) screened from the phage short peptide library to the surface of human umbilical cord mesenchymal stem cell (Huc-MSC)-derived EVs and found that HSTP1-EVs could specifically target the aHSC region after intravenous administration, enhance drug efficacy, and attenuate fibrosis.

3.4.3 Muscle

EVs are widely involved in the physiology and pathology of skeletal muscle and have great clinical potential in the treatment of skeletal muscle diseases. However, the therapeutic cargo contained in natural EVs is limited, and a large number of donors are required to induce sufficient effects in vivo [119]. Therefore, EVs modified with therapeutic cargo and targeting ability are effective and low-cost options for the treatment of muscle diseases. Wang et al. used EVs derived from muscle satellite cells modified with cardiac-targeting peptide to transport therapeutic miR-26a to the myocardium and skeletal muscle in a mouse model of chronic kidney disease (CKD). The results showed that the cross-sectional area of skeletal muscle in mice increased; the myocardial connecting tissue growth factor, fibronectin and collagen type I alpha 1 decreased; and cardiac function improved [120]. The team also used miR-26a-loaded EVs by intramuscular injection to treat muscle atrophy caused by chronic kidney disease and achieved good therapeutic effects [121]. In general, engineered EVs with modified contents and tissue-specific targeting ability have great application prospects in the treatment of muscle diseases.

3.4.4 Lungs

EVs have been reported to be effective in the treatment of lung diseases, such as pulmonary fibrosis, acute lung injury, and pulmonary virus infection. Its targeting to the lungs depends on special injection methods or the homing ability of specific EVs [122]. Sun and his colleagues fused fibroblast-derived EVs with liposomes loaded with clodronate to achieve targeted treatment of pulmonary fibrosis with the help of fibroblast-derived EVs targeting the fibrotic region [123]. The team also used the same method to transport nintedanib to the lungs, enhancing its efficacy in the treatment of pulmonary fibrosis [124]. Kim et al., for the first time, studied engineered EVs to deliver therapeutic short peptides and curcumin to the lungs to treat acute lung injury [125]. They loaded rage-binding peptide (RBP) and curcumin, which have anti-inflammatory effects on the surface and inside of EVs (RBP-EVs/cur), and delivered engineered EVs to the lung through intratracheal instillation, which confirmed that these agents inhibited lung inflammation in a mouse model of acute lung injury and that the therapeutic effect was better than that of the mixture of RBP and curcumin. Holay et al. encapsulated metastatic breast cancer-derived EVs on dexamethasone-loaded polymer cores and used the strong binding ability of metastatic breast cancer-derived EVs to the lung to deliver dexamethasone to the lung, reducing the inflammatory response in a pneumonia model [126]. Although this method is novel, we should pay attention to safety when using tumor-derived EVs. The latest research has designed an acute respiratory syndrome coronavirus 2 (SARS-CoV-2) vaccine using engineered EVs [127]. The vaccine is composed of a recombinant SARS-CoV-2 receptor binding domain (RBD) bound by lung-derived EVs. It enters the body through nebulization and enhances the retention of RBD in both the mucus-lined respiratory airway and the lung parenchyma. The vaccine significantly reduced inflammatory infiltration after a live severe SARS-CoV-2 challenge. Although the effect of these vaccines is significant, there is still a lack of clinical trials verifying the use of these vaccines compared with mRNA vaccines.

4 Future Perspectives

A drug delivery system based on EV surface modification can realize long-term and efficient therapeutic effects in vivo. In the past decade, EVs have been favored in the field of drug delivery and have been widely used in clinical drug transportation and targeted delivery due to their extremely low immunogenicity and strong modifiability. Surface modification of EVs with targeting molecules can support tissue-specific delivery of drugs and is a favorable tool for real precision medicine. In addition, the targeted delivery of fluorescent molecules via EVs is expected to become the most promising real-time diagnostic platform for lesion tracking, helping clinicians quickly determine the specific location of tumors or other lesions to achieve immediate, rapid and accurate intraoperative lesion anchoring and resection. It is believed that with the continuous progress of surface modification technology, simple, rapid, efficient, and stable EV surface modification technology can be developed to help the sustainable development of EVs in clinical disease diagnosis and treatment.

Acknowledgments This study was financially supported by open funding from the Nankai University Eye Institute (NKYKD202203), the Research Project on Skin Injury & Repair (BKJ21J016, 2021XB008), and the Tianjin Natural Science Foundation (22JCZXJC00170, 21JCZDJC00070).

References

1. Kalluri R, LeBleu VS. The biology, function, and biomedical applications of exosomes. Science (New York, NY). 2020;367(6478):eaau6977. https://doi.org/10.1126/science.aau6977.
2. Raposo G, Stahl PD. Extracellular vesicles: a new communication paradigm? Nat Rev Mol Cell Biol. 2019;20(9):509–10. https://doi.org/10.1038/s41580-019-0158-7.
3. Willms E, Cabañas C, Mäger I, Wood MJA, Vader P. Extracellular vesicle heterogeneity: subpopulations, isolation techniques, and diverse functions in cancer progression. Front Immunol. 2018;9:738. https://doi.org/10.3389/fimmu.2018.00738.
4. Xian P, Hei Y, Wang R, Wang T, Yang J, Li J, et al. Mesenchymal stem cell-derived exosomes as a nanotherapeutic agent for amelioration of inflammation-induced astrocyte alterations in mice. Theranostics. 2019;9(20):5956–75. https://doi.org/10.7150/thno.33872.
5. Timmers L, Lim SK, Hoefer IE, Arslan F, Lai RC, van Oorschot AA, et al. Human mesenchymal stem cell-conditioned medium improves cardiac function following myocardial infarction. Stem Cell Res. 2011;6(3):206–14. https://doi.org/10.1016/j.scr.2011.01.001.
6. Xiao C, Wang K, Xu Y, Hu H, Zhang N, Wang Y, et al. Transplanted mesenchymal stem cells reduce autophagic flux in infarcted hearts via the exosomal transfer of miR-125b. Circ Res. 2018;123(5):564–78. https://doi.org/10.1161/circresaha.118.312758.

7. Willis GR, Fernandez-Gonzalez A, Reis M, Yeung V, Liu X, Ericsson M, et al. Mesenchymal stromal cell-derived small extracellular vesicles restore lung architecture and improve exercise capacity in a model of neonatal hyperoxia-induced lung injury. J Extracell Vesicles. 2020;9(1):1790874. https://doi.org/10.1080/20013078.2020.1790874.
8. Wang T, Jian Z, Baskys A, Yang J, Li J, Guo H, et al. MSC-derived exosomes protect against oxidative stress-induced skin injury via adaptive regulation of the NRF2 defense system. Biomaterials. 2020;257:120264. https://doi.org/10.1016/j.biomaterials.2020.120264.
9. Sun Y, Shi H, Yin S, Ji C, Zhang X, Zhang B, et al. Human Mesenchymal stem cell derived Exosomes alleviate type 2 diabetes mellitus by reversing peripheral insulin resistance and relieving β-cell destruction. ACS Nano. 2018;12(8):7613–28. https://doi.org/10.1021/acsnano.7b07643.
10. Venkat P, Zacharek A, Landschoot-Ward J, Wang F, Culmone L, Chen Z, et al. Exosomes derived from bone marrow mesenchymal stem cells harvested from type two diabetes rats promotes neurorestorative effects after stroke in type two diabetes rats. Exp Neurol. 2020;334:113456. https://doi.org/10.1016/j.expneurol.2020.113456.
11. Zhang ZG, Buller B, Chopp M. Exosomes—beyond stem cells for restorative therapy in stroke and neurological injury. Nat Rev Neurol. 2019;15(4):193–203. https://doi.org/10.1038/s41582-018-0126-4.
12. Mathiyalagan P, Liang Y, Kim D, Misener S, Thorne T, Kamide CE, et al. Angiogenic mechanisms of human CD34(+) stem cell Exosomes in the repair of ischemic Hindlimb. Circ Res. 2017;120(9):1466–76. https://doi.org/10.1161/circresaha.116.310557.
13. Wiklander OPB, Brennan M, Lötvall J, Breakefield XO, El Andaloussi S. Advances in therapeutic applications of extracellular vesicles. Sci Transl Med. 2019;11(492):eaav8521. https://doi.org/10.1126/scitranslmed.aav8521.
14. Trials C: Clinical Trials. https://clinicaltrials.gov/ct2/results?cond=exosomes Accessed.
15. Zhang Y, Bi J, Huang J, Tang Y, Du S, Li P. Exosome: a review of its classification, isolation techniques, storage, diagnostic and targeted therapy applications. Int J Nanomedicine. 2020;15:6917–34. https://doi.org/10.2147/ijn.s264498.
16. Salunkhe S, Dheeraj BM, Chitkara D, Mittal A. Surface functionalization of exosomes for target-specific delivery and in vivo imaging & tracking: strategies and significance. J Controll Release. 2020;326:599–614. https://doi.org/10.1016/j.jconrel.2020.07.042.
17. Liang Y, Duan L, Lu J, Xia J. Engineering exosomes for targeted drug delivery. Theranostics. 2021;11(7):3183–95. https://doi.org/10.7150/thno.52570.
18. Langevin C, Jaaro H, Bressanelli S, Fainzilber M, Tuffereau C. Rabies virus glycoprotein (RVG) is a trimeric ligand for the N-terminal cysteine-rich domain of the mammalian p75 neurotrophin receptor. J Biol Chem. 2002;277(40):37655–62. https://doi.org/10.1074/jbc.M201374200.
19. Alvarez-Erviti L, Seow Y, Yin H, Betts C, Lakhal S, Wood MJ. Delivery of siRNA to the mouse brain by systemic injection of targeted exosomes. Nat Biotechnol. 2011;29(4):341–5. https://doi.org/10.1038/nbt.1807.
20. Zhang R, Fu Y, Cheng M, Ma W, Zheng N, Wang Y, et al. sEVs(RVG) selectively delivers antiviral siRNA to fetus brain, inhibits ZIKV infection and mitigates ZIKV-induced microcephaly in mouse model. Mol Ther. 2022;30(5):2078–91. https://doi.org/10.1016/j.ymthe.2021.10.009.
21. Yang J, Wu S, Hou L, Zhu D, Yin S, Yang G, et al. Therapeutic effects of simultaneous delivery of nerve growth factor mRNA and protein via Exosomes on cerebral ischemia. Mol Ther Nucleic Acids. 2020;21:512–22. https://doi.org/10.1016/j.omtn.2020.06.013.
22. Lai N, Wu D, Liang T, Pan P, Yuan G, Li X, et al. Systemic exosomal miR-193b-3p delivery attenuates neuroinflammation in early brain injury after subarachnoid hemorrhage in mice. J Neuroinflammation. 2020;17(1):74. https://doi.org/10.1186/s12974-020-01745-0.
23. Mao L, Li YD, Chen RL, Li G, Zhou XX, Song F, et al. Heart-targeting exosomes from human cardiosphere-derived cells improve the therapeutic effect on cardiac hypertrophy. J Nanobiotechnol. 2022;20(1):435. https://doi.org/10.1186/s12951-022-01630-3.

24. Kim G, Kim M, Lee Y, Byun JW, Hwang DW, Lee M. Systemic delivery of microRNA-21 antisense oligonucleotides to the brain using T7-peptide decorated exosomes. J Controll Release. 2020;317:273–81. https://doi.org/10.1016/j.jconrel.2019.11.009.
25. Liang G, Zhu Y, Ali DJ, Tian T, Xu H, Si K, et al. Engineered exosomes for targeted co-delivery of miR-21 inhibitor and chemotherapeutics to reverse drug resistance in colon cancer. J Nanobiotechnol. 2020;18(1):10. https://doi.org/10.1186/s12951-019-0563-2.
26. Bai J, Duan J, Liu R, Du Y, Luo Q, Cui Y, et al. Engineered targeting tLyp-1 exosomes as gene therapy vectors for efficient delivery of siRNA into lung cancer cells. Asian J Pharm Sci. 2020;15(4):461–71. https://doi.org/10.1016/j.ajps.2019.04.002.
27. Liang Y, Xu X, Li X, Xiong J, Li B, Duan L, et al. Chondrocyte-targeted MicroRNA delivery by engineered Exosomes toward a cell-free osteoarthritis therapy. ACS Appl Mater Interfaces. 2020;12(33):36938–47. https://doi.org/10.1021/acsami.0c10458.
28. Xu X, Liang Y, Li X, Ouyang K, Wang M, Cao T, et al. Exosome-mediated delivery of kartogenin for chondrogenesis of synovial fluid-derived mesenchymal stem cells and cartilage regeneration. Biomaterials. 2021;269:120539. https://doi.org/10.1016/j.biomaterials.2020.120539.
29. Malla RR, Pandrangi S, Kumari S, Gavara MM, Badana AK. Exosomal tetraspanins as regulators of cancer progression and metastasis and novel diagnostic markers. Asia Pac J Clin Oncol. 2018;14(6):383–91. https://doi.org/10.1111/ajco.12869.
30. Théry C, Witwer KW, Aikawa E, Alcaraz MJ, Anderson JD, Andriantsitohaina R, et al. Minimal information for studies of extracellular vesicles 2018 (MISEV2018): a position statement of the International Society for Extracellular Vesicles and update of the MISEV2014 guidelines. J Extracell Vesicles. 2018;7(1):1535750. https://doi.org/10.1080/20013078.2018.1535750.
31. Li Z, Zhou X, Wei M, Gao X, Zhao L, Shi R, et al. In vitro and in vivo RNA inhibition by CD9-HuR functionalized Exosomes encapsulated with miRNA or CRISPR/dCas9. Nano Lett. 2019;19(1):19–28. https://doi.org/10.1021/acs.nanolett.8b02689.
32. Kojima R, Bojar D, Rizzi G, Hamri GC, El-Baba MD, Saxena P, et al. Designer exosomes produced by implanted cells intracerebrally deliver therapeutic cargo for Parkinson's disease treatment. Nat Commun. 2018;9(1):1305. https://doi.org/10.1038/s41467-018-03733-8.
33. Gupta D, Liang X, Pavlova S, Wiklander OPB, Corso G, Zhao Y, et al. Quantification of extracellular vesicles in vitro and in vivo using sensitive bioluminescence imaging. J Extracell Vesicles. 2020;9(1):1800222. https://doi.org/10.1080/20013078.2020.1800222.
34. Joshi BS, de Beer MA, Giepmans BNG, Zuhorn IS. Endocytosis of extracellular vesicles and release of their cargo from endosomes. ACS Nano. 2020;14(4):4444–55. https://doi.org/10.1021/acsnano.9b10033.
35. Luo W, Dai Y, Chen Z, Yue X, Andrade-Powell KC, Chang J. Spatial and temporal tracking of cardiac exosomes in mouse using a nano-luciferase-CD63 fusion protein. Commun Biol. 2020;3(1):114. https://doi.org/10.1038/s42003-020-0830-7.
36. Sung BH, von Lersner A, Guerrero J, Krystofiak ES, Inman D, Pelletier R, et al. A live cell reporter of exosome secretion and uptake reveals pathfinding behavior of migrating cells. Nat Commun. 2020;11(1):2092. https://doi.org/10.1038/s41467-020-15747-2.
37. Curley N, Levy D, Do MA, Brown A, Stickney Z, Marriott G, et al. Sequential deletion of CD63 identifies topologically distinct scaffolds for surface engineering of exosomes in living human cells. Nanoscale. 2020;12(22):12014–26. https://doi.org/10.1039/d0nr00362j.
38. Yim N, Ryu SW, Choi K, Lee KR, Lee S, Choi H, et al. Exosome engineering for efficient intracellular delivery of soluble proteins using optically reversible protein-protein interaction module. Nat Commun. 2016;7:12277. https://doi.org/10.1038/ncomms12277.
39. Choi H, Kim Y, Mirzaaghasi A, Heo J, Kim YN, Shin JH, et al. Exosome-based delivery of super-repressor IκBα relieves sepsis-associated organ damage and mortality. Sci Adv. 2020;6(15):eaaz6980. https://doi.org/10.1126/sciadv.aaz6980.

40. Cheng Q, Shi X, Han M, Smbatyan G, Lenz HJ, Zhang Y. Reprogramming exosomes as nanoscale controllers of cellular immunity. J Am Chem Soc. 2018;140(48):16413–7. https://doi.org/10.1021/jacs.8b10047.
41. Ohno S, Takanashi M, Sudo K, Ueda S, Ishikawa A, Matsuyama N, et al. Systemically injected exosomes targeted to EGFR deliver antitumor microRNA to breast cancer cells. Mol Ther. 2013;21(1):185–91. https://doi.org/10.1038/mt.2012.180.
42. Kamerkar S, LeBleu VS, Sugimoto H, Yang S, Ruivo CF, Melo SA, et al. Exosomes facilitate therapeutic targeting of oncogenic KRAS in pancreatic cancer. Nature. 2017;546(7659):498–503. https://doi.org/10.1038/nature22341.
43. Yang Z, Shi J, Xie J, Wang Y, Sun J, Liu T, et al. Large-scale generation of functional mRNA-encapsulating exosomes via cellular nanoporation. Nat Biomed Eng. 2020;4(1):69–83. https://doi.org/10.1038/s41551-019-0485-1.
44. Barclay AN, Van den Berg TK. The interaction between signal regulatory protein alpha (SIRPα) and CD47: structure, function, and therapeutic target. Annu Rev Immunol. 2014;32:25–50. https://doi.org/10.1146/annurev-immunol-032713-120142.
45. Sedlik C, Vigneron J, Torrieri-Dramard L, Pitoiset F, Denizeau J, Chesneau C, et al. Different immunogenicity but similar antitumor efficacy of two DNA vaccines coding for an antigen secreted in different membrane vesicle-associated forms. J Extracell Vesicles. 2014;2014:3. https://doi.org/10.3402/jev.v3.24646.
46. Rountree RB, Mandl SJ, Nachtwey JM, Dalpozzo K, Do L, Lombardo JR, et al. Exosome targeting of tumor antigens expressed by cancer vaccines can improve antigen immunogenicity and therapeutic efficacy. Cancer Res. 2011;71(15):5235–44. https://doi.org/10.1158/0008-5472.can-10-4076.
47. Bliss CM, Parsons AJ, Nachbagauer R, Hamilton JR, Cappuccini F, Ulaszewska M, et al. Targeting antigen to the surface of EVs improves the in vivo immunogenicity of human and non-human adenoviral vaccines in mice. Mol Ther Methods Clin Dev. 2020;16:108–25. https://doi.org/10.1016/j.omtm.2019.12.003.
48. Morishita M, Takahashi Y, Matsumoto A, Nishikawa M, Takakura Y. Exosome-based tumor antigens-adjuvant co-delivery utilizing genetically engineered tumor cell-derived exosomes with immunostimulatory CpG DNA. Biomaterials. 2016;111:55–65. https://doi.org/10.1016/j.biomaterials.2016.09.031.
49. Liu T, Zhu Y, Zhao R, Wei X, Xin X. Visualization of exosomes from mesenchymal stem cells in vivo by magnetic resonance imaging. Magn Reson Imaging. 2020;68:75–82. https://doi.org/10.1016/j.mri.2020.02.001.
50. Kooijmans SAA, Gitz-Francois J, Schiffelers RM, Vader P. Recombinant phosphatidylserine-binding nanobodies for targeting of extracellular vesicles to tumor cells: a plug-and-play approach. Nanoscale. 2018;10(5):2413–26. https://doi.org/10.1039/c7nr06966a.
51. Hansen MS, Gadegaard ISE, Arnspang EC, Blans K, Nejsum LN, Rasmussen JT. Specific and non-invasive fluorescent Labelling of extracellular vesicles for evaluation of intracellular processing by intestinal epithelial cells. Biomedicines. 2020;8(7):211. https://doi.org/10.3390/biomedicines8070211.
52. Heider S, Dangerfield JA, Metzner C. Biomedical applications of glycosylphosphatidylinositol-anchored proteins. J Lipid Res. 2016;57(10):1778–88. https://doi.org/10.1194/jlr.R070201.
53. Rabesandratana H, Toutant JP, Reggio H, Vidal M. Decay-accelerating factor (CD55) and membrane inhibitor of reactive lysis (CD59) are released within exosomes during in vitro maturation of reticulocytes. Blood. 1998;91(7):2573–80.
54. Kooijmans SA, Aleza CG, Roffler SR, van Solinge WW, Vader P, Schiffelers RM. Display of GPI-anchored anti-EGFR nanobodies on extracellular vesicles promotes tumour cell targeting. J Extracell Vesicles. 2016;5:31053. https://doi.org/10.3402/jev.v5.31053.
55. Lenassi M, Cagney G, Liao M, Vaupotic T, Bartholomeeusen K, Cheng Y, et al. HIV Nef is secreted in exosomes and triggers apoptosis in bystander CD4+ T cells. Traffic (Copenhagen, Denmark). 2010;11(1):110–22. https://doi.org/10.1111/j.1600-0854.2009.01006.x.

56. Lattanzi L, Federico M. A strategy of antigen incorporation into exosomes: comparing cross-presentation levels of antigens delivered by engineered exosomes and by lentiviral virus-like particles. Vaccine. 2012;30(50):7229–37. https://doi.org/10.1016/j.vaccine.2012.10.010.
57. Di Bonito P, Chiozzini C, Arenaccio C, Anticoli S, Manfredi F, Olivetta E, et al. Antitumor HPV E7-specific CTL activity elicited by in vivo engineered exosomes produced through DNA inoculation. Int J Nanomedicine. 2017;12:4579–91. https://doi.org/10.2147/ijn.s131309.
58. Anticoli S, Aricò E, Arenaccio C, Manfredi F, Chiozzini C, Olivetta E, et al. Engineered exosomes emerging from muscle cells break immune tolerance to HER2 in transgenic mice and induce antigen-specific CTLs upon challenge by human dendritic cells. J Mol Med (Berl). 2018;96(2):211–21. https://doi.org/10.1007/s00109-017-1617-2.
59. Ferrantelli F, Manfredi F, Chiozzini C, Anticoli S, Olivetta E, Arenaccio C, et al. DNA vectors generating engineered Exosomes potential CTL vaccine candidates against AIDS, hepatitis B, and tumors. Mol Biotechnol. 2018;60(11):773–82. https://doi.org/10.1007/s12033-018-0114-3.
60. Liu S, Chen X, Bao L, Liu T, Yuan P, Yang X, et al. Treatment of infarcted heart tissue via the capture and local delivery of circulating exosomes through antibody-conjugated magnetic nanoparticles. Nat Biomed Eng. 2020;4(11):1063–75. https://doi.org/10.1038/s41551-020-00637-1.
61. Gao X, Ran N, Dong X, Zuo B, Yang R, Zhou Q, et al. Anchor peptide captures, targets, and loads exosomes of diverse origins for diagnostics and therapy. Sci Transl Med. 2018;10(444):eaat0195. https://doi.org/10.1126/scitranslmed.aat0195.
62. Zuo B, Qi H, Lu Z, Chen L, Sun B, Yang R, et al. Alarmin-painted exosomes elicit persistent antitumor immunity in large established tumors in mice. Nat Commun. 2020;11(1):1790. https://doi.org/10.1038/s41467-020-15569-2.
63. Zuo B, Zhang Y, Zhao K, Wu L, Qi H, Yang R, et al. Universal immunotherapeutic strategy for hepatocellular carcinoma with exosome vaccines that engage adaptive and innate immune responses. J Hematol Oncol. 2022;15(1):46. https://doi.org/10.1186/s13045-022-01266-8.
64. Soma E, Yamayoshi A, Toda Y, Mishima Y, Hosogi S, Ashihara E. Successful incorporation of exosome-capturing antibody-siRNA complexes into multiple myeloma cells and suppression of targeted mRNA transcripts. Cancers. 2022;14(3):566. https://doi.org/10.3390/cancers14030566.
65. Popovic M, Mazzega E, Toffoletto B, de Marco A. Isolation of anti-extra-cellular vesicle single-domain antibodies by direct panning on vesicle-enriched fractions. Microb Cell Factories. 2018;17(1):6. https://doi.org/10.1186/s12934-017-0856-9.
66. Smyth T, Petrova K, Payton NM, Persaud I, Redzic JS, Graner MW, et al. Surface functionalization of exosomes using click chemistry. Bioconjug Chem. 2014;25(10):1777–84. https://doi.org/10.1021/bc500291r.
67. Tian T, Zhang HX, He CP, Fan S, Zhu YL, Qi C, et al. Surface functionalized exosomes as targeted drug delivery vehicles for cerebral ischemia therapy. Biomaterials. 2018;150:137–49. https://doi.org/10.1016/j.biomaterials.2017.10.012.
68. Baskin JM, Prescher JA, Laughlin ST, Agard NJ, Chang PV, Miller IA, et al. Copper-free click chemistry for dynamic in vivo imaging. Proc Natl Acad Sci USA. 2007;104(43):16793–7. https://doi.org/10.1073/pnas.0707090104.
69. Laughlin ST, Baskin JM, Amacher SL, Bertozzi CR. In vivo imaging of membrane-associated glycans in developing zebrafish. Science (New York, NY). 2008;320(5876):664–7. https://doi.org/10.1126/science.1155106.
70. O'Loughlin AJ, Mäger I, de Jong OG, Varela MA, Schiffelers RM, El Andaloussi S, et al. Functional delivery of lipid-conjugated siRNA by extracellular vesicles. Mol Ther. 2017;25(7):1580–7. https://doi.org/10.1016/j.ymthe.2017.03.021.
71. He F, Liu H, Guo X, Yin BC, Ye BC. Direct exosome quantification via bivalent-cholesterol-labeled DNA anchor for signal amplification. Anal Chem. 2017;89(23):12968–75. https://doi.org/10.1021/acs.analchem.7b03919.

72. Yerneni SS, Lathwal S, Shrestha P, Shirwan H, Matyjaszewski K, Weiss L, et al. Rapid on-demand extracellular vesicle augmentation with versatile oligonucleotide tethers. ACS Nano. 2019;13(9):10555–65. https://doi.org/10.1021/acsnano.9b04651.
73. Xia Y, Chen T, Chen G, Weng Y, Zeng L, Liao Y, et al. A nature-inspired colorimetric and fluorescent dual-modal biosensor for exosomes detection. Talanta. 2020;214:120851. https://doi.org/10.1016/j.talanta.2020.120851.
74. Skotland T, Hessvik NP, Sandvig K, Llorente A. Exosomal lipid composition and the role of ether lipids and phosphoinositides in exosome biology. J Lipid Res. 2019;60(1):9–18. https://doi.org/10.1194/jlr.R084343.
75. Laulagnier K, Motta C, Hamdi S, Roy S, Fauvelle F, Pageaux JF, et al. Mast cell—and dendritic cell-derived exosomes display a specific lipid composition and an unusual membrane organization. Biochem J. 2004;380(Pt 1):161–71. https://doi.org/10.1042/bj20031594.
76. Misiak P, Markiewicz KH, Szymczuk D, Wilczewska AZ. Polymeric drug delivery systems bearing cholesterol moieties: a review. Polymers. 2020;12(11):2620. https://doi.org/10.3390/polym12112620.
77. Lu M, Xing H, Xun Z, Yang T, Zhao X, Cai C, et al. Functionalized extracellular vesicles as advanced therapeutic nanodelivery systems. Eur J Pharm Sci. 2018;121:34–46. https://doi.org/10.1016/j.ejps.2018.05.001.
78. Kooijmans SAA, Fliervoet LAL, van der Meel R, Fens M, Heijnen HFG, van Bergen En Henegouwen PMP, et al. PEGylated and targeted extracellular vesicles display enhanced cell specificity and circulation time. J Controll Release. 2016;224:77–85. https://doi.org/10.1016/j.jconrel.2016.01.009.
79. Li C, Guo F, Wang X, Liu D, Wu B, Wang F, et al. Exosome-based targeted RNA delivery for immune tolerance induction in skin transplantation. J Biomed Mater Res A. 2020;108(7):1493–500. https://doi.org/10.1002/jbm.a.36919.
80. Wang J, Li W, Lu Z, Zhang L, Hu Y, Li Q, et al. The use of RGD-engineered exosomes for enhanced targeting ability and synergistic therapy toward angiogenesis. Nanoscale. 2017;9(40):15598–605. https://doi.org/10.1039/c7nr04425a.
81. Wang J, Li W, Zhang L, Ban L, Chen P, Du W, et al. Chemically edited Exosomes with dual ligand purified by microfluidic device for active targeted drug delivery to tumor cells. ACS Appl Mater Interfaces. 2017;9(33):27441–52. https://doi.org/10.1021/acsami.7b06464.
82. Di H, Zeng E, Zhang P, Liu X, Zhang C, Yang J, et al. General approach to engineering extracellular vesicles for biomedical analysis. Anal Chem. 2019;91(20):12752–9. https://doi.org/10.1021/acs.analchem.9b02268.
83. Choi H, Choi Y, Yim HY, Mirzaaghasi A, Yoo JK, Choi C. Biodistribution of exosomes and engineering strategies for targeted delivery of therapeutic exosomes. Tissue Eng Regen Med. 2021;18(4):499–511. https://doi.org/10.1007/s13770-021-00361-0.
84. Shi S, Li T, Wen X, Wu SY, Xiong C, Zhao J, et al. Copper-64 labeled PEGylated Exosomes for in vivo positron emission tomography and enhanced tumor retention. Bioconjug Chem. 2019;30(10):2675–83. https://doi.org/10.1021/acs.bioconjchem.9b00587.
85. Majzoub RN, Chan CL, Ewert KK, Silva BF, Liang KS, Jacovetty EL, et al. Uptake and transfection efficiency of PEGylated cationic liposome-DNA complexes with and without RGD-tagging. Biomaterials. 2014;35(18):4996–5005. https://doi.org/10.1016/j.biomaterials.2014.03.007.
86. Ning B, Huang Z, Youngquist BM, Scott JW, Niu A, Bojanowski CM, et al. Liposome-mediated detection of SARS-CoV-2 RNA-positive extracellular vesicles in plasma. Nat Nanotechnol. 2021;16(9):1039–44. https://doi.org/10.1038/s41565-021-00939-8.
87. Piffoux M, Silva AKA, Wilhelm C, Gazeau F, Tareste D. Modification of extracellular vesicles by fusion with liposomes for the design of personalized biogenic drug delivery systems. ACS Nano. 2018;12(7):6830–42. https://doi.org/10.1021/acsnano.8b02053.
88. Hong JS, Son T, Castro CM, Im H. CRISPR/Cas13a-based MicroRNA detection in tumor-derived extracellular vesicles. Adv Sci. 2023;10(24):e2301766. https://doi.org/10.1002/advs.202301766.

89. Saludes JP, Morton LA, Ghosh N, Beninson LA, Chapman ER, Fleshner M, et al. Detection of highly curved membrane surfaces using a cyclic peptide derived from synaptotagmin-I. ACS Chem Biol. 2012;7(10):1629–35. https://doi.org/10.1021/cb3002705.
90. Saludes JP, Morton LA, Coulup SK, Fiorini Z, Cook BM, Beninson L, et al. Multivalency amplifies the selection and affinity of bradykinin-derived peptides for lipid nanovesicles. Mol BioSyst. 2013;9(8):2005–9. https://doi.org/10.1039/c3mb70109c.
91. de Jesus AJ, White OR, Flynn AD, Yin H. Determinants of curvature-sensing behavior for MARCKS-fragment peptides. Biophys J. 2016;110(9):1980–92. https://doi.org/10.1016/j.bpj.2016.04.007.
92. Battistelli M, Falcieri E. Apoptotic bodies: particular extracellular vesicles involved in intercellular communication. Biology. 2020;9(1):21. https://doi.org/10.3390/biology9010021.
93. Shimaoka M, Kawamoto E, Gaowa A, Okamoto T, Park EJ. Connexins and integrins in exosomes. Cancers. 2019;11:1. https://doi.org/10.3390/cancers11010106.
94. Wang S, Li F, Ye T, Wang J, Lyu C, Qing S, et al. Macrophage-tumor chimeric exosomes accumulate in lymph node and tumor to activate the immune response and the tumor microenvironment. Sci Transl Med. 2021;13(615):eabb6981. https://doi.org/10.1126/scitranslmed.abb6981.
95. Stamp MEM, Halwes M, Nisbet D, Collins DJ. Breaking barriers: exploring mechanisms behind opening the blood-brain barrier. Fluids Barriers CNS. 2023;20(1):87. https://doi.org/10.1186/s12987-023-00489-2.
96. Morad G, Carman CV, Hagedorn EJ, Perlin JR, Zon LI, Mustafaoglu N, et al. Tumor-derived extracellular vesicles breach the intact blood-brain barrier via Transcytosis. ACS Nano. 2019;13(12):13853–65. https://doi.org/10.1021/acsnano.9b04397.
97. Nieland L, Mahjoum S, Grandell E, Breyne K, Breakefield XO. Engineered EVs designed to target diseases of the CNS. J Controll Release. 2023;356:493–506. https://doi.org/10.1016/j.jconrel.2023.03.009.
98. Yu X, Bai Y, Han B, Ju M, Tang T, Shen L, et al. Extracellular vesicle-mediated delivery of circDYM alleviates CUS-induced depressive-like behaviours. J Extracell Vesicles. 2022;11(1):e12185. https://doi.org/10.1002/jev2.12185.
99. Cui GH, Guo HD, Li H, Zhai Y, Gong ZB, Wu J, et al. RVG-modified exosomes derived from mesenchymal stem cells rescue memory deficits by regulating inflammatory responses in a mouse model of Alzheimer's disease. Immun Ageing. 2019;16:10. https://doi.org/10.1186/s12979-019-0150-2.
100. Cooper JM, Wiklander PB, Nordin JZ, Al-Shawi R, Wood MJ, Vithlani M, et al. Systemic exosomal siRNA delivery reduced alpha-synuclein aggregates in brains of transgenic mice. Movement Disord. 2014;29(12):1476–85. https://doi.org/10.1002/mds.25978.
101. Al-Khatib SM, Stevenson WG, Ackerman MJ, Bryant WJ, Callans DJ, Curtis AB, et al. 2017 AHA/ACC/HRS guideline for management of patients with ventricular arrhythmias and the prevention of sudden cardiac death: a report of the American College of Cardiology/American Heart Association task force on clinical practice guidelines and the Heart Rhythm Society. Heart Rhythm. 2018;15(10):e73–e189. https://doi.org/10.1016/j.hrthm.2017.10.036.
102. Vandergriff A, Huang K, Shen D, Hu S, Hensley MT, Caranasos TG, et al. Targeting regenerative exosomes to myocardial infarction using cardiac homing peptide. Theranostics. 2018;8(7):1869–78. https://doi.org/10.7150/thno.20524.
103. Wang H, Maimaitiaili R, Yao J, Xie Y, Qiang S, Hu F, et al. Percutaneous intracoronary delivery of plasma extracellular vesicles protects the myocardium against ischemia-reperfusion injury in Canis. Hypertension (Dallas, Tex: 1979). 2021;78(5):1541–54. https://doi.org/10.1161/hypertensionaha.121.17574.
104. Kim H, Yun N, Mun D, Kang JY, Lee SH, Park H, et al. Cardiac-specific delivery by cardiac tissue-targeting peptide-expressing exosomes. Biochem Biophys Res Commun. 2018;499(4):803–8. https://doi.org/10.1016/j.bbrc.2018.03.227.

105. Wang X, Chen Y, Zhao Z, Meng Q, Yu Y, Sun J, et al. Engineered exosomes with ischemic myocardium-targeting peptide for targeted therapy in myocardial infarction. J Am Heart Assoc. 2018;7(15):e008737. https://doi.org/10.1161/jaha.118.008737.
106. Jiang J, Ni L, Zhang X, Wang H, Liu L, Wei M, et al. Platelet membrane-fused circulating extracellular vesicles protect the heart from ischemia/reperfusion injury. Adv Healthc Mater. 2023;12(21):e2300052. https://doi.org/10.1002/adhm.202300052.
107. Hu S, Wang X, Li Z, Zhu D, Cores J, Wang Z, et al. Platelet membrane and stem cell exosome hybrid enhances cellular uptake and targeting to heart injury. Nano Today. 2021;39:101210. https://doi.org/10.1016/j.nantod.2021.101210.
108. Basak M, Chaudhary DK, Takahashi RU, Yamamoto Y, Tiwari S, Tahara H, et al. Immunocyte derived exosomes: insight into the potential chemo-immunotherapeutic nanocarrier targeting the tumor microenvironment. ACS Biomater Sci Eng. 2023;9(1):20–39. https://doi.org/10.1021/acsbiomaterials.2c00893.
109. Mendt M, Kamerkar S, Sugimoto H, McAndrews KM, Wu CC, Gagea M, et al. Generation and testing of clinical-grade exosomes for pancreatic cancer. JCI Insight. 2018;3(8):e99263. https://doi.org/10.1172/jci.insight.99263.
110. Molavipordanjani S, Khodashenas S, Abedi SM, Moghadam MF, Mardanshahi A, Hosseinimehr SJ. (99m)Tc-radiolabeled HER2 targeted exosome for tumor imaging. Eur J Pharm Sci. 2020;148:105312. https://doi.org/10.1016/j.ejps.2020.105312.
111. Shi X, Cheng Q, Hou T, Han M, Smbatyan G, Lang JE, et al. Genetically engineered cell-derived nanoparticles for targeted breast cancer immunotherapy. Mol Ther. 2020;28(2):536–47. https://doi.org/10.1016/j.ymthe.2019.11.020.
112. Koh E, Lee EJ, Nam GH, Hong Y, Cho E, Yang Y, et al. Exosome-SIRPα, a CD47 blockade increases cancer cell phagocytosis. Biomaterials. 2017;121:121–9. https://doi.org/10.1016/j.biomaterials.2017.01.004.
113. Liu C, Xia C, Xia C. Biology and function of exosomes in tumor immunotherapy. Biomed Pharmacother. 2023;169:115853. https://doi.org/10.1016/j.biopha.2023.115853.
114. Zhao X, Liu Y, Jia P, Cheng H, Wang C, Chen S, et al. Chitosan hydrogel-loaded MSC-derived extracellular vesicles promote skin rejuvenation by ameliorating the senescence of dermal fibroblasts. Stem Cell Res Ther. 2021;12(1):196. https://doi.org/10.1186/s13287-021-02262-4.
115. Tao SC, Guo SC, Li M, Ke QF, Guo YP, Zhang CQ. Chitosan wound dressings incorporating Exosomes derived from MicroRNA-126-overexpressing synovium mesenchymal stem cells provide sustained release of Exosomes and heal full-thickness skin defects in a diabetic rat model. Stem Cells Transl Med. 2017;6(3):736–47. https://doi.org/10.5966/sctm.2016-0275.
116. Kim SH, Bianco NR, Shufesky WJ, Morelli AE, Robbins PD. Effective treatment of inflammatory disease models with exosomes derived from dendritic cells genetically modified to express IL-4. J Immunol. 2007;179(4):2242–9. https://doi.org/10.4049/jimmunol.179.4.2242.
117. Gong L, Zhou H, Zhang S, Wang C CD44-targeting drug delivery system of Exosomes loading forsythiaside a combats liver fibrosis via regulating NLRP3-mediated pyroptosis, Fu K, Ma C, et al. . Adv Healthc Mater 2023;12(11):e2202228. doi: https://doi.org/10.1002/adhm.202202228.
118. Lin Y, Yan M, Bai Z, Xie Y, Ren L, Wei J, et al. Huc-MSC-derived exosomes modified with the targeting peptide of aHSCs for liver fibrosis therapy. J Nanobiotechnol. 2022;20(1):432. https://doi.org/10.1186/s12951-022-01636-x.
119. Ying W, Gao H, Dos Reis FCG, Bandyopadhyay G, Ofrecio JM, Luo Z, et al. MiR-690, an exosomal-derived miRNA from M2-polarized macrophages, improves insulin sensitivity in obese mice. Cell Metab. 2021;33(4):781–90.e5. https://doi.org/10.1016/j.cmet.2020.12.019.
120. Wang B, Zhang A, Wang H, Klein JD, Tan L, Wang ZM, et al. miR-26a limits muscle wasting and cardiac fibrosis through exosome-mediated microRNA transfer in chronic kidney disease. Theranostics. 2019;9(7):1864–77. https://doi.org/10.7150/thno.29579.

121. Wang H, Wang B, Zhang A, Hassounah F, Seow Y, Wood M, et al. Exosome-mediated miR-29 transfer reduces muscle atrophy and kidney fibrosis in mice. Mol Ther. 2019;27(3):571–83. https://doi.org/10.1016/j.ymthe.2019.01.008.
122. Yang Y, Liu Y, Chai Y, Liu K, Hu W, Zhao K, et al. Exosomes in pathogenesis, diagnosis, and treatment of pulmonary fibrosis. Front Pharmacol. 2022;13:927653. https://doi.org/10.3389/fphar.2022.927653.
123. Sun L, Fan M, Huang D, Li B, Xu R, Gao F, et al. Clodronate-loaded liposomal and fibroblast-derived exosomal hybrid system for enhanced drug delivery to pulmonary fibrosis. Biomaterials. 2021;271:120761. https://doi.org/10.1016/j.biomaterials.2021.120761.
124. Ji K, Fan M, Huang D, Sun L, Li B, Xu R, et al. Clodronate-nintedanib-loaded exosome-liposome hybridization enhances the liver fibrosis therapy by inhibiting Kupffer cell activity. Biomater Sci. 2022;10(3):702–13. https://doi.org/10.1039/d1bm01663f.
125. Kim G, Lee Y, Ha J, Han S, Lee M. Engineering exosomes for pulmonary delivery of peptides and drugs to inflammatory lung cells by inhalation. J Controll Release. 2021;330:684–95. https://doi.org/10.1016/j.jconrel.2020.12.053.
126. Holay M, Zhou J, Park JH, Landa I, Ventura CJ, Gao W, et al. Organotropic targeting of biomimetic nanoparticles to treat lung disease. Bioconjug Chem. 2022;33(4):586–93. https://doi.org/10.1021/acs.bioconjchem.1c00599.
127. Wang Z, Popowski KD, Zhu D, de Juan Abad BL, Wang X, Liu M, et al. Exosomes decorated with a recombinant SARS-CoV-2 receptor-binding domain as an inhalable COVID-19 vaccine. Nat Biomed Eng. 2022;6(7):791–805. https://doi.org/10.1038/s41551-022-00902-5.

Chapter 6
Design of Artificial Extracellular Vesicles for Treatments of Cardiovascular Diseases

Yingqi Miao and Xinglu Huang

Abstract Extracellular vesicles (EVs) are naturally nanoparticles released from cells, playing important roles in intercellular communication, cellular signaling, and metabolic waste removal. EVs have acted as a promising treatment option for various diseases, including cardiovascular diseases (CVDs). However, several challenges such as limited mass production, poor stability, and inadequate targeting hindered their widespread use in clinical applications. In recent years, with the advancement in nanotechnology, artificial EVs based on membrane modification are emerging as a solution to overcome the limitations of natural EVs. In this review, we discuss different categories of artificial EVs and the applications of these platforms in CVDs. The review provides new insights into the design of artificial EVs and can facilitate further clinical translation.

Keywords Artificial extracellular vesicles · Membrane · Nanomedicines · Cardiovascular diseases

1 Introduction

Cardiovascular diseases (CVDs) have been the leading cause of death worldwide for decades, accounting for approximately 17.9 million deaths annually [1, 2]. Examples of CVDs encompass a spectrum of conditions, prominently including coronary artery disease, peripheral artery disease, and cerebrovascular disease. The multifaceted etiology of CVDs involves a myriad of risk factors that contribute to the heightened susceptibility to their development. Prominent risk factors encompass modifiable elements such as tobacco smoking, elevated blood pressure,

Y. Miao · X. Huang (✉)
Key Laboratory of Bioactive Materials for the Ministry of Education, and State Key Laboratory of Medicinal Chemical Biology, College of Life Sciences, Nankai University, Tianjin, China
e-mail: huangxinglu@nankai.edu.cn

© The Author(s), under exclusive license to Springer Nature Singapore Pte Ltd. 2024

Z. Li (ed.), *Extracellular Vesicle: Biology and Translational Application*, https://doi.org/10.1007/978-981-97-5536-3_6

dyslipidemia, obesity, and sedentary behavior, alongside non-modifiable factors including age, gender, and genetic predisposition [3]. Despite notable strides in traditional pharmacological therapies, the mortality and morbidity rates linked to CVDs continue to prevail at alarmingly elevated levels [4, 5]. Consequently, an urgent and imperative need arises for the advancement of innovative therapeutic approaches, with the aim of providing more efficacious treatment options that can effectively address the persistent challenges associated with CVDs.

In recent years, the emergence of cell-free therapies centered around extracellular vesicles (EVs) has garnered substantial attention. EVs represent a diverse population of lipid bilayer membrane vesicles that are actively released by cells into the extracellular milieu. This heterogeneous group encompasses exosomes, microvesicles, and apoptotic bodies, each possessing distinct characteristics and functions in intercellular communication [6]. Extensive investigations have unveiled the secretion of EVs by diverse cell types within the cardiovascular system, establishing their crucial involvement in the pathophysiology of CVDs. Notably, endothelial cells, cardiac progenitor cells, cardiac fibroblasts, and cardiomyocytes have been identified as important contributors to the production and release of EVs within this intricate network. These findings underscore the multifaceted roles of EVs in intercellular communication and highlight their implication in the development and progression of various cardiovascular disorders [7]. Benefiting from their exceptional targeting efficiency and capability to traverse tissue barriers, EVs exhibit remarkable proficiency in transporting and delivering endogenous biological cargo, encompassing functional proteins and genetic materials. This inherent ability positions EVs as a promising avenue for targeted drug delivery to the heart, facilitating the precise administration of therapeutic agents to the desired cardiac tissues. By harnessing the inherent properties of EVs, researchers can potentially enhance drug efficacy, minimize off-target effects, and maximize the therapeutic benefits for CVDs [8–10].

Despite the promising clinical potential of native EVs, their extensive implementation faces considerable obstacles, including challenges associated with mass production, quality testing, and bioavailability. These hurdles impede the widespread utilization of native EVs as therapeutic agents for CVDs. Moreover, the field of drug delivery in cardiovascular treatments is frequently hampered by inadequate targeting strategies, leading to suboptimal efficiency in drug delivery. Addressing these limitations is paramount to enhance the clinical translation of EV-based therapies and to maximize their therapeutic efficacy in the context of CVDs [11]. The rapid advancement of nanotechnology in recent years has catalyzed the emergence of biomimetic EVs based on nanomaterials, offering new avenues for therapeutic interventions. Notably, nanomedicines incorporating membrane modifications have demonstrated enhanced targeting capabilities, thereby significantly improving drug delivery efficiency [12–15]. In light of recent advancements in this burgeoning field, we take sight into the evolutionary trajectory of artificial EVs tailored specifically for the treatment of CVDs, with a particular emphasis on membrane-based mimicking EVs. The discussion encompasses the transition from synthetic EVs to modified EVs, delineating the progressive refinement of engineering strategies (Fig. 6.1). Moreover, we elucidate the diverse applications of these artificial EVs in CVD

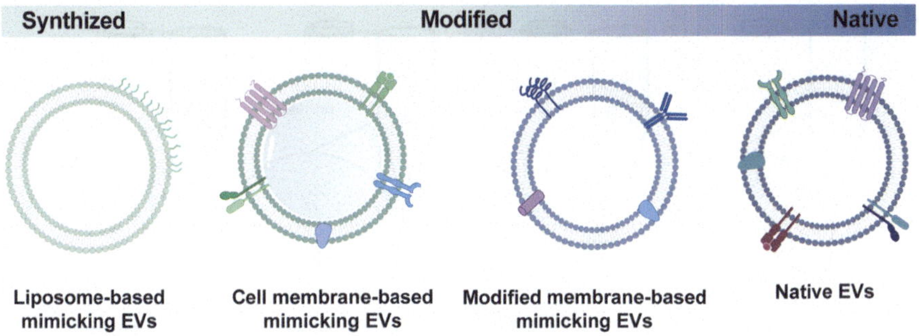

Fig. 6.1 Various designs of artificial EVs, including liposome-based EVs, cell membrane-based EVs, and modified membrane-based EVs

therapies, offering invaluable insights into the potential clinical translation of these emerging nanoplatforms. By providing a comprehensive analysis of the engineering strategies and therapeutic potential of artificial EVs, this review aims to contribute to the advancement and implementation of these innovative approaches in the realm of CVD treatment.

2 Design of Artificial EVs

2.1 Liposome-Based Mimicking EVs

Liposome-based mimicking EVs, alternatively referred to as artificial exosomes or liposome-engineered EVs, represent a distinctive category of synthetic nanoparticles intricately designed to closely emulate both the structure and function of native EVs. Liposomes, comprising a diverse array of lipids or lipidoids, can be purposefully engineered to replicate the key characteristics exhibited by native EVs [16]. A bottom-up approach utilizing liposomes enables precise control over the characteristics of entirely synthetic EVs. Through a bottom-up sequential synthesis process via mechanical emulsification, researchers have successfully developed fully synthetic EVs (fsEVs) with meticulous control over their lipid, protein, and RNA constituents (Fig. 6.2). These synthetic EVs exhibit similar functionality to natural EVs and can be utilized as programmable carriers for targeted therapy in wound healing and neovascularization treatments. Additionally, by analyzing the transcriptome and deciphering the synergistic effects among different EVs components, a comprehensive understanding of EVs signaling between cells has been attained, addressing the technological challenges associated with EV isolation and purification from natural sources [17].

Liposome-based mimicking EVs offer enhanced drug acceptability, allowing for efficient loading of various therapeutic agents, including drugs and nucleic acids. Utilizing liposomes to mimic "artificial EVs" for delivering miRNA, for instance,

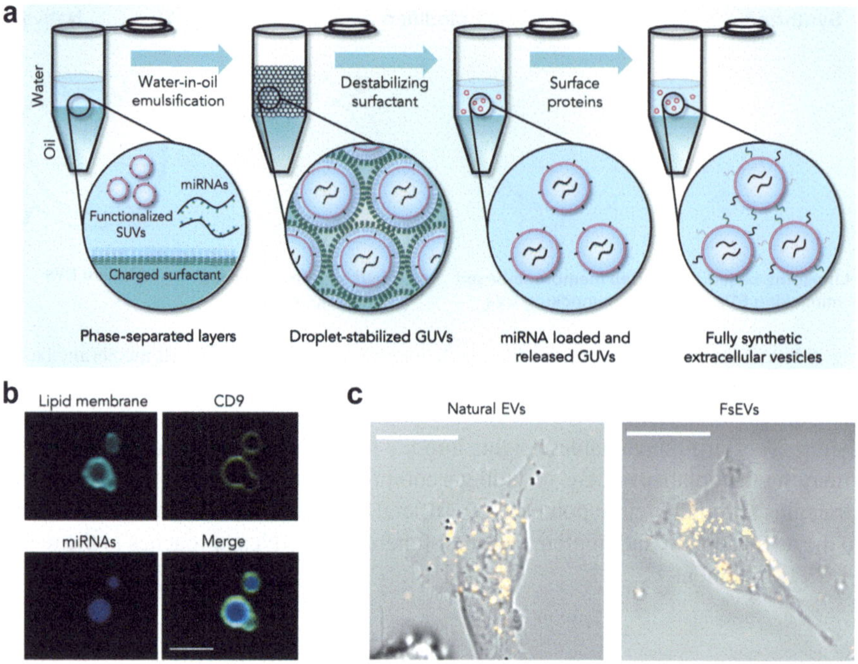

Fig. 6.2 Bottom-up assembly of fully synthetic EVs [17]. (**a**) Schematic illustration of the fsEV formation process inside w/o droplet compartments produced by mechanical emulsification. (**b**) Confocal microscopy image of fsEVs showing surface CD9 and lumen-contained miRNA. (**c**) Fluorescence confocal microscopy image of human dermal fibroblasts incubated with natural EVs and fsEVs. (Reprinted with permission from Staufer et al. 2021)

holds the potential to stimulate angiogenesis and effectively treat ischemic diseases [18]. Indeed, while liposomes have shown promise as delivery vehicles, their efficiency can be limited. Consequently, there is a pressing need to develop strategies that can overcome this challenge. In addressing this issue, Lin et al. introduced a noteworthy solution by developing a hybrid nanoparticle system that combines exosomes and liposomes. This innovative approach enables targeted delivery of the CRISPR/Cas9 gene-editing system to specific receptor cells. By leveraging the advantages of both exosomes and liposomes, this hybrid nanoparticle system offers improved delivery efficiency and enhances the precision of gene-editing therapies [19]. In a similar vein, Evers et al. have recently employed a comparable hybrid system for the delivery of small interfering RNA (siRNA) to various cell types. Remarkably, this hybrid system exhibited distinct functional properties when compared to liposomes alone [20]. By taking advantage of the unique advantages offered by hybrid systems, researchers can unlock new possibilities for efficient and targeted delivery of nucleic acid-based drugs, paving the way for future enhanced clinical treatments.

In order to avoid detection and clearance by the immune system, modification of liposomes with polyethylene glycol (PEG) has proven to be effective in prolonging the circulation time [21, 22]. By coating liposomes with PEG, a hydrophilic polymer, the modified liposomes acquire a stealth-like property, reducing recognition by immune cells and evading rapid clearance. However, it should be noted that PEGylation may induce the production of antibodies against the PEG polymer in vivo. This acquired immune response can lead to accelerated clearance of PEGylated nanoparticles upon repeated dosing, potentially diminishing their efficacy over time. This phenomenon, known as accelerated blood clearance (ABC), poses a challenge in maintaining prolonged circulation and sustained therapeutic levels of PEGylated nanoparticles [16]. Thereby, to better imitate the function of EVs and improve their in vivo biocompatibility and bioavailability, researchers have turned their attention to developing bio-based nanoparticles to mimic native EVs.

2.2 *Cell Membrane-Based Mimicking EVs*

Cell membrane-based mimicking EVs are artificial vesicles designed to closely resemble the structure and functionality of native EVs. These modified vesicles combine the properties of synthetic nanoparticle cores and the biological interface properties of cell membranes, endowing nanoparticles with excellent biocompatibility and immune escape capability [14]. Coating nanoparticles with cell membranes also boosts their ability to interact efficiently with surrounding proteins, cells, and other biological substrates in vivo, allowing them to exhibit similar functionalities to native EVs. The specific choice of membrane components employed in these simulated EVs can impart distinct properties such as circulation time and targeting capabilities (Table 6.1).

Table 6.1 Cell membrane-coated nanoparticles as mimicking EVs

Source		Core	Effect	Reference
Red blood cell		PLGA nanoparticle	Prolong circulation time	[23]
		PLGA nanoparticle	Prolong circulation time	[24]
Platelet		Mesoporous silicon nanoparticle	Target to thrombus site	[25]
		Mesoporous silicon nanoparticle	Remove necrotic cells	[26]
Immune cell	Macrophage	PLGA nanoparticle	Accumulate in atherosclerotic lesions	[27]
	Neutrophil	ZIF-8	Enhance endothelial cells targeting	[28]
Stem cell	MSC	PLGA nanoparticle	Promote cardiomyocyte functions	[29]
	MSC	IONP	Target to ischemic lesion	[30]
Fusion	RBC-EC	PLGA nanoparticle	Prolong circulation time and accumulate in CNV regions	[31]
	RBC-platelet	PLGA nanoparticle	Prolong circulation time and target atherosclerotic lesions	[32]

Red blood cells (RBCs) are naturally long circulating cells that are primarily responsible for transporting oxygen throughout the body. Encapsulating nanoparticles with RBC membranes can lead to an extended half-life in circulation due to the presence of CD47 expressed on their surface, which binds to the inhibitory receptor signaling regulatory protein α (SIRPα) and prevents phagocytosis by macrophages [33]. Consequently, nanoparticles coated with RBC membranes exhibit prolonged retention in vivo, making them highly suitable for applications requiring extended in vivo presence [23]. In contrast to PEG coatings, which may undergo accelerated blood clearance, RBC membrane-coated nanoparticles exhibit consistent performance even following repeated administrations. The stability and extended circulation of RBC membrane-coated nanoparticles contribute to their enhanced therapeutic efficacy and potential for long-term applications.

In addition to RBC membranes, platelets have emerged as a valuable source of cell membranes for surface coating of nanoparticles. Notably, platelets have an inherent capacity to bind to injured vessel walls, facilitated by specific glycoproteins on their membranes such as the glycoprotein IIb/IIIa complex (GPIIb/IIIa), integrin alpha-IIb/beta-3 (αIIbβ3), and P-selectin (CD62P) [34, 35]. These membrane proteins enable platelet membrane-modified nanomedicines to target specific sites of injury or disease, offering a precise and efficient delivery mechanism. By employing biomimetic modifications using platelet membranes, the circulation time of these nanoparticles can be extended as they avoid recognition by the immune system. Consequently, these artificial nanoparticles demonstrate enhanced therapeutic effects in drug delivery [36, 37].

In response to cardiac injury, a robust inflammatory response occurs, leading to the formation of an inflammatory microenvironment at the injury site. Macrophages and neutrophils, possessing specific surface proteins like integrin α4β1 and integrin β2, possess the ability to target sites of inflammation [38]. Utilizing this targeting capability, nanoparticles coated with membranes derived from these immune cells hold promise for specifically targeting cardiac injury sites. For instance, Liu et al. developed a delivery platform consisting of neutrophil-membrane-coated nanoparticles, which effectively transported antisense oligonucleotides (ASOs) against microRNA-155 to endothelial cells in atherosclerotic lesions [39]. Furthermore, immune cell membrane coatings have shown the potential to modulate immune responses within the cardiovascular system. It has been demonstrated that macrophage membrane-coated nanoparticles can regulate inflammation, immune cell recruitment, and immune-modulatory signaling. By promoting a more balanced immune response and mitigating excessive inflammation associated with CVDs, these macrophage cell membrane-coated nanoparticles offer a novel avenue for therapeutic interventions [28].

Currently, numerous studies have provided substantial evidence supporting the involvement of natural stem cells in tissue remodeling and regeneration. Mesenchymal stem cells (MSCs), in particular, play a crucial role in responding to tissue injury and subsequent inflammatory responses. MSCs express various chemokines and cytokine receptors, including CXCR1, CXCR2, CCR1, and CCR2, which facilitate their migration to sites of inflammation. Once at the inflammatory site, MSCs can modulate immune responses and contribute to tissue repair and regeneration [40]. Relying on the specific membrane proteins and inherent homing properties to specifically target inflammation sites of MSCs, researchers have developed stem cell membrane-wrapped nanoparticles to regulate the local microenvironment of inflammation [41]. Furthermore, the paracrine effect of stem cells, which involves the release of numerous pro-angiogenic factors, plays a pivotal role in promoting tissue regeneration. Microparticles encapsulated with MSC membranes and loaded with factors secreted by human bone marrow-derived MSCs have exhibited characteristics akin to genuine MSCs. These bioengineered nanoparticles effectively enhance cardiomyocyte function, offering promising prospects for cardiac tissue regeneration [42].

The fusion of different cell membranes in biomimetic nanoparticles allows for the integration of multiple functions into a single entity. A notable example is the fusion of endothelial cells (EC) and RBC membranes, which has shown promise in the targeted and noninvasive treatment of choroidal neovascularization (CNV). These nanoparticles possess receptors expressed by ECs, such as VEGFR1 and VEGFR2, enabling effective binding to VEGF-A ligands and blocking their effects on host endothelial cells. Simultaneously, the RBC membranes express "don't eat me" protein receptors, endowing the nanoparticles with immune evasion capabilities [32]. In another innovative approach, Zhang et al. proposed the concept of artificial hybrid nanosized cells (Hynocell) with multiple cell functions. This design involves the integration of secretome factors (SF) into PLGA nanoparticles, which are then encapsulated by cell membranes derived from RBCs and platelets fused with copper-containing ferritin (FTn-Cu). The fusion of cell membrane and FTn-Cu facilitated the amalgamation of their distinct properties and functions [43] (Fig. 6.3). Membrane fusion expands the functional repertoire of biomimetic EVs by incorporating diverse membrane properties, thereby enhancing their efficacy and diversifying their functions.

Cell membrane-based mimicking EVs offer a high degree of versatility as platforms for drug delivery, diagnostics, and other biomedical applications due to their ability to transport various cargoes and functional molecules. These modified nanoparticles can be customized to encapsulate and deliver specific therapeutic agents, such as drugs or nucleic acids, to target cells or tissues, enhancing their efficacy and minimizing off-target effects.

2.3 *Modified Membrane-Based Mimicking EVs*

Although nanoparticles coated with natural cell membranes offer significant advantages, there are inherent limitations that restrict their applications. The expression levels and ratios of surface markers on cell membranes may not meet the desired requirements, and the density of certain cognate ligands might not be sufficient for robust targeting [14]. To overcome these challenges, modifying the membranes of mimicking EVs with specific peptides, antibodies, and proteins emerges as a

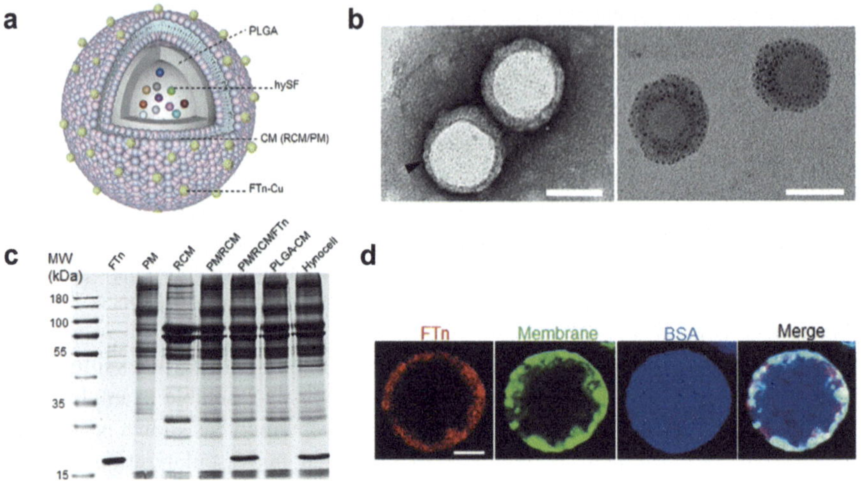

Fig. 6.3 Design of artificial hybrid nanosized cells [43]. (**a**) Schematic illustration of the conceptual design of Hynocell. (**b**) TEM images of PLGA-CM-FTn. (**c**) SDS- PAGE analysis of the protein contents extracted from different cell membranes and fused membranes coated onto the particles. (**d**) Confocal microscopy of the structure of microsized particles. (Reprinted with permission from Zhang et al. 2022)

Table 6.2 Modification of the surface of mimicking EVs

Ligand		Method	Target	Reference
Peptide	IMTP	Click chemistry	Ischemic regions	[44]
	c(RGDyK)	Click chemistry	Ischemic regions	[45]
	CHP	Lipid insertion	Cardiac injury regions	[46]
	HBSP	Lipid insertion	Atherosclerotic lesions	[47]
	IMTP	Genetic engineering	Ischemic regions	[48]
	CMP	Genetic engineering	Cardiac injury regions	[49]
Antibody	GAP43	Chemical conjugation	Damaged neurons	[50]
	VEGF antibodies	Chemical conjugation	Neovascularization lesion	[51]
Protein	CD47	Genetic engineering	–	[52]
	GATA4	Genetic engineering	Ischemic regions	[53]
	CXCR4	Genetic engineering	Ischemic regions	[54]

promising strategy. This approach allows for precise control over the targeting ability and circulation time of nanoparticles in vivo, ultimately enhancing the efficacy of CVDs therapies (Table 6.2).

Attaching peptides to nanoparticles is a straightforward method for creating mimicking EVs with targeting ability to the heart (Fig. 6.4). Several strategies have been developed for modifying membranes with specific peptides, with click chemistry being an emerging technology known for its high reaction efficiency

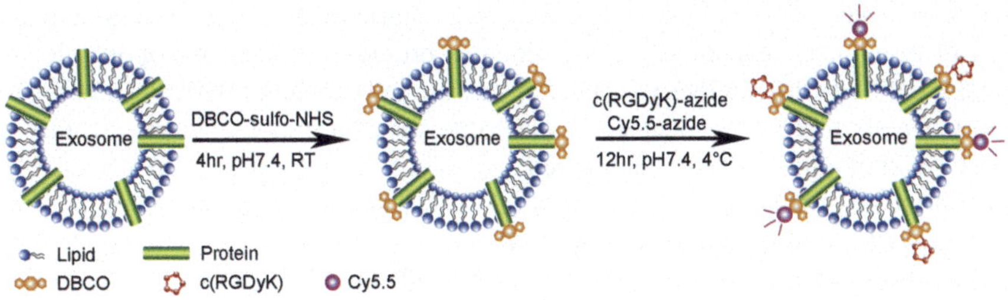

Fig. 6.4 Modified mimicking EVs with specific peptides by various methods. Coupled c(RGDyK) to the surface of MSC-derived EVs via copper-free click chemistry [45]. (Reprinted with permission from Tian et al. 2018)

and mild conditions. Importantly, the conjugation reaction does not alter the size of exosomes and has no impact on their adhesion and uptake by recipient cells. This allows for efficient and effective targeting while maintaining the functional integrity of the exosomes [55]. Coupled ischemic myocardium targeting peptide CSTSMLKAC (IMTP) or cyclic (Arg-Gly-Asp-D-Tyr-Lys) peptide [c(RGDyK)] to mesenchymal stem cell-derived exosomes via copper-free click chemistry showed specific targeting to the site of cardiac ischemic injury [44, 45]. Drawing from the fundamental structure of the phospholipid bilayer, using lipid anchors to attach peptides or functional ligands to the membrane through hydrophobic interactions is a well-established and widely adopted method. Vandergriff et al. conducted a study demonstrating that modifying cardiac stem cell-derived exosomes with a cardiac homing peptide (CHP) using a dioleoylphosphatidylethanolamine N-hydroxysuccinimide (DOPE-NHS) linker improved the targeting of exosomes to regions of cardiac injury [46]. Another effective approach for surface modification of EVs is the incorporation of specific targeting peptides through gene editing. In this method, the targeting ligand is fused to the exosome membrane protein through genetic engineering, resulting in the production of modified EVs with the displayed targeting ligand by the donor cells. For example, the fusion of IMTP or the cardiomyocyte-specific peptide WLSEAGPVVTVRALRGT-GSW (CMP) with lysosome-associated membrane protein (Lamp2b) has been shown to enhance the specificity of exosomes targeting ischemic myocardium [48, 49]. By employing gene-editing techniques, the membranes of biomimetic EVs can be directly modified to optimize their targeting capabilities without compromising their other functions, thereby simplifying the process and eliminating unnecessary steps.

Conjugating antibodies to nanoparticles represents another promising method for enhancing their targeting ability and promoting specific biological activity at desired sites. Guo et al. demonstrated the successful modification of exosome surfaces using the monoclonal antibody GAP43, resulting in increased affinity of the exosomes for damaged neurons. This surface modification enabled precise interaction between the exosomes and the affected cells, facilitating targeted delivery of the neuroprotective compound quercetin [50]. Alternatively, the surface modification of biomimetic exosomes with antibodies provides an effective approach for precise antibody drug delivery. Tian et al. engineered exosomes derived from regulatory T cells (Tregs) to encapsulate cleavable vascular endothelial growth factor (VEGF) antibodies. These engineered exosomes were designed to selectively release the antibodies upon encountering VEGF, thereby targeting the abnormal formation of blood vessels in the eye [51].

Modifying specific proteins on the surface of mimicking EVs can effectively enhance their desired functions, including improved cellular uptake and enhanced targeting effects. One such protein of significance is CD47, which plays a crucial role in regulating the immune system by sending a signal that prevents immune attack and destruction of cells [56]. Wei et al. used lentivirus to overexpress CD47 on the membrane of MSCs, and then extracted CD47-EVs to improve the efficacy and circulation time of the modified EVs [52]. In the context of targeted delivery of therapeutic agents to ischemic tissue, researchers have tackled this challenge by exploiting the natural homing response of MSCs. This response is mediated by the interaction between C-X-C chemokine receptor 4 (CXCR4) and stromal-derived factor (SDF) present in injured tissues. Bose et al. designed bioengineered stem cell membrane-functionalized nanocarriers that were membrane-modified with the CXCR4 receptor for targeted delivery of therapeutic agents to the ischemic hindlimb, resulting in enhanced delivery and therapeutic efficacy [54].

In general, the utilization of different strategies to modify EVs offers several notable advantages for the treatment of CVDs. These modifications significantly enhance the targeting capabilities of EVs, enabling precise delivery of therapeutic cargo to specific sites within the cardiovascular system. Furthermore, modified EVs exhibit immune modulation abilities, allowing for the regulation of the immune response associated with CVDs. The modifications also contribute to improved stability and prolonged circulation of EVs in the body, ensuring sustained therapeutic effects. Additionally, the scalability in the production of modified EVs further enhances their potential for widespread clinical applications. Overall, the use of modified EVs represents a promising approach in the field of CVD treatment, offering the potential for more precise and effective therapeutic interventions.

3 Applications of Artificial EVs in Cardiovascular Diseases

Various engineered EVs have shown great promise in the treatment of CVDs. These artificial EVs not only replicate the essential features of native EVs but also utilize nanomaterial engineering to enhance their therapeutic capabilities. In this section, we present a comprehensive overview of the advancements in utilizing artificial EVs for the treatment of prevalent CVDs such as atherosclerosis, thrombus, myocardial infarction, and stroke (Table 6.3).

Table 6.3 Application of artificial EVs in cardiovascular diseases

Disease	Cargo	Efficiency	Reference
Atherosclerosis	RAP	Target to inflammatory lesions and increase drug availability	[57]
	Anti-CXCR2 antibody	Reduce macrophage accumulation and plaque size in atherosclerotic lesions	[58]
	Far-red AS fluorescent dye	Target to vulnerable plaque and detect site of disease	[59]
	HAL	Alleviate the inflammation and visualize progression of disease	[60]
Thrombus	Argatroban	Target to thrombus and protect cells from H_2O_2-induced cellular injury	[61]
	Uk and ICG	Target and accumulate at blood clot sites	[62]
	Janus polymer motors	Ablation of thrombus by photothermal therapy	[63]
Myocardial infarction	Anti-IL-1β antibodies	Alleviate the inflammation and improve cardiac function	[64]
	Secreted factors	Enhance cardiac function without T cell infiltration	[65]
	Secreted factors	Regenerate blood vessels	[66]
	MicroRNA	Modulate the macrophage phenotype and promote cardiac repair	[67]
Stroke	Curcumin	Alleviate the inflammation and inhibit cellular apoptosis	[45]
	tPA	Accumulate in ischemic region and alleviate the inflammation	[68]
	Hb and methoxatin	Modulate the metabolic microenvironment	[69]

3.1 Atherosclerosis

Atherosclerosis, which underlies CVDs such as coronary artery disease, heart attacks, and strokes, is primarily characterized by endothelial damage, chronic inflammation, and lipid accumulation in the arterial wall [70]. The compromised endothelium within atherosclerotic plaques exhibits increased permeability, enabling nanoparticles to passively penetrate the plaque through the endothelial barrier. This phenomenon, known as the enhanced permeability and retention (EPR) effect, is commonly observed in diseased tissues, including atherosclerosis. Moreover, nanoparticles can be engineered to carry ligands or antibodies that specifically bind to receptors expressed on cells within the plaque. These ligands interact with molecules that are overexpressed in the diseased tissue, promoting the accumulation and retention of nanoparticles within the plaque [71]. The targeted approach employed in treating atherosclerosis enables the enhancement and optimization of therapy, leading to greater efficacy in its treatment.

To ensure effective treatment of atherosclerosis, optimizing the circulation time of drugs in the body plays a vital role. Prolonged drug circulation enables sustained

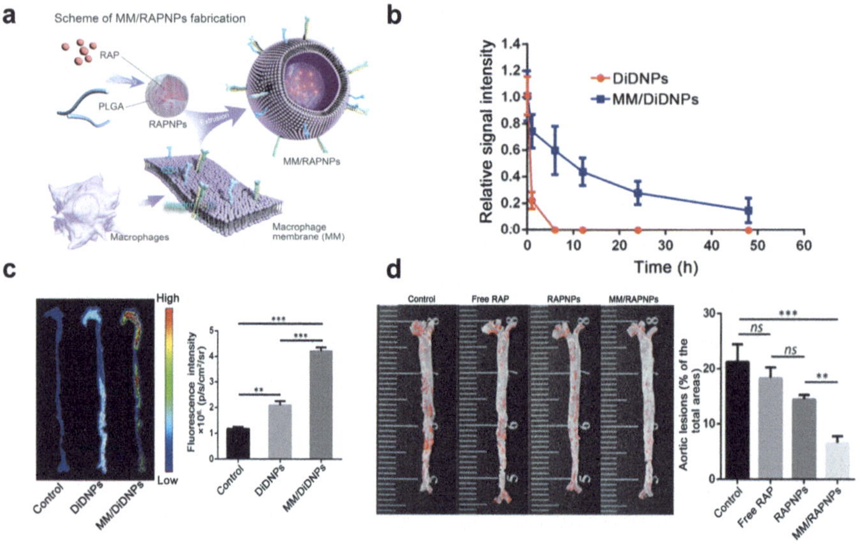

Fig. 6.5 PLGA nanoparticles coated with macrophage membranes for treatment of atherosclerosis [57]. (**a**) Schematic illustration of macrophage membrane-coated PLGA nanoparticles loaded with rapamycin (MM/RAPNPs). (**b**) Comparison of in vivo circulation time between MM/NPs and PLGA nanoparticles. (**c**) Targeting ability to atherosclerotic region of MM/NPs in vivo. (**d**) The effect of MM/RAPNPs in treating atherosclerosis. (Reprinted with permission from Wang et al. 2021)

therapeutic levels and improved delivery to affected areas. Wang et al. proposed a strategy utilizing PLGA nanoparticles coated with red blood cell membranes, aiming to enhance biocompatibility and camouflage the nanoparticles from the immune system. This innovative approach prolongs the presence of nanoparticles in the bloodstream, thereby improving drug delivery and treatment outcomes in atherosclerosis [24]. Another investigation explored the potential of macrophage membrane (MM)-coated biomimetic nanoparticles, which effectively evaded macrophage phagocytosis and exhibited an extended circulation time. These nanoparticles harnessed the unique homing ability of macrophages, enabling selective targeting and accumulation within atherosclerotic plaques. The incorporation of these nanoparticles for the delivery of rapamycin (RAP), an inhibitor that targets the mTOR pathway known for its anti-inflammatory and anti-proliferative properties, facilitated localized drug release, effectively hindering the progression of atherosclerosis [57] (Fig. 6.5).

Early detection and intervention are crucial in managing atherosclerosis and reducing the risk of complications. Biomimetic nanoparticles loaded with MRI imaging agents can be prepared by coating platelet membranes. These nanoparticles have been successfully detected in live animal models of atherosclerosis, enabling visualization and monitoring of the disease progression [59]. In addition, M2 macrophage-derived exosomes engineered with hexyl 5-aminolevulinate hydrochloride (HAL) have demonstrated anti-inflammatory properties, offering potential relief from the inflammation associated with atherosclerosis. The use of intermediate molecule IX (PpIX) for imaging and tracking further enhances the ability to visualize and evaluate the effectiveness of treatment in managing atherosclerosis [60].

3.2 *Thrombus*

Thrombosis plays a crucial role in the advancement of atherosclerosis. The rupture of an atherosclerotic plaque exposes tissue factors and other molecules that trigger the coagulation process. This activation leads to the aggregation of platelets at the site of plaque rupture, ultimately resulting in the formation of a thrombus. The thrombus comprises platelets, fibrin, and other blood components that together obstruct or partially obstruct the normal flow of blood. Due to the complex cellular environment at the thrombus site, engineered EVs can be designed to specifically target and accumulate at the thrombus site, potentially enhancing diagnostics, drug delivery, and therapeutic interventions for thrombotic disorders.

Platelets play a vital role in maintaining hemostasis and facilitating blood clotting. In the event of a bleeding injury, platelets rapidly gather at the site of the wound to initiate the formation of a clot, thereby preventing excessive blood loss [72]. Wan et al. have devised a targeted drug delivery system utilizing mesoporous/macroporous silica/platinum nanomotors coated with platelet membranes which could interact with damaged blood vessels in a manner akin to platelets. By this approach, thrombolytic and anticoagulant drugs can be delivered directly to the site of thrombus formation. The system further allows for controlled drug release upon exposure to near-infrared radiation, thereby facilitating an optimal thrombolytic effect [25] (Fig. 6.6). In thrombosis, insufficient blood flow triggers hypoxia, reactive oxygen species (ROS) production, and tissue damage. Zhao et al. developed H2O2-responsive polymeric nanoparticles coated with platelet membranes. The nanoparticles scavenge excessive ROS and deliver argatroban, an anticoagulant, synergistically treating thrombotic diseases [61].

Some artificial EVs utilizing photothermal therapy have also been involved in the treatment of thrombus. Chang et al. developed a nanocomposite designed for the thrombus-targeted co-delivery of urokinase-type plasminogen activator (Uk) and indocyanine green (ICG), which improves both the diagnostic and therapeutic efficacy for thrombosis. The nanocomposite is P-selectin-specific and has multiple functions including thrombus targeting and imaging, photothermal thrombolysis, and near-infrared (NIR) fluorescence imaging [62]. Shao et al. constructed an erythrocyte membrane-masked Janus polymer motors (JPMs) propelled by NIR laser irradiation and are successfully applied in thrombus ablation. These JPMs, consisting of chitosan and heparin with a gold (Au) coating, demonstrate effective mobility in biological environments [63].

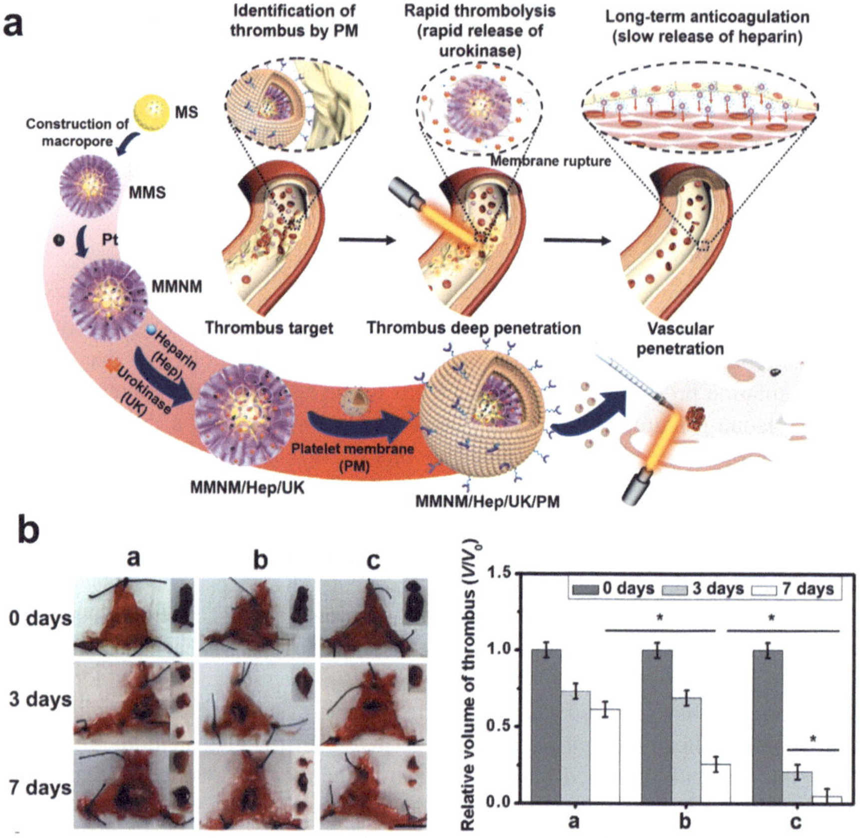

Fig. 6.6 Platelet-derived porous nanomotor for thrombus therapy [25]. (**a**) Schematic illustration of the fabrication of MMNM/Hep/UK/PM nanomotors and their application for thrombus targeting and thrombolytic treatment. (**b**) The effect of MMNM/Hep/UK/PM nanomotors for treating thrombus. (Reprinted with permission from Wan et al. 2021)

3.3 Myocardial Infarction

Myocardial infarction (MI), commonly known as a heart attack, occurs when the coronary arteries, which are responsible for supplying oxygen and nutrients to the heart muscle, become blocked by a thrombus or blood clot. The ability of exosomes to promote healing, reduce inflammation, stimulate angiogenesis and enhance cardiac function offers a promising avenue for cell therapy of MI [73]. In this regard, nanotechnology-based artificial exosomes can be engineered to take advantage of various factors that contribute to their accumulation in damaged tissues. They can benefit from the EPR effect, allowing them to passively accumulate in areas with increased vascularization and inflammation. Additionally, these artificial exosomes can be designed with targeting ligands that actively bind to receptors or biomarkers overexpressed in the lesions, facilitating their selective accumulation.

MI triggers a severe inflammatory response, which adversely affects cardiac remodeling. However, conventional anti-inflammatory drugs lack targeting precision and can lead to adverse effects [74, 75]. To overcome this challenge, researchers have devised a targeted drug delivery strategy utilizing platelet microparticles conjugated with interleukin-1 beta (IL-1β) antibodies. These microparticles selectively bind to IL-1β, mitigating inflammation, and resulting in a significant reduction in infarct size and improved cardiac function. The homing ability of platelets enables a highly specific and effective approach for treating CVDs [64]. In another research, Xue et al. conducted a study where they developed a targeted approach for managing MI by constructing nanoparticles that were enveloped by macrophage membranes and encapsulated the anti-MI agent microRNA (miR) 199a-3p. These macrophage membrane-coated nanoparticles, known as MMNPmiR199a-3p, demonstrated efficient uptake by cardiovascular cells. By reducing the levels of inflammatory cytokines and promoting a shift toward an M2 phenotype in the peri-infarct zone, MMNPmiR199a-3p effectively inhibited inflammatory responses, improved left ventricular remodeling, and enhanced cardiac function [27].

Although stem cell-based therapy has been recognized as a promising approach for treating CVDs, it poses several challenges such as low delivery efficiency and low survival rates at the injury site. To address these problems, researchers have developed mimicking EVs loaded with cytokines secreted by stem cells to enhance stem cell-based therapy. Tang et al. firstly reported that therapeutic microspheres with a core-shell structure were used to simulate stem cells. These synthetic stem cell particles, containing the conditioned medium of stem cells and modified stem cell membranes on the surface, were injected into a mouse model of MI. The results demonstrated enhanced cardiac function without T cell infiltration, indicating good biological safety in clinical trials [65]. Another approach, described by Zhang et al., involved the construction of 3D stem cell spheroids (SSPs) using cell adhesion-modifying chemicals to induce cell aggregation. These SSPs were then incorporated into micro/nanoparticles with a surface coating derived from various cell membranes, resulting in artificially reconstituted 3D spheroids (ASSPs). This strategy exhibited superior revascularization effects in hindlimb ischemia and myocardial infarction models [66] (Fig. 6.7). This cell membrane coating strategy proves to be a valuable approach in overcoming the aforementioned deficiencies associated with stem cell delivery, thereby improving the efficacy and specificity of therapeutic interventions.

Moreover, due to the characteristics of platelets targeting injured hearts, employing platelet membranes to decorate MSC-exosomes can enhance the therapeutic effects [76, 77]. Li et al. designed a platelet membrane-engineered MSC-EVs loaded with microRNA for targeting cardiac repair and immune regulation. Based on the increased binding ability of platelets to circulating monocytes after MI/R, the engineered MSC-EVs could bind to Ly6C^{high} monocytes in peripheral blood and be carried by monocytes into ischemic myocardium. They were subsequently endocytosed by monocyte-differentiated macrophages, releasing miRNAs to modulate the

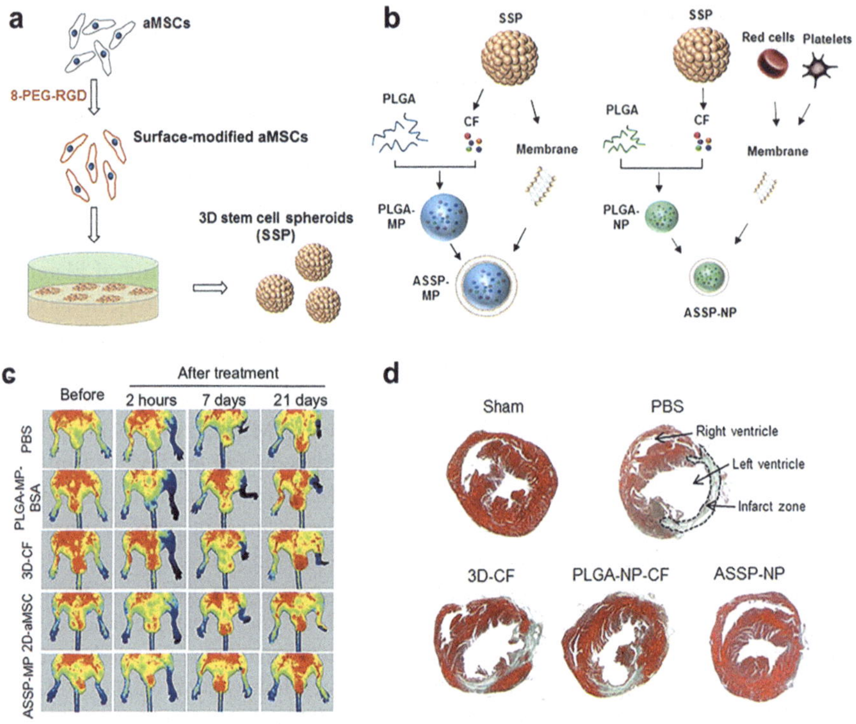

Fig. 6.7 Artificially reconstituted 3D spheroids for revascularization of ischemic diseases [66]. (**a**) Schematic illustration of 3D SSPs. (**b**) Schematic illustration of ASSP-MPs. ASSP-NPs were prepared by encapsulating CF secreted from SSPs into PLGA-NPs and featured a surface coating derived from various cell membranes. (**c**) The therapeutic effect of MM/RAPNPs in treating hindlimb ischemia. (**d**) The therapeutic effect of MM/RAPNPs in treating MI. (Reprinted with permission from Zhang et al. 2020)

macrophage phenotype, thus promoting cardiac repair [67]. Another research has proposed a design in which platelet nanovesicles (PNVs) were decorated onto the surface of cardio sphere-derived cardiac stem cells (CSCs). The natural MI-homing ability of platelets is exploited to enhance CSC vascular delivery to the site of MI injury. Modified CSCs selectively bind collagen-coated surfaces and endothelialized rat aorta, increasing their retention in the heart and reducing infarct size [78].

3.4 Stroke

Stroke is a prevalent cardiovascular disease characterized by the formation of a blood clot that obstructs a cerebral blood vessel. This occlusion disrupts the regular circulation of blood and vital oxygen supply to the brain, resulting in the demise or impairment of brain cells within the affected region. The occurrence of thrombosis plays a significant role in the pathogenesis of stroke, underscoring the criticality of immediate medical attention to reestablish blood flow and mitigate the extent of cerebral injury.

EVs possess the inherent property of efficiently traversing the blood–brain barrier (BBB), making them promising delivery platforms for therapeutic interventions in the brain. Nevertheless, the effective delivery of EVs to the ischemic brain is confronted with notable challenges stemming from the limited targeting capabilities of EVs. To enhance the brain-targeting properties, it is essential to modify EVs by binding specific surface molecules, such as brain-targeting peptides [79, 80]. The surface conjugation of c(RGDyK) peptide onto exosomes via click chemistry provides a means for the targeted delivery of exosomes to areas affected by ischemic brain injury. This surface modification strategy imparts exosomes with the capability to selectively recognize and bind to specific receptors present in ischemic brain regions. Furthermore, the integration of curcumin into cRGD-Exo harnesses the synergistic anti-inflammatory and anti-apoptotic properties of curcumin, thereby amplifying the therapeutic potential for the treatment of ischemic brain injury [45]. Tian et al. employed a deliberate modification strategy by fusing the RGD-4C peptide to the phosphatidylserine-binding domain of lactadherin. This specific modification resulted in the facilitated self-association of the fusion protein with the membrane of the EVs, thereby enhancing their ability to specifically target the ischemic brain region in vivo. Moreover, the RGD-EVs demonstrated a robust anti-inflammatory effect, effectively attenuating the inflammatory response triggered by reperfusion and resulting in a substantial reduction in reperfusion-induced brain damage [81].

Current stroke treatment methods involve thrombolysis and antiplatelet therapy to improve blood flow in the brain. However, the use of tissue plasminogen activator (tPA) as a thrombolytic drug can lead to detrimental effects such as oxidative damage and increased BBB permeability. To address this challenge, Yu et al. developed thrombus targeting nanovesicles (tMP) by simple encapsulating melanin nanoparticles (MNP) and tPA with a platelet membrane vesicle (PM). These nanovesicles achieved precise release of tPA at the thrombus site, enhanced thrombolytic activity. The released MNP particles, with a size of 4.5 nm, could traverse the BBB, accumulate in the region affected by cerebral ischemia, and provide neuroprotection by scavenging free radicals, reducing inflammation, and modulating immune responses [68]. Moreover, the treatment of acute ischemic stroke is hindered by the challenges presented by reperfusion injury and the associated poor long-term prognosis following thrombolytic therapy. The susceptibility of the brain to metabolic changes, particularly disruptions in oxygen and glucose regulation, significantly contributes to the progression of acute ischemic stroke. Thus, restoring metabolic homeostasis is of paramount importance in facilitating effective long-term recovery. Liu et al. developed nanored blood cells that aim to reshape the metabolic microenvironment during different phases of stroke. These nanored blood cells utilized oxygen-carrying mechanism of natural hemoglobin, incorporating hemoglobin and

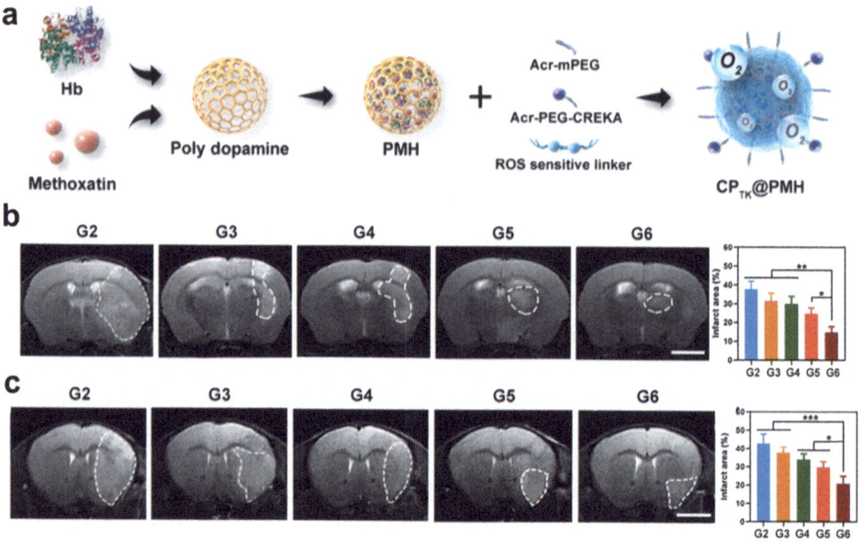

Fig. 6.8 Artificially nanored blood cells for acute ischemic stroke treatment [69]. (**a**) Schematic illustration of the nanoerythrocyte formation. (**b**) The therapeutic effect of nanoerythrocyte in pMCAO mice model. (**c**) The therapeutic effect of nanoerythrocyte in tMCAO mice model. (Reprinted with permission from Liu et al. 2023)

methotrexate into polydopamine nanocages coated with a polymer layer. By actively accumulating in the ischemic core and responding to the metabolic microenvironment, these artificial cells restore cellular metabolic homeostasis, offering a potential solution for acute ischemic stroke treatment [69] (Fig. 6.8).

Overall, artificial EVs hold great promise in CVD treatment and diagnosis. They offer targeted drug delivery, cardioprotective effects, and immune modulation and serve as diagnostic tools in various CVD-related conditions. However, further research and clinical studies are necessary to optimize their design, standardize production, and assess the safety and efficacy of biomimetic EVs for clinical use.

4 Conclusion and Perspectives

In recent years, native EVs have been recognized as potential vehicles for novel cardiovascular disease therapies due to their ability to regulate cell apoptosis, angiogenesis, and inflammation. There are currently around 20 clinical studies investigating the use of EVs in CVDs, with most of them focusing on cardiovascular diagnostics [82]. Preclinical studies have demonstrated the therapeutic potential of EVs in various CVDs, such as myocardial infarction [83], hindlimb ischemia [84], and tissue regeneration [85]. Despite the immense promise, the application of native EVs as therapies for CVDs does present several challenges, including off-target effects, low yield, and inconsistent cargo loading. Additionally, removing endogenous cargo without disrupting the intrinsic structure of EVs poses a significant challenge, as it may lead to unintended side effects. However, it is worth noting that endogenous cargo may contribute to augmentation or synergistic effects, particularly when EVs are derived from cells with inherent therapeutic properties [86]. To address these obstacles and enhance treatment effectiveness, the development of mimicking EVs has emerged as a promising solution. Artificial EVs offer distinct advantages over their natural counterparts, including increased stability and specificity, improved cargo loading and delivery efficiency, and the potential for mass production. By harnessing the benefits of biomimicry, these artificial EVs hold significant potential in overcoming the limitations associated with native EVs, thus paving the way for improved therapeutic outcomes in CVDs.

In this review, our primary focus revolves around membrane-based mimicking EVs, ranging from fully synthetic to modified versions. One such example is liposome-based nanomedicines, which exhibit biocompatibility and versatility. The ease of modification empowers liposomes with enhanced targeting capabilities as drug delivery systems, thereby minimizing in vivo side effects. These synthetic EVs can be produced on a larger scale and manufactured in accordance with regulatory guidelines. However, certain limitations such as poor penetration, short circulation time, and high production costs hinder the broader applications of liposomes. To overcome these challenges, cell membrane coating-based artificial EVs offer a unique combination of bio-membrane and nano-core characteristics [87]. Nonetheless, the biggest challenge for such biomimetic EVs is production constraints. While certain technologies have been developed for large-scale industrial production of cell membranes [88–90], ensuring consistent quality across different batches remains an unresolved issue [91]. Furthermore, when considering cell membrane-based nanomedicines, it is crucial to address the issues of immune-compatibility. Red blood cell and platelet membrane-encased nanoparticles can be readily administered by matching donor and recipient blood types, while other cell types require more care due to possible immunogenicity caused by MHC mismatch [14, 92]. Additionally, a recent study has highlighted a significant concern regarding membrane-coated nanoparticles, revealing that a majority of these particles are only partially coated. This partial coating can have implications for the internalization of nanoparticles within the body. The incomplete coverage of the nanoparticle surface with the desired cell membrane may hinder efficient interactions with target cells and affect the overall efficacy of the nanomedicine [93]. Consequently, it is necessary to address this issue and develop strategies to improve the uniformity and completeness of the membrane coating on nanoparticles to optimize their in vivo behavior and therapeutic outcomes. Moreover, investigating the intricate interaction between membrane-modified nanoparticles and cells is of utmost importance as it lays the foundation for understanding crucial processes such as blood circulation, pharmacokinetics, biodistribution, cellular uptake, and intracellular transport. By delving into these dynamics, researchers can gain valuable insights into how these nanoparticles behave within the body and design more effective therapeutic regimens.

In general, nanotechnology-based artificial EVs hold great promise for the treatment of CVDs. Extensive preclinical studies have been conducted, laying the foundation for its future clinical application. Sustained research on these bioinspired nanoplatforms is critical to tackle CVDs, which remains an important global health problem. By advancing the development and understanding of these technologies, significant progress can be made in addressing CVDs and improving patient outcomes worldwide.

References

1. Anon. Global, regional, and national life expectancy, all-cause mortality, and cause-specific mortality for 249 causes of death, 1980-2015: a systematic analysis for the global burden of disease study 2015. Lancet. 2016;388(10053):1459–544. https://doi.org/10.1016/s0140-6736(16)31012-1.
2. Anon. Global, regional, and national comparative risk assessment of 84 behavioural, environmental and occupational, and metabolic risks or clusters of risks for 195 countries and territories, 1990-2017: a systematic analysis for the global burden of disease study 2017. Lancet. 2018;392(10159):1923–94. https://doi.org/10.1016/s0140-6736(18)32225-6.
3. Yusuf S, Reddy S, Ounpuu S, Anand S. Global burden of cardiovascular diseases: part I: general considerations, the epidemiologic transition, risk factors, and impact of urbanization. Circulation. 2001;104(22):2746–53. https://doi.org/10.1161/hc4601.099487.
4. Park JH, Dehaini D, Zhou J, Holay M, Fang RH, Zhang L. Biomimetic nanoparticle technology for cardiovascular disease detection and treatment. Nanoscale Horiz. 2020;5(1):25–42. https://doi.org/10.1039/c9nh00291j.
5. de Abreu RC, Fernandes H, da Costa Martins PA, Sahoo S, Emanueli C, Ferreira L. Native and bioengineered extracellular vesicles for cardiovascular therapeutics. Nat Rev Cardiol. 2020;17(11):685–97. https://doi.org/10.1038/s41569-020-0389-5.
6. Walker S, Busatto S, Pham A, Tian M, Suh A, Carson K, et al. Extracellular vesicle-based drug delivery systems for cancer treatment. Theranostics. 2019;9(26):8001–17. https://doi.org/10.7150/thno.37097.
7. Barile L, Moccetti T, Marbán E, Vassalli G. Roles of exosomes in cardioprotection. Eur Heart J. 2017;38(18):1372–9. https://doi.org/10.1093/eurheartj/ehw304.
8. Garikipati VNS, Shoja-Taheri F, Davis ME, Kishore R. Extracellular vesicles and the application of system biology and computational modeling in cardiac repair. Circ Res. 2018;123(2):188–204. https://doi.org/10.1161/circresaha.117.311215.
9. Hu W, Liu C, Bi ZY, Zhou Q, Zhang H, Li LL, et al. Comprehensive landscape of extracellular vesicle-derived RNAs in cancer initiation, progression, metastasis and cancer immunology. Mol Cancer. 2020;19(1):102. https://doi.org/10.1186/s12943-020-01199-1.
10. Han C, Yang J, Sun J, Qin G. Extracellular vesicles in cardiovascular disease: biological functions and therapeutic implications. Pharmacol Ther. 2022;233:108025. https://doi.org/10.1016/j.pharmthera.2021.108025.
11. Singh AP, Biswas A, Shukla A, Maiti P. Targeted therapy in chronic diseases using nanomaterial-based drug delivery vehicles. Signal Transduct Target Ther. 2019;4:33. https://doi.org/10.1038/s41392-019-0068-3.
12. Fang RH, Kroll AV, Gao W, Zhang L. Cell membrane coating nanotechnology. Adv Mater. 2018;30(23):e1706759. https://doi.org/10.1002/adma.201706759.
13. Fang RH, Jiang Y, Fang JC, Zhang L. Cell membrane-derived nanomaterials for biomedical applications. Biomaterials. 2017;128:69–83. https://doi.org/10.1016/j.biomaterials.2017.02.041.

14. Fang RH, Gao W, Zhang L. Targeting drugs to tumours using cell membrane-coated nanoparticles. Nat Rev Clin Oncol. 2023;20(1):33–48. https://doi.org/10.1038/s41571-022-00699-x.
15. Saeed S, Ud Din SR, Khan SU, Gul R, Kiani FA, Wahab A, et al. Nanoparticle: a promising player in nanomedicine and its Theranostic applications for the treatment of cardiovascular diseases. Curr Probl Cardiol. 2023;48(5):101599. https://doi.org/10.1016/j.cpcardiol.2023.101599.
16. Li YJ, Wu JY, Liu J, Xu W, Qiu X, Huang S, et al. Artificial exosomes for translational nanomedicine. J Nanobiotechnol. 2021;19(1):242. https://doi.org/10.1186/s12951-021-00986-2.
17. Staufer O, Dietrich F, Rimal R, Schröter M, Fabritz S, Boehm H, et al. Bottom-up assembly of biomedical relevant fully synthetic extracellular vesicles. Sci Adv. 2021;7(36):eabg6666. https://doi.org/10.1126/sciadv.abg6666.
18. Aday S, Hazan-Halevy I, Chamorro-Jorganes A, Anwar M, Goldsmith M, Beazley-Long N, et al. Bioinspired artificial exosomes based on lipid nanoparticles carrying let-7b-5p promote angiogenesis in vitro and in vivo. Mol Ther. 2021;29(7):2239–52. https://doi.org/10.1016/j.ymthe.2021.03.015.
19. Lin Y, Wu J, Gu W, Huang Y, Tong Z, Huang L, et al. Exosome-liposome hybrid nanoparticles deliver CRISPR/Cas9 system in MSCs. Adv Sci (Weinh). 2018;5(4):1700611. https://doi.org/10.1002/advs.201700611.
20. Evers MJW, van de Wakker SI, de Groot EM, de Jong OG, Gitz-François JJJ, Seinen CS, et al. Functional siRNA delivery by extracellular vesicle-liposome hybrid nanoparticles. Adv Healthc Mater. 2022;11(5):e2101202. https://doi.org/10.1002/adhm.202101202.
21. Wolfram J, Ferrari M. Clinical Cancer Nanomedicine. Nano Today. 2019;25:85–98. https://doi.org/10.1016/j.nantod.2019.02.005.
22. Knop K, Hoogenboom R, Fischer D, Schubert US. Poly(ethylene glycol) in drug delivery: pros and cons as well as potential alternatives. Angew Chem Int Ed Engl. 2010;49(36):6288–308. https://doi.org/10.1002/anie.200902672.
23. Hu CM, Zhang L, Aryal S, Cheung C, Fang RH, Zhang L. Erythrocyte membrane-camouflaged polymeric nanoparticles as a biomimetic delivery platform. Proc Natl Acad Sci USA. 2011;108(27):10980–5. https://doi.org/10.1073/pnas.1106634108.
24. Wang Y, Zhang K, Qin X, Li T, Qiu J, Yin T, et al. Biomimetic Nanotherapies: red blood cell based core-shell structured nanocomplexes for atherosclerosis management. Adv Sci (Weinh). 2019;6(12):1900172. https://doi.org/10.1002/advs.201900172.
25. Wan M, Wang Q, Wang R, Wu R, Li T, Fang D, et al. Platelet-derived porous nanomotor for thrombus therapy. Sci Adv. 2020;6(22):eaaz9014. https://doi.org/10.1126/sciadv.aaz9014.
26. Chen L, Zhou Z, Hu C, Maitz MF, Yang L, Luo R, et al. Platelet membrane-coated Nanocarriers targeting plaques to deliver anti-CD47 antibody for atherosclerotic therapy. Research (Wash D C). 2022;2022:9845459. https://doi.org/10.34133/2022/9845459.
27. Xue Y, Zeng G, Cheng J, Hu J, Zhang M, Li Y. Engineered macrophage membrane-enveloped nanomedicine for ameliorating myocardial infarction in a mouse model. Bioeng Transl Med. 2021;6(2):e10197. https://doi.org/10.1002/btm2.10197.
28. Li C, Zhao Z, Luo Y, Ning T, Liu P, Chen Q, et al. Macrophage-disguised manganese dioxide nanoparticles for neuroprotection by reducing oxidative stress and modulating inflammatory microenvironment in acute ischemic stroke. Adv Sci (Weinh). 2021;8(20):e2101526. https://doi.org/10.1002/advs.202101526.
29. Feng L, Dou C, Xia Y, Li B, Zhao M, Yu P, et al. Neutrophil-like cell-membrane-coated nanozyme therapy for ischemic brain damage and long-term neurological functional recovery. ACS Nano. 2021;15(2):2263–80. https://doi.org/10.1021/acsnano.0c07973.
30. Yao C, Wu W, Tang H, Jia X, Tang J, Ruan X, et al. Self-assembly of stem cell membrane-camouflaged nanocomplex for microRNA-mediated repair of myocardial infarction injury. Biomaterials. 2020;257:120256. https://doi.org/10.1016/j.biomaterials.2020.120256.
31. Kim HY, Kim TJ, Kang L, Kim YJ, Kang MK, Kim J, et al. Mesenchymal stem cell-derived magnetic extracellular nanovesicles for targeting and treatment of ischemic stroke. Biomaterials. 2020;243:119942. https://doi.org/10.1016/j.biomaterials.2020.119942.

32. Li M, Xu Z, Zhang L, Cui M, Zhu M, Guo Y, et al. Targeted noninvasive treatment of choroidal neovascularization by hybrid cell-membrane-cloaked biomimetic nanoparticles. ACS Nano. 2021;15(6):9808–19. https://doi.org/10.1021/acsnano.1c00680.
33. Oldenborg PA, Zheleznyak A, Fang YF, Lagenaur CF, Gresham HD, Lindberg FP. Role of CD47 as a marker of self on red blood cells. Science. 2000;288(5473):2051–4. https://doi.org/10.1126/science.288.5473.2051.
34. van der Meijden PEJ, Heemskerk JWM. Platelet biology and functions: new concepts and clinical perspectives. Nat Rev Cardiol. 2019;16(3):166–79. https://doi.org/10.1038/s41569-018-0110-0.
35. Lippi G, Franchini M, Targher G. Arterial thrombus formation in cardiovascular disease. Nat Rev Cardiol. 2011;8(9):502–12. https://doi.org/10.1038/nrcardio.2011.91.
36. Hu CM, Fang RH, Wang KC, Luk BT, Thamphiwatana S, Dehaini D, et al. Nanoparticle biointerfacing by platelet membrane cloaking. Nature. 2015;526(7571):118–21. https://doi.org/10.1038/nature15373.
37. Li J, Angsantikul P, Liu W, Esteban-Fernández de Ávila B, Chang X, Sandraz E, et al. Biomimetic platelet-camouflaged Nanorobots for binding and isolation of biological threats. Adv Mater. 2018;30(2):1. https://doi.org/10.1002/adma.201704800.
38. Swirski FK, Nahrendorf M. Cardioimmunology: the immune system in cardiac homeostasis and disease. Nat Rev Immunol. 2018;18(12):733–44. https://doi.org/10.1038/s41577-018-0065-8.
39. Liu Y, He M, Yuan Y, Nie C, Wei K, Zhang T, et al. Neutrophil-membrane-coated biomineralized metal-organic framework nanoparticles for atherosclerosis treatment by targeting gene silencing. ACS Nano. 2023;17:7721. https://doi.org/10.1021/acsnano.3c00288.
40. Karp JM, Leng Teo GS. Mesenchymal stem cell homing: the devil is in the details. Cell Stem Cell. 2009;4(3):206–16. https://doi.org/10.1016/j.stem.2009.02.001.
41. Thanuja MY, Anupama C, Ranganath SH. Bioengineered cellular and cell membrane-derived vehicles for actively targeted drug delivery: so near and yet so far. Adv Drug Deliv Rev. 2018;132:57–80. https://doi.org/10.1016/j.addr.2018.06.012.
42. Luo L, Tang J, Nishi K, Yan C, Dinh PU, Cores J, et al. Fabrication of synthetic mesenchymal stem cells for the treatment of acute myocardial infarction in mice. Circ Res. 2017;120(11):1768–75. https://doi.org/10.1161/circresaha.116.310374.
43. Zhang X, Zhang Y, Zhang R, Jiang X, Midgley AC, Liu Q, et al. Biomimetic design of artificial hybrid nanocells for boosted vascular regeneration in ischemic tissues. Adv Mater. 2022;34(14):e2110352. https://doi.org/10.1002/adma.202110352.
44. Zhu LP, Tian T, Wang JY, He JN, Chen T, Pan M, et al. Hypoxia-elicited mesenchymal stem cell-derived exosomes facilitates cardiac repair through miR-125b-mediated prevention of cell death in myocardial infarction. Theranostics. 2018;8(22):6163–77. https://doi.org/10.7150/thno.28021.
45. Tian T, Zhang HX, He CP, Fan S, Zhu YL, Qi C, et al. Surface functionalized exosomes as targeted drug delivery vehicles for cerebral ischemia therapy. Biomaterials. 2018;150:137–49. https://doi.org/10.1016/j.biomaterials.2017.10.012.
46. Vandergriff A, Huang K, Shen D, Hu S, Hensley MT, Caranasos TG, et al. Targeting regenerative exosomes to myocardial infarction using cardiac homing peptide. Theranostics. 2018;8(7):1869–78. https://doi.org/10.7150/thno.20524.
47. Guo L, Miao Y, Wang Y, Zhang Y, Zhou E, Wang J, et al. Biomimetic macrophage membrane and Lipidated peptide hybrid Nanovesicles for atherosclerosis therapy. Adv Funct Mater. 2022;32(52):2204822. https://doi.org/10.1002/adfm.202204822.
48. Wang X, Chen Y, Zhao Z, Meng Q, Yu Y, Sun J, et al. Engineered exosomes with ischemic myocardium-targeting peptide for targeted therapy in myocardial infarction. J Am Heart Assoc. 2018;7(15):e008737. https://doi.org/10.1161/jaha.118.008737.
49. Mentkowski KI, Lang JK. Exosomes engineered to express a cardiomyocyte binding peptide demonstrate improved cardiac retention in vivo. Sci Rep. 2019;9(1):10041. https://doi.org/10.1038/s41598-019-46407-1.

50. Guo L, Huang Z, Huang L, Liang J, Wang P, Zhao L, et al. Surface-modified engineered exosomes attenuated cerebral ischemia/reperfusion injury by targeting the delivery of quercetin towards impaired neurons. J Nanobiotechnol. 2021;19(1):141. https://doi.org/10.1186/s12951-021-00879-4.
51. Tian Y, Zhang F, Qiu Y, Wang S, Li F, Zhao J, et al. Reduction of choroidal neovascularization via cleavable VEGF antibodies conjugated to exosomes derived from regulatory T cells. Nat Biomed Eng. 2021;5(9):968–82. https://doi.org/10.1038/s41551-021-00764-3.
52. Wei Z, Chen Z, Zhao Y, Fan F, Xiong W, Song S, et al. Mononuclear phagocyte system blockade using extracellular vesicles modified with CD47 on membrane surface for myocardial infarction reperfusion injury treatment. Biomaterials. 2021;275:121000. https://doi.org/10.1016/j.biomaterials.2021.121000.
53. Yu B, Kim HW, Gong M, Wang J, Millard RW, Wang Y, et al. Exosomes secreted from GATA-4 overexpressing mesenchymal stem cells serve as a reservoir of anti-apoptotic microRNAs for cardioprotection. Int J Cardiol. 2015;182:349–60. https://doi.org/10.1016/j.ijcard.2014.12.043.
54. Bose RJ, Kim BJ, Arai Y, Han IB, Moon JJ, Paulmurugan R, et al. Bioengineered stem cell membrane functionalized nanocarriers for therapeutic targeting of severe hindlimb ischemia. Biomaterials. 2018;185:360–70. https://doi.org/10.1016/j.biomaterials.2018.08.018.
55. Salunkhe S, Dheeraj BM, Chitkara D, Mittal A. Surface functionalization of exosomes for target-specific delivery and in vivo imaging & tracking: strategies and significance. J Control Release. 2020;326:599–614. https://doi.org/10.1016/j.jconrel.2020.07.042.
56. Logtenberg MEW, Scheeren FA, Schumacher TN. The CD47-SIRPα immune checkpoint. Immunity. 2020;52(5):742–52. https://doi.org/10.1016/j.immuni.2020.04.011.
57. Wang Y, Zhang K, Li T, Maruf A, Qin X, Luo L, et al. Macrophage membrane functionalized biomimetic nanoparticles for targeted anti-atherosclerosis applications. Theranostics. 2021;11(1):164–80. https://doi.org/10.7150/thno.47841.
58. Huang R, Zhang L, Li X, Liu F, Cheng X, Ran H, et al. Anti-CXCR2 antibody-coated nanoparticles with an erythrocyte-platelet hybrid membrane layer for atherosclerosis therapy. J Control Release. 2023;356:610–22. https://doi.org/10.1016/j.jconrel.2023.02.036.
59. Wei X, Ying M, Dehaini D, Su Y, Kroll AV, Zhou J, et al. Nanoparticle functionalization with platelet membrane enables multifactored biological targeting and detection of atherosclerosis. ACS Nano. 2018;12(1):109–16. https://doi.org/10.1021/acsnano.7b07720.
60. Wu G, Zhang J, Zhao Q, Zhuang W, Ding J, Zhang C, et al. Molecularly engineered macrophage-derived exosomes with inflammation tropism and intrinsic Heme biosynthesis for atherosclerosis treatment. Angew Chem Int Ed Engl. 2020;59(10):4068–74. https://doi.org/10.1002/anie.201913700.
61. Zhao Y, Xie R, Yodsanit N, Ye M, Wang Y, Wang B, et al. Hydrogen peroxide-responsive platelet membrane-coated nanoparticles for thrombus therapy. Biomater Sci. 2021;9(7):2696–708. https://doi.org/10.1039/d0bm02125c.
62. Chang LH, Chuang EY, Cheng TM, Lin C, Shih CM, Wu AT, et al. Thrombus-specific theranostic nanocomposite for codelivery of thrombolytic drug, algae-derived anticoagulant and NIR fluorescent contrast agent. Acta Biomater. 2021;134:686–701. https://doi.org/10.1016/j.actbio.2021.07.072.
63. Shao J, Abdelghani M, Shen G, Cao S, Williams DS, van Hest JCM. Erythrocyte membrane modified Janus polymeric motors for thrombus therapy. ACS Nano. 2018;12(5):4877–85. https://doi.org/10.1021/acsnano.8b01772.
64. Li Z, Hu S, Huang K, Su T, Cores J, Cheng K. Targeted anti-IL-1β platelet microparticles for cardiac detoxing and repair. Sci Adv. 2020;6(6):eaay0589. https://doi.org/10.1126/sciadv.aay0589.
65. Tang J, Shen D, Caranasos TG, Wang Z, Vandergriff AC, Allen TA, et al. Therapeutic microparticles functionalized with biomimetic cardiac stem cell membranes and secretome. Nat Commun. 2017;8:13724. https://doi.org/10.1038/ncomms13724.

66. Zhang R, Luo W, Zhang Y, Zhu D, Midgley AC, Song H, et al. Particle-based artificial three-dimensional stem cell spheroids for revascularization of ischemic diseases. Sci Adv. 2020;6(19):eaaz8011. https://doi.org/10.1126/sciadv.aaz8011.
67. Li Q, Huang Z, Wang Q, Gao J, Chen J, Tan H, et al. Targeted immunomodulation therapy for cardiac repair by platelet membrane engineering extracellular vesicles via hitching peripheral monocytes. Biomaterials. 2022;284:121529. https://doi.org/10.1016/j.biomaterials.2022.121529.
68. Yu W, Yin N, Yang Y, Xuan C, Liu X, Liu W, et al. Rescuing ischemic stroke by biomimetic nanovesicles through accelerated thrombolysis and sequential ischemia-reperfusion protection. Acta Biomater. 2022;140:625–40. https://doi.org/10.1016/j.actbio.2021.12.009.
69. Liu P, Zhang T, Li C, Zhang Y, Zhou Z, Zhao Z, et al. Bioinspired nanoerythrocytes for metabolic microenvironment remodeling and long-term prognosis promoting of acute ischemic stroke. Nano Today. 2023;49:101806. https://doi.org/10.1016/j.nantod.2023.101806.
70. Charo IF, Taub R. Anti-inflammatory therapeutics for the treatment of atherosclerosis. Nat Rev Drug Discov. 2011;10(5):365–76. https://doi.org/10.1038/nrd3444.
71. Lobatto ME, Fuster V, Fayad ZA, Mulder WJ. Perspectives and opportunities for nanomedicine in the management of atherosclerosis. Nat Rev Drug Discov. 2011;10(11):835–52. https://doi.org/10.1038/nrd3578.
72. Zhu C, Ma J, Ji Z, Shen J, Wang Q. Recent advances of cell membrane coated nanoparticles in treating cardiovascular disorders. Molecules. 2021;26(11):3428. https://doi.org/10.3390/molecules26113428.
73. Sahoo S, Losordo DW. Exosomes and cardiac repair after myocardial infarction. Circ Res. 2014;114(2):333–44. https://doi.org/10.1161/circresaha.114.300639.
74. Lutgens E, Atzler D, Döring Y, Duchene J, Steffens S, Weber C. Immunotherapy for cardiovascular disease. Eur Heart J. 2019;40(48):3937–46. https://doi.org/10.1093/eurheartj/ehz283.
75. Everett BM, Cornel JH, Lainscak M, Anker SD, Abbate A, Thuren T, et al. Anti-inflammatory therapy with Canakinumab for the prevention of hospitalization for heart failure. Circulation. 2019;139(10):1289–99. https://doi.org/10.1161/circulationaha.118.038010.
76. Li Q, Song Y, Wang Q, Chen J, Gao J, Tan H, et al. Engineering extracellular vesicles with platelet membranes fusion enhanced targeted therapeutic angiogenesis in a mouse model of myocardial ischemia reperfusion. Theranostics. 2021;11(8):3916–31. https://doi.org/10.7150/thno.52496.
77. Hu S, Wang X, Li Z, Zhu D, Cores J, Wang Z, et al. Platelet membrane and stem cell exosome hybrid enhances cellular uptake and targeting to heart injury. Nano Today. 2021;39:39. https://doi.org/10.1016/j.nantod.2021.101210.
78. Tang J, Su T, Huang K, Dinh PU, Wang Z, Vandergriff A, et al. Targeted repair of heart injury by stem cells fused with platelet nanovesicles. Nat Biomed Eng. 2018;2:17–26. https://doi.org/10.1038/s41551-017-0182-x.
79. Alvarez-Erviti L, Seow Y, Yin H, Betts C, Lakhal S, Wood MJ. Delivery of siRNA to the mouse brain by systemic injection of targeted exosomes. Nat Biotechnol. 2011;29(4):341–5. https://doi.org/10.1038/nbt.1807.
80. Liu L, Li Y, Peng H, Liu R, Ji W, Shi Z, et al. Targeted exosome coating gene-chem nanocomplex as "nanoscavenger" for clearing α-synuclein and immune activation of Parkinson's disease. Sci Adv. 2020;6(50):eaba3967. https://doi.org/10.1126/sciadv.aba3967.
81. Tian T, Cao L, He C, Ye Q, Liang R, You W, et al. Targeted delivery of neural progenitor cell-derived extracellular vesicles for anti-inflammation after cerebral ischemia. Theranostics. 2021;11(13):6507–21. https://doi.org/10.7150/thno.56367.
82. Sahoo S, Kariya T, Ishikawa K. Targeted delivery of therapeutic agents to the heart. Nat Rev Cardiol. 2021;18(6):389–99. https://doi.org/10.1038/s41569-020-00499-9.
83. Gallet R, Dawkins J, Valle J, Simsolo E, de Couto G, Middleton R, et al. Exosomes secreted by cardiosphere-derived cells reduce scarring, attenuate adverse remodelling, and improve function in acute and chronic porcine myocardial infarction. Eur Heart J. 2017;38(3):201–11. https://doi.org/10.1093/eurheartj/ehw240.

84. Mathiyalagan P, Liang Y, Kim D, Misener S, Thorne T, Kamide CE, et al. Angiogenic mechanisms of human CD34(+) stem cell exosomes in the repair of ischemic Hindlimb. Circ Res. 2017;120(9):1466–76. https://doi.org/10.1161/circresaha.116.310557.
85. Henriques-Antunes H, Cardoso RMS, Zonari A, Correia J, Leal EC, Jiménez-Balsa A, et al. The kinetics of small extracellular vesicle delivery impacts skin tissue regeneration. ACS Nano. 2019;13(8):8694–707. https://doi.org/10.1021/acsnano.9b00376.
86. Witwer KW, Wolfram J. Extracellular vesicles versus synthetic nanoparticles for drug delivery. Nat Rev Mater. 2021;6(2):103–6. https://doi.org/10.1038/s41578-020-00277-6.
87. Molinaro R, Corbo C, Martinez JO, Taraballi F, Evangelopoulos M, Minardi S, et al. Biomimetic proteolipid vesicles for targeting inflamed tissues. Nat Mater. 2016;15(9):1037–46. https://doi.org/10.1038/nmat4644.
88. Busatto S, Vilanilam G, Ticer T, Lin WL, Dickson DW, Shapiro S, et al. Tangential flow filtration for highly efficient concentration of extracellular vesicles from large volumes of fluid. Cells. 2018;7(12):273. https://doi.org/10.3390/cells7120273.
89. Ganesan P, Karthivashan G, Park SY, Kim J, Choi DK. Microfluidization trends in the development of nanodelivery systems and applications in chronic disease treatments. Int J Nanomedicine. 2018;13:6109–21. https://doi.org/10.2147/ijn.S178077.
90. Zhou Z, Luo R, Chen L, Hu C, Chen C, Maitz MF, et al. Dressing blood-contacting devices with platelet membrane enables large-scale multifunctional biointerfacing. Matter. 2022;5(7):2334–51. https://doi.org/10.1016/j.matt.2022.05.018.
91. Herrmann IK, Wood MJA, Fuhrmann G. Extracellular vesicles as a next-generation drug delivery platform. Nat Nanotechnol. 2021;16(7):748–59. https://doi.org/10.1038/s41565-021-00931-2.
92. Shlomchik WD. Graft-versus-host disease. Nat Rev Immunol. 2007;7(5):340–52. https://doi.org/10.1038/nri2000.
93. Liu L, Bai X, Martikainen MV, Kårlund A, Roponen M, Xu W, et al. Cell membrane coating integrity affects the internalization mechanism of biomimetic nanoparticles. Nat Commun. 2021;12(1):5726. https://doi.org/10.1038/s41467-021-26052-x.

Chapter 7
EV Tetraspanins in Regenerative Medicine

Jun Li and Shijing Yue

Abstract Extracellular vesicles (EVs) are biological nano-sized vesicles secreted by all cell types and can be found in different species. EVs show promise in diagnostics and therapeutics due to their ability to act as intercellular messengers. The therapeutic potential of EVs in regenerative medicine has been increasing attention in recent years. Many studies have demonstrated the important role of EVs in wound healing and tissue regeneration, inhibition of inflammation, and regulation of the immune system. Tetraspanins are a family of four transmembrane proteins that participate in various physiological and pathological processes including cell proliferation, migration, adhesion, stemness, and immune regulation. Tetraspanins also serve as the characteristic molecular markers and membranous components of EVs. In this chapter, we summarize the current understanding of the structure and function of tetraspanins, the effects of tetraspanins on EVs' biogenesis, component sorting, transportation, and biological distribution, and discuss their potential mechanisms for influencing EVs therapeutic effects. Finally, we focus on the recent research that has shown the multiple functions of EV tetraspanins in regenerative medicine. These findings will provide suggestions for accelerating the effective implementation of EVs therapy in regenerative medicine.

Keyword Extracellular vesicles (EVs) · Tetraspanins · Four transmembrane proteins · Membranous components · Regenerative medicine

J. Li
Tianjin Key Lab of Ophthalmology and Vision Science, Tianjin Eye Institute, Tianjin Eye Hospital, Tianjin, China

Nankai University Affiliated Eye Hospital, Nankai University, Tianjin, China

S. Yue (✉)
School of Medicine, Nankai University, Tianjin, China
e-mail: shijingyue@nankai.edu.cn

© The Author(s), under exclusive license to Springer Nature Singapore Pte Ltd. 2024

Z. Li (ed.), *Extracellular Vesicle: Biology and Translational Application*,
https://doi.org/10.1007/978-981-97-5536-3_7

1 Introduction

Currently, regenerative medicine is developed into an interdisciplinary field of research and clinical applications focused on developing new therapies to repair or replace damaged or diseased tissues and organs [1]. Regenerative medicine involves the use of stem cells, tissue engineering, and other advanced techniques to cure a disease by regenerating damaged tissue and to improve the quality of life for patients suffering from a wide range of conditions, including heart disease, diabetes, neurological disorders, and cancer [2–4]. Mesenchymal stem cells (MSCs) are the major resources of cell-based therapy, which are pluripotent stem cells due to their ability to differentiate into specialized cells and repair damaged tissues [5–7]. Further studies have demonstrated that the beneficial effects of MSCs applications in regenerative medicine are attributable to paracrine signaling, which includes MSCs-derived EVs [8–10].

In recent years, many researchers have paid attention to EVs-based regenerative medicine. EVs are lipid bilayer membrane nano-sized particles, which have been confirmed to be secreted by most cell types including MSCs, epithelial cells, endothelial cells, immune cells, neuronal cells, etc. EVs mainly include exosomes and microvesicles (MVs). MVs range from 50 to 2000 nm, and exosomes are in smaller sizes, from 30 to 150 nm [11]. As the natural nanovesicles, EVs carry much information from the parent cells, containing a variety of molecules including proteins, lipids, and nucleic acids, which play a crucial role in the regulation of cell-extracellular matrix (ECM) and cell-cell interaction [12, 13]. Moreover, EVs also have been found to serve as a delivery system for proteins, nucleic acids and even synthetic or natural therapeutic agents to the target cells [14]. In the last decade, studies have demonstrated that EVs have advantageous effects for therapy in various processes including neurological [15], respiratory [16], cartilage [17], kidney [18], cardiac [19], and liver diseases [20], bone repair [21], and cancer [22].

Tetraspanins, the classical molecular markers of EVs, are a family of conserved proteins with 33 members in humans, which contain four transmembrane domains. Tetraspanins are localized on the cell membrane system, including the plasma membrane, various intracellular organelles membrane and vesicle membrane. They are widely expressed in nearly all cell types [23]. A fundamental role of tetraspanins is to organize other proteins into a network to form tetraspanin-enriched microdomains (TEMs), which can influence cell migration, cell adhesion, invasion, and survival and regulate downstream signaling [24]. EV tetraspanins (CD63, CD81, CD9) are involved in the regulation of EV biogenesis, EV component sorting, EV transportation, the biological distribution of EVs, and cell biological functions [13, 25, 26]. Moreover, recent studies have discussed that tetraspanins can also contribute to isolating EVs using the specific antibody. Furthermore, EVs may be modified through genetic engineering, covalent modification, and non-covalent modification [27–29].

Thus, we will describe the structure and function of tetraspanins in the process of EV biogenesis and EV uptake by recipient cells. Moreover, we will discuss the contribution of EV tetraspanins to regenerative medicine. Finally, we will summarize the therapeutic applications of EV tetraspanins in regenerative medicine, such as bone regeneration, heart disease, neurological disorders, wound healing, and future trends.

1.1 Tetraspanins

Tetraspanins are small transmembrane proteins that a family of molecules has been found by the efforts in cloning membrane antigens in the early 1980s. Tetraspanins are expressed in all metazoans with the number of family members varies between species. They have been studied and found 34 members in mammals, 37 in Drosophila melanogaster, 20 in Caenorhabditis elegans, and 17 in Arabidopsis thaliana. The name of tetraspanin refers to the four membrane-spanning segments of protein and flanks a highly conserved topology of two extracellular loops. The naming of a part of tetraspanins is based on the cluster of differentiation (CD) classification that originated the early study by identification with several different monoclonal antibodies and are currently known as tetraspanins CD9, CD37, CD53, CD63, CD81, CD82, and CD151. Apart from the "CD," systematic names also exist, like Tspan1–33 for the human tetraspanins [23].

1.1.1 Structure of Tetraspanins

Tetraspanins are a family of conserved four transmembrane glycoproteins with one small and one large extracellular loop (SEL and LEL), four transmembrane domains (TMs), one short cytoplasmic amino- and one carboxy-terminal domain, and one short intracellular loop (ICL) (see Fig. 7.1). Recent studies have revealed the crystal structure of several tetraspanins. Especially, CD81, the crystal structure of CD81 LEL is firstly reported in 2001 [30]. Up to now, many tetraspanin structures have been identified, including CD9 [31], CD53 [32], and Tspan15 [33], etc. Most tetraspanins share 20–30% homology, they contain a cysteine-cysteine-glycine (CCG) motif in LEL and several polar residues in TM, which also is the specific signature of tetraspanins compared with other four transmembrane glycoproteins [34, 35].

Extracellular loops are the most variable regions of tetraspanins. LEL, a 69–132 amino acids region, contains several conserved cysteine residues (such as Cys-Cys-Gly (CCG) motif) to form 2–4 disulfide bonds. The secondary structure of tetraspanins LEL is determined and stabilized by the variable disulfide bonds. These elements are crucial for the correct folding of LEL and the whole molecule. Due to the specificity of LEL, the monoclonal antibody is designed based on this region. Importantly, LEL is the functional region of tetraspanins, which is the fundamental structure of homodimerization formation and protein–protein interactions. In addition, LEL is glycosylated in most tetraspanins. SEL, a 13–31 amino acids region, is

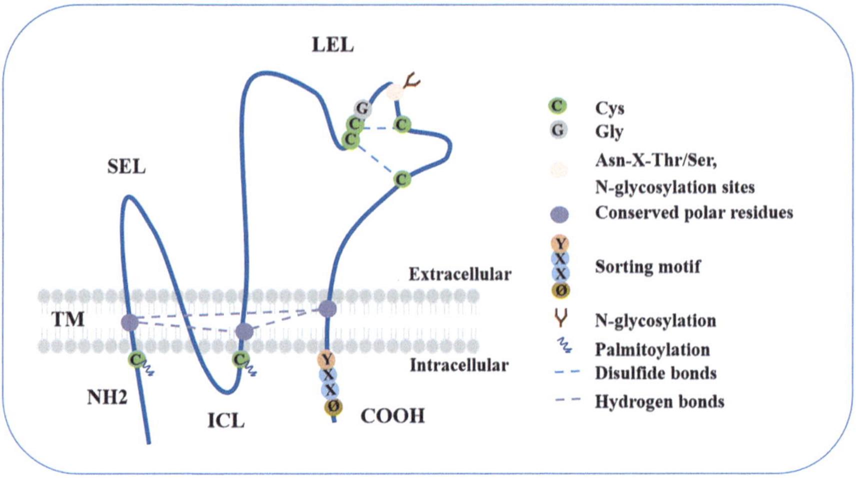

Fig. 7.1 The overall structure of tetraspanins

rarely glycosylated. It usually was considered as the structure supporting domain. However, the more details about SEL are not clear [36].

The sequences of four TMs are highly conserved in tetraspanins. The crystal structures of CD81, CD9, and CD53 show the cone-shaped architecture in the TMs, where are the binding pocket of molecular partners. In addition, the polar residues in TMs help to form hydrogen bonds to stabilize the cone-shaped structure and the interaction with partners [37].

The cytoplasmic domains of tetraspanins usually contain functional motifs: YXXØ motif, PDZ-binding motif, and ITIM/ITAM-like motif, which are involved in protein trafficking and signaling transduction. YXXØ motif drives the trafficking of tetraspanins and partners into EVs. The Gly-Tyr-Glu-Val-Met (GYEVM) motif contributes to driving CD63 into the clathrin-coated vesicles. The PDZ-binding motif of CD53 contributes to participate in PKC pathway. Meanwhile, the palmitoylation of cytoplasmic cysteine residues in cytoplasmic domains is the crucial signature of tetraspanins, contributing to organizing the interactions between different tetraspanins and stable the structure [38] (see Fig. 7.1).

The unique conformation of LEL is mainly dependent on the disulfide bonds between conserved cysteine residues. The N-glycosylation sites (Asn-X-Thr/Ser) on LEL endow tetraspanins with more specificity. In the TMs, the polar residues form hydrogen bonds to maintain the conformation and lipids binding cavity (lipids not shown in this figure). In the cytoplasmic domains, there are several sorting motifs, like YXXØ, which contribute to protein trafficking. The cysteines on membrane-proximal are usually palmitoylated, which are considered to organize the interactions between different tetraspanins [39, 40].

1.1.2 Tetraspanin Webs

Tetraspanin webs refer to the network formed by the interactions of tetraspanin-tetraspanin and tetraspanin partners, which are also described as TEMs (see Fig. 7.2). As mentioned earlier, the structure of tetraspanins is the basis for these interactions that contribute to the formation of TEMs by recruiting multiple partners and secondary partners.

Literature shows that the partners of tetraspanins can be categorized into four categories: extracellular enzymes, adhesion proteins, lipids, and intracellular signaling proteins. The LEL and TM are the key elements of tetraspanin interaction with molecular partners. For example, the association of CD81-CD19 and CD81-hepatitis C virus (HCV) E2 complex is dependent on the structure of LEL [41, 42]. The interaction of CD151 and integrins is mediated by LEL [43]. LEL is also essential for the interaction between CD82 and integrin α5 [44]. The binding sites of CD9-EWI2/F and Tspan15/ADAM10 are within TM [36]. The binding pocket of cholesterol in CD81 is formed by four TM helices and several polar residues in TM [37]. The cytoplasmic domains are essential for the recruitment of intracellular signaling proteins, like PKC, Rac, and ezrin–radixin–moesin (ERM) family [45] (see Fig. 7.2).

Meanwhile, the two major post-translational modifications of tetraspanins are palmitoylation and glycosylation, which also play an important role in the interaction of tetraspanin-tetraspanin and tetraspanin-partner. Palmitoylation is involved in the binding of different tetraspanins like CD9 with CD53, CD81, and even with integrin β4 [46]. The function of glycosylation has become more clear in recent years. The N-glycosylation of CD82 is essential for the interaction with integrins [44]. The localization and stability of tetraspanins are also regulated by glycosylation.

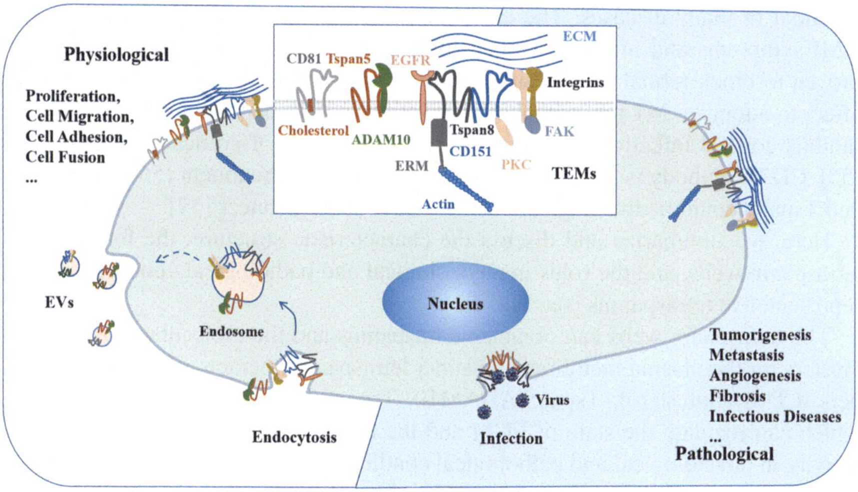

Fig. 7.2 The overview of tetraspanin webs in biological processes

Taken together, the tetraspanin web organizes the tetraspanins and extracellular, transmembrane, and intracellular partners into these TEMs to regulate cell function and arrange EV cargos.

1.1.3 Physiological and Pathological Roles of Tetraspanins

Tetraspanins widely expressed in all cell types, participate in many biological processes including cell growth, migration, cell adhesion, and cell fusion. Moreover, tetraspanins are also involved in the pathogenesis of some human diseases like inflammation, virus infection, neovascularization, and tumorigenesis (see Fig. 7.2).

CD81 is firstly found in the immune system that involves B cell and T cell activation [47]. The similar function of CD81, CD9, and Tspan7 is recovered in brain development [48], retinal pigment epithelium development [49, 50], and mammalian reproduction [51]. Furthermore, CD81, CD9, and Tspan8 also involved in viruses (HIV, COVID-19, HPV, and HCV, etc.) or other pathogens' infection process as an entry helper on the cell membrane [52], tumorigenesis, and metastasis [53]. CD151 and Tspan8 have been reported to play an important role in skin wound healing and blood vessel development. Meanwhile, CD151 and Tspan8 are also highly expressed in many kinds of tumor tissues, as the potential targeted therapeutic markers. The aberrant CD151 and Tspan8 also promote fibrosis, pathological angiogenesis, tumorigenesis, and metastasis by triggering excessive growth and migration through the partners and downstream signal pathways in tetraspanin webs, like integrins, EGFR, PKC, cytoskeleton, CD44, and β-catenin, and so on. Even though CD82 belongs to the tetraspanins family, CD82 develops an opposite effect with CD151 and Tspan8, inhibiting fibrosis, angiogenesis, and metastasis. These processes are partially implemented through EVs.

Based on these functions, tetraspanins have huge therapeutic and diagnostic potential in many diseases. The exogenous expression of CD151 can relieve the AMI symptoms and improve cardiac function. The CD82 LEL peptide has been proven to block retinal neovascularization in vivo and show an anti-inflammatory effect to attenuate M1 macrophage polarization [54]. A peptide derived from CD9 inhibits corneal infection and promotes wound healing of corneal epithelial cells [55]. CD37 antibody is a potential choice for lymphoma treatment [56]. EV CD151 and Tspan8 could be the diagnostic biomarkers of lung cancer [57].

Here, we summarize and discuss the characteristic structure, the formation of tetraspanin webs, and the roles in physiological and pathological conditions of the representative tetraspanins (see Fig. 7.2).

The tetraspanin webs can organize tetraspanins and the molecular partners into clusters on the plasma membrane. Distinct tetraspanin interacts with specific partners, CD81/cholesterol, Tspan5/ADAM10, Tspan8/EGFR, and CD151/integrins, which can regulate the state of ECM and the downstream signal to influence cell activity in physiological and pathological conditions. At the same time, Tetraspanin webs are involved in the biogenesis of EVs through endocytosis. Moreover, multiple pathogens infection is mediated by tetraspanins [23].

1.2 *Tetraspanins in EVs Journey*

The specific structure and domains of tetraspanins are fundamental for interacting with a large variety of direct or indirect molecular partners, which contribute to the biogenesis, cargo sorting, and biodistribution of EVs [13].

1.2.1 EVs Biogenesis

Currently, the mechanisms of EV biogenesis have been almost revealed. Multiple evidences support that exosome biogenesis originates from membrane invaginations. The first invagination occurs on the membrane surface by endocytosis. The invagination of the cell membrane forms the early endosomes and integrates into late endosomes in cytoplasm. The second invagination, part of the late endosomes membrane invades the inner endosome and invaginates to generate the multivesicular bodies (MVBs). Subsequently, part of MVBs fuse with lysosomes, leading to the degradation. Alternatively, part of MVBs contain CD63, and LAMP1/2 on the surface, which mediate MVBs fusion with the plasma membrane to release EVs into the extracellular space [58]. MVs are generated through plasma membrane budding, and tetraspanins, TSG101, and ESCRT complex are involved in the progress [59]. (see Fig. 7.3).

Multiple evidences support that the biogenesis and secretion of MVBs are driven by the ESCRT-dependent and ESCRT-independent progress. ESCRT is a protein complex that is mainly a regulator to sort specific components into MVBs, which is composed of nearly 20 proteins, including four complexes (ESCRT-0, I, II, III) and various associated proteins (VTA1, ALIX, VPS4, etc.). In the ESCRT-dependent

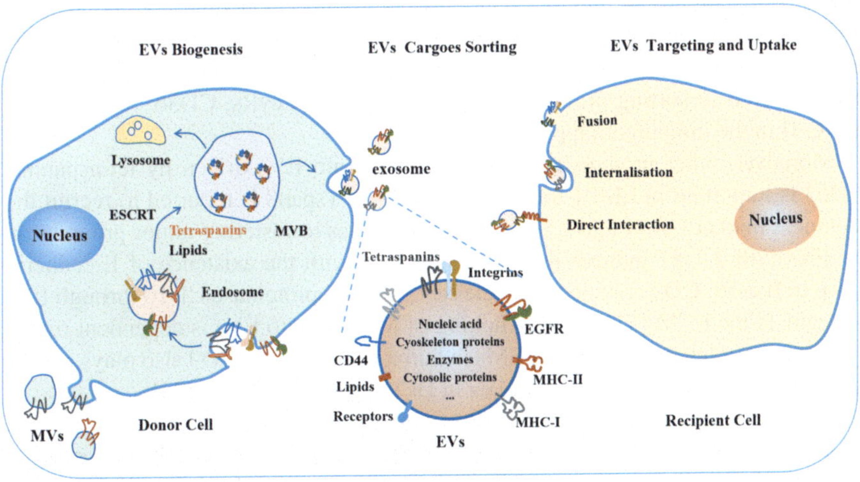

Fig. 7.3 Biogenesis and molecular composition of EVs and interaction with recipient cell

process, ESCRT complexes recognize the ubiquitinated proteins on the endosomal membrane and sequester specific components into MVBs. The members of the ESCRT complexes are dependent on cell types [60, 61]. It has been found that ALIX and ESCRT-III are involved in the sorting of CD9, CD81, and CD63 into EVs in Hela cells [62]. However, in the absence of ESCRT cells, CD63-positive EVs are also found [63].

The ESCRT-independent processes are also involved in the biogenesis and secretion of EVs. Tetraspanins seem to play an important role in these processes [13]. For instance, in CD9 knockout mice, the EVs secretion is decreased compared with wild-type mice [64]. Consistently, enhanced levels of CD9 could significantly increase the number of EVs in several human cells [65]. CD63 contributes to the increasing release of EVs when cells are infected with HSV-1 [66]. The decreased expression of CD63 reduced MVB formation and EV biogenesis. In contrast, EV secretion is negatively regulated by tspan-6, which directs the MVBs cargos for lysosomal degradation through SDC4-syntenin [67]. Lipids are also involved in the biogenesis of EVs [12]. However, the more detailed mechanism of how tetraspanins influence EV biogenesis still needs further evidence to be confirmed.

1.2.2 Biological Composition of EVs

The increasing evidence has shown that tetraspanins contribute to ILVs sorting of various cargos. Tetraspanins form the specific TEMs on the EV membrane, playing a role in the recruitment of associated partners, and influencing the compositions of EVs, including receptors, protease, and possibly nucleic acids.

Protein cargos are the major partners of EV tetraspanins. In CD151 and Tspan8 knockout rats, CD151 and Tspan8 have been confirmed to contribute to integrin and protease recruitment, such as integrin $\alpha6\beta1$, MMP2, and MMP9 [68]. Distinct integrin subunits are involved in different tetraspanin sorting machinery. CD82 recruits $\alpha5$ and $\beta1$ into EVs by direct interaction in ovarian cancer cells [44]. The CD63-αIIbβ3-CD9 complex is located on the surface of activated platelets [69]. CD9 is also involved in sorting proteins into EVs, including FcγRs, CD36, integrin, and MHC-II in the inflammation process.

Moreover, other proteins can also be sorted into EVs driven by tetraspanins. CD63 is important in LMP1 trafficking into EVs. Tspan8 is reported to recruit the stemness marker CD44 into EVs, which contributes to fostering tumor progression by EVs [70]. CD82 induces β-catenin into EVs with the existence of E-cadherin [64]. In B cells, CD9 conduces the release of metalloproteinase CD10 through EVs to regulate the ECM [71]. The trafficking of MHC-II into EVs is dependent on the recruitment of tetraspanins into TEMs in human B cells [72]. CD9 also plays a similar role in recruiting MHC-II into EVs in DCs to activate T cells [73].

Nucleic acids are the important compositions of EV cargo. Recent research has found that tetraspanins may be involved in the recruitment of mRNA and miRNA into EVs. In ovarian cancer cells, the finding demonstrated that micronuclei served as a source of nuclear content shuttled into MVBs via tetraspanins [74]. The analysis of Tspan8 overexpression cell-derived EVs showed that Tspan8 contributed to the selective recruitment of mRNA into EVs [75].

Therefore, tetraspanins are enriched on EVs and regulate the associated partners through TEMs, contributing to sorting and recruiting selective cargoes into EVs, which also have the potential to carry drugs as natural nanovectors (see Fig. 7.3).

1.2.3 Targeting and Uptake of EVs

As the natural nanovesicles, it is a complicated process that EVs mediate the cell-cell communication. By the way, it is a long journey to target cells since EVs are derived from the donor cells. EVs are uptake by the recipient cells in several possible ways: (1) directly interacting with receptors; (2) direct fusion with the plasma membrane; (3) internalization (see Fig. 7.3).

The non-specific uptake is ubiquitous in various cell types. However, the specific targeting of EVs is also existed and mysterious, which is possibly dependent on direct interaction. For instance, CD63 positive EVs derived from neuroblastoma cells can be uptake by neuronal dendrites, while the CD63 negative EVs are uptake by whole neurons and glial cells [76]. Tumor EVs have the organotropic ability, which is useful for predicting metastatic propensity and plays a role in determining the organ sites of metastasis. The organotropic of EVs was confirmed to be caused by the different compositions of integrin that bind to specific proteins on target cells. Tetraspanins may be involved in the recruitment of these integrins [77]. CD81 positive EVs were shown to target laminin-secreting cells by the interaction between CD81 and laminin [78]. CD151 and Tspan8 positive EVs derived from the highly metastatic pancreatic cancer cells are rich in $\alpha6\beta4$, and preferentially target lung and lymph node stromal cells. In our latest research, Tspan8 positive EVs derived from epithelial cancer cells can promote confined diffusion and pronounce the uptake of EVs by recipient cells [79].

Besides, the targeting of EVs can be also mediated by other proteins and lipids, like nervous system-specific rabies viral glycoprotein (RVG), LFA-1 interacting with ICAM1 on vascular endothelial cells, MHC-I/II, Apolipoprotein E targeting to liver, E7 peptide designed to target to MSCs, Sphingomyelin [80–84]. All of these have the potential to drive EVs uptaken by target cells and will facilitate tailoring EVs for drug delivery.

Exosomes originate from membrane invagination, following the formation of endosomes, the selection of cargos into MVBs through ESCRT-dependent and ESCRT-independent (tetraspanins, lipids), the fusion of MVBs with the plasma membrane, and finally releasing EVs into the extracellular environment. A part of the cargo is sorted into lysosomes for degradation. The cargos are sorted into EVs including tetraspanins, integrins, MHC-I/II, receptors, adhesion proteins, cytoskeletal proteins, enzymes, other cytosolic proteins, and nucleic acids, which are attributed partly to TEMs. MVs are generated by direct budding of the plasma membrane [13, 59]. The uptake of EVs into recipient cells mainly relies on three ways: 1) directly interacting with receptors; (2) direct fusion with the plasma membrane; (3) internalization.

1.3 Contribution of EV Tetraspanins in Regenerative Medicine

Multiple evidences have confirmed the potential of EVs in regenerative medicine. Especially, EVs from MSCs as a choice of MSCs play a significant role in cell-free therapy, which promotes wound healing, tissue regeneration, neurological damage repair, inflammation, etc. As mentioned above, tetraspanins as the classical markers of EVs are involved in the biogenesis, sorting associated partners (integrins, ICAM-1, CD44, MMPs, MHC-II, β-catenin, etc), targeting and uptake of EVs [85, 86]. Therefore, EV tetraspanins contribute to regenerative medicine on different planes.

Studies demonstrated that EVs, especially EVs derived from MSCs, contribute to regenerative medicine through their natural cargos, containing receptors, protease, mRNA, miRNA, etc. [9]. Further studies identified EVs as natural nanovectors, which could carry multiple endogenous and exogenous substances. Compared to natural EVs, the engineered EVs will be endowed with more functions in regenerative medicine therapy [87]. Considering the role of tetraspanins in EVs, EV tetraspanins have the potential to be the targets of modification for engineered EVs [13]. We will discuss the applications of EV tetraspanins on different planes.

1.3.1 Drug Delivery

EVs have lipid bilayer membranes to encapsulate and protect the cargo during circulation or extracellular space, which is an excellent drug delivery system for therapy. Although the natural EVs exhibit poor targeting and lack therapeutic cargos, the engineered EVs can enhance the functional content through the use of biochemical and biologic engineering techniques. EV membrane-embedded protein scaffolds are widely used in the engineering of EVs for loading therapeutic cargo [88]. EV tetraspanins are excellent targets for modification. Previous studies have examined the use of EV tetraspanins, including CD9, CD63, and CD81, to generate fusion proteins to enrich cargo or enable site-specific drug delivery.

A fusion protein CD9-HuR was designed by CD9 and HuR, which is an RNA binding protein. The CD9-HuR fusion protein could effectively enrich miR-155 into EVs when miR-155 was overexpressed in donor cells. In addition, if the RNAs contained the AU-rich elements, CRISPR/dCas9 could also be enriched in EVs. This is a novel and effective strategy to enrich more RNAs or related cargos into EVs compared to the traditional way [89].

Another fusion protein is to combine a peptide from Apo-A1 with CD63, which is a targeted delivery system. Apo-A1 helps EVs target scavenger receptor class B type 1-expressing liver cancer cells. After the purification, miR-26a was loaded into EVs via electroporation [90]. Lamp2b-RVG EVs carried drugs are used to target the brain in many articles [91]. RVG also can be selected to fuse with tetraspanins. This system can be extended to multiple applications, delivering all kinds of drugs to target disease sites.

Besides fusion proteins, tetraspanins or their specific binding peptide also can mediate drug delivery. CD81-positive EVs have the potential to target cells and enhance cellular uptake. A study designed a system to enhance the ability of targeted delivery and identified human placental laminin as the target of CD81 in a mammalian display system [78]. The specific binding peptide of CD63, CP05 is widely used to bind to CD63 on EVs, which can label synthetic drugs to EVs or PEG2000 via thioketal (TK) to avoid unnecessary uptake by mononuclear phagocytic system [92].

The potential of tetraspanins in loading cargo and targeting delivery will enhance the application and function of EVs in regenerative medicine. Furthermore, the discovery of new targets for diseases will improve much more efficiency of EVs-mediated regenerative medicine therapy.

1.3.2 Monitor System

Except for drug delivery, in vivo imaging is an emerging area of EVs-mediated regenerative medicine therapy. The monitor system is essential for understanding the uptake and biodistribution of EVs in vivo. The EV membrane proteins including tetraspanins, Lamp-2b, and GPI can be the reporter proteins modified on the EV surface [87].

Genetic engineering is the major way to build the monitor system to image and track EVs. The EV CD63-luciferase fusion protein makes it possible to directly analyze the EVs in a culture medium and monitor EV production, uptake, and long-term distribution in vivo. A study showed that both NanoLuc luciferase and ThermoLuc luciferase fused with the N- or C- terminal of CD63 were efficient for EV labeling. The study illustrated that the biodistribution of CD9 positive and CD63 positive EVs was slightly different. Compared to CD9 positive EVs, CD63 positive EVs were significantly enriched in the gastrointestinal tract and reduced in the lungs [93]. To allow dynamic monitoring of EV secretion in living cells, a pH-sensitive luciferase, pHluorin, was designed to fuse with CD63, which could only detect the luciferase signaling when EVs were secreted into the weakly acidic extracellular environment, not in the acidic MVB lumen [94].

Besides bioluminescent strategies, the Cy3/Cy5 labeled aptamer [95], anchoring CP05 peptide [96], EGFP [97], photocleavable protein [98], tetracycline transactivator (tTA) [99], and loxP [100, 101] were also designed into monitor system fused with tetraspanins to accurately interpret the cargos delivery process of EVs, which provide more opportunities to create EVs with new functions and properties for regenerative medicine research (see Table 7.1).

Table 7.1 Modification of EV tetraspanins

Strategy of modification	Cell source	Model	Function	Ref.
CD9 fused with HuR (RNA binding protein) and cotransfected with miR-155	293 T	THP1/mice	Enhance miR-155 loading into EVs	[89]
CD9 fused with HuR and cotransfected with CRISPR/dCas9	AML12	Mouse model of liver injury	Enhance CRISPR/dCas9 loading into EVs	[89]
CD9 fused with AGO2(RNA binding protein)	293 T	ARPE-19	Enhance RNAs loading into EVs	[102]
CD63 fused with EGFP-APO-A1 and loaded with miR-26a	293 T	HepG2	Target to liver cells and transfer miRNA cargos	[90]
CD63 modified by CP05-TK-mPEG and cotransfected with Bmp7 cDNA	293 T	RAW264.7 cells and mice omental adipose tissue	Transfer Bmp7-mRNA into omental adipose tissue and avoid phagocytosis of macrophages through mPEG	[92]
CD63 fused with NanoLuc/ThermoLuc luciferase	293 T	Huh7/B16-F10/mice	Monitor the biodistribution of CD63 EVs in vitro and in vivo	[93]
CD9 fused with NanoLuc/ThermoLuc luciferase	293 T	Huh7/B16-F10/mice	Monitor the biodistribution of CD9 EVs in vitro and in vivo	[93]
CD63 fused with a pH-sensitive luciferase (pHluorin)	HT1080	HT1080 in 2D and 3D culture	Reveal the role of EVs in autocrine and paracrine	[94]
CD63 fused with pHluorin	Primary mouse hippocampal neurons	Primary mouse hippocampal neurons	Monitor the fusion of MVB-PM and the release of EVs	[103]
CD63 fused with emGFP$^{loxP/stop/loxP}$	Any cell type	Lineage-specific Cre recombinase driver mice	Generate cell type-specific, CD63^{+} EVs	[101]
CD9/CD63/CD81 fused with tetracycline transactivator (tTA)	293 T	Tetracycline responsive element reporter 293 T	Evaluate cytoplasmic cargo delivery efficiency of EVs	[99]
CD63 labeled with CP05-TBP and loaded with anti-miR-214	Red blood cell	Mouse model of osteoporosis	Deliver anti-miR-214 into specific targeting tissue, osteoclasts in vivo	[104]
CD63 labeled with CP05-M12 and CP05-PMO	Murine myotube C2C12 cell	Mystrophin -deficient mdx mice	Deliver PMO into muscle tissue through M12	[105]

(continued)

Table 7.1 (continued)

CD cluster of differentiation, *HuR* Human antigen R, *THP1* Tohoku Hospital Pediatrics-1, *EVs* extracellular vesicles, *CRISPR* Clustered Regularly Interspaced Short Palindromic Repeats, *dCas9* dead CRISPR-associated protein 9, *AML12* alpha mouse liver 12, *AGO2* Argonaute2, *ARPE-19* Adult Retinal Pigment Epithelial cell line-19, *EGFP* Enhanced Green Fluorescent Protein, *APO-A1* Apolipoprotein A-1, *TK* thioketal, *PEG* polyethylene glycol, *Bmp7* bone morphogenetic protein 7, *MVB* multivesicular bodies, *PM* plasma membrane, *emGFP* emerald green fluorescent protein, *tTA* transactivator, *M12* muscle-targeting peptide 12, *PMO* morpholino oligomer

1.3.3 Signaling Transduction

As the organizers of TEMs, EV tetraspanins might serve as the scaffolds to carry cargo and imaging substances and be involved in the regulation of cell proliferation, migration, adhesion, stemness, and immune process by signaling transduction.

The role of tetraspanins in cancer development is more in-depth. In prostate cancer, EV CD151 could affect cell migration and invasion by activating the integrin pathway and promoting the cleavage of MMP9 in recipient cells [106]. In endothelial cells, EV Tspan8-CD49d complex could induce the expression of several angiogenesis related genes, including VEGF, CXCL5, MIF, and CXCR1 to enhance EC activity, even the maturation of EC progenitors [107]. CD9-positive EVs derived from cancerassociated fibroblasts might regulate the progression of melanoma and gastric cancer through MMP2.

In our previous study, we focused on the role of tumor EVs on cancer metastasis and identified both EV CD151 and Tspan8 contributing to ECM remodeling due to the interactions of tetraspanin-integrin and tetraspanin-protease [108]. In ovarian cancer, we demonstrated that CD82, located on the surface of EVs, could block fibronectin-integrin-FAK pathway to inhibit cell migration [44]. Moreover, EV CD82 participated in the development and metastasis of many kinds of cancers, like esophageal squamous cell carcinoma, pancreatic cancer, and breast cancer.

1.4 Potential Therapeutic Use of EV Tetraspanins in Regenerative Medicine

EV tetraspanins have many functions in EV-mediated therapy. Along with the understanding of EV tetraspanins and the application of EVs in regenerative medicine, researchers pay attention to the role of EV tetraspanins in regenerative medicine. As the classical markers of EVs, tetraspanins can contribute to regenerative medicine in many diseases.

1.4.1 Bone Regeneration

Bone-associated disorders are the leading cause of disability worldwide, which can seriously influence the life quality of people. Although bone has a self-repair ability, poor bone repair is still an increasing problem in aging, diabetes, and smoking people. Bone regeneration is a slow and complex process that requires the cooperation of multiple cells. Currently, bone grafts are primarily chosen to treat serious fractures or large bone damage. However, infection, insufficient bone source and insufficient vascularization may contribute to the failure of transplantation [109, 110]. With the rapid development of cell therapy, MSCs have become one of the methods used in bone regeneration. According to the paracrine effect of MSCs, EVs have gradually received increasing attention.

Human perivascular stem cells (PSCs) are pluripotent stem cells, which are necessary for angiogenesis. Studies have demonstrated EVs from PSCs stimulated bone repair by EV tetraspanins. Through EVs trypsinization, antibodies neutralizing for tetraspanins or shRNA knockdown demonstrated that CD9/CD81 on the surface of EVs could bind to the molecular partners IGSF8 and PTGFRN on the membrane surface of recipient cells. The binding is the basis for EV bioactivity to stimulate skeletal cell proliferation, migration, and osteodifferentiation (see Fig. 7.4) [111]. In a mouse femur fracture model, the $CD9^{-/-}$mice were used to explore the role of CD9 and bone-marrow-derived mesenchymal stem cells (BM-MSCs) EVs in fracture healing. The study demonstrated that the delayed fracture healing in $CD9^{-/-}$mice may be linked to the impaired release of EVs and could be rescued by BM-MSCs-derived EVs [112].

As the markers of EVs, modified tetraspanins are involved in bone regeneration. For example, in osteoclasts, the existence of miR-214 can lead to bone absorption

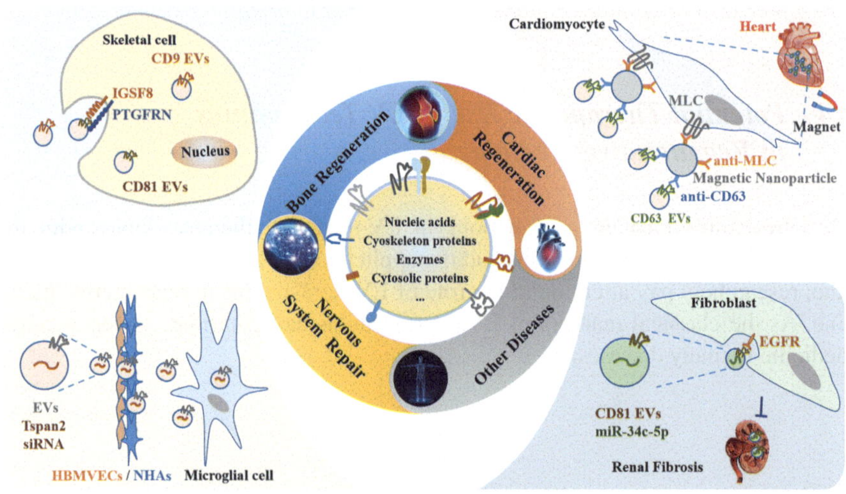

Fig. 7.4 The application of EV tetraspanins in regenerative medicine

inducing osteoporosis. So, the use of anti-miRNA oligonucleotide (AMO) by silencing the function of miR-214 can become a potential treatment. To target osteoclasts, studies reported a peptide-guided delivery strategy to realize osteoclast-targeted delivery of AMO. CP05, a peptide can specifically bind to the CD63 on EVs. TRAP-binding peptide (TBP) can link the receptors on osteoclasts. The TBP-CP05 complex can act as a guide. TBP-CP05-modified EVs can target osteoclasts both in vitro and in vivo. CD63 indirectly contributes to bone regeneration [104].

These results implicate that EV tetraspanins play an essential role in EVs treatment of bone regeneration and give more directions for target therapy for further studies.

1.4.2 Cardiac Regeneration

Adult mammalian have no ability to regenerate cardiomyocytes. The loss of cardiomyocytes in the cardiac injury will be replaced by myocardial scarring, resulting in a reduction in ventricular function, heart failure, and ultimately death [113]. In recent years, MSCs, especially the BM-MSCs, have been widely used in cardiac regenerative medicine. The potential of BM-MSCs has been confirmed to differentiate into cardiomyocytes and revascularize the arteries and veins [114]. Recent evidence shows that the paracrine signaling of MSCs can be mediated by EVs. EV cargos derived from MSCs are promising therapeutic tools for myocardial infarction, including miRNAs and proteins [115].

The effective concentration of EVs in local sites is the limitation of therapy. To accumulate EVs, the researcher designed antibody-conjugated magnetic nanoparticles consisting of a Fe_3O_4 core and a silica shell. The silica shell was designed to link two antibodies, recognizing CD63 on the surface of EVs and myosin-light-chain surface markers on injured cardiomyocytes, respectively. The link between the silica shell and with antibody is hydrazone bonds which can be cleaved under an acidic environment. The application of a magnetic field will accumulate nanoparticles into injured cardiac tissue and then the EVs will be locally released through the break of the hydrazone bonds under the acidic environment of injured cardiac tissue. In rabbit and rat models of myocardial infarction, these CD63-modified EVs could accumulate into infarcted tissue, leading to reduce the infarct size and improve the ejection fraction of the left-ventricle (see Fig. 7.4), [116]. The design is a promising method to manipulate the biodistribution of EVs for application in regenerative medicine.

Studies have demonstrated that CD82 is involved in the differentiation of human-induced pluripotent stem cells (hiPSC) into cardiomyocytes. The hiPSC-derived EVs contain CD82 increasing the secretion of β-catenin, indicating that CD82 is involved in differentiation through Wnt signaling. EV CD82 may serve as a therapeutic target for cardiac regeneration [117].

1.4.3 Nervous System Repair

Neurological disorders are the major causes of disability and death in older people, including Alzheimer's disease, Parkinson's disease, ischemic reperfusion stroke, infectious encephalopathy, etc. [118]. Lack of sustained repair and regeneration capacity is the major obstacle to the cure of neurological disorders. Recent studies have suggested that EVs, as a cell-free therapy, have the potential to treat neurological disorders [119–121]. Especially, the ability to cross the blood–brain barrier (BBB) provides more possibilities for EV therapy [122].

Neurological disorders are the most common complications after HIV infection [123]. Evidences have demonstrated that cells are involved in the pathological process and act as a viral storage pool of HIV, which can be the effective target of neurological disorder repair [124]. Tspan2 is a member of tetraspains, which has been involved in cell fusion and adhesion [125, 126]. Researchers used the microglia derived EVs to deliver Cy5-labeled Tspan2 siRNA to evaluate the effect of Tspan2 in modulating the neuroinflammation and track the process of EVs across the BBB. The study demonstrated the engineered EVs could deliver siRNA across the BBB to microglial cells infected by HIV and modulate the neuroinflammatory response (see Fig. 7.4), [127].

In a spinal cord injury model, researchers found the population of $CD81^+$ EVs was significantly increased at 1–7 days post-injury, which indicated that EV CD81 may serve as a therapeutic target of posttraumatic brain disorders [128]. Moreover, EV tetraspanins are also involved in nervous system repair, acting as a scaffold. CD63-pHluorin fusion protein can visualize the secretion of EVs in primary neurons [103].

1.4.4 Other Diseases

In recent years, multiple studies have demonstrated that the potential of EV tetraspanins in regenerative medicine by acting as the scaffolds, or regulating the EV cargos in many pathological processes, would repair, renal fibrosis dystrophy, and liver repair.

Tspan8 and CD151 have been confirmed to regulate cell proliferation and migration [129–131]. In the Tspan8-ko and CD151-ko mouse models, they indicated Tspan8-positive and CD151-positive EVs to promote skin wound healing [108]. Moreover, tetraspanins also can promote the uptake of EVs. In a renal fibrosis model, they found that CD81 on the surface of EVs could interact with EGFR located on the membrane surface of fibroblasts and promote the uptake of EVs into target cells, providing a potential therapeutic strategy for renal fibrosis (see Fig. 7.4) [132]. TM4SF5 positive EVs could target brown adipose tissues to clear high extracellular glucose levels in diabetes [133].

As the classical markers of EVs, CD63 and CD9 are usually used to be the scaffolds to load cargos to monitor EVs or target sites. EVs loaded with phosphorodiamidate morpholino oligomer (PMO) are modified by CP05-muscle-targeting peptide, which can help to target muscle tissue and enhance the expression of dystrophin in quadriceps of dystrophin-deficient mdx mice [105]. Another EVs delivering and tracking system is based on CD9 with two plasmids. One plasmid is CD9-PhoCl-catalase, expressing CD9, a photocleavable protein and an antioxidant enzyme. Another plasmid is coding apolipoprotein A-I and transmembrane domain of PDGFR. The co-expression of these two plasmids in donor cells will get an engineered EV which contributes to liver-specific delivery of antioxidant enzymes with the help of CD9 and apolipoprotein A-I [98].

In bone generation, CD9 and CD81 localized on the surface of EVs contribute to the uptake of EVs through the interaction between CD9/CD81 and IGSF8/PTGFRN on skeletal cells, improving the therapeutic efficiency of MSCs-derived EVs. In myocardial infarction, the magnetic nanoparticle modified with one MLC and one CD63 antibody can bind to both $CD63^+$ EVs and MLC^+ injured cardiomyocytes together. Upon the application of a magnetic field, EVs will be accumulated in injured cardiac tissue. In the nervous system, EVs loaded with Tspan2 siRNA can deliver siRNA across the HBMVECs layer and NHAs layer (BBB) to microglial cells. In renal fibrosis, $CD81^+$ EVs promote the uptake of fibroblast through binding with EGFR and efficiently transfer miR-34c-5p into fibroblast to alleviate renal fibrosis.

1.5 Future Challenges and Perspectives

Despite the recent evidence that has shown the functions of tetraspanins in EVs much more clear, the application of EV tetraspanins in regenerative medicine still needs further exploration. Up to now, the application of EV tetraspanins in regenerative medicine mainly utilizes to modify the functions of EVs. Considering the effects of tetraspanins on membrane organization and downstream signal pathways, the exogenous overexpression of tetraspanins fusion proteins in donor cells can alter the cargos and influence the innate functions of EVs. The effect may be a double-edged sword, which can enhance the function or give side effects of EVs.

Full consideration is needed for the selection of tetraspanins in clinical application. Studies demonstrate that distinct tetraspanins play different roles in the various cell types or tissues and also have unique functions in specific conditions. CD82 has the potential to inhibit angiogenesis, which cannot be the target of modification for EVs for the ischemic disease but can be the candidate for abnormal hemorrhagic disease, like diabetic retinopathy [54]. Due to the interaction between EV CD81 and EGFR, the CD81 positive EVs could be an efficient carrier to target EGFR positive tissues [132]. CD9 plays an important role in bone generation and promotes the uptake of EVs, which can be a target for modification of EVs to improve the therapeutic effect of bone-associated disorders [112]. CD151 and Tspan8 promote wound healing and angiogenesis in vitro and in vivo, which is a considerable candidate to enhance the function of MSCs EVs in regenerative medicine [108]. In addition, some tetraspanins have efficient organ tropism. Although the molecular mechanism remains to be addressed, these advantages of EV tetraspanins can be considered in regenerative medicine.

Moreover, the low production yield of EVs is a major limitation of EVs for clinical applications. There are several methods available for isolating EVs, including ultracentrifugation, ultrafiltration, size exclusion chromatography, PEG precipitation, immunoaffinity capture, and microfluidic technology [134]. EV tetraspanins may contribute to improving the efficiency of EVs enrichment through immunomagnetic beads or microfluidic technology based on immunoaffinity. Although these methods have high specificity, are expensive and not suitable for large scale. The development of novel strategies for modifying donor cells and the EVs helps to further improve the efficiency.

Combining EVs with biomaterials may be a promising method for improving EV stability and achieving controlled release in vivo. However, it should be noted that the ability of biomaterials to carry EVs varies, which may depend on the binding efficiency between EVs and biomaterials [135]. The TEMs play an important role in the interaction between EVs and extracellular matrix (ECM), which means that tetraspanins have the potential to improve the binding rate of EVs to biomaterials. However, the function of EV tetraspanins in binding with biomaterials remains to be further studied.

2 Conclusion

In recent years, EVs have been widely used to alleviate diseases in vitro, in vivo, and even in many clinical trials. EVs have been demonstrated with enormous potential for regenerative medicine, especially EVs derived from MSCs. The modification of EVs has been performed widely and aims to improve therapeutic activities. The role of EV tetraspanins in EVs biogenesis, cargo sorting, and biodistribution gives more possibility to the application of EVs. Along with the deep understanding of the biological structure and functions of EV tetraspanins, the application of EV tetraspanins in regenerative medicine will be further expanded.

In this chapter, we describe the function of tetraspanins based on TEMs, discuss the role of tetraspanins in the journey and modification of EVs, and summarize the application of EV tetraspanins in recent studies. The loading of exogenous cargos, the targeting of specific tissue, and the tracking of delivery based on EV tetraspanins, all of these will give more possibilities and drive more attention to the application of EV tetraspanins in regenerative medicine.

References

1. Jacques E, Suuronen EJ. The progression of regenerative medicine and its impact on therapy translation. Clin Transl Sci. 2020;13(3):440–50. https://doi.org/10.1111/cts.12736.

2. Berthiaume F, Maguire TJ, Yarmush ML. Tissue engineering and regenerative medicine: history, progress, and challenges. Annu Rev Chem Biomol Eng. 2011;2:403–30. https://doi.org/10.1146/annurev-chembioeng-061010-114257.
3. Verma S, Domb AJ, Kumar N. Nanomaterials for regenerative medicine. Nanomedicine (Lond). 2011;6(1):157–81. https://doi.org/10.2217/nnm.10.146.
4. Cossu G, Fears R, Griffin G, Ter Meulen V. Regenerative medicine: challenges and opportunities. Lancet. 2020;395(10239):1746–7. https://doi.org/10.1016/s0140-6736(20)31250-2.
5. Kolios G, Moodley Y. Introduction to stem cells and regenerative medicine. Respiration. 2013;85(1):3–10. https://doi.org/10.1159/000345615.
6. Suman S, Domingues A, Ratajczak J, Ratajczak MZ. Potential clinical applications of stem cells in regenerative medicine. Adv Exp Med Biol. 2019;1201:1–22. https://doi.org/10.1007/978-3-030-31206-0_1.
7. Munda M, Velnar T. Stem cell therapy for degenerative disc disease: bridging the gap between preclinical promise and clinical potential. Biomol Biomed. 2023;24:210. https://doi.org/10.17305/bb.2023.9518.
8. Burdon TJ, Paul A, Noiseux N, Prakash S, Shum-Tim D. Bone marrow stem cell derived paracrine factors for regenerative medicine: current perspectives and therapeutic potential. Bone Marrow Res. 2011;2011:207326. https://doi.org/10.1155/2011/207326.
9. Hade MD, Suire CN, Suo Z. Mesenchymal stem cell-derived exosomes: applications in regenerative medicine. Cells. 2021;10(8):1959. https://doi.org/10.3390/cells10081959.
10. Cui L, Saeed Y, Li H, Yang J. Regenerative medicine and traumatic brain injury: from stem cell to cell-free therapeutic strategies. Regen Med. 2022;17(1):37–53. https://doi.org/10.2217/rme-2021-0069.
11. Raposo G, Stoorvogel W. Extracellular vesicles: exosomes, microvesicles, and friends. J Cell Biol. 2013;200(4):373–83. https://doi.org/10.1083/jcb.201211138.
12. Kalluri R, LeBleu VS. The biology, function, and biomedical applications of exosomes. Science. 2020;367(6478):eaau6977. https://doi.org/10.1126/science.aau6977.
13. Gurung S, Perocheau D, Touramanidou L, Baruteau J. The exosome journey: from biogenesis to uptake and intracellular signalling. Cell Commun Signal. 2021;19(1):47. https://doi.org/10.1186/s12964-021-00730-1.
14. Xi XM, Xia SJ, Lu R. Drug loading techniques for exosome-based drug delivery systems. Pharmazie. 2021;76(2):61–7. https://doi.org/10.1691/ph.2021.0128.
15. Iranpanah A, Kooshki L, Moradi SZ, Saso L, Fakhri S, Khan H. The exosome-mediated PI3K/Akt/mTOR signaling pathway in neurological diseases. Pharmaceutics. 2023;15(3):1006. https://doi.org/10.3390/pharmaceutics15031006.
16. Purghè B, Manfredi M, Ragnoli B, Baldanzi G, Malerba M. Exosomes in chronic respiratory diseases. Biomed Pharmacother. 2021;144:112270. https://doi.org/10.1016/j.biopha.2021.112270.
17. Xu X, Liang Y, Li X, Ouyang K, Wang M, Cao T, et al. Exosome-mediated delivery of kartogenin for chondrogenesis of synovial fluid-derived mesenchymal stem cells and cartilage regeneration. Biomaterials. 2021;269:120539. https://doi.org/10.1016/j.biomaterials.2020.120539.
18. Thongboonkerd V, Kanlaya R. The divergent roles of exosomes in kidney diseases: pathogenesis, diagnostics, prognostics and therapeutics. Int J Biochem Cell Biol. 2022;149:106262. https://doi.org/10.1016/j.biocel.2022.106262.
19. Nasser MI, Masood M, Adlat S, Gang D, Zhu S, Li G, et al. Mesenchymal stem cell-derived exosome microRNA as therapy for cardiac ischemic injury. Biomed Pharmacother. 2021;143:112118. https://doi.org/10.1016/j.biopha.2021.112118.
20. Luo N, Li J, Dong R, Lu J. Exosome-based Theranostics for liver diseases. Dis Markers. 2022;2022:7888906. https://doi.org/10.1155/2022/7888906.
21. Ma S, Zhang Y, Li S, Li A, Li Y, Pei D. Engineering exosomes for bone defect repair. Front Bioeng Biotechnol. 2022;10:1091360. https://doi.org/10.3389/fbioe.2022.1091360.

22. Zhang L, Yu D. Exosomes in cancer development, metastasis, and immunity. Biochim Biophys Acta Rev Cancer. 2019;1871(2):455–68. https://doi.org/10.1016/j.bbcan.2019.04.004.
23. Lang T, Hochheimer N. Tetraspanins. Curr Biol. 2020;30(5):R204–r6. https://doi.org/10.1016/j.cub.2020.01.007.
24. Huang C, Fu C, Wren JD, Wang X, Zhang F, Zhang YH, et al. Tetraspanin-enriched microdomains regulate digitation junctions. Cell Mol Life Sci. 2018;75(18):3423–39. https://doi.org/10.1007/s00018-018-2803-2.
25. Berditchevski F, Odintsova E. Tetraspanins as regulators of protein trafficking. Traffic. 2007;8(2):89–96. https://doi.org/10.1111/j.1600-0854.2006.00515.x.
26. Kang T, Atukorala I, Mathivanan S. Biogenesis of extracellular vesicles. Subcell Biochem. 2021;97:19–43. https://doi.org/10.1007/978-3-030-67171-6_2.
27. Barile L, Vassalli G. Exosomes: therapy delivery tools and biomarkers of diseases. Pharmacol Ther. 2017;174:63–78. https://doi.org/10.1016/j.pharmthera.2017.02.020.
28. Liang Y, Duan L, Lu J, Xia J. Engineering exosomes for targeted drug delivery. Theranostics. 2021;11(7):3183–95. https://doi.org/10.7150/thno.52570.
29. Panwar D, Shrivastava D, Kumar A, Gupta LK, Kumar NSS, Chintagunta AD. Efficient strategy to isolate exosomes using anti-CD63 antibodies conjugated to gold nanoparticles. AMB Express. 2023;13(1):90. https://doi.org/10.1186/s13568-023-01592-1.
30. Kitadokoro K, Bordo D, Galli G, Petracca R, Falugi F, Abrignani S, et al. CD81 extracellular domain 3D structure: insight into the tetraspanin superfamily structural motifs. EMBO J. 2001;20(1–2):12–8. https://doi.org/10.1093/emboj/20.1.12.
31. Umeda R, Nishizawa T, Nureki O. Crystallization of the human tetraspanin protein CD9. Acta Crystallogr F Struct Biol Commun. 2019;75(Pt 4):254–9. https://doi.org/10.1107/s2053230x1801840x.
32. Yang Y, Liu XR, Greenberg ZJ, Zhou F, He P, Fan L, et al. Open conformation of tetraspanins shapes interaction partner networks on cell membranes. EMBO J. 2020;39(18):e105246. https://doi.org/10.15252/embj.2020105246.
33. Lipper CH, Gabriel KH, Seegar TCM, Dürr KL, Tomlinson MG, Blacklow SC. Crystal structure of the Tspan15 LEL domain reveals a conserved ADAM10 binding site. Structure. 2022;30(2):206–14.e4. https://doi.org/10.1016/j.str.2021.10.007.
34. Susa K, Kruse A, Blacklow S. Tetraspanins: structure, dynamics, and principles of partner-protein recognition. Trends Cell Biol. 2023;S0962–8924(23):00195–2. https://doi.org/10.1016/j.tcb.2023.09.003.
35. Querol Cano L, Dunlock V-ME, Schwerdtfeger F, van Spriel AB. Membrane organization by tetraspanins and galectins shapes lymphocyte function. Nat Rev Immunol. 2023;24:193. https://doi.org/10.1038/s41577-023-00935-0.
36. Tetraspanins_ structure, dynamics, andprinciples of partner-protein recognition.pdf. https://doi.org/10.1016/j.tcb.2023.09.003.
37. Zimmerman B, Kelly B, McMillan BJ, Seegar TCM, Dror RO, Kruse AC, et al. Crystal structure of a full-length human Tetraspanin reveals a cholesterol-binding pocket. Cell. 2016;167(4):1041–51.e11. https://doi.org/10.1016/j.cell.2016.09.056.
38. María YM, Andreu Z. Tetraspanins in extracellular vesicle formation and function. Front Immunol. 2014;5:442. https://doi.org/10.3389/fimmu.2014.00442.
39. Lu J, Li J, Liu S, Wang T, Ianni A, Bober E, et al. Exosomal tetraspanins mediate cancer metastasis by altering host microenvironment. Oncotarget. 2017;8(37):62803–15. https://doi.org/10.18632/oncotarget.19119.
40. van Deventer SJ. Molecular interactions shaping the tetraspanin web. Biochem Soc Trans. 2017;45:741–50. https://doi.org/10.1042/BST20160284.
41. Susa KJ, Seegar TC, Blacklow SC, Kruse AC. A dynamic interaction between CD19 and the tetraspanin CD81 controls B cell co-receptor trafficking. elife. 2020;9:9. https://doi.org/10.7554/eLife.52337.

42. Grigorov B, Molle J, Rubinstein E, Zoulim F, Bartosch B. CD81 large extracellular loop-containing fusion proteins with a dominant negative effect on HCV cell spread and replication. J Gen Virol. 2017;98(7):1646–57. https://doi.org/10.1099/jgv.0.000850.
43. Yu J, Lee CY, Changou CA, Cedano-Prieto DM, Takada YK, Takada Y. The CD9, CD81, and CD151 EC2 domains bind to the classical RGD-binding site of integrin αvβ3. Biochem J. 2017;474(4):589–96. https://doi.org/10.1042/bcj20160998.
44. Li J, Xu J, Li L, Ianni A, Kumari P, Liu S, et al. MGAT3-mediated glycosylation of tetraspanin CD82 at asparagine 157 suppresses ovarian cancer metastasis by inhibiting the integrin signaling pathway. Theranostics. 2020;10(14):6467–82. https://doi.org/10.7150/thno.43865.
45. Hemler ME. Tetraspanin proteins promote multiple cancer stages. Nat Rev Cancer. 2014;14(1):49–60. https://doi.org/10.1038/nrc3640.
46. Termini CM, Gillette JM. Tetraspanins function as regulators of cellular signaling. Front Cell Dev Biol. 2017;5:34. https://doi.org/10.3389/fcell.2017.00034.
47. Oren R, Takahashi S, Doss C, Levy R, Levy S. TAPA-1, the target of an antiproliferative antibody, defines a new family of transmembrane proteins. Mol Cell Biol. 1990;10(8):4007–15. https://doi.org/10.1128/mcb.10.8.4007-4015.1990.
48. Geisert EE Jr, Williams RW, Geisert GR, Fan L, Asbury AM, Maecker HT, et al. Increased brain size and glial cell number in CD81-null mice. J Comp Neurol. 2002;453(1):22–32. https://doi.org/10.1002/cne.10364.
49. Pan Y, Geisert DF, Orr WE, Geisert EE. The effects of a CD81 null mutation on retinal pigment epithelium in mice. Neurochem Res. 2011;36(4):569–73. https://doi.org/10.1007/s11064-010-0268-2.
50. Chang Y, Finnemann SC. Tetraspanin CD81 is required for the alpha v beta5-integrin-dependent particle-binding step of RPE phagocytosis. J Cell Sci. 2007;120(Pt 17):3053–63. https://doi.org/10.1242/jcs.006361.
51. Jankovičová J, Neuerová Z, Sečová P, Bartóková M, Bubeníčková F, Komrsková K, et al. Tetraspanins in mammalian reproduction: spermatozoa, oocytes and embryos. Med Microbiol Immunol. 2020;209(4):407–25. https://doi.org/10.1007/s00430-020-00676-0.
52. van Spriel AB, Figdor CG. The role of tetraspanins in the pathogenesis of infectious diseases. Microbes Infect. 2010;12(2):106–12. https://doi.org/10.1016/j.micinf.2009.11.001.
53. Vences-Catalán F, Duault C, Kuo CC, Rajapaksa R, Levy R, Levy S. CD81 as a tumor target. Biochem Soc Trans. 2017;45(2):531–5. https://doi.org/10.1042/bst20160478.
54. Lee JW, Hur J, Kwon YW, Chae CW, Choi JI, Hwang I, et al. KAI1(CD82) is a key molecule to control angiogenesis and switch angiogenic milieu to quiescent state. J Hematol Oncol. 2021;14(1):148. https://doi.org/10.1186/s13045-021-01147-6.
55. Jadi PK, Dave A, Issa R, Tabbasum K, Okurowska K, Samarth A, et al. Tetraspanin CD9-derived peptides inhibit Pseudomonas aeruginosa corneal infection and aid in wound healing of corneal epithelial cells. Ocul Surf. 2023;32:211. https://doi.org/10.1016/j.jtos.2023.07.001.
56. Payandeh Z, Noori E, Khalesi B, Mard-Soltani M, Abdolalizadeh J, Khalili S. Anti-CD37 targeted immunotherapy of B-cell malignancies. Biotechnol Lett. 2018;40(11–12):1459–66. https://doi.org/10.1007/s10529-018-2612-6.
57. Sandfeld-Paulsen B, Jakobsen KR, Bæk R, Folkersen BH, Rasmussen TR, Meldgaard P, et al. Exosomal proteins as diagnostic biomarkers in lung cancer. J Thorac Oncol. 2016;11(10):1701–10. https://doi.org/10.1016/j.jtho.2016.05.034.
58. Han QF, Li WJ, Hu KS, Gao J, Zhai WL, Yang JH, et al. Exosome biogenesis: machinery, regulation, and therapeutic implications in cancer. Mol Cancer. 2022;21(1):207. https://doi.org/10.1186/s12943-022-01671-0.
59. Han C, Yang J, Sun J, Qin G. Extracellular vesicles in cardiovascular disease: biological functions and therapeutic implications. Pharmacol Ther. 2022;233:108025. https://doi.org/10.1016/j.pharmthera.2021.108025.
60. Juan T, Fürthauer M. Biogenesis and function of ESCRT-dependent extracellular vesicles. Semin Cell Dev Biol. 2018;74:66–77. https://doi.org/10.1016/j.semcdb.2017.08.022.

61. Baietti MF, Zhang Z, Mortier E, Melchior A, Degeest G, Geeraerts A, et al. Syndecan-syntenin-ALIX regulates the biogenesis of exosomes. Nat Cell Biol. 2012;14(7):677–85. https://doi.org/10.1038/ncb2502.
62. Manohar S, Harlow M, Nguyen H, Li J, Hankins GR, Park M. Chromatin modifying protein 1A (Chmp1A) of the endosomal sorting complex required for transport (ESCRT)-III family activates ataxia telangiectasia mutated (ATM) for PanC-1 cell growth inhibition. Cell Cycle. 2011;10(15):2529–39. https://doi.org/10.4161/cc.10.15.15926.
63. Stuffers S, Sem Wegner C, Stenmark H, Brech A. Multivesicular endosome biogenesis in the absence of ESCRTs. Traffic. 2009;10(7):925–37. https://doi.org/10.1111/j.1600-0854.2009.00920.x.
64. Chairoungdua A, Smith DL, Pochard P, Hull M, Caplan MJ. Exosome release of β-catenin: a novel mechanism that antagonizes Wnt signaling. J Cell Biol. 2010;190(6):1079–91. https://doi.org/10.1083/jcb.201002049.
65. Böker KO, Lemus-Diaz N, Rinaldi Ferreira R, Schiller L, Schneider S, Gruber J. The impact of the CD9 Tetraspanin on lentivirus infectivity and exosome secretion. Mol Ther. 2018;26(2):634–47. https://doi.org/10.1016/j.ymthe.2017.11.008.
66. Waisner H, Lasnier S, Suma SM, Kalamvoki M. Effects on exocytosis by two HSV-1 mutants unable to block autophagy. J Virol. 2023;97:e0075723. https://doi.org/10.1128/jvi.00757-23.
67. Ghossoub R, Chéry M, Audebert S, Leblanc R, Egea-Jimenez AL, Lembo F, et al. Tetraspanin-6 negatively regulates exosome production. Proc Natl Acad Sci USA. 2020;117(11):5913–22. https://doi.org/10.1073/pnas.1922447117.
68. Yue S, Mu W, Erb U, Zöller M. The tetraspanins CD151 and Tspan8 are essential exosome components for the crosstalk between cancer initiating cells and their surrounding. Oncotarget. 2015;6(4):2366–84. https://doi.org/10.18632/oncotarget.2958.
69. Pols MS, Klumperman J. Trafficking and function of the tetraspanin CD63. Exp Cell Res. 2009;315(9):1584–92. https://doi.org/10.1016/j.yexcr.2008.09.020.
70. Wang Z, Sun H, Provaznik J, Hackert T, Zöller M. Pancreatic cancer-initiating cell exosome message transfer into noncancer-initiating cells: the importance of CD44v6 in reprogramming. J Exp Clin Cancer Res. 2019;38(1):132. https://doi.org/10.1186/s13046-019-1129-8.
71. Mazurov D, Barbashova L, Filatov A. Tetraspanin protein CD9 interacts with metalloprotease CD10 and enhances its release via exosomes. FEBS J. 2013;280(5):1200–13. https://doi.org/10.1111/febs.12110.
72. Petersen SH, Odintsova E, Haigh TA, Rickinson AB, Taylor GS, Berditchevski F. The role of tetraspanin CD63 in antigen presentation via MHC class II. Eur J Immunol. 2011;41(9):2556–61. https://doi.org/10.1002/eji.201141438.
73. Buschow SI, Nolte-'t Hoen EN, van Niel G, Pols MS, ten Broeke T, Lauwen M, et al. MHC II in dendritic cells is targeted to lysosomes or T cell-induced exosomes via distinct multivesicular body pathways. Traffic. 2009;10(10):1528–42. https://doi.org/10.1111/j.1600-0854.2009.00963.x.
74. Yokoi A, Villar-Prados A, Oliphint PA, Zhang J, Song X, De Hoff P, et al. Mechanisms of nuclear content loading to exosomes. Sci Adv. 2019;5(11):eaax8849. https://doi.org/10.1126/sciadv.aax8849.
75. Nazarenko I, Rana S, Baumann A, McAlear J, Hellwig A, Trendelenburg M, et al. Cell surface tetraspanin Tspan8 contributes to molecular pathways of exosome-induced endothelial cell activation. Cancer Res. 2010;70(4):1668–78. https://doi.org/10.1158/0008-5472.CAN-09-2470.
76. Laulagnier K, Javalet C, Hemming FJ, Chivet M, Lachenal G, Blot B, et al. Amyloid precursor protein products concentrate in a subset of exosomes specifically endocytosed by neurons. Cell Mol Life Sci. 2018;75(4):757–73. https://doi.org/10.1007/s00018-017-2664-0.
77. Hoshino A, Costa-Silva B, Shen TL, Rodrigues G, Hashimoto A, Tesic Mark M, et al. Tumour exosome integrins determine organotropic metastasis. Nature. 2015;527(7578):329–35. https://doi.org/10.1038/nature15756.

78. Vogt S, Bobbili MR, Stadlmayr G, Stadlbauer K, Kjems J, Rüker F, et al. An engineered CD81-based combinatorial library for selecting recombinant binders to cell surface proteins: laminin binding CD81 enhances cellular uptake of extracellular vesicles. J Extracell Vesicles. 2021;10(11):e12139. https://doi.org/10.1002/jev2.12139.
79. Wang T, Wang X, Wang H, Li L, Zhang C, Xiang R, et al. High TSPAN8 expression in epithelial cancer cell-derived small extracellular vesicles promote confined diffusion and pronounced uptake. J Extracell Vesicles. 2021;10(13):e12167. https://doi.org/10.1002/jev2.12167.
80. Man Z, Li T, Zhang L, Yuan L, Wu C, Li P, et al. E7 peptide-functionalized Ti6Al4V alloy for BMSC enrichment in bone tissue engineering. Am J Transl Res. 2018;10(8):2480–90.
81. Zhang W, Zhong W, Wang B, Yang J, Yang J, Yu Z, et al. ICAM-1-mediated adhesion is a prerequisite for exosome-induced T cell suppression. Dev Cell. 2022;57(3):329–43.e7. https://doi.org/10.1016/j.devcel.2022.01.002.
82. Toda Y, Takata K, Nakagawa Y, Kawakami H, Fujioka S, Kobayashi K, et al. Effective internalization of U251-MG-secreted exosomes into cancer cells and characterization of their lipid components. Biochem Biophys Res Commun. 2015;456(3):768–73. https://doi.org/10.1016/j.bbrc.2014.12.015.
83. He J, Du Z, Zhang H, Wang B, Xia J. Exosomes derived from human umbilical cord mesenchymal stem cells loaded with RVG-Lamp2b and Netrin-1 promotes Schwann cell invasion and migration. Tissue Cell. 2023;85:102219. https://doi.org/10.1016/j.tice.2023.102219.
84. Jin Y, Chung YW, Jung MK, Lee JH, Ko KY, Jang JK, et al. Apolipoprotein E-mediated regulation of selenoprotein P transportation via exosomes. Cell Mol Life Sci. 2020;77(12):2367–86. https://doi.org/10.1007/s00018-019-03287-y.
85. Rana S, Zöller M. Exosome target cell selection and the importance of exosomal tetraspanins: a hypothesis. Biochem Soc Trans. 2011;39(2):559–62. https://doi.org/10.1042/bst0390559.
86. Rana S, Yue S, Stadel D, Zöller M. Toward tailored exosomes: the exosomal tetraspanin web contributes to target cell selection. Int J Biochem Cell Biol. 2012;44(9):1574–84. https://doi.org/10.1016/j.biocel.2012.06.018.
87. Kim HY, Kwon S, Um W, Shin S, Kim CH, Park JH, et al. Functional extracellular vesicles for regenerative medicine. Small. 2022;18(36):e2106569. https://doi.org/10.1002/smll.202106569.
88. Salunkhe S, Dheeraj BM, Chitkara D, Mittal A. Surface functionalization of exosomes for target-specific delivery and in vivo imaging & tracking: strategies and significance. J Control Release. 2020;326:599–614. https://doi.org/10.1016/j.jconrel.2020.07.042.
89. Li Z, Zhou X, Wei M, Gao X, Zhao L, Shi R, et al. In vitro and in vivo RNA inhibition by CD9-HuR functionalized exosomes encapsulated with miRNA or CRISPR/dCas9. Nano Lett. 2019;19(1):19–28. https://doi.org/10.1021/acs.nanolett.8b02689.
90. Liang G, Kan S, Zhu Y, Feng S, Feng W, Gao S. Engineered exosome-mediated delivery of functionally active miR-26a and its enhanced suppression effect in HepG2 cells. Int J Nanomedicine. 2018;13:585–99. https://doi.org/10.2147/ijn.S154458.
91. Zhang R, Fu Y, Cheng M, Ma W, Zheng N, Wang Y, et al. sEVs(RVG) selectively delivers antiviral siRNA to fetus brain, inhibits ZIKV infection and mitigates ZIKV-induced microcephaly in mouse model. Mol Ther. 2022;30(5):2078–91. https://doi.org/10.1016/j.ymthe.2021.10.009.
92. Guo Y, Wan Z, Zhao P, Wei M, Liu Y, Bu T, et al. Ultrasound triggered topical delivery of Bmp7 mRNA for white fat browning induction via engineered smart exosomes. J Nanobiotechnol. 2021;19(1):402. https://doi.org/10.1186/s12951-021-01145-3.
93. Gupta D, Liang X, Pavlova S, Wiklander OPB, Corso G, Zhao Y, et al. Quantification of extracellular vesicles in vitro and in vivo using sensitive bioluminescence imaging. J Extracell Vesicles. 2020;9(1):1800222. https://doi.org/10.1080/20013078.2020.1800222.
94. Sung BH, von Lersner A, Guerrero J, Krystofiak ES, Inman D, Pelletier R, et al. A live cell reporter of exosome secretion and uptake reveals pathfinding behavior of migrating cells. Nat Commun. 2020;11(1):2092. https://doi.org/10.1038/s41467-020-15747-2.

95. Wei J, Zhu K, Chen Z, Yang Z, Yang K, Wang Z, et al. Triple-color fluorescence co-localization of PD-L1-overexpressing cancer exosomes. Mikrochim Acta. 2022;189(5):182. https://doi.org/10.1007/s00604-022-05278-6.
96. Wan Z, Gan X, Mei R, Du J, Fan W, Wei M, et al. ROS triggered local delivery of stealth exosomes to tumors for enhanced chemo/photodynamic therapy. J Nanobiotechnol. 2022;20(1):385. https://doi.org/10.1186/s12951-022-01591-7.
97. Li W, Wang J, Yin X, Shi H, Sun B, Ji M, et al. Construction of a mouse model that can be used for tissue-specific EV screening and tracing in vivo. Front Cell Dev Biol. 2022;10:1015841. https://doi.org/10.3389/fcell.2022.1015841.
98. Cheng Q, Dai Z, Shi X, Duan X, Wang Y, Hou T, et al. Expanding the toolbox of exosome-based modulators of cell functions. Biomaterials. 2021;277:121129. https://doi.org/10.1016/j.biomaterials.2021.121129.
99. Somiya M, Kuroda S. Reporter gene assay for membrane fusion of extracellular vesicles. J Extracell Vesicles. 2021;10(13):e12171. https://doi.org/10.1002/jev2.12171.
100. Luo W, Dai Y, Chen Z, Yue X, Andrade-Powell KC, Chang J. Spatial and temporal tracking of cardiac exosomes in mouse using a nano-luciferase-CD63 fusion protein. Commun Biol. 2020;3(1):114. https://doi.org/10.1038/s42003-020-0830-7.
101. McCann JV, Bischoff SR, Zhang Y, Cowley DO, Sanchez-Gonzalez V, Daaboul GD, et al. Reporter mice for isolating and auditing cell type-specific extracellular vesicles in vivo. Genesis. 2020;58(7):e23369. https://doi.org/10.1002/dvg.23369.
102. Es-Haghi M, Neustroeva O, Chowdhury I, Laitinen P, Väänänen MA, Korvenlaita N, et al. Construction of fusion protein for enhanced small RNA loading to extracellular vesicles. Genes (Basel). 2023;14(2):261. https://doi.org/10.3390/genes14020261.
103. Pescosolido MF, Ouyang Q, Liu JS, Morrow EM. Live-imaging detection of multivesicular body-plasma membrane fusion and exosome release in cultured primary neurons. Methods Mol Biol. 2023;2683:213–20. https://doi.org/10.1007/978-1-0716-3287-1_17.
104. Xu L, Xu X, Liang Y, Wen C, Ouyang K, Huang J, et al. Osteoclast-targeted delivery of anti-miRNA oligonucleotides by red blood cell extracellular vesicles. J Control Release. 2023;358:259–72. https://doi.org/10.1016/j.jconrel.2023.04.043.
105. Gao X, Ran N, Dong X, Zuo B, Yang R, Zhou Q, et al. Anchor peptide captures, targets, and loads exosomes of diverse origins for diagnostics and therapy. Sci Transl Med. 2018;10(444):eaat0195. https://doi.org/10.1126/scitranslmed.aat0195.
106. Malla RR, Pandrangi S, Kumari S, Gavara MM, Badana AK. Exosomal tetraspanins as regulators of cancer progression and metastasis and novel diagnostic markers. Asia Pac J Clin Oncol. 2018;14(6):383–91. https://doi.org/10.1111/ajco.12869.
107. Nazarenko I, Rana S, Baumann A, McAlear J, Hellwig A, Trendelenburg M, et al. Cell surface tetraspanin Tspan8 contributes to molecular pathways of exosome-induced endothelial cell activation. Cancer Res. 2010;70(4):1668–78. https://doi.org/10.1158/0008-5472.Can-09-2470.
108. Zhao K, Erb U, Hackert T, Zöller M, Yue S. Distorted leukocyte migration, angiogenesis, wound repair and metastasis in Tspan8 and Tspan8/CD151 double knockout mice indicate complementary activities of Tspan8 and CD51. Biochim Biophys Acta, Mol Cell Res. 2018;1865(2):379–91. https://doi.org/10.1016/j.bbamcr.2017.11.007.
109. Wang W, Liang X, Zheng K, Ge G, Chen X, Xu Y, et al. Horizon of exosome-mediated bone tissue regeneration: the all-rounder role in biomaterial engineering. Mater Today Bio. 2022;16:100355. https://doi.org/10.1016/j.mtbio.2022.100355.
110. Elgali I, Omar O, Dahlin C, Thomsen P. Guided bone regeneration: materials and biological mechanisms revisited. Eur J Oral Sci. 2017;125(5):315–37. https://doi.org/10.1111/eos.12364.
111. Xu J, Wang Y, Hsu CY, Gao Y, Meyers CA, Chang L, et al. Human perivascular stem cell-derived extracellular vesicles mediate bone repair. elife. 2019;8:8. https://doi.org/10.7554/eLife.48191.

112. Furuta T, Miyaki S, Ishitobi H, Ogura T, Kato Y, Kamei N, et al. Mesenchymal stem cell-derived exosomes promote fracture healing in a mouse model. Stem Cells Transl Med. 2016;5(12):1620–30. https://doi.org/10.5966/sctm.2015-0285.
113. Saleh M, Ambrose JA. Understanding myocardial infarction. F1000Res. 2018;7:F1000 Faculty Rev-1378. https://doi.org/10.12688/f1000research.15096.1.
114. Shafei AE, Ali MA, Ghanem HG, Shehata AI, Abdelgawad AA, Handal HR, et al. Mesenchymal stem cell therapy: a promising cell-based therapy for treatment of myocardial infarction. J Gene Med. 2017;19(12):1. https://doi.org/10.1002/jgm.2995.
115. Ma ZJ, Yang JJ, Lu YB, Liu ZY, Wang XX. Mesenchymal stem cell-derived exosomes: toward cell-free therapeutic strategies in regenerative medicine. World J Stem Cells. 2020;12(8):814–40. https://doi.org/10.4252/wjsc.v12.i8.814.
116. Liu S, Chen X, Bao L, Liu T, Yuan P, Yang X, et al. Treatment of infarcted heart tissue via the capture and local delivery of circulating exosomes through antibody-conjugated magnetic nanoparticles. Nat Biomed Eng. 2020;4(11):1063–75. https://doi.org/10.1038/s41551-020-00637-1.
117. Takeda M, Kanki Y, Masumoto H, Funakoshi S, Hatani T, Fukushima H, et al. Identification of cardiomyocyte-fated progenitors from human-induced pluripotent stem cells marked with CD82. Cell Rep. 2018;22(2):546–56. https://doi.org/10.1016/j.celrep.2017.12.057.
118. Pena SA, Iyengar R, Eshraghi RS, Bencie N, Mittal J, Aljohani A, et al. Gene therapy for neurological disorders: challenges and recent advancements. J Drug Target. 2020;28(2):111–28. https://doi.org/10.1080/1061186x.2019.1630415.
119. Namini MS, Daneshimehr F, Beheshtizadeh N, Mansouri V, Ai J, Jahromi HK, et al. Cell-free therapy based on extracellular vesicles: a promising therapeutic strategy for peripheral nerve injury. Stem Cell Res Ther. 2023;14(1):254. https://doi.org/10.1186/s13287-023-03467-5.
120. Yu Z, Teng Y, Yang J, Yang L. The role of exosomes in adult neurogenesis: implications for neurodegenerative diseases. Neural Regen Res. 2024;19(2):282–8. https://doi.org/10.4103/1673-5374.379036.
121. Huber CC, Wang H. Pathogenic and therapeutic role of exosomes in neurodegenerative disorders. Neural Regen Res. 2024;19(1):75–9. https://doi.org/10.4103/1673-5374.375320.
122. Wang J, Yang L. The role of exosomes in central nervous system tissue regeneration and repair. Biomed Mater. 2023;18(5):052003. https://doi.org/10.1088/1748-605X/ace39c.
123. Uwishema O, Ayoub G, Badri R, Onyeaka H, Berjaoui C, Karabulut E, et al. Neurological disorders in HIV: Hope despite challenges. Immun Inflamm Dis. 2022;10(3):e591. https://doi.org/10.1002/iid3.591.
124. Borrajo López A, Penedo MA, Rivera-Baltanas T, Pérez-Rodríguez D, Alonso-Crespo D, Fernández-Pereira C, et al. Microglia: the real foe in HIV-1-associated neurocognitive disorders? Biomedicines. 2021;9(8):925. https://doi.org/10.3390/biomedicines9080925.
125. Otsubo C, Otomo R, Miyazaki M, Matsushima-Hibiya Y, Kohno T, Iwakawa R, et al. TSPAN2 is involved in cell invasion and motility during lung cancer progression. Cell Rep. 2014;7(2):527–38. https://doi.org/10.1016/j.celrep.2014.03.027.
126. de Monasterio-Schrader P, Patzig J, Möbius W, Barrette B, Wagner TL, Kusch K, et al. Uncoupling of neuroinflammation from axonal degeneration in mice lacking the myelin protein tetraspanin-2. Glia. 2013;61(11):1832–47. https://doi.org/10.1002/glia.22561.
127. Reynolds JL, Mahajan SD. Transmigration of Tetraspanin 2 (Tspan2) siRNA via microglia derived exosomes across the blood brain barrier modifies the production of immune mediators by microglia cells. J Neuroimmune Pharmacol. 2020;15(3):554–63. https://doi.org/10.1007/s11481-019-09895-6.
128. Khan NZ, Cao T, He J, Ritzel RM, Li Y, Henry RJ, et al. Spinal cord injury alters microRNA and CD81+ exosome levels in plasma extracellular nanoparticles with neuroinflammatory potential. Brain Behav Immun. 2021;92:165–83. https://doi.org/10.1016/j.bbi.2020.12.007.
129. Zhao K, Wang Z, Hackert T, Pitzer C, Zöller M. Tspan8 and Tspan8/CD151 knockout mice unravel the contribution of tumor and host exosomes to tumor progression. J Exp Clin Cancer Res. 2018;37(1):312. https://doi.org/10.1186/s13046-018-0961-6.

130. Yue S, Mu W, Zöller M. Tspan8 and CD151 promote metastasis by distinct mechanisms. Eur J Cancer. 2013;49(13):2934–48. https://doi.org/10.1016/j.ejca.2013.03.032.
131. Zhang L, Xu Y, Cai E, Zheng M, Liu L, Wang Q, et al. TSPAN8 regulates EGFR/AKT pathway to enhance metastasis in gastric cancer. Mol Biol Rep. 2023;50(10):7955–65. https://doi.org/10.1007/s11033-023-08662-4.
132. Hu X, Shen N, Liu A, Wang W, Zhang L, Sui Z, et al. Bone marrow mesenchymal stem cell-derived exosomal miR-34c-5p ameliorates RIF by inhibiting the core fucosylation of multiple proteins. Mol Ther. 2022;30(2):763–81. https://doi.org/10.1016/j.ymthe.2021.10.012.
133. Jung JW, Kim JE, Kim E, Lee H, Lee H, Shin EA, et al. Liver-originated small extracellular vesicles with TM4SF5 target brown adipose tissue for homeostatic glucose clearance. J Extracell Vesicles. 2022;11(9):e12262. https://doi.org/10.1002/jev2.12262.
134. Willms E, Cabañas C, Mäger I, Wood MJA, Vader P. Extracellular vesicle heterogeneity: subpopulations, isolation techniques, and diverse functions in cancer progression. Front Immunol. 2018;9:738. https://doi.org/10.3389/fimmu.2018.00738.
135. Zhai Y, Wang Q, Zhu Z, Zheng W, Ma S, Hao Y, et al. Cell-derived extracellular matrix enhanced by collagen-binding domain-decorated exosomes to promote neural stem cells neurogenesis. Biomed Mater. 2021;17(1):014104. https://doi.org/10.1088/1748-605X/ac4089.

Chapter 8
The Mechanisms and Applications of Exosomes in Dermatology

Yinjiao Fang and Hong Cai

Abstract Exosomes, small membrane-surrounded vesicles secreted by cells, play significant roles in processes such as wound healing, skin rejuvenation, hair growth, and immune-mediated dermatoses. This chapter explores the mechanisms by which exosomes contribute to these processes, their potential therapeutic applications, and the challenges in translating research findings into clinical practice. Key areas of focus include the use of exosomes derived from different cell types for enhancing wound healing, reducing signs of aging on the skin, promoting hair growth, and treating immune-related skin conditions.

Keywords Exosomes · Dermatology · Wound Healing · Skin Rejuvenation · Hair Growth · Immunomediated Dermatoses · Therapeutic Applications · Cell Communication · Research Challenges

1 Introduction

Exosomes, which are cell-secreted nanovesicles encapsulated by a membrane, typically measure between 40 and 150 nm in diameter and have a density ranging from 1.09 to 1.18 g/mL. These vesicles transport proteins, lipids, polysaccharides, and RNA and form part of the broader category of extracellular vesicles (EVs). Exosomes can be extracted from various species and types of cell culture supernatants, such as neuronal cells, mesenchymal stem cells, fibroblasts, endothelial cells, megakaryocytes, etc., and exosomes have also been found to exist in various body fluids of the human body, such as blood, saliva, urine, and milk. Studies in recent

Y. Fang
China Medical University, Shenyang, China

Meishan People's Hospital, Meishan, China

H. Cai (✉)
Department of Dermatology, Air Force Medical Center, PLA, Beijing, China

© The Author(s), under exclusive license to Springer Nature Singapore Pte Ltd. 2024
Z. Li (ed.), *Extracellular Vesicle: Biology and Translational Application*,
https://doi.org/10.1007/978-981-97-5536-3_8

years have shown that exosomes may play important roles in a variety of pathophysiological processes and disease processes by participating in paracrine signaling as a communication tool between cells and have gradually become a research hotspot in the field of tumor immunity and regenerative medicine. In the field of skin diseases, there have also been many advances in research on exosomes. Herein, we summarize recent updates about the pathogenesis and applications of exosomes in dermatology, including wound healing, skin rejuvenation, hair growth, and immune-mediated dermatoses.

2 Exosomes in Dermatology

2.1 Exosomes and Wound Healing

Skin wound healing is an intricate and dynamic activity aimed at restoring the structural integrity and functionality of damaged tissue. This process encompasses multiple, well-orchestrated phases including inflammation, cellular migration and proliferation, angiogenesis, and the remodeling of the extracellular matrix [1]. However, under pathological conditions, such as infection, diabetes, ischemia, and hypoxia, it may lead to poor or even nonhealing behavior as a result of disruption and prolongation of the wound healing process [2]. Additionally, compromised wound healing often results in scar development. Pathologic scars mainly manifest as atrophic, hypertrophic, or keloids with variable degrees of redness or hyperpigmentation, which can affect esthetics or even affect normal function when scarring on specific areas.

Exosomes are reported to be derived mainly from mesenchymal stem cells (MCSs) and adipose-derived stem cells (ADSCs), which promote wound healing by participating in the afore-mentioned stages. For example, ADSCs derived exosomes (ADSC-Exos) can modulate immune responses and inflammation in the early wound healing stage. They facilitate this by enabling activated T regulatory cells to decrease both the production of interferon alpha (IFN-α) and the clustering ability of M1 macrophages [3]. ADSC-Exos serve an immunosuppressive function by curtailing IFN-α secretion, thereby preventing T cell activation [4]. They can also enhance angiogenesis by boosting the proliferation and migration of vascular endothelial cells, which provide sufficient blood to transport nutrients and metabolic waste. Angiogenesis is crucial for wound repair as it affects multiple processes, including fibroblast proliferation, collagen synthesis, and re-epithelialization [2]. In the proliferative phase of wound healing, fibroblasts multiply to form the extracellular matrix (ECM), while epithelial cells also grow and move toward the wound to promote healing. ADSC-Exos are internalized by fibroblasts, stimulating their proliferation, migration, and collagen synthesis in a dose-responsive manner [5]. The partial properties of fibroblasts are also optimized by the delivery of miR-31 through exosomes, as shown in in vivo experiments [6]. In addition, ADSC-Exos also play a role in modulating collagen remodeling, which helps prevent scar hyperplasia. In the early stage, exosomes aid in collagen remodeling by synthesizing types I and III collagen, whereas in the late stage, they contribute to scar reduction by inhibiting collagen formation [5].

The clinical use of exosomes in wound healing and treatment is promising based on the multiple preclinical studies mentioned earlier. To maintain exosome bioactivity and ensure the controlled release of their effective concentrations during extended wound care, developing an advanced delivery system is crucial. A variety of biomaterials, such as membranes, electrospun nanofibers, colloidal nanoparticles, and hydrogels, have been employed together with exosomes in fabricating wound dressings or scaffolds. Among these materials, hydrogels have become a popular option because of their versatile properties, including antibacterial effects, hemostatic capabilities, tissue adhesion, UV protection, injectability, and self-healing features [7]. In animal model experiments by Safi et al., wounds treated with seaweed gel exosomes were observed to heal faster and be scar free [8].

2.2 *Exosomes and Skin Rejuvenation*

Skin aging is driven by a combination of internal and external factors, leading to reduced structural integrity and impaired physiological functions. External aggressive factors, such as environmental pollution, lifestyle habits, and notably ultraviolet (UV) exposure, accelerate skin aging. Specifically, exposure to long-wavelength ultraviolet radiation (UVA) and medium-wavelength ultraviolet radiation (UVB) primarily contribute to external skin aging, commonly referred to as photoaging. This form of aging is responsible for approximately 80% of the aging changes observed in facial skin and is characterized by sunburn, irregular pigmentation, dryness, sallowness, roughness, and fine and coarse wrinkles.

Research on the role of exosomes in skin rejuvenation has focused mainly on the mechanism of anti-aging in vitro. The reports of exosomes for anti-aging are mainly derived from various stem cells, including umbilical cord mesenchymal stem cells (UCMSCs), adipose-derived stem cells (ADSCs), and induced pluripotent stem cells (iPSCs). Beyond exosomes sourced from stem cells, those extracted from dermal fibroblasts—which are fully differentiated cells—have also been researched and found to possess regenerative properties similar to UCMSC-Exos [9]. In vitro studies have demonstrated that exosomes can increase cell proliferation and mobility, modify the expression of genes and proteins involved in synthesizing the extracellular matrix, and boost the production of growth factors. These findings indicate that exosomes hold the potential as an effective treatment for reducing the appearance of fine lines and wrinkles.

For example, administering UCMSC-Exos to human dermal fibroblasts (HDFs) resulted in significant improvements in HDF migration, enhanced gene expression of collagen 1, fibronectin, and elastin, increased levels of procollagen type 1 C protein, and decreased matrix metalloproteinase (MMP) 1 protein levels, all observed within 24 hours [10]. Furthermore, UCMSC-Exos also promoted both the proliferation and migration of healthy keratinocytes. Following exposure of keratinocytes to UVB radiation, subsequent treatment with UCMSC-Exos enhanced cell survival, decreased the levels of intracellular reactive oxygen species (ROS), and reduced the number of senescence-associated beta-galactosidase (SA-beta-Gal) positive cells. These results indicate that UCMSC-Exos facilitate the repair processes in photoaged keratinocytes during in vitro studies [11]. In vivo studies mainly use murine models and support the application of exosomes in rejuvenation. In such studies, applying UCMSC-Exos to a photoaging mouse model every other day for 2 weeks has been shown to significantly reduce the appearance of micro-wrinkles, as observed through clinical, histological, or molecular analyses [12]. Another in vivo study explored the impact of the origin and culture conditions of exosomes on their anti-aging efficacy. This study utilized 3-dimensional (3D) spheroid cultures of HDFs as a source of exosomes, where 3D HDF-Exos demonstrated greater effectiveness in mitigating signs of photoaging compared to both MSC-Exos and 2D HDF-Exos 3 weeks post-treatment, as observed both clinically and histologically [13] (Fig. 8.1).

Although the initial preclinical results appear promising, the advancement of exosome research into clinical application faces challenges due to the diverse sources of exosomes, varying isolation methods, and the complexities of quality assessment. Presently, no exosome-based therapies have secured FDA approval, and the clinical data available on their efficacy in skin rejuvenation remains sparse.

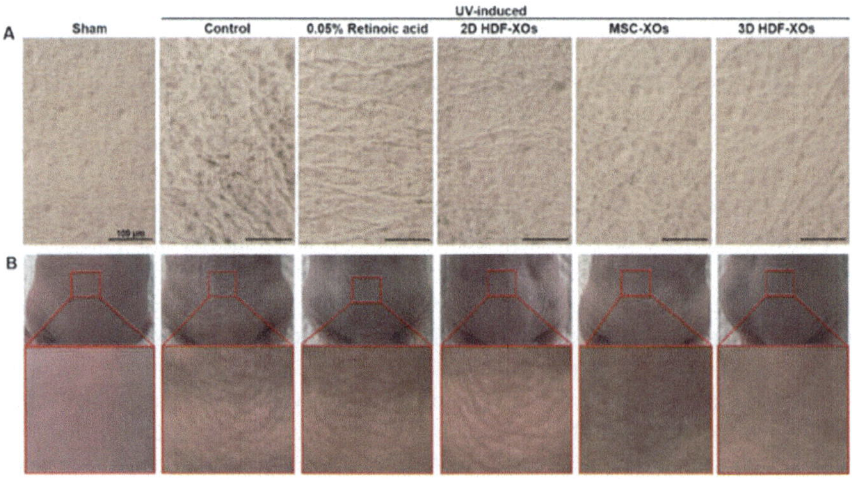

Fig. 8.1 Impact of retinoic acid (RA) and various cellular exosomes on wrinkle development in UVB-irradiated nude mice. (**a**) Microscopic examination of the replicas (scale bar: 100 μm) and (**b**) images of the dorsal skin from different experimental groups ($n = 3$). (Reprinted with permission from ACS Nano. 2019;13 (10):11273–11282)

2.3 Exosomes and Hair Growth

Alopecia, also known as hair loss or baldness, is the most readily visible human feature for various reasons—such as age, disease, trauma, and medication—and is usually divided into two categories: congenital disorders and acquired disorders. Intervention strategies include medications, natural ingredients, acupuncture therapy, low-laser therapy, and hair follicle transplantation. However, these therapies have potential problems such as adverse drug reactions, unfavorable outcomes and impermanent results. Hair loss is significantly influenced by the cyclical phases of hair follicles, namely the growth (anagen) and resting (telogen) stages, which are punctuated by periods of intense tissue transformation, known as the catagen and neogen phases. Hair follicle stem cells (HFSCs) and dermal papilla cells (DPCs) are crucial in promoting hair growth, particularly in sustaining the follicles through the anagen phase. Recent research has thus been directed toward modifying the cellular composition and managing the secretions of these stem cells, particularly through exosomes, to enhance or restore the signaling necessary for hair regeneration [14].

Contemporary research highlights the potential of exosomes as promising treatments for hair loss. Various cell types, including dermal papilla cells, the outer root sheath of hair, dermal fibroblasts, adipose-derived stem cells, immune cells, and bone marrow, have been studied for their role in hair regrowth via exosome production. Among these, exosomes derived from DPCs (DPCs-Exos) have yielded substantial preclinical evidence supporting their effectiveness in treating alopecia. DPCs-Exos mediate hair regrowth mainly in the following ways: 1. up-regulation of β-catenin and sonic hedgehog (Shh) levels; 2. Promoting the differentiation of HFSCs. First, research demonstrates that these exosomes present Wnt proteins on their exterior, triggering β-catenin activation—a crucial pathway for initiating and sustaining the growth phase of hair follicles. By the in vitro study, injection of DPCs-Exos accelerated the entry into HF anagen, extended catagen, and resulted in thicker hair shafts and more pronounced follicular bulges in the skin of mice, as shown by H&E-stained tissue sections [15]. Furthermore, studies by Yan et al. have shown that during the coculture of HFSCs with DPCs, HFSCs differentiation was induced, and transmission electron microscopy (TEM) was used to visualize DPC exosomes attached to the surface of HFSCs during coculture. This research also identified 111 differentially expressed microRNAs in DPC exosomes compared to DPCs alone, underscoring the significant role of DPC-derived exosomes in hair follicle regeneration [16].

Traditional subcutaneous injections of exosomes typically necessitate frequent applications due to their brief retention time in the body. To enhance clinical effectiveness and increase patient satisfaction, various studies have pursued alternative methods of delivery. For instance, Shiekh [17] et al. developed an oxygen releasing, antioxidant wound dressing named OxOBand (Fig. 8.2), made from polyurethane, that encapsulates ADSC-Exos. Similarly, Yang [18] et al. created a microneedle device (Fig. 8.3) that utilizes a water-soluble hyaluronic acid (HA) base. This device is equipped to deliver MSC-Exos combined with the small-molecule drug UK5099

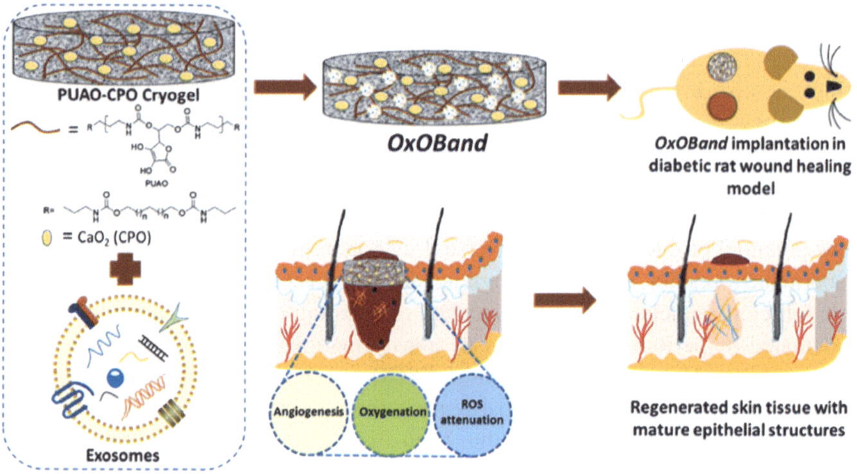

Fig. 8.2 Schematic representation of the formation of OxOBand with oxygen releasing antioxidant PUAO-CPO cryogels and adipose-derived stem cell exosomes. (Reprinted with permission from Biomaterials. 2020; 249:120020)

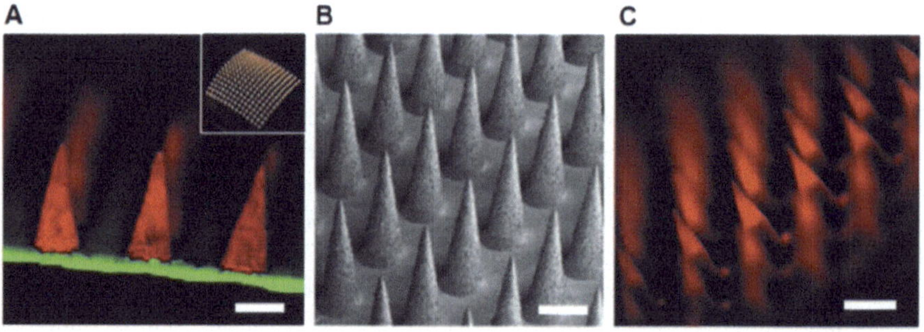

Fig. 8.3 The microneedle system—a removable microneedle patch identified as HMN. (**a**) Fluorescence images. Scale bar: 250 μm. Inset: Photograph of a representative image of an HMN patch. (**b**) Scanning electron microscopy (SEM) image. Scale bar: 250 μm. (**c**) Fluorescence image. Scale bar: 250 μm. Reprinted with permission from ACS Nano. 2019;13 (4):4354–4360

directly into the hair follicle's microenvironment, thus optimizing the treatment for hair growth by reducing the required dose.

Currently, there are no completed clinical trials that evaluate the efficacy of exosomes in treating hair loss. Nonetheless, numerous patents are in progress targeting the use of exosomes for alopecia treatment [19–21]. Furthermore, initial pilot studies have shown promising results in using exosomes to address hair loss [22–24]. Collectively, these early findings are encouraging, yet further clinical and preclinical research is essential to fully understand both the effectiveness and the underlying mechanisms of exosome-based therapies for alopecia.

2.4 Exosomes and Immune-Mediated Dermatoses

Immune-mediated skin disorders such as psoriasis, atopic dermatitis (AD), and systemic sclerosis (SSc) are characterized by their inflammatory nature and disruptions in immune system functionality. Exosomes play a crucial role in facilitating communication among immune cells, essential for orchestrating an effective immune response. Exosomes execute several immunoregulatory roles, such as altering antigen presentation, modulating immune response activation and suppression, conducting immune surveillance, and enhancing intercellular communication1. Due to these properties, exosomes hold potential as diagnostic or prognostic biomarkers for these diseases and could serve as therapeutic targets in the future.

2.4.1 Psoriasis

Psoriasis is a chronic, immune-mediated inflammatory dermatological disease characterized by red, scaly patches typically appearing on the elbows, knees, scalp, and lower back. Beyond its skin manifestations, psoriasis is associated with systemic implications, including psychological distress, metabolic issues, arthritic symptoms, and cardiovascular complications. The underlying mechanisms of psoriasis have long been a hotspot in dermatological investigations. Originally, it was theorized that exosomes derived from TH1 cells could transport cytokines like TNF-α to adjacent antigen-presenting dendritic cells, thus facilitating the initial stages of psoriasis. Recent findings, however, highlight that Th-17 cells and IL-17 (not exclusively produced by Th17 cells) have more significant roles in the pathogenesis of this prevalent condition. The release of IL-17 is partially regulated by IL-23, which dermal dendritic cells emit via exosomes. Given the effectiveness of therapies targeting IL-17 and IL-23, adjusting the function of exosomes from dendritic cells might represent a viable strategy for addressing this widespread chronic disease [25]. Furthermore, the abnormal proliferation and differentiation of keratinocytes are fundamental aspects of psoriasis. Research by Jiang [26] et al. has shown that exosomes from psoriatic keratinocytes activate the NF-κB and p38 MAPK pathways, leading to the production of pro-inflammatory agents like IL-6, IL-8, and TNF-α. These observations imply that targeting exosomes from keratinocytes could offer new therapeutic possibilities for managing psoriasis.

Taken together, these findings reveal that exosomes originating from both immune and non-immune cells are involved in the development of psoriasis, indicating a potential new approach for the treatment of psoriasis.

2.4.2 Atopic Dermatitis

Atopic dermatitis (AD) manifests as a chronic inflammatory skin disorder with a complex origin involving compromised skin barriers, imbalanced immune function, and altered skin microbial composition. Management strategies encompass a spectrum of treatments like the application of topical anti-inflammatory agents, the utilization of light therapy, the administration of systemic immunosuppressive medication, or the employment of drugs like dupilumab. These interventions are designed to mitigate itchiness and secure disease management. However, overseeing AD poses a sustained challenge due to the complexity of its care, underscoring the critical demand for treatments that are both effective and safe.

Exosomes present robust paracrine capabilities by releasing bioactive cytokines and agents that influence inflammatory pathways, thereby offering a prospective option for AD treatment. A critical element in managing AD is the integrity of the skin's epidermal barrier. Research by Shin [27] et al. has revealed that hAT-MSC derived exosomes can enhance this barrier by bolstering epidermal permeability, coupled with a marked rise in ceramide levels and a suppression of immunological reactions as AD advances. Moreover, MSC exosomes are known to induce the synthesis of ceramides within the epidermis and the development of structured layers within the stratum granulosum and stratum corneum, aiding in keratinocyte differentiation and fortifying the epidermal barrier. These MSC exosomes also appear to activate genes that are crucial to keratinocyte maturation. Conversely, the modulation of anomalous immune activity and reduction of inflammation are essential in lessening AD symptoms. Studies by Cho [28] et al. have shown that ADSC-derived exosomes were effective in reducing symptoms of atopic dermatitis in NC/Nga mice exposed to house dust mite antigens by lowering serum IgE levels, reducing eosinophil counts, and decreasing the levels of inflammatory cytokines such as IL-4, IL-23, IL-31, and TNF-α.

In addition, exosomes may be useful as a serologic monitoring indicator of disease changes and are convenient. A recent report revealed that circulating CD3 + HLA-DR+ EVs can serve as indicators for pathological conditions related to T-cells, such as AD [16].

2.4.3 Systemic Sclerosis

Systemic sclerosis (SSc) is an autoimmune condition with an unknown cause, characterized by vascular irregularities, amplified immune reactions, and functional disruptions of fibroblasts in both the skin and internal organs. The underlying mechanisms of SSc remain enigmatic, complicating the stratification of patients and the selection of optimal treatments. This complexity often results in pronounced disability and can be fatal for those afflicted with SSc. In addition, there is a pronounced gap in the development and validation of non-invasive biomarkers that could reliably diagnose or track the progression of this condition.

Exosomes participate in the pathogenesis of SSc by functioning as messengers between immune, vascular, and fibrotic components (Fig. 8.4). Diverse cellular sources, including immune, endothelial, mesenchymal stem (MSCs), fibroblast, and vascular smooth muscle cells, are involved in the production and secretion of exosomes. These exosomes transport a variety of molecules such as DNA, microRNAs,

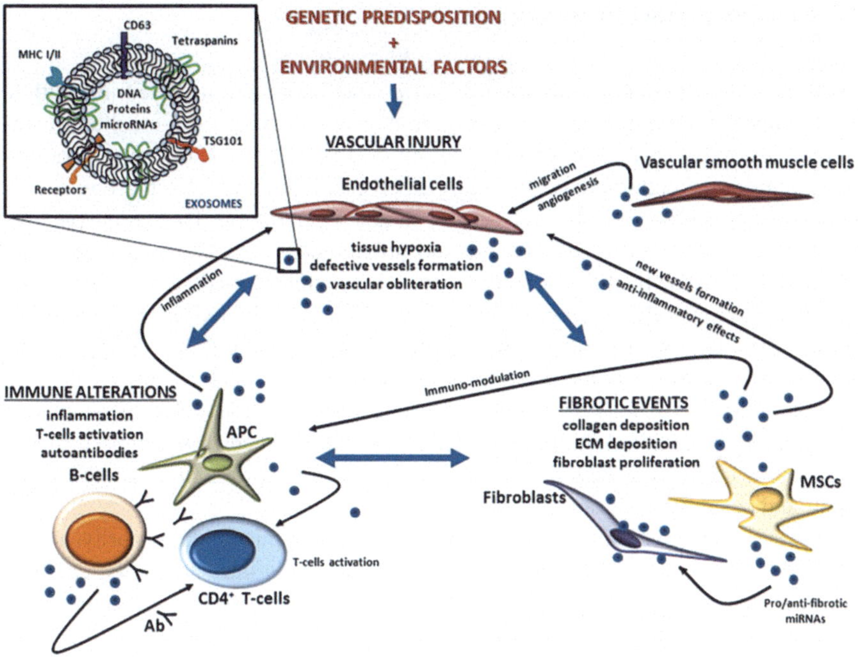

Fig. 8.4 Potential functions of exosomes in SSc. The Blue arrows demonstrate the connections among vasculopathy, autoimmunity, and fibrosis; the black arrows show how exosomes influence target cells. Reproduced under a Creative Commons Attribution 4.0 International License from Int J Mol Sci. 2019 Sep 4;20 (18):4337

and proteins, pivotal in modulating activities in recipient cells and thus affecting cellular functions. Exosomes originating from immune cells can target similar or different tissue cells, like endothelial cells, to regulate processes, particularly inflammation. Exosomes derived from MSCs are known for their immune-modulating capabilities and contain microRNAs that can both suppress and promote fibrosis and are instrumental in vasculogenesis. Vascular smooth muscle cell-released exosomes, once internalized by endothelial cells, can alter their angiogenic and migratory capabilities. Such activities contribute to the clinical manifestations of SSc [29]. In this light, exosomes are also being examined for their potential as carriers of microRNAs and proteins, which might serve as novel biomarkers for the clinical oversight of SSc. For example, Wermuth et al. reported that, compared with normal serum exosomes, SSc serum exosomes upregulate six profibrotic miRNAs and downregulate ten antifibrotic miRNAs [17].

SSc is a systemic and complex disease that affects multiple regions and various organs, involves diverse cell types, and includes intricate interactions with the microenvironment and extracellular matrix. Exosomes shed light on the underlying mechanisms and provide possible biomarkers for assessing disease severity. However, further research is still needed to determine the appropriate therapy.

3 Prospects and Perspectives

Exosomes, which contain important paracrine mediators, are involved in multiple physiological and pathological processes in the skin. Serving as intercellular communicators, these vesicles transport between diverse cell types including keratinocytes, fibroblasts, endothelial cells, adipose cells, and immune cells, thereby having a dual impact on the processes of tissue repair and regeneration. The unique patterns of expression, the medley of immune and inflammatory reactions, as well as the facilitation of tissue healing and other control mechanisms mediated by exosomes, are central to the body's response to certain dermatological conditions and injuries. Due to their precise and responsive nature, exosomes naturally occurring in the body hold promise as markers for diagnosing dermatological conditions that are immune-related, such as psoriasis, AD, and systemic SSc. Moreover, current research has illuminated the potential of exosomes, sourced from external means like iPSCs, UMMSCs, HDFs, BMSCs, and ADSCs, as promising agents for the repair and cosmetic restoration of skin.

However, several notable obstacles merit attention in this field. Primarily, a myriad of skin cell types give rise to exosomes, each carrying intricate biological messages; as a result, the exact process by which these exosomes are transported remains elusive within the realm of dermatological research, necessitating more detailed exploration. Additionally, while early-stage research shows promise, translating these results from controlled lab environments and animal subjects to human applications is challenging due to the complex interplay of human biological responses that could alter exosome efficacy. Moreover, the advancement of exosome research toward clinical application is hindered by the diversity of exosomal origins, the methods used to isolate them, and the need for comprehensive analysis. To date, no exosome-based therapies have received FDA approval. Consequently, in-depth research into exosome-centric approaches is essential to augment clinical treatments in dermatology.

References

1. Gao S, Chen T, Hao Y, Zhang F, Tang X, Wang D, Wei Z, Qi J. Exosomal miR-135a derived from human amnion mesenchymal stem cells promotes cutaneous wound healing in rats and fibroblast migration by directly inhibiting LATS2 expression. Stem Cell Res Ther. 2020;11(1):56. https://doi.org/10.1186/s13287-020-1570-9.
2. Liu J, Yan Z, Yang F, Huang Y, Yu Y, Zhou L, Sun Z, Cui D, Yan Y. Exosomes derived from human umbilical cord Mesenchymal stem cells accelerate cutaneous wound healing by enhancing angiogenesis through delivering Angiopoietin-2. Stem Cell Rev Rep. 2021;17(2):305–17. https://doi.org/10.1007/s12015-020-09992-7.
3. Lipsky PE. Systemic lupus erythematosus: an autoimmune disease of B cell hyperactivity. Nat Immunol. 2001;2(9):764–6. https://doi.org/10.1038/ni0901-764.
4. Blazquez R, Sanchez-Margallo FM, de la Rosa O, Dalemans W, Alvarez V, Tarazona R, Casado JG. Immunomodulatory potential of human adipose Mesenchymal stem cells derived

Exosomes on in vitro stimulated T cells. Front Immunol. 2014;5:556. https://doi.org/10.3389/fimmu.2014.00556.
5. Hu L, Wang J, Zhou X, Xiong Z, Zhao J, Yu R, Huang F, Zhang H, Chen L. Exosomes derived from human adipose mensenchymal stem cells accelerates cutaneous wound healing via optimizing the characteristics of fibroblasts. Sci Rep. 2016;6:32993. https://doi.org/10.1038/srep32993.
6. Kang T, Jones TM, Naddell C, Bacanamwo M, Calvert JW, Thompson WE, Bond VC, Chen YE, Liu D. Adipose-derived stem cells induce angiogenesis via microvesicle transport of miRNA-31. Stem Cells Transl Med. 2016;5(4):440–50. https://doi.org/10.5966/sctm.
7. Safari B, Aghazadeh M, Davaran S, Roshangar L. Exosome-loaded hydrogels: a new cell-free therapeutic approach for skin regeneration. Eur J Pharm Biopharm. 2022;171:50–9. https://doi.org/10.1016/j.ejpb.2021.11.002. Epub 2021 Nov 15
8. Shafei S, Khanmohammadi M, Heidari R, Ghanbari H, et al. Exosome loaded alginate hydrogel promotes tissue regeneration in full-thickness skin wounds: an in vivo study. J Biomed Mater Res A. 2020;108(3):545–56.
9. Deng M, Yu TZ, Li D, Wang X, Zhou G, Liu W, Cao Y, Xia W, Li W, Zhang WJ. Human umbilical cord mesenchymal stem cell-derived and dermal fibroblast-derived extracellular vesicles protect dermal fibroblasts from ultraviolet radiation-induced photoaging in vitro. Photochem Photobiol Sci. 2020;19(3):406–14.
10. Kim YJ, Yoo SM, Park HH, Lim HJ, Kim YL, Lee S, Seo KW, Kang KS. Exosomes derived from human umbilical cord blood mesenchymal stem cells stimulates rejuvenation of human skin. Biochem Biophys Res Commun. 2017;493(2):1102–8.
11. Liu SJ, Meng MY, Han S, Gao H, Zhao YY, Yang Y, Lin ZY, Yang LR, Zhu K, Han R, Huang WW, Wang RQ, Yang LL, Wang WJ, Li L, Wang XD, Hou ZL, Liao LW, Yang L. Umbilical cord Mesenchymal stem cell-derived Exosomes ameliorate HaCaT cell photo-aging. Rejuvenation Res. 2021;24(4):283–93.
12. Zhang K, Yu L, Li FR, Li X, Wang Z, Zou X, Zhang C, Lv K, Zhou B, Mitragotri S, Chen M. Topical application of Exosomes derived from human umbilical cord Mesenchymal stem cells in combination with sponge spicules for treatment of Photoaging. Int J Nanomedicine. 2020;15:2859–72.
13. Hu S, Li Z, Cores J, Huang K, Su T, Dinh PU, Cheng K. Needle-free injection of Exosomes derived from human dermal fibroblast spheroids ameliorates skin Photoaging. ACS Nano. 2019;13(10):11273–82.
14. Xiong M, Zhang Q, Hu W, Zhao C, Lv W, Yi Y, Wang Y, Tang H, Wu M, Wu Y. The novel mechanisms and applications of exosomes in dermatology and cutaneous medical aesthetics. Pharmacol Res. 2021;166:105490.
15. Zhou L, Wang H, Jing J, Yu L, Wu X, Lu Z. Regulation of hair follicle development by exosomes derived from dermal papilla cells. Biochem Biophys Res Commun. 2018;500(2):325–32.
16. Yan H, Gao Y, Ding Q, Liu J, Li Y, Jin M, Xu H, Ma S, Wang X, Zeng W, Chen Y. Exosomal micro RNAs derived from dermal papilla cells mediate hair follicle stem cell proliferation and differentiation. Int J Biol Sci. 2019;15(7):1368–82.
17. Shiekh PA, Singh A, Kumar A. Exosome laden oxygen releasing antioxidant and antibacterial cryogel wound dressing OxOBand alleviate diabetic and infectious wound healing. Biomaterials. 2020;249:120020.
18. Yang G, Chen Q, Wen D, Chen Z, Wang J, Chen G, Wang Z, Zhang X, Zhang Y, Hu Q, Zhang L, Gu Z. A therapeutic microneedle patch made from hair-derived keratin for promoting hair regrowth. ACS Nano. 2019;13(4):4354–60.
19. Lim SK, Yeo MSW, Chen TS, Lai RC. Use of eexosomes to promote or enhance hair growth. Google Patents. 2015.
20. Fikrettin Sahin OSS, Pakize Neslihan TASLI, Demir O. inventor; patent application WO2017078641A1.
21. ライ，ルエン キリサリサウイマウイマチテチテラル. JP5911874B2.

22. Cooley J. Bone Marrow Derived MSC Exosomes for AA: The First 100 Patients. presented at: International Society of Hair Restoration Surgeons World Congress; 2020; Virtual.
23. Zari S. The Use of Exosomes in the Management of Androgenetic Alopecia/Female Pattern Hair Loss. presented at: International Society of Hair Restoration Surgeons World Congress; 2020; Virtual.
24. Huh CHKS. Exosome for hair regeneration: from bench to bedside. J Am Acad Dermatol. 2019;81(4):AB62.
25. Jacquin-Porretaz C, Cordonnier M, Nardin C, Boullerot L, Chanteloup G, Vautrot V, Adotevi O, Garrido C, Gobbo J, Aubin F. Increased levels of interleukin-17A Exosomes in psoriasis. Acta Derm Venereol. 2019;99(12):1143–7.
26. Jiang M, Fang H, Shao S, Dang E, Zhang J, Qiao P, Yang A, Wang G. Keratinocyte exosomes activate neutrophils and enhance skin inflammation in psoriasis. FASEB J. 2019;33(12):13241–53.
27. Shin KO, Ha DH, Kim JO, Crumrine DA, Meyer JM, Wakefield JS, Lee Y, Kim B, Kim S, Kim HK, Lee J, Kwon HH, Park GH, Lee JH, Lim J, Park S, Elias PM, Park K, Yi YW, Cho BS. Exosomes from human adipose tissue-derived Mesenchymal stem cells promote epidermal barrier repair by inducing de novo synthesis of Ceramides in atopic dermatitis. Cells. 2020;9(3):680.
28. Cho BS, Kim JO, Ha DH, Yi YW. Exosomes derived from human adipose tissue-derived mesenchymal stem cells alleviate atopic dermatitis. Stem Cell Res Ther. 2018;9(1):187.
29. Colletti M, Galardi A, De Santis M, Guidelli GM, Di Giannatale A, Di Luigi L, Antinozzi C. Exosomes in systemic sclerosis: messengers between immune, vascular and fibrotic components? Int J Mol Sci. 2019;20(18):4337.

Chapter 9
Stem Cell-Derived Extracellular Vesicles as Therapeutics Against Renal Diseases

Chuyue Zhang, Fei Peng, Tingting Qi, Xijun Sun, Jin Yao, Liang Ma, Xiangmei Chen, and Ping Fu

Abstract Mesenchymal stem cells are critical for maintaining steady-state organ homeostasis and regenerating injured tissues. Due to the natural property of extracellular vesicles being of cellular origin, stem cell-derived extracellular vesicles (EVs) have regenerative and immunomodulatory properties, making them attractive innate biotherapeutics for renal disorders. In this chapter, we summarize the current applications and therapeutic molecular mechanisms of EVs derived from natural stem cells in the treatment of renal diseases, with a focus on an updated overview of the key engineering approaches for better exploiting EVs in renal disease intervention. In addition, we present recent advances in utilizing natural membrane nanoparticles (NPs) made from stem cell exosomes with excellent delivery capacity as drug carriers to alleviate kidney injury and delve into harnessing their full potential in renal therapeutics.

Keywords Kidney · Stem cell · Acute kidney injury · Extracellular vesicles (EVs) · Therapy

---

Chuyue Zhang and Fei Peng contributed equally to this work.

C. Zhang (✉)
Kidney Research Institute, Division of Nephrology, West China Hospital of Sichuan University, Chengdu, China

Department of Nephrology, State Key Laboratory of Kidney Diseases, First Medical Center of Chinese PLA General Hospital, Beijing, China
e-mail: zhangcycywch@wchscu.edu.cn

F. Peng · T. Qi · X. Sun · J. Yao · X. Chen
taNational Key Laboratory of Kidney Diseases, Chinese PLA General Hospital, Beijing, China

L. Ma · P. Fu
Kidney Research Institute, Division of Nephrology, West China Hospital of Sichuan University, Chengdu, China
e-mail: fupinghx@scu.edu.cn

© The Author(s), under exclusive license to Springer Nature Singapore Pte Ltd. 2024

Z. Li (ed.), *Extracellular Vesicle: Biology and Translational Application*,
https://doi.org/10.1007/978-981-97-5536-3_9

1 Introduction

Extracellular vesicles (EVs) contain thousands of biomolecules, including RNA, DNA, proteins, lipids, and carbohydrates [1]. Due to the natural property of extracellular vesicles being of cellular origin, EVs play a key role in intercellular communication by transmitting signals to recipient cells through triggering a signaling cascade by binding to receptors on the surface of recipient cells [2] or by being endocytosed by cells [3, 4]. Based on this property, EVs have been developed as therapeutic agents or drug delivery tools with promising applications in renal diseases [5].

Growing evidence indicates that EVs derived from stem cells transport active molecules within them to recipient cells to modulate multiple signaling pathways, thereby alleviating a variety of diseases [6–9]. Because EVs have low immunogenicity, low tumorigenicity, and low risk of teratomas [10–12], there is no need to be concerned about the risk of direct implantation of these cells [13–15]. Additionally, because the preparation of EVs is simple, easy to store, and able to cross biological barriers [16, 17], all of these factors make EVs derived from MSCs promising candidates for cell-free therapies and promising treatments for kidney diseases [18–22]. In this section, we will review the progress and the latest frontiers of the study of stem cell-derived EVs in renal diseases.

1.1 Stem Cell-Derived EVs as Natural Renal Therapeutics

1.1.1 Stem Cell-Derived EVs in Combating AKI

Acute kidney injury (AKI) is a common clinical critical condition. Currently, there is a lack of effective treatments for this disease [23]. Stem cells are widely used in AKI therapy due to their self-renewal ability and potential to differentiate into any cell type. EVs or exosomes secreted from different sources of stem cells have been shown to play important roles in AKI (Table 9.1).

1.1.2 Bone Marrow MSC-Derived EVs for the Treatment of AKI

Several studies have shown that bone marrow mesenchymal stem cell-derived EVs (BM-MSC-EVs) can significantly ameliorate acute kidney injury (AKI). BM-MSC-EVs can accelerate renal self-repair in IRI mice by activating M2 macrophage polarization [24]. It has also been reported that miR-223-3p delivered by BM-MSC-EVs can directly target NLRP3 to inhibit inflammatory activation, promote mitosis, and ameliorate the function of the injured kidney in ischemia–reperfusion-induced AKI [25]. BM-MSC-EVs restore the stability of TFAM proteins and TFAM-mtDNA complexes, thereby reversing mtDNA deletion and mitochondrial oxidative

Table 9.1 Use of stem cell-derived EVs in combating AKI

MSC source	Animal model	Therapeutic cargo	Mechanism of action	Pathophysiological effects	Ref.
BM-MSC	IRI	/	Activating M2 macrophage polarization	Accelerate repair process	[24]
BM-MSC	IRI	miR-223-3p	miR-223-3p/ NLRP3 Axis	Inhibit inflammatory activation, promote mitosis, and ameliorate the function	[25]
BM-MSC	IRI	/	Restore the stability of TFAM proteins and TFAM-mtDNA complexes	Attenuate mitochondrial damage and inflammation	[26]
BM-MSC	Sepsis	miR-21a-5p	Targeting toll-like receptor 4 and programmed cell death 4	Inhibit inflammation and improve renal function	[27]
BM-MSC	Cisplatin	Pulsed focused ultrasound-stimulated	Upregulating MAPK/ERK, PI3K/Akt and eNOS, SIRT3	Upregulating proliferative signaling pathways, regenerative pathways and inhibiting inflammation	[28]
UC-MSC	IRI	miR-125b-5p	miR-125b-5p/ p53 pathway	Targeting cell cycle arrest and apoptosis	[29]
UC-MSC	IRI	miR-874-3p	Regulate necroptosis through miR-874-3p	Attenuate renal tubular epithelial cell injury and enhance repair	[30]
UC-MSC	IRI	/	Ameliorating systemic inflammation	Attenuate renal inflammation and oxidative stress	[31]
UC-MSC	IRI	/	Upregulation of klotho and bone morphogenetic protein 7	Accelerate renal tubular cell repair and regeneration	[32]

(continued)

Table 9.1 (continued)

MSC source	Animal model	Therapeutic cargo	Mechanism of action	Pathophysiological effects	Ref.
UC-MSC	Glycerol	Platelet-rich plasma-stimulated	Inducing YAP nuclear expression; activated AKT/Rab27 pathway	Reduced renal tubular cell apoptosis; reverse the histopathological kidney damage	[33]
Placenta-MSC	IRI	miR-200a-3p	Activating the Keap1-Nrf2 signaling pathway and enhancing mitochondrial function	Promoted the recovery of kidney function	[36]
Placenta-MSC	IRI	Let-7a-5p	Boosting autophagy and inhibiting apoptosis	Ameliorate renal function, attenuated histopathological damage in early phases of AKI, and augment antifibrotic effects in chronic stages	[37]
iPSC	IRI	SP1	Delivering SP1 to activate the expression of SK1 and the generation of S1P	Anti-necroptosis	[39]
iPSC	IRI	/	Activated ERK 1/2 signaling	Improved kidney function and tissue damage	[40]
iPSC	IRI	/	Reducing cell death and inflammatory response	Protected functional mitochondria and regulated several genes associated with oxidative stress	[41]
AD-MSC	Lipopolysaccharide	/	Inhibiting oxidative stress and inflammation	Improved renal function and structure	[34]
AD-MSC	IRI	/	Decrease of mitochondrial O2•-formation with the aid of the upregulated antioxidant HO-1/NRF2 system	Improved renal function and structure	[35]
Embryonic MSC	IRI	/	Activating $Sox9^+$ cells	Physiological repair and inhibit the pathological repair	[38]

phosphorylation defects in injured renal tubular cells [26]. MiR-21a-5p is highly enriched in exosomes secreted by bone marrow mesenchymal stem cells and can inhibit inflammation and improve renal function by targeting toll-like receptor 4 and programmed cell death 4 in a mouse model of sepsis due to cecum ligation [27]. In addition, pulsed focused ultrasound (pFUS) can improve the efficacy of BM-MSC-EVs in the treatment of AKI by upregulating proliferative signaling pathways (MAPK/ERK, PI3K/Akt) and regenerative pathways (eNOS, SIRT3) as well as inhibiting inflammation [28].

1.1.3 Umbilical Cord MSC-Derived EVs Against AKI

Human umbilical cord-derived mesenchymal stem cell (UC-MSC-EV) exosomes play an important role in AKI. UC-MSC-EVs ameliorate ischemic AKI and promote tubular repair by targeting cell cycle arrest and apoptosis in TECs via the miR-125b-5p/p53 pathway [29]. Additionally, UC-MSC-EVs can regulate necroptosis through miR-874-3p to attenuate renal tubular epithelial cell injury and enhance repair, providing new therapeutic modalities and ideas for the treatment of AKI and the progression of AKI to CKD [30]. UC-MSC-EVs also attenuate renal inflammation and oxidative stress by ameliorating systemic inflammation, thereby reducing renal injury and improving renal function [31]. Moreover, UC-MSC-EVs play an important role in promoting renal tubular cell proliferation and angiogenesis in a porcine ischemia–reperfusion model [32]. Interestingly, platelet-rich plasma-stimulated human umbilical cord-derived mesenchymal stem cells effectively ameliorated glycerol-induced AKI in a mouse model [33].

1.1.4 Adipose MSC-Derived EVs Against AKI

Adipose MSC-derived EVs (AD-MSC-EVs) are considered ideal candidate cells due to their low immunogenicity, ease of extraction, and lack of ethical concerns. Compared with BM-MSC-EVs, AD-MSC-EV exosomes protect against lipopolysaccharide-induced AKI by inhibiting oxidative stress and inflammation and significantly improving renal function and structure [34]. AD-MSC-EVs protect against I/R-induced AKI by upregulating the antioxidant HO-1/nuclear factor erythroid 2-associated factor 2 system and reducing the production of the intense anion superoxide (O2•-) [35].

1.1.5 Placental MSC-Derived EVs Against AKI

Placental MSC-derived EVs also have therapeutic effects on AKI. It has been reported that placenta MSC-derived EVs promote the recovery of kidney function by activating the Keap1-Nrf2 signaling pathway and enhancing the mitochondrial function of tubular epithelial cells [36]. Facilitated by RGD hydrogel-mediated local delivery, EVs provide superior rescue effects on renal function, attenuate histopathological damage, decrease tubular injury, and promote cell proliferation in the early phases of AKI; additionally, they augment the antifibrotic effects of MSC-EVs in chronic stages via let-7a-5p by boosting autophagy and inhibiting apoptosis [37].

1.1.6 Embryonic Stem Cell-Derived EVs Against AKI

Embryonic stem cell-derived EVs (ESC-EVs) increase the proliferation of renal tubular epithelial cells, facilitate renal angiogenesis, inhibit the progression of renal fibrosis, and reverse DNA damage caused by ischemia and reperfusion after AKI by activating Sox9+ cells [38].

1.1.7 Induction of AKI with Pluripotent Stem Cell-Derived EVs

Induced pluripotent stem cell (iPSC)-derived exosomes (iPSC-EVs) exert powerful therapeutic effects on renal ischemia–reperfusion. An anti-necroptotic effect of induced pluripotent stem cell (iPSC)-derived mesenchymal stem cell exosomes (hiPSC-MSC-Exos) was shown to occur against renal I/R injury via the delivery of SP1 to target renal cells and the intracellular activation of SK1 expression and S1P generation [39]. Compared with those in the I/R injury group, the levels of activated ERK 1/2 signaling molecules were significantly greater in mice treated with iPSC-MSCs-EVs, which significantly improved kidney function and tissue damage [40]. Other studies have shown that iPSC-EVs protect mitochondrial function and play an important role in regulation and oxidative stress [41].

1.2 Stem Cell-Derived EVs in Combating CKD

Chronic kidney disease (CKD) was defined as kidney injury and/or a decline in kidney function lasting more than 3 months. CKD has become an important public health problem that jeopardizes human health worldwide due to its high prevalence, high disability, high healthcare costs, and low awareness. Exosomes secreted by stem cells can eliminate pathogenic damage in CKD by targeting renal fibrosis, reducing tubular atrophy and inflammation, and promoting angiogenesis, thereby promoting tissue regeneration (Table 9.2).

1.2.1 Bone Marrow MSC-Derived EVs for the Treatment of CKD

BM-MSC-EV therapy is also a promising strategy in chronic kidney disease. Many studies have found that in a unilateral ureteral obstruction (UUO)-induced renal fibrosis model, BM-MSC-EVs can activate the tuberous sclerosis complex subunit

Table 9.2 MSC-derived EVs in combating CKD

MSC source	Animal model	Therapeutic cargo	Mechanism of action	Pathophysiological effects	Ref.
BM-MSC	UUO	Let-7i-5p	Activate TSC1/mTOR pathway	Improving renal fibrosis	[42]
BM-MSC	UUO	miR-133b	Targeting and inhibiting connective tissue growth factor (CTGF) expression	Reducing renal interstitial fibrosis	[43]
BM-MSC	UUO	miR-122a	Regulates mTOR signaling and autophagy	Alleviating renal fibrosis	[44]
BM-MSC	UUO	/	Activating EP2 receptors to reduce the polarization of M1 and M2 macrophages	Reducing renal fibrosis	[45]
BM-MSC	UUO	/	Inhibiting the glycolysis of renal tubular epithelial cells	Reducing renal fibrosis	[46]
BM-MSC	UUO	miR-296	Pro-angiogenic	Renal protective effects	[47]
BM-MSC	5/6 subtotal nephrotomy	/	Regulating the Smurf 2/Smad 7 axis	Inhibited renal fibrosis	[48]
BM-MSC	UUO	miR-204-5p	Modulating PI3K/Akt signaling	Ameliorates renal fibrosis	[49]
BM-MSC	IRI	miR-34c-5p	Inhibiting the core fucosylation of multiple proteins	Ameliorates RIF	[50]
BM-MSC	Diabetic nephropathy	miR-125b	Downregulation of TRAF6	Induced autophagy and inhibited apoptosis	[51]
BM-MSC	Diabetic nephropathy	miR-30e-5p	Targets ELAVL1	Inhibit caspase-1-mediated pyroptosis	[52]
BM-MSC	5/6 subtotal nephrotomy	/	Lowering blood glucose levels and inhibiting JAK2/STAT3 expression	Improve renal function	[53]
BM-MSC	5/6 subtotal nephrotomy	/	SIRT6-HMGB1 deacetylation	Suppressing phosphate-induced aortic calcification	[54]
BM-MSC	Diabetic peripheral neuropathy	/	Paracrine effect in controlling hyperglycemia and loss in body weigh	Attenuation of damage to the tissues such as the pancreas, kidney, and liver	[55]
BM-MSC	Bilateral ovariectomy	/	Reducing inflammation and degeneration	Reducing chronic inflammatory changes in the kidney	[56]

(continued)

Table 9.2 (continued)

MSC source	Animal model	Therapeutic cargo	Mechanism of action	Pathophysiological effects	Ref.
UC-MSC	Diabetic nephropathy	miR-146a-5p	M2 macrophage polarization by targeting TRAF6	Restoring renal function	[57]
UC-MSC	Partial bladder outlet obstruction	/	Inhibiting Wnt/β--catenin signaling pathway	Ameliorated renal fibrosis, and reversed renal injury	[58]
UC-MSC	Diabetic nephropathy	miR-451a	Inhibit cell cycle inhibitors P15 and P19 to restart the blocked cell cycle and reverse EMT	Promote structural and functional improvements of injured kidneys in vivo	[59]
UC-MSC	UUO	/	CK1δ/β-TRCP inhibited YAP activity	Attenuates renal fibrosis	[60]
AD-MSC	Diabetic nephropathy	miR-15b-5p	Downregulating VEGF/PDK4 axis	Protects podocytes	[61]
AD-MSC	Diabetic nephropathy	miR-125a	Targeting histone deacetylase 1 and downregulating Endothelin-1	Inhibit DN progression and alleviate the symptoms	[62]
AD-MSC	Adenine-containing diet to induce CKD	Melatonin-stimulated	Upregulating the levels of aquaporins 2 and 5	Regulating inflammation and fibrosis	[64]
Placenta MSC	IRI	/	Restoring CPT1A mediated fatty acid oxidation	Ameliorate renal fibrosis	[65]

1/mammalian target of rapamycin (TSC1/mTOR) pathway and improve renal fibrosis by delivering antilet-7i-5p [42]; reducing renal interstitial fibrosis by targeting and inhibiting connective tissue growth factor (CTGF) expression through secretion of miR-133b [43]; reducing renal fibrosis by mediating miRNA-122a, which regulates mTOR signaling and autophagy to alleviate renal fibrosis in HK-2 cells [44]; reducing renal fibrosis by activating E2 receptor 2 (EP2) receptors to reduce the polarization of M1 and M2 macrophages [45]; reducing renal fibrosis by inhibiting the glycolysis of renal tubular epithelial cells (TECs) [46]; and BM-MSC-EVs derived from trained animals can produce miR-296 with pro-angiogenic effects and thus exert renoprotective effects [47].

In addition to their role in the UUO-induced renal fibrosis model, BM-MSC-EVs play an important protective role in other CKD models. BM-MSC-EVs inhibited renal fibrosis both in vivo and in vitro by regulating the Smurf 2/Smad 7 axis in 5/6 subtotal nephrotomy (SNx) rat models and in TGF-β1-induced human renal proximal tubular epithelial cells (HRPTEpiCs) [48]. Sinomenine ameliorates renal fibrosis by affecting miR-204-5p expression and modulating the phosphatidylinositol 3-kinase (PI3K)/protein kinase B (Akt) signaling pathway in BM-MSC-EVs [49].

BM-MSC-EV-secreted miR-34c-5p ameliorates RIF by inhibiting the core fucosylation of multiple proteins [50]. In a diabetic nephropathy model, BM-MSC-EVs secreted miR-125b induced autophagy and inhibited apoptosis in HG-treated human embryonic kidney epithelial cells (HKCs) via the downregulation of tumor necrosis factor receptor-associated factor 6 (TRAF6), [51]; miR-30e-5p secreted by BM-MSC-EVs targeted ELAVL1 in HG-treated HK-2 cells to inhibit caspase-1-mediated pyroptosis [52]; and BM-MSC-EVs also improved renal function by lowering blood glucose levels and inhibiting JAK2/STAT3 expression [53] or suppressing phosphate-induced aortic calcification via SIRT6-HMGB1 deacetylation in a 5/6 nephrectomized mouse CKD model [54]. Transplantation of engineered BM-MSC-EVs ameliorates diabetic peripheral neuropathy under electrical stimulation [55]. Interestingly, BM-MSC-EVs are renoprotective against postmenopausal chronic kidney injury by reducing inflammation and degeneration [56].

1.2.2 Umbilical Cord MSC-Derived EVs Against CKD

UC-MSC-EV-derived miR-146a-5p promotes M2 macrophage polarization by targeting TRAF6, thereby restoring renal function in diabetic nephropathy rats [57]. The administration of UC-MSC-EVs to mice with partial bladder outlet obstruction (PBOO) inhibited the expression of proteins related to the Wnt/β-catenin signaling pathway, ameliorated renal fibrosis, and reversed renal injury [58]. UC-MSC-EV-derived miR-451a could specifically inhibit cell cycle inhibitors to restart the blocked cell cycle and reverse epithelial mesenchymal transformation (EMT) in vivo and in vitro [59]. Additionally, UC-MSC-EVs attenuate renal fibrosis through CK1δ/β-TRCP-mediated inhibition of YAP activity in a UUO-induced renal fibrosis model [60].

1.2.3 Other Kinds of EVs Against CKD

MicroRNA-15b-5p secreted by AD-MSC-Exos protects podocytes from diabetic nephropathy by downregulating the vascular endothelial growth factor (VEGF)/pyruvate dehydrogenase kinase 4 (PDK4) axis [61]. AD-MSC-Exos Carry microRNA-125a to Protect Against Diabetic Nephropathy by Targeting Histone Deacetylase 1 and Downregulating Endothelin-1 [62]. Autophagy-induced AD-MSC-Exos also significantly ameliorate renal fibrosis [63]. Exosomes isolated from melatonin-stimulated mesenchymal stem cells improve kidney function by upregulating the levels of aquaporin 2 and 5 and reducing blood urea nitrogen and creatinine in a chronic kidney disease mouse model [64].

Hypoxia-pretreated human placenta-derived MSC extracellular vesicles ameliorate renal fibrosis after ischemia–reperfusion injury by restoring CPT1A-mediated fatty acid oxidation [65].

1.3 Stem Cell-Derived EVs for Other Kidney-Related Diseases

Exosomes secreted by stem cells also play an important protective role in kidney-induced models, except for AKI and CKD models. UC-MSC-EVs antagonize senescence in primary tubular epithelial cells (PTECs) of the mouse kidney by decreasing the transcription of senescence markers and senescence-associated secretory phenotype (SASP) factors, reducing DNA damage foci and increasing the number of proliferating cells [66]; additionally, intravenously injected UC-MSC-EVs produce an antiaging phenotype in aging mice, including reduced bone and kidney degeneration [67]. AD-MSC-Exos prevent frailty, improve healthspan, and decrease epigenetic age in old mice [68].

BM-MSC-EVs promoted the anti-inflammatory polarization of macrophages by secreting miR-16 and miR-21. These macrophages alleviate systemic lupus erythematosus nephritis in lupus mice by phagocytosing apoptotic debris and recruiting Treg cells [69]. Donor BM-MSC-EVs alleviated the inflammatory response by inhibiting the location of mature DCs (mDCs) in kidney grafts and increasing the number of regulatory T (Treg) cells in the spleens of kidney allograft rats [70].

2 Engineering Approaches for Effective Therapeutic Applications Based on Stem Cell-Derived Extracellular Vesicles

Rather than being satisfied with the strategy of using only extracellular vesicles as therapeutic agents, recent studies have explored the possibility of engineering or functionalizing parental cells to alter the secretion of their extracellular vesicles, thereby enhancing the therapeutic potential of these vesicles [2, 71–73]. In this section, we will provide an overview of strategies for the therapeutic optimization of extracellular vesicles (Fig. 9.1).

2.1 Cargo Loading Strategies

A major advantage of extracellular vesicles over cellular or synthetic materials is the ability to deliver active ingredients (RNA, proteins, lipids, etc.) across biological barriers [16, 17]. However, issues such as poor cellular uptake and stability can limit the efficacy of EVs. Thus, increasing the activity of EVs can improve their therapeutic potential. The loading of component molecules into EVs can be categorized into two basic approaches: endogenous loading and exogenous loading. Endogenous loading is performed by genetically modifying the parental cells or by altering the culture conditions of the parental cells (means of preconditioning) to increase their active content during the biogenesis of the extracellular vesicles, whereas

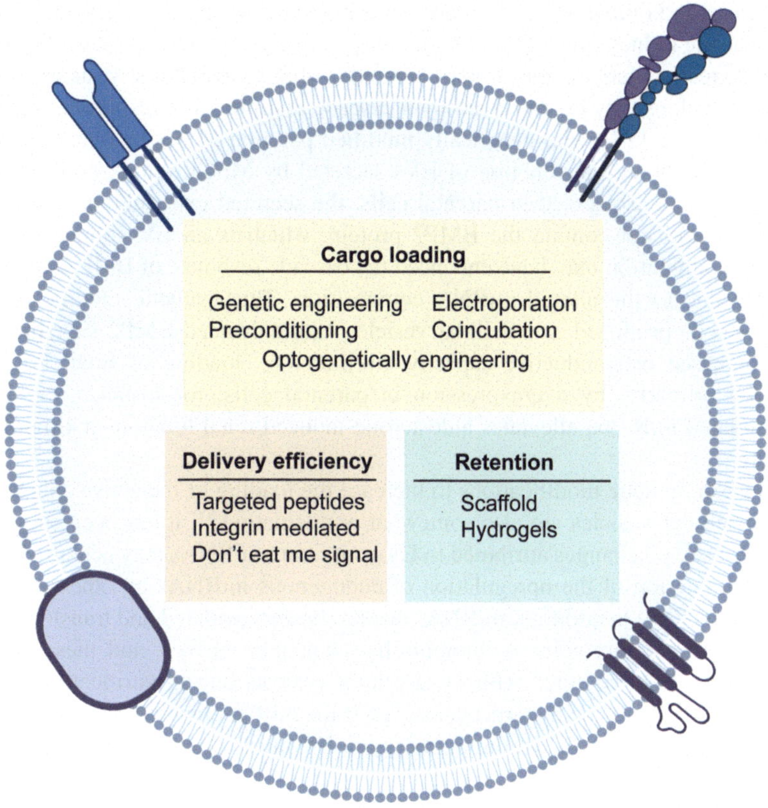

Fig. 9.1 Engineering approaches for effective therapeutic applications based on stem cell-derived extracellular vesicles (produced via biorender)

exogenous reprinting is performed by incorporating additional content after the generation of the extracellular vesicles utilizing electroporation, chemical incorporation, sonication, and so on.

2.1.1 Genetic Engineering

Genetic engineering allows for the incorporation of target active components during extracellular vesicle biogenesis, the content of beneficial components of EVs can be increased by means of genetic engineering, and the overexpression of genes involved in repair in parental cells can be used to ameliorate tissue damage [74, 75]. Bone marrow mesenchymal stem cell-derived exosomes (MSC-Exos) with indoleamine 2,3-dioxygenase (IDO) overexpression accelerated renal self-repair in IRI mice by activating M2 macrophage polarization, and these effects were amplified by IDO overexpression compared with those of the control [24]. Similarly, lentivirus-mediated modification of human umbilical cord MSCs with ACE2 has been shown to have anti-inflammatory, antioxidative stress, and antiapoptotic effects that protect against acute kidney injury [76]. On the other hand, even if overexpressed proteins do not directly increase the protein content within extracellular vesicles, downstream signaling may be enhanced by a cascade amplification response, as shown by Huang et al. [77], who genetically modified parental MSCs by overexpressing BMP2 to modulate the function of EVs secreted by MSCs. Although the BMP2 protein was overexpressed in parental cells, the secreted extracellular vesicles of these cells did not contain the BMP2 protein, which is an EV component, and genetic modification may have enhanced the cascade response of BMP2 signaling, possibly due to the altered miRNA composition. Thus, genetic modification of parental cells produced extracellular vesicles with enhanced BMP2 signaling and thus increased osteoinductive capacity. Furthermore, loading of small RNAs is achieved primarily by overexpression in parental cells; for instance, exosome-encapsulated miR-26a alleviates aldosterone-induced renal tubulointerstitial fibrosis [78].

However, genetic modifications to increase the loading of the active component of extracellular vesicles are also somewhat problematic [2]; it has been suggested that the observed changes attributed to EV-containing miRNAs may actually be due to the stimulation of the upregulation of endogenous miRNAs by other vesicular components [79]. In addition, miRNAs need to be encapsulated and translocated to target cells by lipoproteins or protein-oligonucleotide affixes, and these protein complexes can be coeluted with extracellular vesicles during purification, which prevents miRNA uptake by target cells even if the miRNA load within extracellular vesicles increases [79]. Therefore, a thorough EV purification step should be used with caution [80].

2.1.2 Preconditioning of Parent Cells

The physiological state of the cells and their culture conditions are crucial for influencing the content and type of components of the EVs secreted by them [80]. Many scholars have explored the possibility of enhancing the content of EVs in parental cells by prestimulation, thereby enhancing the efficacy of EVs. The most common is the utilization of inflammatory factors. There is growing evidence that the immunomodulatory capacity of MSCs is not static and that an inflammatory environment may enhance the therapeutic efficacy of MSCs [81–84]. EVs secreted by MSCs activated by different inflammatory cytokines may give rise to extracellular vesicles with heightened immunomodulatory properties [82].

For example, extracellular vesicles derived from MSCs prestimulated with TNF-α/IFN-γ were found to play a significant role in regenerative processes, including angiogenesis and fibrosis, and inflammation was found to play a more pronounced role in the effective recovery of renal function; additionally, anti-inflammatory exosome-assembled integrated bioactive scaffolds could be better used as advanced tissue engineering platforms for kidney tissue regeneration [85]. Extracellular vesicles from interleukin 1β (IL-1β)-pretreated mouse MSCs were applied to a mouse model of sepsis, and the results revealed that the extracellular vesicles obtained from prestimulation efficiently induced M2-like polarization of macrophages and thus more effectively attenuated the symptoms of sepsis in septic mice than exosomes released from MSCs alone and improved survival [86]. Likewise, hexyl 5-aminolevulinate hydrochloride (HAL)-loaded engineered cytokine-primed MSC-EVs (H@TI-EVs) with high expression of the immune checkpoint molecule programmed death-legend 1 (PD-L1) exert excellent inflammatory targeting and immunosuppressive effects on T1D imaging and therapy [87].

MSCs cultured under hypoxic conditions also exhibit altered secretion of extracellular vesicles. Compared to normoxic secreted extracellular vesicles, adipose-derived MSCs secrete more EVs in hypoxic conditions, which trigger different energy supplies and antiapoptotic, immunomodulatory, angiogenic, and antioxidant stress responses in renal tissues [88]. A recent study reported that pretreatment of MSCs with extracellular vesicles (Hypo-EVs) derived from hypoxia has improved therapeutic efficacy in repairing renal structure damage, restoring renal function, and reducing renal fibrosis. This effect is achieved by improving mitochondrial homeostasis and thus restoring CPT1A-mediated mitochondrial fatty acid oxidation, which reduces renal fibrosis [65]. Another study indicated that the culture of MSCs under hypoxic conditions is a promising in vitro preconditioning regimen that alters the protein cargo and increases the antioxidative potential of EVs [89]. In addition, increased levels of angiogenic extracellular vesicular proteins, including FGF2, PDGF, and epidermal growth factor (EGF), were found in EVs secreted by MSCs exposed to ischemia-mimicking conditions, where nuclear factor-κB (NF-κB) signaling was identified as a major mediator of EV-induced angiogenesis [90].

In addition to common preconditions involving inflammatory factors and hypoxic prestimulation, in recent years, EVs have been reported to prime specific parental cells to obtain greater efficacy. Exosomes derived from MSCs preconditioned with melatonin had a better protective effect against renal ischemia–reperfusion injury than did those derived from nonpreconditioned MSCs or from their exosomes [91]. Furthermore, platelet-rich plasma promotes the paracrine activity of MSC-derived exosomes to repair acute kidney injury via the AKT/Rab27 pathway [33]; interestingly, in nonrenal diseases, platelet preconditioning also enhances efficacy but by a different mechanism, in which platelets facilitate the wound-healing capability of mesenchymal stem cells by mitochondrial transfer and metabolic reprogramming [92].

It is worth noting, however, that this strategy of varying the culture conditions of the parental cells also has limitations, as sometimes the culture conditions alone may not be sufficient to address the limitations on the efficacy of extracellular vesicles [93] or because their therapeutic efficacy may not be sufficiently stable [94].

2.1.3 Exogenous Cargo Loading

Following extracellular vesicle biogenesis, the active contents of EVs can be enhanced by exogenous methods. To increase the levels of specific miRNAs known for their pro-regenerative properties in extracellular vesicles (EVs) for the treatment of AKI, mesenchymal stem cells (MSCs) were transiently transfected with miRNA mimics to enrich the EVs with the desired miRNAs [95]. In vitro and in vivo experiments demonstrated that engineered EVs retained their pro-regenerative effects and exhibited greater efficacy than EVs from naïve MSCs, even at lower doses, suggesting the potential use of a minimal amount of EVs for renal regeneration. Likewise, electroporation was used to generate transient membrane pores to load exogenous siRNA into purified EVs for delivery to the mouse brain against Alzheimer's disease [96]. However, there is also disagreement that loading siRNA into EVs by electroporation is accompanied by the formation of large amounts of siRNA aggregates, which may lead to an overestimation of the amount of siRNA actually loaded into EVs (the retention of siRNA is less than 0.05%), and the efficiency of electroporation is much lower than that previously described [97]. In addition, electroporation may interfere with the membrane structure of EVs required for fusion with the target cell membrane. Therefore, loading siRNA by electroporation also requires caution [98]. Coincubation is also an optimized strategy for enhancing the content of extracellular vesicles by exogenous loading. Didiot et al. [99] showed that hydrophobically modified small interfering RNAs were efficiently loaded into extracellular vesicles after coincubation without altering vesicle integrity. When extracellular vesicles containing oligonucleotide therapeutic agents were taken up by mouse primary cortical neurons, Huntington mRNA and protein silencing were promoted, thus effectively alleviating Huntington's disease.

Despite the conventional operations, such as electroporation and coincubation, there are several novel loading techniques, such as EXPLOR, a novel, optogenetically engineered exosome technology. We successfully delivered the exosomal suppressor inhibitor of NF-κB into B6 wild-type mice before/after kidney ischemia–reperfusion surgery, which resulted in an enhanced anti-inflammatory effect compared with that of a control exosome [100].

2.2 *Delivery Efficiency Enhancement Strategies*

2.2.1 Targeted Delivery

Extracellular vesicles are used as therapeutic agents and are mainly injected intravenously, with unavoidable losses during delivery to the target organ [19–22]. Thus, enhanced targeting of extracellular vesicles can more effectively utilize the therapeutic advantages of EVs. The utilization of targeted peptides for exosome modification represents a highly direct renal targeting strategy. This is exemplified in a novel RNA interference (RNAi) therapy in which Kim-1-targeted extracellular vesicles were loaded with therapeutic siRNAs, which demonstrated efficient homing to the injured kidney tubules and subsequent delivery of siRNAs to suppress the transcription factors P65 and Snai1, thereby alleviating inflammation and fibrosis in AKI [101]. Furthermore, P-selectin-binding peptide (PBP)-engineered extracellular vesicles (PBP-EVs) with imaging and therapeutic functions have a selective targeting propensity for injured kidneys, as does quantify P-selectin expression in kidneys by molecular imaging, providing spatiotemporal information for the early diagnosis of AKI, and displaying superior renoprotective functions in renal repair and inhibition of renal fibrosis, which reduces inflammatory infiltrates, improves reparative angiogenesis, and improves poor renal parenchymal repair [102].

In addition, the targeting peptides were fused to extracellular vesicle membrane proteins for target organ enrichment. In a pioneering study by Alvarez-Erviti et al., the targeting peptide was fused to an extracellular vesicular membrane protein, i.e., a chimeric protein consisting of a fusion of the neural-targeting rabies virus glycoprotein (RVG) to the EV membrane protein Lamp2b was expressed in parental cells. The produced EVs containing the targeting peptide facilitated brain-targeted delivery of siRNAs, which were subsequently used for Alzheimer's disease treatment in mice [96]. However, while they achieved neuronal targeting of extracellular vesicles by demonstrating the RVG peptide, they were unable to achieve muscle targeting by fusing Lamp2b to a muscle-specific peptide [96]. This finding suggested that this method of combining a targeting peptide with Lamp2b is not sufficiently stable, and other methods (e.g., engineering glycosylated chimeric proteins) are needed to stabilize the foreign body-targeting peptide [103]. In addition to Lamp proteins, other proteins genetically inherited on the membrane of extracellular vesicles can be fused to a targeting ligand or receptor to facilitate tissue-specific targeting of EVs. Cheng et al. [104] focused on glycosylphosphatidylinositols (GPIs) on EV membranes, anchored anti-EGFR nanoantibodies anchored to GPI into chimeric proteins and expressed them on EV membranes; the authors showed that EVs engineered to express EGFR and CD3 antibodies induced T cells to cross-link with EGFR-expressing breast cancer cells, thereby triggering effective antitumor immunity. In addition, there are also CD63 proteins on EV membranes that can be fused to targeting peptides; for example, Gao et al. [105] bound the CP05 peptide to CD63, after which the drug was loaded to deliver engineered EVs that target the muscles of dystrophin-deficient mdx mice and improve the recovery of the dystrophin phenotype.

2.2.2 Don't Eat me Signals

Previous studies have focused on engineering modified targeting peptides to increase the lifetime of EVs in addition to altering the targeting efficacy of EVs. Kamerkar et al. adopted this approach based on the finding that CD47-SIRPα binding inhibits phagocytosis by initiating "do not-eat-me" signaling; thus, the expression of CD47 on EVs (termed "Exosomes") on EVs prolongs their half-life [106].

2.3 Stability and Retention Improvement Strategies

The low stability and retention of extracellular vesicles in vivo greatly limit their efficacy [107]; moreover, free extracellular vesicles are rapidly cleared in vivo [16, 108]. Therefore, strategies that can improve the retention of extracellular vesicles in target organs are urgently needed, and a variety of bioscaffolds, including hydrogels, are being increasingly utilized to increase the retention time of extracellular vesicles [107].

Hydrogels are three-dimensional polymers that undergo physical or chemical cross-linking to form biocompatible scaffolds and exhibit a strong affinity for water, mimicking natural tissues [109]. Many studies have utilized hydrogels to improve the efficacy of EVs in treating renal disease [110–112]. Hydrogels containing RGD (Arg-Gly-Asp) peptides have been shown to enhance the binding of extracellular vesicles to injured kidney cells that express RGD-binding integrins on their surface [37]. In a mouse AKI model, extracellular vesicles that were entrapped within hydrogels and injected under the renal capsule were more effective than free extracellular vesicles at limiting tubular injury in the early stages of AKI and limiting fibrosis in the later stages. In addition, hydrogels based on hyaluronic acid [113, 114] and peptides [115] have been used to encapsulate EVs from different cellular sources to improve the retention time of EVs in vivo.

Recent studies have not been satisfactory at simply prolonging the retention of extracellular vesicles but have also explored the development of functional or responsive hydrogels. A matrix metalloproteinase-2 (MMP2)-sensitive self-assembling peptide (KMP2) hydrogel was developed for the local delivery of MSC-EVs. In vitro and in vivo experiments demonstrated that this system effectively increased the therapeutic efficacy of EVs by promoting angiogenesis, reducing inflammation, and inhibiting apoptosis and fibrosis in renal ischemia–reperfusion (I/R) injury models [116]. Another hydrogel-based dual-drug delivery platform was developed by combining a self-assembling peptide/heparin (SAP/Hep) hydrogel with TNF-α neutralizing antibody (anti-TNF-α) and hepatocyte growth factor (HGF), which demonstrated that treatment with the SAP drug (anti-TNF-α/HGF in the SAP hydrogel) effectively reduced renal injury by inhibiting inflammation, promoting cell proliferation, and reducing chronic fibrosis, surpassing the benefits of individual drug administration or SAP alone [117].

2.4 Summary

In summary, various strategies have been developed to enhance the therapeutic efficacy of extracellular vesicles: (1) increasing the content of critical miRNAs, proteins, and other components of extracellular vesicles to mediate tissue and organ repair; this can be achieved by endogenous measures of genetic engineering modifications, prestimulation of parental cells or three-dimensional cultures or exogenous means of electroporation and coincubation. (2) Tissue targeting of extracellular vesicles can also be enhanced to allow greater delivery to target organs to improve EV efficacy; this can be achieved through optimized strategies of fusing extracellular vesicle membrane proteins to target peptides. (3) Biomaterials for enhancing the in vivo stability and retention of extracellular vesicles can be developed to effectively improve the bioavailability of extracellular vesicles; this can be achieved by having optimized strategies to develop bioscaffolds, such as biocompatible and functional hydrogels, to improve the efficacy of EVs.

3 The Role of EVs as Delivery Systems in the Treatment of Renal Disease

3.1 EV Delivery of Nucleic Acids

EVs can transport microRNAs (miRNAs), circular RNAs (circRNAs), and long noncoding RNAs (lncRNAs) to corresponding target cells, mediate intercellular information exchange, and play important roles in the treatment of kidney-related diseases (Fig. 9.2).

3.1.1 EV Delivery of miRNAs

MicroRNAs (miRNAs) are a class of noncoding single-stranded RNA molecules encoded by endogenous genes of approximately 22 nucleotides in length that inhibit gene expression at the posttranscriptional level by binding to the 3′-untranslated region (3′-UTR) of target mRNAs. miRNAs play important roles in a variety of biological processes, such as the cell cycle, apoptosis, cell proliferation and differentiation. Many studies have shown that exosomes exert significant therapeutic effects on a variety of renal diseases by delivering miRNAs.

EV Delivery of miRNA for AKI Treatment

Mesenchymal stem cell-derived EVs improved renal function in ischemic AKI by (1) delivering miR-223-3p to directly target NLRP3, which inhibited the activation of inflammasomes and enhanced mitophagy [25], (2) delivering miR-148b-3p to significantly reduce endoplasmic reticulum stress [118]; and (3) delivering miR-125b-5p to activate the p53 pathway, which targeted cell cycle arrest and apoptosis in TECs [29]. In addition to MSC-EVs, miR-486-5p-enriched cord blood endothelial colony-forming cell-derived exosomes were also effective at alleviating ischemic AKI in mice [119].

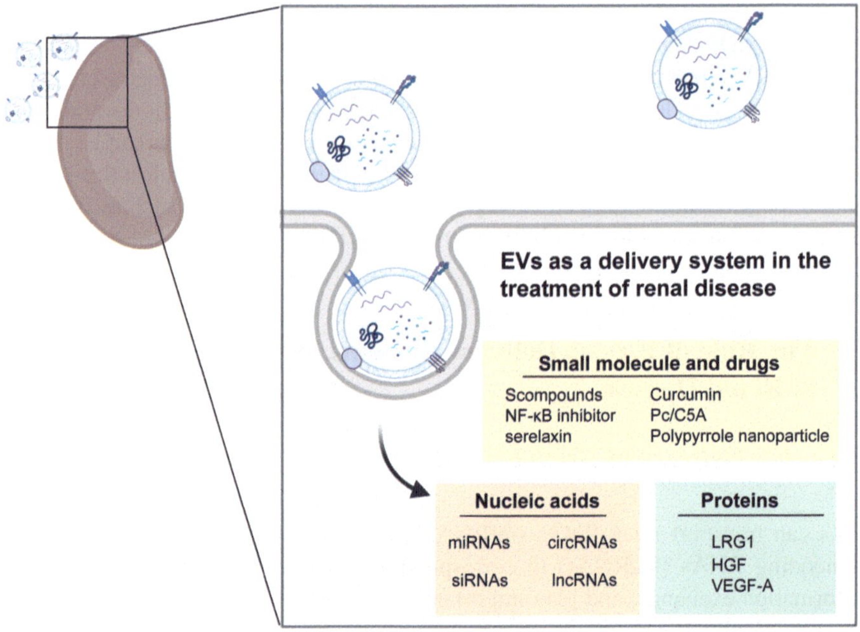

Fig. 9.2 EVs as a delivery system in the treatment of renal disease (produced via biorender)

In a lipopolysaccharide (LPS)-induced AKI model, MSC-EVs inhibited AKI-related inflammation and focal death by regulating miR-21-5p, which targets TLR4 and inhibits the NF-κB/NLRP3 pathway [120], and attenuated AKI-induced inflammation and focal death by delivering miR-223-3p, targeting HDAC2 and promoting SNRK transcription [121]. In a cisplatin-induced AKI model, urinary EVs from preterm infants inhibited HK-2-related apoptosis by targeting MAPK8 through the secretion of miR-30a-5p [122], and miR-1184 secreted by exosomes derived from mesenchymal stem cells attenuated renal injury by mediating IL-1β and TNF-α [123]. Endothelial progenitor cell-derived EVs can silence RUNX1 by releasing miR-21-5p to alleviate sepsis-induced AKI [124].

EV-delivered miRNAs may also cause damage to renal tissues. In a sepsis-induced acute kidney injury model, macrophage-isolated EVs directly regulate TXNIP by delivering miR-93-5p, which affects the thermal apoptosis of renal epithelial cells [125]. Macrophage-derived EVs exacerbate renal tubular injury by delivering miR-155, which targets the suppressor of cytokine signaling 1 (SOCS-1), [126].

EV Delivery of miRNA for CKD Treatment

In a model of unilateral ureteral obstruction (UUO)-induced renal fibrosis, MSC-EVs attenuated renal fibrosis in vitro and in vivo by downregulating Smad5 through the delivery of miR-186-5p [127] or inhibiting the expression of phosphofructokinase muscle isoforms (PFKMs) through the secretion of miR-21a-5p, which suppressed glycolysis in TECs and significantly ameliorated renal fibrosis [46]; modulating mTOR signaling and autophagy through the delivery of miRNA-122a [44]; or attenuating renal tubular epithelial cell injury via miR-874-3p [30]. Interestingly, compared with untrained vesicles, bone marrow MSC vesicles from trained animals promoted angiogenesis through the secretion of miR-296, which in turn exerted a renoprotective effect [47]. Renal tubular epithelial cell exosomes can accelerate the development of renal fibrosis by secreting miR-21 and activating the PTEN/Akt pathway [128]. MSC-EVs also ameliorate renal interstitial fibrosis by delivering miR-34c-5p [50], miR-335-5p [129] or miR-181d [130]. Furthermore, EV-encapsulated miR-26a attenuates aldosterone-induced renal fibrosis by directly binding to its 3′-UTR and inhibiting the activation of SMAD3 [78].

MSC-derived EVs can play an important role in the treatment of diabetic nephropathy by carrying miR-125a [62], miR-181, miR-29b, Let-7b [131], miR-374a-5p [132], miR-22-3p [133], miR-424-5p [134], miR-30e-5p [52], miR-146a-5p [57], miR-15b-5p [61], and miR-125b [51] to alleviate renal injury. miRNA-93-5p in exosomes derived from M2 macrophages improves lipopolysaccharide-induced podocyte apoptosis by targeting Toll-like receptor 4 [135], and satellite cell-derived EV-mediated delivery of the microRNA-23a/27a/26a cluster ameliorates renal tubulointerstitial fibrosis in mice with diabetic nephropathy [136].

EV-delivered miRNAs can also cause damage to renal tissues. Exosomal miR-7002-5p derived from high glucose-induced macrophages suppresses autophagy in tubular epithelial cells by targeting Atg9b [137], and proximal tubule-derived exosomes contribute to mesangial cell injury in diabetic nephropathy via miR-92a-1-5p transfer [138].

EV Delivery of miRNA for Other Kidney Diseases

Circulating EV miR-1-3p from myocardial infarction rats has a protective effect on contrast-induced nephropathy via targeting ATG13 and activating the AKT signaling pathway in contrast-induced nephropathy [139]. EVs secreted by liver stem cells combined with the antitumor miRNA miR-145 could reduce the invasiveness of renal cancer more effectively than other EVs [140]. EVs from BMMSCs alleviate systemic lupus erythematosus nephritis in lupus mice by delivering miR-16 and miR-21, inducing Treg cell recruitment and promoting macrophage anti-inflammatory polarization [69].

3.1.2 EV Delivery of circRNA

Circular RNAs (circRNAs) are a special class of noncoding RNA molecules with a closed loop structure that are not affected by RNA exonucleases, are more stable in expression and are less prone to degradation. Recent studies have shown that circRNA molecules are rich in microRNA (miRNA) binding sites, which act as miRNA sponges in cells, thus enhancing the inhibitory effect of miRNAs on their target genes and elevating the expression level of target genes; this mechanism of action is known as the competitive endogenous RNA (ceRNA) mechanism. By interacting with disease-associated miRNAs, circRNAs play an important regulatory role in diseases.

circRNA is a key regulator of therapeutic progression in kidney disease. Exosomal hsa_circ_0125310 promotes cell proliferation and fibrosis by sponging miR-422a and targeting the IGF1R/p38 axis [141], and Circ_0008529 contributes to renal tubular cell dysfunction in high-glucose stress via the miR-185-5p/SMAD2 pathway in diabetic nephropathy [142]. The urinary stem cell-derived exocrine circRNA ATG7 regulates the SOCS1/STAT3 signaling pathway through miR-4500, inhibits M1 macrophage polarization, and alleviates the progression of diabetes nephropathy [143]. Interestingly, exosome-derived circ_0001806 was found to exacerbate LPS-induced HK2 cell injury by targeting miR-942-5p to regulate TXNIP expression [144]. Similarly, it was also found that exosome-derived circ_0001818 could activate TXNIP expression by targeting miR-136–5p, leading to LPS-induced HK2 cell injury [145].

3.1.3 EV Delivery of lncRNA or siRNA

Long noncoding RNAs (lncRNAs) are noncoding RNAs with a length of more than 200 nucleotides that are involved in a variety of intracellular signaling pathways. BM-MSC-EVs transfer the functional lncRNA Loc108349490 into dendritic cells, which in turn alleviates acute rejection after renal allograft transplantation [70]. The eTUG1 lncRNA derived from human urine-derived stem cells attenuates renal ischemia/reperfusion injury by interacting with SRSF1 to regulate ASCL4-mediated ferroptosis [146]. Four lncRNAs (LINC01015, LINC01986, AC087257.1, and AC022596.1) and two miRNAs (miR-16-5p and miR-101-3p) were found to treat systemic lupus erythematosus renal tissue by targeting key pathways associated with inflammation, fibrosis, epithelial–mesenchymal transition, and actin cytoskeleton injury [147]. EVs from MSCs can effectively target Gli1+ cells, the progenitors of smooth muscle cells, by delivering lncRNA-ANCR, which promotes the rapid reconstruction of smooth muscle cells, significantly inhibits the differentiation of $Gli1^+$ cells to osteoblast-like cells, and ameliorates vascular calcification caused by chronic kidney disease [148].

Small interfering RNA (siRNA) binds to target-specific mRNAs and mediates mRNA degradation through the RNA-induced silencing complex (RISC). EVs can attenuate renal injury by delivering siRNAs. An extracellular vesicle derived from erythrocytes targets Kim-1 in the kidneys of acutely injured mice and delivers siRNAs targeting the transcription factors P65 and Snai1, thereby attenuating tubular inflammation and fibrosis [101].

3.2 EV Delivery of Drugs/Monomers or Small Molecule Compounds

EV-encapsulated drugs or small molecule compounds have also been found to play important protective roles in kidney disease. A novel optogenetically engineered exosome technology, "EXPLOR," was utilized to deliver nuclear factor-κB inhibitor exosomes to mice before and after renal ischemia–reperfusion. These exosomes downregulated renal nuclear factor-κB signaling and ameliorated acute kidney injury [100]. A porous pneumatic microextrusion (PME) composite scaffold consisting of poly (lactic-co-glycolic acid) (PLGA, P), magnesium hydroxide (MH, M), and decellularized porcine kidney extracellular matrix (kECM, E) is functionalized with bioactive compounds, polydeoxyribonucleotide (PDRN), and tumor necrosis factor-α (TNF-α)/interferon-γ (IFN-γ)-primed mesenchymal stem-cell-derived extracellular vesicles (TI-EVs) to improve the regeneration and maintenance of functional kidney tissue [149]. The macrocyclic amphiphile C5A, the commercial dye sulfonated aluminum phthalocyanine (Pc), and extracellular vesicles secreted by mesenchymal stem cells (MSC-EVs) were prepared into a unique nanoscale hypoxia-sensitive coassembly (Pc/C5A@EVs) by molecular recognition and self-assembly techniques. The results showed that Pc/C5A coassembly could exert regenerative effects by inhibiting the expression of HIF-1α and the downstream NF-κB signaling pathway in hypoxic renal tubular epithelial cells (TECs) and inducing the transformation of M1-M2 macrophages in a unilateral or bilateral ischemia/reperfusion injury mouse model; additionally, Pc/C5A coassembly significantly restored renal function by decreasing apoptosis, inhibiting the progression of renal fibrosis, and attenuating renal tubulointerstitial inflammation, thus significantly restoring renal function [150]. Tubular epithelial cell-derived exosomes containing CD26 protect mice against renal ischemia/reperfusion injury by maintaining proliferation and dissipating inflammation [151]. Polypyrrole nanoparticles (PpyNps) containing liposomes encapsulated in exosomes secreted by bone marrow mesenchymal stem cells were found to ameliorate diabetic peripheral neuropathy under electrical stimulation [55]. The combination of the antifibrotic drug serelaxin (RLX) with bone marrow mesenchymal stem cells and their exosomes enhanced renoprotection in preclinical CKD models [152]. Transcript multifunctional FOXP3 gene-engineered extracellular vesicles attenuate the dynamics of the allogeneic response in kidney transplantation [153]. Exosomes derived from renal tubular epithelial cells promote peritubular capillary repair in ischemic kidney injury by delivering renal tubular vascular endothelial growth factor-A (VEGF-A) [154]. Additionally, human adult renal progenitor cells prevent cisplatin nephrotoxicity by inducing CYP1B1 overexpression and miR-27b-3p downregulation through extracellular vesicles [155].

3.3 EV Delivery Protein/Antibodies

EV-encapsulated proteins or antibodies have also been found to play an important protective role in kidney disease. Surface albumin-binding peptide (ABP)-displayed EVs (ABP-EVs) were produced by transfecting parent cells with the ABP-Lamp2b fusion plasmid. Compared with unmodified EVs (NC-EVs), small molecule drugs (e.g., curcumin) or proteins (e.g., hepatocyte growth factor) delivered by ABP-EVs have superior therapeutic (e.g., antiapoptotic, antioxidant, and anti-inflammatory) effects both in vitro and in vivo [156]. In addition to playing a protective role, proteins or antibodies secreted by exosomes may also cause damage to the kidneys. β-catenin in renal tubular cells induces exosome secretion. Osteoprotegerin (OPN) is encapsulated in renal tubular cell-derived exosomes controlled by β-catenin and subsequently delivered to fibroblasts. By binding to CD44, exosomal OPN promotes the proliferation and activation of fibroblasts and accelerates renal fibrosis [157]. Leucine alpha-2-glycoprotein 1 (LRG1)-rich EVs activate macrophages through a TGFβR1-dependent process, creating a negative feedback loop between renal tubular epithelial cells and macrophages that promotes renal inflammation and apoptosis in diabetic nephropathy [158], and macrophage-derived, LRG1-enriched extracellular vesicles exacerbate aristolochic acid nephropathy in a TGFβR1-dependent manner [159].

4 Conclusion

Stem cell-derived extracellular vesicles (EVs) have great potential as treatments for kidney diseases. With the increasing incidence of kidney disease, the need for new drug development has not been urgent. As a novel personalized nanomedicine, EVs have the potential to change the trajectory of kidney disease. While the fabrication and engineering of EVs are indeed complex and challenging, they are not insurmountable. The development of EV products will evolve as the understanding of the nature of disease deepens and methods for producing, characterizing, and functionalizing EVs continue to be refined.

As the functions and correlates carried out in the regulation or pathologic intercellular communication of EVs have been revealed, they have been actively explored as therapeutics, tools and targets for a range of diseases. Engineering approaches have evolved in the boom of clinically translational EV research, seeking to fully utilize the overall therapeutic potential of EVs. In particular, various strategies have been proposed to improve overall therapeutic efficacy, including enriching the desired biomolecules with EVs, efficiently loading EVs with exogenous vectors, increasing their accumulation at preferred sites of action, and controlling their pathological effects. These engineering approaches offer new possibilities for the therapeutic application of EVs as well as associated challenges and future research directions.

However, the challenges are equally great. Most relevant clinical trials have not yet been completed. We believe that the advances in engineering approaches presented in this review will help more therapeutic applications of engineered EVs reach the clinical trial stage. However, other challenges need to be addressed on many fronts, ranging from explicit methods for clearly defining, isolating, and characterizing EV subtypes to optimized methods for clearly defining, identifying, and reducing EV doses in clinical trials. Therefore, there is an urgent need for interdisciplinary collaboration aimed at enabling EV-based therapeutic applications in the clinic.

The above description confirms the potential application of stem cell-derived extracellular vesicles in the treatment of renal diseases and emphasizes the importance of an engineering approach that will help advance the field and provide insights for the development of additional EV therapeutic applications in the future.

Declaration of Competing Interests The authors declare the following competing interests.

Acknowledgments The study was supported by the following grants: National Natural Science Foundation of China (82200764, 62250001, 82303288), Young Elite Scientists Sponsorship Program by CAST (2022QNRC001), Sichuan Science and Technology Program (2022YFS0328, 2023YFG0211), China Postdoctoral Science Foundation (2022 M722276, 2023 M733986 and 2023 T160741), Postdoctoral Research Fund from West China Hospital of Sichuan University (2023HXBH092), 1.3.5 project for disciplines of excellence from West China Hospital of Sichuan University (ZYGD2023015), and Beijing Natural Science Foundation (L232122).

References

1. Witwer KW, Van Balkom BWM, Bruno S, Choo A, Dominici M, Gimona M, et al. Defining mesenchymal stromal cell (MSC)-derived small extracellular vesicles for therapeutic applications. J Extracell Vesicles. 2019;8(1):1609206. https://doi.org/10.1080/20013078.2019.1609206.
2. Armstrong JP, Holme MN, Stevens MM. Re-engineering extracellular vesicles as smart nanoscale therapeutics. ACS Nano. 2017;11(1):69–83. https://doi.org/10.1021/acsnano.6b07607.
3. Zhang Y, Xu J, Liu S, Lim M, Zhao S, Cui K, et al. Embryonic stem cell-derived extracellular vesicles enhance the therapeutic effect of mesenchymal stem cells. Theranostics. 2019;9(23):6976–90. https://doi.org/10.7150/thno.35305.
4. Wang M, Wang C, Chen M, Xi Y, Cheng W, Mao C, et al. Efficient angiogenesis-based diabetic wound healing/skin reconstruction through bioactive antibacterial adhesive ultraviolet shielding Nanodressing with exosome release. ACS Nano. 2019;13(9):10279–93. https://doi.org/10.1021/acsnano.9b03656.

5. Grange C, Bussolati B. Extracellular vesicles in kidney disease. Nat Rev Nephrol. 2022;18(8):499–513. https://doi.org/10.1038/s41581-022-00586-9.
6. Nargesi AA, Lerman LO, Eirin A. Mesenchymal stem cell-derived extracellular vesicles for kidney repair: current status and looming challenges. Stem Cell Res Ther. 2017;8:8. https://doi.org/10.1186/s13287-017-0727-7.
7. Zhang S, Teo KYW, Chuah SJ, Lai RC, Lim SK, Toh WS. MSC exosomes alleviate temporomandibular joint osteoarthritis by attenuating inflammation and restoring matrix homeostasis. Biomaterials. 2019;200:35–47. https://doi.org/10.1016/j.biomaterials.2019.02.006.
8. Wu J, Kuang L, Chen C, Yang J, Zeng W-N, Li T, et al. miR-100-5p-abundant exosomes derived from infrapatellar fat pad MSCs protect articular cartilage and ameliorate gait abnormalities via inhibition of mTOR in osteoarthritis. Biomaterials. 2019;206(undefined):87–100.
9. Wei Y, Wu Y, Zhao R, Zhang K, Midgley AC, Kong D, et al. MSC-derived sEVs enhance patency and inhibit calcification of synthetic vascular grafts by immunomodulation in a rat model of hyperlipidemia. Biomaterials. 2019;204(undefined):13–24.
10. Liao Z, Luo R, Li G, Song Y, Zhan S, Zhao K, et al. Exosomes from mesenchymal stem cells modulate endoplasmic reticulum stress to protect against nucleus pulposus cell death and ameliorate intervertebral disc degeneration in vivo. Theranostics. 2019;9(14):4084–100. https://doi.org/10.7150/thno.33638.
11. Bei Y, Das S, Rodosthenous RS, Holvoet P, Vanhaverbeke M, Monteiro MC, et al. Extracellular vesicles in cardiovascular Theranostics. Theranostics. 2017;7(17):4168–82. https://doi.org/10.7150/thno.21274.
12. Mihajlovic M, Wever KE, van der Made TK, de Vries RBM, Hilbrands LB, Masereeuw R. Are cell-based therapies for kidney disease safe? A systematic review of preclinical evidence. Pharmacol Ther. 2019;197:191–211. https://doi.org/10.1016/j.pharmthera.2019.01.004.
13. Dickhut A, Pelttari K, Janicki P, Wagner W, Eckstein V, Egermann M, et al. Calcification or dedifferentiation: requirement to lock mesenchymal stem cells in a desired differentiation stage. J Cell Physiol. 2009;219(1):219–26. https://doi.org/10.1002/jcp.21673.
14. Savkovic V, Li H, Seon JK, Hacker M, Franz S, Simon JC. Mesenchymal stem cells in cartilage regeneration. Curr Stem Cell Res Ther. 2014;9(6):469–88.
15. Karpman D, Stahl AL, Arvidsson I. Extracellular vesicles in renal disease. Nat Rev Nephrol. 2017;13(9):545–62. https://doi.org/10.1038/nrneph.2017.98.
16. Liu Y, Cui J, Wang H, Hezam K, Zhao X, Huang H, et al. Enhanced therapeutic effects of MSC-derived extracellular vesicles with an injectable collagen matrix for experimental acute kidney injury treatment. Stem Cell Res Ther. 2020;11(1):161. https://doi.org/10.1186/s13287-020-01668-w.
17. Zhang K, Chen S, Sun H, Wang L, Li H, Zhao J, et al. In vivo two-photon microscopy reveals the contribution of Sox9+ cell to kidney regeneration in a mouse model with extracellular vesicle treatment. J Biol Chem. 2020;295:12203. https://doi.org/10.1074/jbc.RA120.012732.
18. Zhang G, Wang D, Miao S, Zou X, Liu G, Zhu Y. Extracellular vesicles derived from mesenchymal stromal cells may possess increased therapeutic potential for acute kidney injury compared with conditioned medium in rodent models: a meta-analysis. Exp Ther Med. 2016;11(4):1519–25. https://doi.org/10.3892/etm.2016.3076.
19. Zou X, Gu D, Xing X, Cheng Z, Gong D, Zhang G, et al. Human mesenchymal stromal cell-derived extracellular vesicles alleviate renal ischemic reperfusion injury and enhance angiogenesis in rats. Am J Transl Res. 2016;8(10):4289–99.
20. Zhang G, Zou X, Huang Y, Wang F, Miao S, Liu G, et al. Mesenchymal stromal cell-derived extracellular vesicles protect against acute kidney injury through anti-oxidation by enhancing Nrf2/ARE activation in rats. Kidney Blood Press Res. 2016;41(2):119–28. https://doi.org/10.1159/000443413.
21. Zou X, Zhang G, Cheng Z, Yin D, Du T, Ju G, et al. Microvesicles derived from human Wharton's jelly mesenchymal stromal cells ameliorate renal ischemia-reperfusion injury in rats by suppressing CX3CL1. Stem Cell Res Ther. 2014;5(2):40. https://doi.org/10.1186/scrt428.

22. Gatti S, Bruno S, Deregibus MC, Sordi A, Cantaluppi V, Tetta C, et al. Microvesicles derived from human adult mesenchymal stem cells protect against ischaemia-reperfusion-induced acute and chronic kidney injury. Nephrol Dial Transplant. 2011;26(5):1474–83. https://doi.org/10.1093/ndt/gfr015.
23. Guo C, Dong G, Liang X, Dong Z. Epigenetic regulation in AKI and kidney repair: mechanisms and therapeutic implications. Nat Rev Nephrol. 2019;15(4):220–39. https://doi.org/10.1038/s41581-018-0103-6.
24. Xie X, Yang X, Wu J, Tang S, Yang L, Fei X, et al. Exosome from indoleamine 2,3-dioxygenase-overexpressing bone marrow mesenchymal stem cells accelerates repair process of ischemia/reperfusion-induced acute kidney injury by regulating macrophages polarization. Stem Cell Res Ther. 2022;13(1):367. https://doi.org/10.1186/s13287-022-03075-9.
25. Sun Z, Gao Z, Wu J, Zheng X, Jing S, Wang W. MSC-derived extracellular vesicles activate Mitophagy to alleviate renal ischemia/reperfusion injury via the miR-223-3p/NLRP3 Axis. Stem Cells Int. 2022;2022:6852661. https://doi.org/10.1155/2022/6852661.
26. Zhao M, Liu S, Wang C, Wang Y, Wan M, Liu F, et al. Mesenchymal stem cell-derived extracellular vesicles attenuate mitochondrial damage and inflammation by stabilizing mitochondrial DNA. ACS Nano. 2021;15(1):1519–38. https://doi.org/10.1021/acsnano.0c08947.
27. Niu R, Pan P, Li C, Luo B, Ma H, Hao H, et al. Bone mesenchymal stromal cell-derived small extracellular vesicles inhibit inflammation and ameliorate sepsis via delivery of microRNA-21a-5p. Cytotherapy. 2023;25(6):625–39. https://doi.org/10.1016/j.jcyt.2023.02.002.
28. Ullah M, Liu DD, Rai S, Razavi M, Concepcion W, Thakor AS. Pulsed focused ultrasound enhances the therapeutic effect of mesenchymal stromal cell-derived extracellular vesicles in acute kidney injury. Stem Cell Res Ther. 2020;11(1):398. https://doi.org/10.1186/s13287-020-01922-1.
29. Cao JY, Wang B, Tang TT, Wen Y, Li ZL, Feng ST, et al. Exosomal miR-125b-5p deriving from mesenchymal stem cells promotes tubular repair by suppression of p53 in ischemic acute kidney injury. Theranostics. 2021;11(11):5248–66. https://doi.org/10.7150/thno.54550.
30. Yu Y, Chen M, Guo Q, Shen L, Liu X, Pan J, et al. Human umbilical cord mesenchymal stem cell exosome-derived miR-874-3p targeting RIPK1/PGAM5 attenuates kidney tubular epithelial cell damage. Cell Mol Biol Lett. 2023;28(1):12. https://doi.org/10.1186/s11658-023-00425-0.
31. Zhang Y, Wang C, Bai Z, Li P. Umbilical cord mesenchymal stem cell exosomes alleviate the progression of kidney failure by modulating inflammatory responses and oxidative stress in an ischemia-reperfusion mice model. J Biomed Nanotechnol. 2021;17(9):1874–81. https://doi.org/10.1166/jbn.2021.3155.
32. Huang J, Cao H, Cui B, Ma X, Gao L, Yu C, et al. Mesenchymal stem cells-derived exosomes ameliorate ischemia/reperfusion induced acute kidney injury in a porcine model. Front Cell Dev Biol. 2022;10:899869. https://doi.org/10.3389/fcell.2022.899869.
33. Ji C, Zhang J, Zhou Z, Shi H, Liu W, Sun F, et al. Platelet-rich plasma promotes MSCs exosomes paracrine to repair acute kidney injury via AKT/Rab27 pathway. Am J Transl Res. 2021;13(3):1445–57.
34. Zhang W, Zhang J, Huang H. Exosomes from adipose-derived stem cells inhibit inflammation and oxidative stress in LPS-acute kidney injury. Exp Cell Res. 2022;420(1):113332. https://doi.org/10.1016/j.yexcr.2022.113332.
35. Lopes JA, Collino F, Rodrigues-Ferreira C, Sampaio LDS, Costa-Sarmento G, Wendt CHC, et al. Early effects of extracellular vesicles secreted by adipose tissue mesenchymal cells in renal ischemia followed by reperfusion: mechanisms rely on a decrease in mitochondrial anion superoxide production. Int J Mol Sci. 2022;23(6):2906. https://doi.org/10.3390/ijms23062906.
36. Cao H, Cheng Y, Gao H, Zhuang J, Zhang W, Bian Q, et al. In vivo tracking of mesenchymal stem cell-derived extracellular vesicles improving mitochondrial function in renal

ischemia-reperfusion injury. ACS Nano. 2020;14(4):4014–26. https://doi.org/10.1021/acsnano.9b08207.
37. Zhang C, Shang Y, Chen X, Midgley AC, Wang Z, Zhu D, et al. Supramolecular nanofibers containing arginine-glycine-aspartate (RGD) peptides boost therapeutic efficacy of extracellular vesicles in kidney repair. ACS Nano. 2020;14(9):12133–47. https://doi.org/10.1021/acsnano.0c05681.
38. Yu L, Liu S, Wang C, Zhang C, Wen Y, Zhang K, et al. Embryonic stem cell-derived extracellular vesicles promote the recovery of kidney injury. Stem Cell Res Ther. 2021;12(1):379. https://doi.org/10.1186/s13287-021-02460-0.
39. Yuan X, Li D, Chen X, Han C, Xu L, Huang T, et al. Extracellular vesicles from human-induced pluripotent stem cell-derived mesenchymal stromal cells (hiPSC-MSCs) protect against renal ischemia/reperfusion injury via delivering specificity protein (SP1) and transcriptional activating of sphingosine kinase 1 and inhibiting necroptosis. Cell Death Dis. 2017;8(12):3200. https://doi.org/10.1038/s41419-017-0041-4.
40. Lim SW, Kim KW, Kim BM, Shin YJ, Luo K, Quan Y, et al. Alleviation of renal ischemia/reperfusion injury by exosomes from induced pluripotent stem cell-derived mesenchymal stem cells. Korean J Intern Med. 2022;37(2):411–24. https://doi.org/10.3904/kjim.2020.438.
41. Collino F, Lopes JA, Tapparo M, Tortelote GG, Kasai-Brunswick TH, Lopes GMC, et al. Extracellular vesicles derived from induced pluripotent stem cells promote Renoprotection in acute kidney injury model. Cells. 2020;9(2):453. https://doi.org/10.3390/cells9020453.
42. Jin J, Qian F, Zheng D, He W, Gong J, He Q. Mesenchymal stem cells attenuate renal fibrosis via exosomes-mediated delivery of microRNA let-7i-5p Antagomir. Int J Nanomedicine. 2021;16:3565–78. https://doi.org/10.2147/ijn.S299969.
43. Cao D, Wang Y, Zhang Y, Zhang Y, Huang Q, Yin Z, et al. Regulation of connective tissue growth factor expression by miR-133b for the treatment of renal interstitial fibrosis in aged mice with unilateral ureteral obstruction. Stem Cell Res Ther. 2021;12(1):171. https://doi.org/10.1186/s13287-021-02210-2.
44. Li D, Qu J, Yuan X, Zhuang S, Wu H, Chen R, et al. Mesenchymal stem cells alleviate renal fibrosis and inhibit autophagy via exosome transfer of miRNA-122a. Stem Cells Int. 2022;2022:1981798. https://doi.org/10.1155/2022/1981798.
45. Lu Y, Yang L, Chen X, Liu J, Nie A, Chen X. Bone marrow mesenchymal stem cell-derived exosomes improve renal fibrosis by reducing the polarisation of M1 and M2 macrophages through the activation of EP2 receptors. IET Nanobiotechnol. 2022;16(1):14–24. https://doi.org/10.1049/nbt2.12071.
46. Xu S, Cheuk YC, Jia Y, Chen T, Chen J, Luo Y, et al. Bone marrow mesenchymal stem cell-derived exosomal miR-21a-5p alleviates renal fibrosis by attenuating glycolysis by targeting PFKM. Cell Death Dis. 2022;13(10):876. https://doi.org/10.1038/s41419-022-05305-7.
47. Luiz RDS, Rampaso RR, Dos Santos AAC, Convento MB, Barbosa DA, da Fonseca CD, et al. BM-MSC-derived small extracellular vesicles (sEV) from trained animals presented nephroprotective potential in unilateralureteral obstruction model. J Venom Anim Toxins Incl Trop Dis. 2021;27:e20200187. https://doi.org/10.1590/1678-9199-jvatitd-2020-0187.
48. Liu Y, Guo W, Guo Y, Chen X, Liu W. Bone marrow mesenchymal stem cell-derived exosomes improve renal fibrosis via regulating Smurf 2/Smad 7. Front Biosci (Landmark Ed). 2022;27(1):17. https://doi.org/10.31083/j.fbl2701017.
49. Gu H, Li J, Ni Y. Sinomenine improves renal fibrosis by regulating mesenchymal stem cell-derived exosomes and affecting autophagy levels. Environ Toxicol. 2023;38(10):2524–37. https://doi.org/10.1002/tox.23890.
50. Hu X, Shen N, Liu A, Wang W, Zhang L, Sui Z, et al. Bone marrow mesenchymal stem cell-derived exosomal miR-34c-5p ameliorates RIF by inhibiting the core fucosylation of multiple proteins. Mol Ther. 2022;30(2):763–81. https://doi.org/10.1016/j.ymthe.2021.10.012.
51. Cai X, Zou F, Xuan R, Lai XY. Exosomes from mesenchymal stem cells expressing microribonucleic acid-125b inhibit the progression of diabetic nephropathy via the tumour necrosis

factor receptor-associated factor 6/Akt axis. Endocr J. 2021;68(7):817–28. https://doi.org/10.1507/endocrj.EJ20-0619.
52. Lv J, Hao YN, Wang XP, Lu WH, Xie LY, Niu D. Bone marrow mesenchymal stem cell-derived exosomal miR-30e-5p ameliorates high-glucose induced renal proximal tubular cell pyroptosis by inhibiting ELAVL1. Ren Fail. 2023;45(1):2177082. https://doi.org/10.1080/0886022x.2023.2177082.
53. Wang S, Bao L, Fu W, Deng L, Ran J. Protective effect of exosomes derived from bone marrow mesenchymal stem cells on rats with diabetic nephropathy and its possible mechanism. Am J Transl Res. 2021;13(6):6423–30.
54. Wei W, Guo X, Gu L, Jia J, Yang M, Yuan W, et al. Bone marrow mesenchymal stem cell exosomes suppress phosphate-induced aortic calcification via SIRT6-HMGB1 deacetylation. Stem Cell Res Ther. 2021;12(1):235. https://doi.org/10.1186/s13287-021-02307-8.
55. Singh A, Raghav A, Shiekh PA, Kumar A. Transplantation of engineered exosomes derived from bone marrow mesenchymal stromal cells ameliorate diabetic peripheral neuropathy under electrical stimulation. Bioact Mater. 2021;6(8):2231–49. https://doi.org/10.1016/j.bioactmat.2021.01.008.
56. Alasmari WA, El-Shetry ES, Ibrahim D, ElSawy NA, Eldoumani H, Metwally AS, et al. Mesenchymal stem-cells' exosomes are renoprotective in postmenopausal chronic kidney injury via reducing inflammation and degeneration. Free Radic Biol Med. 2022;182:150–9. https://doi.org/10.1016/j.freeradbiomed.2022.02.024.
57. Zhang Y, Le X, Zheng S, Zhang K, He J, Liu M, et al. MicroRNA-146a-5p-modified human umbilical cord mesenchymal stem cells enhance protection against diabetic nephropathy in rats through facilitating M2 macrophage polarization. Stem Cell Res Ther. 2022;13(1):171. https://doi.org/10.1186/s13287-022-02855-7.
58. Wang Z, Yu Y, Jin L, Tan X, Liu B, Zhang Z, et al. HucMSC exosomes attenuate partial bladder outlet obstruction-induced renal injury and cell proliferation via the Wnt/β-catenin pathway. Eur J Pharmacol. 2023;952:175523. https://doi.org/10.1016/j.ejphar.2023.175523.
59. Zhong L, Liao G, Wang X, Li L, Zhang J, Chen Y, et al. Mesenchymal stem cells-microvesicle-miR-451a ameliorate early diabetic kidney injury by negative regulation of P15 and P19. Exp Biol Med (Maywood). 2018;243(15–16):1233–42. https://doi.org/10.1177/1535370218819726.
60. Ji C, Zhang J, Zhu Y, Shi H, Yin S, Sun F, et al. Exosomes derived from hucMSC attenuate renal fibrosis through CK1δ/β-TRCP-mediated YAP degradation. Cell Death Dis. 2020;11(5):327. https://doi.org/10.1038/s41419-020-2510-4.
61. Zhao T, Jin Q, Kong L, Zhang D, Teng Y, Lin L, et al. microRNA-15b-5p shuttled by mesenchymal stem cell-derived extracellular vesicles protects podocytes from diabetic nephropathy via downregulation of VEGF/PDK4 axis. J Bioenerg Biomembr. 2022;54(1):17–30. https://doi.org/10.1007/s10863-021-09919-y.
62. Hao Y, Miao J, Liu W, Cai K, Huang X, Peng L. Mesenchymal stem cell-derived exosomes carry MicroRNA-125a to protect against diabetic nephropathy by targeting histone deacetylase 1 and downregulating Endothelin-1. Diabetes Metab Syndr Obes. 2021;14:1405–18. https://doi.org/10.2147/dmso.S286191.
63. Ahrabi B, Abbaszadeh HA, Piryaei A, Shekari F, Ahmady Roozbahany N, Rouhollahi M, et al. Autophagy-induced mesenchymal stem cell-derived extracellular vesicles ameliorated renal fibrosis in an in vitro model. Bioimpacts. 2023;13(5):359–72. https://doi.org/10.34172/bi.2022.24256.
64. Yea JH, Yoon YM, Lee JH, Yun CW, Lee SH. Exosomes isolated from melatonin-stimulated mesenchymal stem cells improve kidney function by regulating inflammation and fibrosis in a chronic kidney disease mouse model. J Tissue Eng. 2021;12:20417314211059624. https://doi.org/10.1177/20417314211059624.
65. Gao Z, Zhang C, Peng F, Chen Q, Zhao Y, Chen L, et al. Hypoxic mesenchymal stem cell-derived extracellular vesicles ameliorate renal fibrosis after ischemia-reperfusion injure by

restoring CPT1A mediated fatty acid oxidation. Stem Cell Res Ther. 2022;13(1):191. https://doi.org/10.1186/s13287-022-02861-9.
66. Liao CM, Luo T, von der Ohe J, de Juan MB, Schmitt R, Hass R. Human MSC-derived exosomes reduce cellular senescence in renal epithelial cells. Int J Mol Sci. 2021;22(24):13562. https://doi.org/10.3390/ijms222413562.
67. Lei Q, Gao F, Liu T, Ren W, Chen L, Cao Y, et al. Extracellular vesicles deposit PCNA to rejuvenate aged bone marrow-derived mesenchymal stem cells and slow age-related degeneration. Sci Transl Med. 2021;13(578):eaaz8697. https://doi.org/10.1126/scitranslmed.aaz8697.
68. Sanz-Ros J, Romero-García N, Mas-Bargues C, Monleón D, Gordevicius J, Brooke RT, et al. Small extracellular vesicles from young adipose-derived stem cells prevent frailty, improve health span, and decrease epigenetic age in old mice. Sci Adv. 2022;8(42):eabq2226. https://doi.org/10.1126/sciadv.abq2226.
69. Zhang M, Johnson-Stephenson TK, Wang W, Wang Y, Li J, Li L, et al. Mesenchymal stem cell-derived exosome-educated macrophages alleviate systemic lupus erythematosus by promoting efferocytosis and recruitment of IL-17(+) regulatory T cell. Stem Cell Res Ther. 2022;13(1):484. https://doi.org/10.1186/s13287-022-03174-7.
70. Wang ZG, Xu HE, Cheng FM, Zhang J, Feng YH, Liu DH, et al. Donor BMSC-derived small extracellular vesicles relieve acute rejection post-renal allograft through transmitting Loc108349490 to dendritic cells. Aging Cell. 2021;20(10):e13461. https://doi.org/10.1111/acel.13461.
71. Teng F, Fussenegger M. Shedding light on extracellular vesicle biogenesis and bioengineering. Adv Sci. 2020;8(1):2003505. https://doi.org/10.1002/advs.202003505.
72. Khayambashi P, Iyer J, Pillai S, Upadhyay A, Zhang Y, Tran SD. Hydrogel encapsulation of mesenchymal stem cells and their derived exosomes for tissue engineering. Int J Mol Sci. 2021;22(2):684. https://doi.org/10.3390/ijms22020684.
73. Di H, Zeng E, Zhang P, Liu X, Zhang C, Yang J, et al. General approach to engineering extracellular vesicles for biomedical analysis. Anal Chem. 2019;91(20):12752–9. https://doi.org/10.1021/acs.analchem.9b02268.
74. Huang F, Zhu X, Hu XQ, Fang ZF, Tang L, Lu XL, et al. Mesenchymal stem cells modified with miR-126 release angiogenic factors and activate notch ligand Delta-like-4, enhancing ischemic angiogenesis and cell survival. Int J Mol Med. 2013;31(2):484–92. https://doi.org/10.3892/ijmm.2012.1200.
75. Tao SC, Huang JY, Li ZX, Zhan S, Guo SC. Small extracellular vesicles with LncRNA H19 "overload": YAP regulation as a tendon repair therapeutic tactic. iScience. 2021;24(3):102200. https://doi.org/10.1016/j.isci.2021.102200.
76. Zhang J, Su R, Wang Y, Wang H, Li S, Yang X, et al. Protective effect of small extracellular vesicles (EVs) derived from ACE2-modified human umbilical cord mesenchymal stem cells against renal ischemia-reperfusion injury. Nephrology (Carlton). 2023;29:5. https://doi.org/10.1111/nep.14237.
77. Huang CC, Kang M, Lu Y, Shirazi S, Diaz JI, Cooper LF, et al. Functionally engineered extracellular vesicles improve bone regeneration. Acta Biomater. 2020;109:182–94. https://doi.org/10.1016/j.actbio.2020.04.017.
78. Zheng H, Ji J, Zhao T, Wang E, Zhang A. Exosome-encapsulated miR-26a attenuates aldosterone-induced tubulointerstitial fibrosis by inhibiting the CTGF/SMAD3 signaling pathway. Int J Mol Med. 2023;51(2):11. https://doi.org/10.3892/ijmm.2022.5214.
79. Tkach M, Théry C. Communication by extracellular vesicles: where we are and where we need to go. Cell. 2016;164(6):1226–32. https://doi.org/10.1016/j.cell.2016.01.043.
80. Thery C, Witwer KW, Aikawa E, Alcaraz MJ, Anderson JD, Andriantsitohaina R, et al. Minimal information for studies of extracellular vesicles 2018 (MISEV2018): a position statement of the International Society for Extracellular Vesicles and update of the MISEV2014 guidelines. J Extracell Vesicles. 2018;7(1):1535750. https://doi.org/10.1080/20013078.2018.1535750.

81. Yin JQ, Zhu J, Ankrum JA. Manufacturing of primed mesenchymal stromal cells for therapy. Nat Biomedi Eng. 2019;3(2):90–104. https://doi.org/10.1038/s41551-018-0325-8.
82. Shi Y, Wang Y, Li Q, Liu K, Hou J, Shao C, et al. Immunoregulatory mechanisms of mesenchymal stem and stromal cells in inflammatory diseases. Nat Rev Nephrol. 2018;14(8):493–507. https://doi.org/10.1038/s41581-018-0023-5.
83. Zhao L, Hu C, Zhang P, Jiang H, Chen J. Novel preconditioning strategies for enhancing the migratory ability of mesenchymal stem cells in acute kidney injury. Stem Cell Res Ther. 2018;9(1):225. https://doi.org/10.1186/s13287-018-0973-3.
84. Brennan MÁ, Layrolle P, Mooney DJ. Biomaterials functionalized with MSC secreted extracellular vesicles and soluble factors for tissue regeneration. Adv Funct Mater. 2020;30(37):1909125. https://doi.org/10.1002/adfm.201909125.
85. Ko KW, Park SY, Lee EH, Yoo YI, Kim DS, Kim JY, et al. Integrated bioactive scaffold with Polydeoxyribonucleotide and stem-cell-derived extracellular vesicles for kidney regeneration. ACS Nano. 2021;15:7575. https://doi.org/10.1021/acsnano.1c01098.
86. Yao M, Cui B, Zhang W, Ma W, Zhao G, Xing L. Exosomal miR-21 secreted by IL-1β-primed-mesenchymal stem cells induces macrophage M2 polarization and ameliorates sepsis. Life Sci. 2021;264:118658. https://doi.org/10.1016/j.lfs.2020.118658.
87. Wang L, Qi C, Cao H, Zhang Y, Liu X, Qiu L, et al. Engineered cytokine-primed extracellular vesicles with high PD-L1 expression ameliorate type 1 diabetes. Small. 2023;19(38):e2301019. https://doi.org/10.1002/smll.202301019.
88. Collino F, Lopes JA, Corrêa S, Abdelhay E, Takiya CM, Wendt CHC, et al. Adipose-derived mesenchymal stromal cells under hypoxia: changes in extracellular vesicles secretion and improvement of renal recovery after ischemic injury. Cell Physiol Biochem. 2019;52(6):1463–83. https://doi.org/10.33594/000000102.
89. Geßner A, Koch B, Klann K, Fuhrmann DC, Farmand S, Schubert R, et al. Characterization of extracellular vesicles from preconditioned human adipose-derived stromal/stem cells. Int J Mol Sci. 2021;22(6):2873. https://doi.org/10.3390/ijms22062873.
90. Anderson JD, Johansson HJ, Graham CS, Vesterlund M, Pham MT, Bramlett CS, et al. Comprehensive proteomic analysis of mesenchymal stem cell exosomes reveals modulation of angiogenesis via nuclear factor-KappaB signaling. Stem Cells. 2016;34(3):601–13. https://doi.org/10.1002/stem.2298.
91. Alzahrani FA. Melatonin improves therapeutic potential of mesenchymal stem cells-derived exosomes against renal ischemia-reperfusion injury in rats. Am J Transl Res. 2019;11(5):2887–907.
92. Levoux J, Prola A, Lafuste P, Gervais M, Chevallier N, Koumaiha Z, et al. Platelets facilitate the wound-healing capability of mesenchymal stem cells by mitochondrial transfer and metabolic reprogramming. Cell Metab. 2021;33(2):283–99.e9. https://doi.org/10.1016/j.cmet.2020.12.006.
93. Phan J, Kumar P, Hao D, Gao K, Farmer D, Wang A. Engineering mesenchymal stem cells to improve their exosome efficacy and yield for cell-free therapy. J Extracell Vesicles. 2018;7(1):1522236. https://doi.org/10.1080/20013078.2018.1522236.
94. Zhu L, Kalimuthu S, Oh JM, Gangadaran P, Baek SH, Jeong SY, et al. Enhancement of antitumor potency of extracellular vesicles derived from natural killer cells by IL-15 priming. Biomaterials. 2018;190-191:38–50. https://doi.org/10.1016/j.biomaterials.2018.10.034.
95. Tapparo M, Bruno S, Collino F, Togliatto G, Deregibus MC, Provero P, et al. Renal regenerative potential of extracellular vesicles derived from miRNA-engineered mesenchymal stromal cells. Int J Mol Sci. 2019;20(10):2381. https://doi.org/10.3390/ijms20102381.
96. Alvarez-Erviti L, Seow Y, Yin H, Betts C, Lakhal S, Wood MJ. Delivery of siRNA to the mouse brain by systemic injection of targeted exosomes. Nat Biotechnol. 2011;29(4):341–5. https://doi.org/10.1038/nbt.1807.
97. Kooijmans SAA, Stremersch S, Braeckmans K, de Smedt SC, Hendrix A, Wood MJA, et al. Electroporation-induced siRNA precipitation obscures the efficiency of siRNA loading into

extracellular vesicles. J Control Release. 2013;172(1):229–38. https://doi.org/10.1016/j.jconrel.2013.08.014.

98. Reshke R, Taylor JA, Savard A, Guo H, Rhym LH, Kowalski PS, et al. Reduction of the therapeutic dose of silencing RNA by packaging it in extracellular vesicles via a pre-microRNA backbone. Nat Biomed Eng. 2020;4(1):52–68. https://doi.org/10.1038/s41551-019-0502-4.
99. Didiot MC, Hall LM, Coles AH, Haraszti RA, Godinho BM, Chase K, et al. Exosome-mediated delivery of Hydrophobically modified siRNA for huntingtin mRNA silencing. Mol Ther. 2016;24(10):1836–47. https://doi.org/10.1038/mt.2016.126.
100. Kim S, Lee SA, Yoon H, Kim MY, Yoo JK, Ahn SH, et al. Exosome-based delivery of super-repressor IκBα ameliorates kidney ischemia-reperfusion injury. Kidney Int. 2021;100(3):570–84. https://doi.org/10.1016/j.kint.2021.04.039.
101. Tang TT, Wang B, Li ZL, Wen Y, Feng ST, Wu M, et al. Kim-1 targeted extracellular vesicles: a new therapeutic platform for RNAi to treat AKI. J Am Soc Nephrol. 2021;32(10):2467–83. https://doi.org/10.1681/ASN.2020111561.
102. Zhang K, Li R, Chen X, Yan H, Li H, Zhao X, et al. Renal endothelial cell-targeted extracellular vesicles protect the kidney from ischemic injury. Adv Sci (Weinh). 2023;10(3):e2204626. https://doi.org/10.1002/advs.202204626.
103. Hung ME, Leonard JN. Stabilization of exosome-targeting peptides via engineered glycosylation. J Biol Chem. 2015;290(13):8166–72. https://doi.org/10.1074/jbc.M114.621383.
104. Cheng Q, Shi X, Han M, Smbatyan G, Lenz HJ, Zhang Y. Reprogramming exosomes as nanoscale controllers of cellular immunity. J Am Chem Soc. 2018;140(48):16413–7. https://doi.org/10.1021/jacs.8b10047.
105. Gao X, Ran N, Dong X, Zuo B, Yang R, Zhou Q, et al. Anchor peptide captures, targets, and loads exosomes of diverse origins for diagnostics and therapy. Sci Transl Med. 2018;10(444):eaat0195. https://doi.org/10.1126/scitranslmed.aat0195.
106. Kamerkar S, LeBleu VS, Sugimoto H, Yang S, Ruivo CF, Melo SA, et al. Exosomes facilitate therapeutic targeting of oncogenic KRAS in pancreatic cancer. Nature. 2017;546(7659):498–503. https://doi.org/10.1038/nature22341.
107. Katyal P, Mahmoudinobar F, Montclare JK. Recent trends in peptide and protein-based hydrogels. Curr Opin Struct Biol. 2020;63:97–105. https://doi.org/10.1016/j.sbi.2020.04.007.
108. Zhang K, Zhao X, Chen X, Wei Y, Du W, Wang Y, et al. Enhanced therapeutic effects of mesenchymal stem cell-derived exosomes with an injectable hydrogel for Hindlimb ischemia treatment. ACS Appl Mater Interfaces. 2018;10(36):30081–91. https://doi.org/10.1021/acsami.8b08449.
109. Mantha S, Pillai S, Khayambashi P, Upadhyay A, Zhang Y, Tao O, et al. Smart hydrogels in tissue engineering and regenerative medicine. Materials (Basel, Switzerland). 2019;12(20):3323. https://doi.org/10.3390/ma12203323.
110. Luo L, Tang J, Nishi K, Yan C, Dinh PU, Cores J, et al. Fabrication of synthetic mesenchymal stem cells for the treatment of acute myocardial infarction in mice. Circ Res. 2017;120(11):1768–75. https://doi.org/10.1161/CIRCRESAHA.116.310374.
111. Liu X, Yang Y, Li Y, Niu X, Zhao B, Wang Y, et al. Integration of stem cell-derived exosomes with in situ hydrogel glue as a promising tissue patch for articular cartilage regeneration. Nanoscale. 2017;9(13):4430–8. https://doi.org/10.1039/c7nr00352h.
112. Han C, Zhou J, Liang C, Liu B, Pan X, Zhang Y, et al. Human umbilical cord mesenchymal stem cell derived exosomes encapsulated in functional peptide hydrogels promote cardiac repair. Biomater Sci. 2019;7(7):2920–33. https://doi.org/10.1039/c9bm00101h.
113. Derkus B. Human cardiomyocyte-derived exosomes induce cardiac gene expressions in mesenchymal stromal cells within 3D hyaluronic acid hydrogels and in dose-dependent manner. J Mater Sci Mater Med. 2021;32(1):2. https://doi.org/10.1007/s10856-020-06474-7.
114. Chen CW, Wang LL, Zaman S, Gordon J, Arisi MF, Venkataraman CM, et al. Sustained release of endothelial progenitor cell-derived extracellular vesicles from shear-thinning hydrogels improves angiogenesis and promotes function after myocardial infarction. Cardiovasc Res. 2018;114(7):1029–40. https://doi.org/10.1093/cvr/cvy067.

115. Huang CC, Kang M, Shirazi S, Lu Y, Cooper LF, Gajendrareddy P, et al. 3D encapsulation and tethering of functionally engineered extracellular vesicles to hydrogels. Acta Biomater. 2021;126:199. https://doi.org/10.1016/j.actbio.2021.03.030.
116. Zhou Y, Liu S, Zhao M, Wang C, Li L, Yuan Y, et al. Injectable extracellular vesicle-released self-assembling peptide nanofiber hydrogel as an enhanced cell-free therapy for tissue regeneration. J Control Release. 2019;316:93–104. https://doi.org/10.1016/j.jconrel.2019.11.003.
117. Liu S, Zhao M, Zhou Y, Li L, Wang C, Yuan Y, et al. A self-assembling peptide hydrogel-based drug co-delivery platform to improve tissue repair after ischemia-reperfusion injury. Acta Biomater. 2020;103:102–14. https://doi.org/10.1016/j.actbio.2019.12.011.
118. Shi W, Zhou X, Li X, Peng X, Chen G, Li Y, et al. Human umbilical cord mesenchymal stem cells protect against renal ischemia-reperfusion injury by secreting extracellular vesicles loaded with miR-148b-3p that target pyruvate dehydrogenase kinase 4 to inhibit endoplasmic reticulum stress at the reperfusion stages. Int J Mol Sci. 2023;24(10):8899. https://doi.org/10.3390/ijms24108899.
119. Viñas JL, Spence M, Porter CJ, Douvris A, Gutsol A, Zimpelmann JA, et al. MicroRNA-486-5p protects against kidney ischemic injury and modifies the apoptotic transcriptome in proximal tubules. Kidney Int. 2021;100(3):597–612. https://doi.org/10.1016/j.kint.2021.05.034.
120. Bian Z, Wang X, Zhu R, Chen S. miR-21-5p in extracellular vesicles obtained from adipose tissue-derived stromal cells facilitates tubular epithelial cell repair in acute kidney injury. Cytotherapy. 2023;25(3):310–22. https://doi.org/10.1016/j.jcyt.2022.08.002.
121. Xie Z, Tang J, Chen Z, Wei L, Chen J, Liu Q. Human bone marrow mesenchymal stem cell-derived extracellular vesicles reduce inflammation and pyroptosis in acute kidney injury via miR-223-3p/HDAC2/SNRK. Inflamm Res. 2023;72(3):553–76. https://doi.org/10.1007/s00011-022-01653-4.
122. Ma M, Luo Q, Fan L, Li W, Li Q, Meng Y, et al. The urinary exosomes derived from premature infants attenuate cisplatin-induced acute kidney injury in mice via microRNA-30a-5p/mitogen-activated protein kinase 8 (MAPK8). Bioengineered. 2022;13(1):1650–65. https://doi.org/10.1080/21655979.2021.2021686.
123. Zhang J, He W, Zheng D, He Q, Tan M, Jin J. Exosomal-miR-1184 derived from mesenchymal stem cells alleviates cisplatin-associated acute kidney injury. Mol Med Rep. 2021;24(5):795. https://doi.org/10.3892/mmr.2021.12435.
124. Zhang Y, Huang H, Liu W, Liu S, Wang XY, Diao ZL, et al. Endothelial progenitor cells-derived exosomal microRNA-21-5p alleviates sepsis-induced acute kidney injury by inhibiting RUNX1 expression. Cell Death Dis. 2021;12(4):335. https://doi.org/10.1038/s41419-021-03578-y.
125. Juan CX, Mao Y, Cao Q, Chen Y, Zhou LB, Li S, et al. Exosome-mediated pyroptosis of miR-93-TXNIP-NLRP3 leads to functional difference between M1 and M2 macrophages in sepsis-induced acute kidney injury. J Cell Mol Med. 2021;25(10):4786–99. https://doi.org/10.1111/jcmm.16449.
126. Zhang Z, Chen H, Zhou L, Li C, Lu G, Wang L. Macrophage-derived exosomal miRNA-155 promotes tubular injury in ischemia-induced acute kidney injury. Int J Mol Med. 2022;50(3):116. https://doi.org/10.3892/ijmm.2022.5172.
127. Yang Y, Wang J, Zhang Y, Hu X, Li L, Chen P. Exosomes derived from mesenchymal stem cells ameliorate renal fibrosis via delivery of miR-186-5p. Hum Cell. 2022;35(1):83–97. https://doi.org/10.1007/s13577-021-00617-w.
128. Zhao S, Li W, Yu W, Rao T, Li H, Ruan Y, et al. Exosomal miR-21 from tubular cells contributes to renal fibrosis by activating fibroblasts via targeting PTEN in obstructed kidneys. Theranostics. 2021;11(18):8660–73. https://doi.org/10.7150/thno.62820.
129. Qiu Z, Zhong Z, Zhang Y, Tan H, Deng B, Meng G. Human umbilical cord mesenchymal stem cell-derived exosomal miR-335-5p attenuates the inflammation and tubular epithelial-myofibroblast transdifferentiation of renal tubular epithelial cells by reducing ADAM19 protein levels. Stem Cell Res Ther. 2022;13(1):373. https://doi.org/10.1186/s13287-022-03071-z.

130. Wang SJ, Qiu ZZ, Chen FW, Mao AL, Bai JC, Hong YJ, et al. Bone marrow mesenchymal stem cell-derived extracellular vesicles containing miR-181d protect rats against renal fibrosis by inhibiting KLF6 and the NF-κB signaling pathway. Cell Death Dis. 2022;13(6):535. https://doi.org/10.1038/s41419-022-04875-w.
131. Khamis T, Alsemeh AE, Alanazi A, Eltaweel AM, Abdel-Ghany HM, Hendawy DM, et al. Breast Milk mesenchymal stem cells and/or derived exosomes mitigated adenine-induced nephropathy via modulating renal autophagy and fibrotic signaling pathways and their epigenetic regulations. Pharmaceutics. 2023;15(8):2149. https://doi.org/10.3390/pharmaceutics15082149.
132. Liang M, Zhang D, Zheng D, He W, Jin J. Exosomes from miR-374a-5p-modified mesenchymal stem cells inhibit the progression of renal fibrosis by regulating MAPK6/MK5/YAP axis. Bioengineered. 2022;13(2):4517–27. https://doi.org/10.1080/21655979.2022.2033465.
133. Wang Y, Liu J, Wang H, Lv S, Liu Q, Li S, et al. Mesenchymal stem cell-derived exosomes ameliorate diabetic kidney disease through the NLRP3 signaling pathway. Stem Cells. 2023;41(4):368–83. https://doi.org/10.1093/stmcls/sxad010.
134. Cui C, Zang N, Song J, Guo X, He Q, Hu H, et al. Exosomes derived from mesenchymal stem cells attenuate diabetic kidney disease by inhibiting cell apoptosis and epithelial-to-mesenchymal transition via miR-424-5p. FASEB J. 2022;36(10):e22517. https://doi.org/10.1096/fj.202200488R.
135. Wang Z, Sun W, Li R, Liu Y. miRNA-93-5p in exosomes derived from M2 macrophages improves lipopolysaccharide-induced podocyte apoptosis by targeting toll-like receptor 4. Bioengineered. 2022;13(3):7683–96. https://doi.org/10.1080/21655979.2021.2023794.
136. Ji JL, Shi HM, Li ZL, Jin R, Qu GT, Zheng H, et al. Satellite cell-derived exosome-mediated delivery of microRNA-23a/27a/26a cluster ameliorates the renal tubulointerstitial fibrosis in mouse diabetic nephropathy. Acta Pharmacol Sin. 2023;44:2455. https://doi.org/10.1038/s41401-023-01140-4.
137. Zhao J, Chen J, Zhu W, Qi XM, Wu YG. Exosomal miR-7002-5p derived from highglucose-induced macrophages suppresses autophagy in tubular epithelial cells by targeting Atg9b. FASEB J. 2022;36(9):e22501. https://doi.org/10.1096/fj.202200550RR.
138. Tsai YC, Kuo MC, Hung WW, Wu PH, Chang WA, Wu LY, et al. Proximal tubule-derived exosomes contribute to mesangial cell injury in diabetic nephropathy via miR-92a-1-5p transfer. Cell Commun Signal. 2023;21(1):10. https://doi.org/10.1186/s12964-022-00997-y.
139. Zhao P, Zhu Y, Sun L, Zhu W, Lu Y, Zhang J, et al. Circulating Exosomal miR-1-3p from rats with myocardial infarction plays a protective effect on contrast-induced nephropathy via targeting ATG13 and activating the AKT signaling pathway. Int J Biol Sci. 2021;17(4):972–85. https://doi.org/10.7150/ijbs.55887.
140. Brossa A, Tapparo M, Fonsato V, Papadimitriou E, Delena M, Camussi G, et al. Coincubation as miR-loading strategy to improve the anti-tumor effect of stem cell-derived EVs. Pharmaceutics. 2021;13(1):76. https://doi.org/10.3390/pharmaceutics13010076.
141. Zhu Y, Zha F, Tang B, Ji TT, Li XY, Feng L, et al. Exosomal hsa_circ_0125310 promotes cell proliferation and fibrosis in diabetic nephropathy via sponging miR-422a and targeting the IGF1R/p38 axis. J Cell Mol Med. 2022;26(1):151–62. https://doi.org/10.1111/jcmm.17065.
142. Niu Z, Ren G, Huang L, Mu L. Circ_0008529 contributes to renal tubular cell dysfunction in high glucose stress via miR-185-5p/SMAD2 pathway in diabetic nephropathy. Biochem Genet. 2023;61(3):963–78. https://doi.org/10.1007/s10528-022-10296-3.
143. Sun Y, Zhao Y, Lu Y, Li H, Xiang J, Yang D, et al. Urinary stem cell-derived exocrine circRNA ATG7 regulates the SOCS1/STAT3 signaling pathway through miR-4500, inhibits M1 macrophage polarization, and alleviates the progression of diabetes nephropathy. Int Urol Nephrol. 2023;56:1449. https://doi.org/10.1007/s11255-023-03819-3.
144. Chen M, Zhang L. Circ_0001806 relieves LPS-induced HK2 cell injury by regulating the expression of miR-942-5p and TXNIP. J Bioenerg Biomembr. 2023;55(4):301–12. https://doi.org/10.1007/s10863-023-09978-3.

145. Kuang F, Wang B, You T, Liu Y, Li P, Wang J, et al. CIRC_0001818 targets MIR-136-5P to increase lipopolysaccharide-induced HK2 cell injuries by activating TXNIP/NLRP3 INFLAMMASOME pathway. Shock. 2023;60(1):110–20. https://doi.org/10.1097/shk.0000000000002140.
146. Sun Z, Wu J, Bi Q, Wang W. Exosomal lncRNA TUG1 derived from human urine-derived stem cells attenuates renal ischemia/reperfusion injury by interacting with SRSF1 to regulate ASCL4-mediated ferroptosis. Stem Cell Res Ther. 2022;13(1):297. https://doi.org/10.1186/s13287-022-02986-x.
147. Flores-Chova A, Martinez-Arroyo O, Riffo-Campos AL, Ortega A, Forner MJ, Cortes R. Plasma Exosomal non-coding RNA profile associated with renal damage reveals potential therapeutic targets in lupus nephritis. Int J Mol Sci. 2023;24(8):7088. https://doi.org/10.3390/ijms24087088.
148. Yan J, Xiao H, Zhou X, Li Y, Zhao S, Zhao X, et al. Engineered exosomes reprogram Gli1(+) cells in vivo to prevent calcification of vascular grafts and autologous pathological vessels. Sci Adv. 2023;9(29):eadf7858. https://doi.org/10.1126/sciadv.adf7858.
149. Ko KW, Park SY, Lee EH, Yoo YI, Kim DS, Kim JY, et al. Integrated bioactive scaffold with Polydeoxyribonucleotide and stem-cell-derived extracellular vesicles for kidney regeneration. ACS Nano. 2021;15(4):7575–85. https://doi.org/10.1021/acsnano.1c01098.
150. Cheng YQ, Yue YX, Cao HM, Geng WC, Wang LX, Hu XY, et al. Coassembly of hypoxia-sensitive macrocyclic amphiphiles and extracellular vesicles for targeted kidney injury imaging and therapy. J Nanobiotechnol. 2021;19(1):451. https://doi.org/10.1186/s12951-021-01192-w.
151. Du J, Sun Q, Wang Z, Wang F, Chen F, Wang H, et al. Tubular epithelial cells derived-exosomes containing CD26 protects mice against renal ischemia/reperfusion injury by maintaining proliferation and dissipating inflammation. Biochem Biophys Res Commun. 2021;553:134–40. https://doi.org/10.1016/j.bbrc.2021.03.057.
152. Li Y, Chakraborty A, Broughton BRS, Ferens D, Widdop RE, Ricardo SD, et al. Comparing the renoprotective effects of BM-MSCs versus BM-MSC-exosomes, when combined with an anti-fibrotic drug, in hypertensive mice. Biomed Pharmacother. 2021;144:112256. https://doi.org/10.1016/j.biopha.2021.112256.
153. Lin J, Lv J, Yu S, Chen Y, Wang H, Chen J. Transcript engineered extracellular vesicles alleviate Alloreactive dynamics in renal transplantation. Adv Sci (Weinh). 2022;9(31):e2202633. https://doi.org/10.1002/advs.202202633.
154. Zhong X, Tang TT, Shen AR, Cao JY, Jing J, Wang C, et al. Tubular epithelial cells-derived small extracellular vesicle-VEGF-A promotes peritubular capillary repair in ischemic kidney injury. NPJ Regen Med. 2022;7(1):73. https://doi.org/10.1038/s41536-022-00268-x.
155. Franzin R, Stasi A, De Palma G, Picerno A, Curci C, Sebastiano S, et al. Human adult renal progenitor cells prevent cisplatin-nephrotoxicity by inducing CYP1B1 overexpression and miR-27b-3p Down-regulation through extracellular vesicles. Cells. 2023;12(12):1655. https://doi.org/10.3390/cells12121655.
156. Liu S, Lv K, Wang Y, Lou P, Zhou P, Wang C, et al. Improving the circulation time and renal therapeutic potency of extracellular vesicles using an endogenous ligand binding strategy. J Control Release. 2022;352:1009–23. https://doi.org/10.1016/j.jconrel.2022.11.017.
157. Chen S, Zhang M, Li J, Huang J, Zhou S, Hou X, et al. β-Catenin-controlled tubular cell-derived exosomes play a key role in fibroblast activation via the OPN-CD44 axis. J Extracell Vesicles. 2022;11(3):e12203. https://doi.org/10.1002/jev2.12203.
158. Jiang WJ, Xu CT, Du CL, Dong JH, Xu SB, Hu BF, et al. Tubular epithelial cell-to-macrophage communication forms a negative feedback loop via extracellular vesicle transfer to promote renal inflammation and apoptosis in diabetic nephropathy. Theranostics. 2022;12(1):324–39. https://doi.org/10.7150/thno.63735.
159. Jiang W, Xu C, Xu S, Su W, Du C, Dong J, et al. Macrophage-derived, LRG1-enriched extracellular vesicles exacerbate aristolochic acid nephropathy in a TGFβR1-dependent manner. Cell Biol Toxicol. 2022;38(4):629–48. https://doi.org/10.1007/s10565-021-09666-1.

Chapter 10
Exosomes for Ophthalmic Therapeutics

Hongyan Tao and Zongjin Li

Abstract Exosomes, as key mediators of cell-to-cell communication, are emerging as promising therapeutic agents in ophthalmology, particularly for conditions with diagnostic and treatment challenges such as age-related macular degeneration, glaucoma, infectious endophthalmitis, diabetic retinopathy, and autoimmune uveitis. These challenges include delayed symptom onset, difficulties in microbial identification, and the complexities of masquerade syndromes. This chapter reviews the roles of exosomes in treating various ophthalmic conditions, highlighting their potential as novel vectors for drug and gene therapy delivery. It explores recent advancements in exosome-based therapies and other innovative therapeutic delivery approaches, including small drug molecules, gene therapy, nanomedicine, stem cells, and trophic factors. These techniques demonstrated to be promising in preclinical animal models for conditions such as corneal scar/haze, inflammation, and angiogenesis are on the cusp of human clinical application. This discussion emphasizes the potential of exosomal and bioengineering approaches in revolutionizing ocular disease treatment, moving toward more effective, targeted, and less invasive therapies.

Keywords Exosomes · Stem cells · Ophthalmic therapeutics · Bioengineering

H. Tao
MRC Weatherall Institute of Molecular Medicine (Oxford University & John Radcliffe Hospital), University of Oxford, Oxford, UK

School of Medicine, Nankai University, Tianjin, China
e-mail: hongyan.tao@rdm.ox.ac.uk

Z. Li (✉)
School of Medicine, Nankai University, Tianjin, China
e-mail: zongjinli@nankai.edu.cn

© The Author(s), under exclusive license to Springer Nature Singapore Pte Ltd. 2024
Z. Li (ed.), *Extracellular Vesicle: Biology and Translational Application*,
https://doi.org/10.1007/978-981-97-5536-3_10

1 Introduction

Visual impairment, affecting over 2.2 billion people globally, is a major health concern and the third leading contributor to reduced quality of life after tumors and cardiovascular diseases [1, 2]. Primary treatments for eye diseases are predominantly surgery and pharmacotherapy [3]. Advanced-stage eye diseases often necessitate surgical interventions such as vitrectomy, corneal transplantation, cataract extraction, and artificial lens implantation [4–6]. However, these invasive procedures can lead to suboptimal visual outcomes, particularly in conditions like advanced proliferative diabetic retinopathy (DR), where a significant proportion of patients experience no improvement or worsening of vision post-surgery [7]. This has shifted the focus increasingly toward pharmacotherapy, even for conditions traditionally managed surgically [8].

Ophthalmology currently emphasizes early intervention to preserve visual function and comfort, but conventional drug delivery methods face challenges such as inefficiency and poor targeting [9, 10]. Biological drugs, although having a longer half-life than chemical drugs, often require frequent administrations, exemplified by the monthly intravitreal injections needed for ranibizumab in age-related macular degeneration (AMD) treatment [11].

Recent advancements in novel therapeutic methods, particularly cell-based therapies, offer sustained effectiveness and superior biocompatibility [12]. A notable advancement is the intravitreal administration of mesenchymal stromal cells (MSCs) during pars plana vitrectomy (PPV) for large macular holes (MHs), which significantly improved closure rates compared to standard PPV [13]. Yet, challenges such as reduced cell survival and risks of undesired differentiation persist [14, 15].

Exosomes are gaining recognition as potential therapeutics in various diseases due to their biocompatibility, immune neutrality, and ability to cross biological barriers [16–21]. In ophthalmic therapeutics, they are being explored both as direct therapeutic agents and as drug delivery vehicles [22, 23].

As drug delivery platforms, exosomes offer several advantages, such as effective targeting, minimal immunogenicity, substantial drug loading capacity, synergistic therapeutic effects, and high stability [24–30]. This chapter provides a comprehensive exploration of exosomes in ophthalmic therapy, discussing their roles as delivery vehicles, recent progress, challenges, and future prospects.

2 Anatomy of the Cornea and Corneal Response to Injury

In adults, the cornea typically spans 11.5–12.0 mm in horizontal diameter, which is marginally greater than its vertical diameter, and has a central thickness of about 0.5 mm, increasing toward its periphery [31]. Its prolate configuration, being flatter at the edges and steeper at the center, creates an aspheric optical system. The shape and curvature of the cornea are determined by its biomechanical structure and external influences. The human cornea is structurally composed of five distinct layers, each playing a critical role in maintaining corneal function and integrity. These layers include three cellular layers—the epithelium, stroma, and endothelium—and two acellular interface layers, namely the Bowman membrane and Descemet membrane [32]. The epithelium, the outermost layer, serves as a barrier against environmental insults and contributes to optical clarity. The Bowman membrane, situated beneath the epithelium, provides structural support and protects against corneal trauma. The stroma, the thickest layer, is crucial for corneal transparency and shape, while the Descemet membrane, located between the stroma and endothelium, functions as a supportive structure. The rigidity of its anterior stromal layer plays a crucial role in sustaining its curvature, with the variation in collagen bundles providing structural integrity and resistance to hydration-related alterations. Additionally, the cornea's response to mechanical stress is affected by the hydration of the stroma. Finally, the endothelium, the innermost layer, is essential for maintaining corneal hydration and transparency.

2.1 *Anatomy of the Cornea*

2.1.1 Epithelium

The corneal epithelial surface, a primary barrier against external factors, plays a crucial role in the tear film–cornea interface, significantly enhancing the eye's refractive capability. Originating from the surface ectoderm in the 5th–6th weeks of gestation, it consists of 4–6 layers of nonkeratinized, stratified squamous epithelial cells [33]. Collaborating with the tear film, it evens out anterior surface irregularities, accounting for two-thirds of the eye's total refractive power.

Positioned between the ocular surface epithelium and the external environment, the tear film has a close anatomical and functional relationship with the epithelium. The tear film's mucinous layer, produced by conjunctival goblet cells, combines with the corneal epithelial cells' glycocalyx, aiding tear film spreading with each blink. This synergy is crucial for maintaining the ocular optical system's integrity, aiding wound healing, and ensuring comfort and high-quality vision [34].

Corneal epithelial cells, with a 7- to 10-day lifespan, are constantly renewing [35]. The outermost 2–3 layers, consisting of superficial cells, feature microvilli and microplicae, coated with a glycocalyceal layer vital for tear film adherence and a clear optical system. Additionally, these cells form tight junctions, enhancing the barrier against external elements [36].

2.1.2 Bowman's Layer

Bowman's layer, positioned directly anterior to the stroma, is an acellular structure formed by condensed stromal elements, with randomly arranged collagen fibers [37]. With a thickness of 13.7–19.6 μm, this smooth layer is crucial in maintaining the cornea's shape, and it cannot regenerate post-injury [38].

2.1.3 Stroma

The corneal stroma constitutes about 80–85% of the cornea's total thickness [39]. The primary reason for its transparency is the orderly arrangement of collagen fibers within the extracellular matrix (ECM) [40]. The stroma comprises 200–250 distinct lamellae, each positioned at right angles to the fibers in adjacent layers. Notably, the stroma is thicker at the periphery, with fibrils oriented circumferentially near the limbus [41]. The stromal collagen fibrils consist of highly aligned layers of Type I and Type V collagen lamellae, surrounded by proteoglycans such as keratan sulfate and chondroitin/dermatan sulfate. These elements are crucial for regulating the stroma's hydration and structural integrity. Keratocytes, the primary cells in the stroma, play an integral role in maintaining the ECM. They synthesize collagen, glycosaminoglycans, and matrix metalloproteases (MMPs), crucial for maintaining stromal homeostasis. Additionally, the organized structure of these cells and components contains corneal "crystallins," pivotal in minimizing light backscatter and ensuring corneal transparency [42, 43].

2.1.4 Descemet Membrane (DM)

The development of Descemet's membrane (DM) is divided into two phases—prenatal and postnatal—each exhibiting distinct characteristics over time. Forming from the 8th week of gestation, the prenatal DM, remaining unchanged with aging, is approximately 3 μm thick and exhibits a banded pattern under electron microscopy. In contrast, the postnatal DM, which grows continuously after birth, exhibits an amorphous ultrastructure and can thicken to as much as 10 μm. These variations in DM are crucial for understanding the onset and progression of corneal endothelial diseases [44].

2.1.5 Endothelium

Originating early in embryogenesis as a neural crest-derived monolayer, the corneal endothelial layer, crucial for maintaining corneal clarity through dehydration, evolves into a 10-μm thick, tightly adherent endothelium with a honeycomb structure at birth, integrating with trabecular meshwork cells above the developing Descemet membrane. As aging progresses, these cells, interconnected by gap and tight junctions and featuring Na^+, K^+-ATPase pumps and hemidesmosomes, flatten to approximately 4 μm in adulthood. With age, endothelial cell density decreases, leading to altered shapes and a reduction in hexagonal cells, which impacts dehydration efficiency. Although these cells lack mitotic activity and decrease in number over time, they adapt by enlarging and altering their shape to compensate for cell loss, potentially leading to reduced effectiveness in corneal dehydration, particularly in cases of aging, trauma, or disease [45, 46].

2.1.6 Blood Supply and Nerve Supply of the Cornea

Despite its avascularity, the human cornea relies on blood-derived nutrients, delivered through peripheral vessels and from facial and ophthalmic arteries via the aqueous humor and tear film [47]. As one of the most densely innervated tissues, corneal nerves, primarily from the trigeminal nerve's nasociliary branch, enter the stroma in thick trunks, forming complex networks beneath the Bowman membrane and the basal epithelial layer, contributing to its high sensitivity. This structure is further supplemented by autonomic sympathetic fibers [48, 49].

3 Characterization of Exosomes

Exosomes, vesicles measuring 30–150 nm, have gained prominence in organ physiology and pathology, including in the eye. Identified in 1983 and named by Johnstone in 1987 [50], they rose to prominence following Schekman, Rothman, and Südhof's Nobel Prize-winning research on vesicle transport in the late twentieth century. Present in a variety of biological fluids and released by many cell types, exosomes transport active components such as miRNA, impacting diseases ranging from cardiac to neurologic disorders and cancer [51]. Detected in blood [52], urine [53], and tears [54] and released from cells such as lymphocytes and tumor cells [55, 56], exosomes are potential biomarkers and drug delivery systems owing to their stability and compatibility [57, 58]. They are formed through endosomal processes [59], and their composition varies based on their origin [60]. Isolation techniques for exosomes include ultracentrifugation and size-exclusion chromatography, while characterization methods vary from Western blotting to electron microscopy [61, 62]. Engineered exosomes, genetically or chemically modified for specific functions, present advanced therapeutic potential, albeit with efficiency challenges [63]. Engineered exosomes have significant advantages over traditional nanocarriers in drug and gene delivery primarily because of their enhanced stability and greater affinity for nucleic acids [3].

4 Application of Exosome for Ophthalmic Therapeutics

The human eyeball consists of three distinct layers: the outer layer (cornea and sclera), the middle layer (housing the eye's primary blood supply, including the iris and pupil), and the inner layer (retina). Among the numerous eye diseases and disorders, the four most common conditions causing vision loss or blindness are cataracts, diabetic retinopathy, glaucoma, and age-related macular degeneration. However, the range of ophthalmic diseases is extensive, covering hundreds of different conditions (Fig. 10.1). In this section, we will explore the use of exosomes in treating these common ophthalmic diseases, focusing on their potential roles and therapeutic benefits.

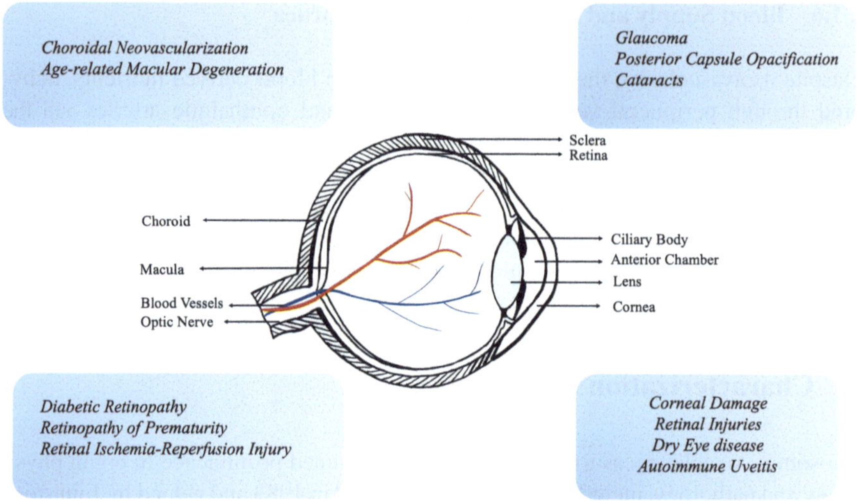

Fig. 10.1 Anatomy of the human eye and related ocular conditions This diagram illustrates the anatomical features of the human eye, including the choroid, macula, blood vessels, optic nerve, sclera, ciliary body, anterior chamber, lens, and cornea. Adjacent to each anatomical feature are related ocular conditions. Near the choroid and macula, conditions such as Choroidal Neovascularization and Age-Related Macular Degeneration are highlighted. In the context of the retina, Glaucoma, Posterior Capsule Opacification, and Cataracts are indicated. Below the diagram, conditions including Diabetic Retinopathy, Retinopathy of Prematurity, and Retinal Ischemia-Reperfusion Injury are noted. Conditions related to the cornea, such as Corneal Damage, Retinal Injuries, Dry Eye Disease, and Autoimmune Uveitis

4.1 Traumatic Diseases

4.1.1 Corneal Damage

Corneal blindness, impacting approximately 1.9 million people globally, accounts for about 5% of total blindness cases, predominantly due to corneal opacification [64, 65]. This epidemiological variance in visual impairment and blindness is largely attributed to a variety of infectious and inflammatory eye diseases. These conditions often result in corneal scarring, a common pathway in the progression of many ocular disorders, which can ultimately lead to blindness. Geographically, the incidence of corneal diseases differs significantly, with major contributing factors being cataracts, trachoma (affecting an estimated 4.9 million individuals through corneal scarring), ocular trauma, and corneal ulceration. In pediatric populations, conditions such as xerophthalmia and ophthalmia neonatorum and infections like herpes simplex play a substantial role in visual impairment [66]. This section aims to delve into the advancements and therapeutic potential of exosomes in treating corneal damage.

Addressing the limitations of traditional treatments, like artificial tears, which show minimal effectiveness in conditions such as neurotrophic keratopathy [23], recent developments in cell-based therapies, especially those utilizing stem cells, have opened new avenues. In particular, the application of exosomes in corneal regeneration has emerged as a promising area. MSC-derived exosomes, for example, have been proven effective in promoting corneal epithelial cell proliferation and aiding in wound healing, signaling a significant breakthrough in treating corneal damage [67]. Additionally, research has demonstrated that exosomes from mouse corneal epithelial cells can induce myofibroblastic transformation in the cornea. Intriguingly, these exosomal vesicles have been detected intercalating between epithelial cells and the stroma in damaged rat corneas via electron microscopy, highlighting their role in cellular communication and response to injury [68].

Moreover, the capabilities of stem cell-derived exosomes extend to addressing corneal stroma opacity, improving cell viability, and facilitating the remodeling of the extracellular matrix. Notably, specific exosomal miRNAs, such as miR-19a, miR-432-5p, and miR-159a, have been identified as crucial in suppressing undesirable cell differentiation and promoting transparent tissue regeneration after corneal injuries [42, 69, 70]. Furthermore, innovative approaches such as hydrogels that release exosomes have been developed, enhancing their therapeutic efficacy in ocular applications [71].

Exosomes are increasingly recognized as a viable and innovative solution to the challenges presented by ocular barriers in ophthalmic therapeutics. The particular interest in exosomes derived from mesenchymal stem cells (MSCs) is due to their accessibility, multifunctionality, and favorable safety profile [72]. While there are a limited number of clinical trials currently underway (see Table 10.1), it is crucial to recognize that the yield and specific functions of therapeutic-grade exosomes can vary based on their cellular origin. This variation means that safety considerations might differ depending on the type of exosome utilized [73]. Therefore, ongoing research and meticulous evaluation are essential in their application to ensure safety and efficacy.

4.1.2 Retinal Injuries

Retinal injuries, often caused by accidents or environmental factors, can lead to significant vision loss. While traditional treatments have limitations, with surgeries typically reserved for severe cases [74], recent advancements have revealed the potential of MSC exosomes in this domain. Notably, studies have demonstrated their neuroprotective effects in conditions such as traumatic optic neuropathy, aiding in the survival and regeneration of retinal ganglion cells [75]. Mead et al. showed that bone marrow-derived MSC (BMSC) exosomes exert a substantial neuroprotective and neurogenic effect in a rat optic nerve crush (ONC) model. Intriguingly, the study found that inhibition of Ago2, a key protein in miRNA regulation, diminishes these beneficial effects, suggesting that miRNAs in BMSC-derived exosomes play a more significant role than proteins [76].

Table 10.1 Current clinical trials involving exosome therapies in ophthalmology

ClinicalTrial (gov ID)	Study	Exosomes resources	Diseases	Study type	Phase	Status
NCT04213248	Effect of UMSCs Derived Exosomes on Dry Eye in Patients with cGVHD	Umbilical Mesenchymal Stem Cells (UMSCs)	Chronic Graft Versus Host Diseases (cGVHD)	Interventional	Phase 1/2	RECRUITING
NCT05738629	Safety and Efficacy of PSC-MSC-Exo Eye Drops for Dry Eye Post Refractive Surgery and Blepharospasm	Pluripotent Stem Cell-derived Mesenchymal Stem Cell (PSC-MSC)	Dry eye disease (DED)	Interventional	Phase 1/2	NOT YET RECRUITING
NCT03437759	MSC-Exos Promote Healing of MHs	Mesenchymal Stem Cells (MSCs)	Large and refractory macular holes (MHs)	Interventional	Early Phase 1	UNKNOWN STATUS
NCT05888558	Serum Exosomal miRNA as a Biomarker for Ocular Myasthenia Gravis	Serum Exosomal miRNA Screening	Ocular muscle myasthenia gravis (OMG)	Observational [Patient Registry]	NA	ENROLLING BY INVITATION
NCT05413148	Effect of Stem Cells and Exosomes on Visual Functions in RP	Mesenchymal Stem Cells (MSCs)	Retinitis pigmentosa (RP)	Interventional	Phase 2/3	RECRUITING
NCT03264976	Role of Serum Exosomal miRNA in DR	Serum Exosomal miRNA Screening	Diabetic retinopathy (DR)	Observational	NA	UNKNOWN STATUS

In addressing refractory and large MHs, human umbilical cord tissue-derived MSC exosomes have shown promise [13]. Since the introduction of pars plana vitrectomy (PPV) by Kelly and Wendel in 1991, the technique has evolved to achieve a 98% closure rate for recent onset MHs less than 400 μm [77, 78]. However, larger (>400 μm) and long-standing (>6 months) MHs, particularly in highly myopic eyes, present a challenge [79, 80]. A study involving five patients with MHs larger than 400 μm showed improved anatomic and visual outcomes after intravitreal injection of MSCs and MSC-Exos following standard PPV. This MSC-Exo therapy, being safer than cell therapy due to reduced risk of cell proliferation and requiring no additional surgery, offers a promising alternative.

Further investigations have explored the use of exosomes with specific properties for treating retinal detachment (RD), presenting new potential treatments. Ma et al.

investigated the effects of MSC-derived exosomes in a rat-induced retinal detachment model, observing that these exosomes suppress inflammatory cytokines and enhance autophagy, thereby improving photoreceptor survival [81]. Specifically, they noted a significant reduction in TNF-α and IL-1β expression, an increased LC3-II to LC3-I ratio, and decreased Atg5 cleavage following MSC-Exo treatment. This inhibition of TNF-α and consequent autophagy increase plays a crucial role in reducing apoptosis and enhancing photoreceptor survival, aligning with findings from previous studies.

4.2 Autoimmune Diseases

4.2.1 Dry Eye Disease

The eye's intricate anatomy, comprising the outer fibrous layer (cornea/sclera), middle vascular layer (uvea: iris, ciliary body, choroid), and inner sensorineural layer (retina), is essential for immune protection. The outer layer supports ocular health with its blood-ocular barrier and immunosuppressive neuropeptides. The middle layer induces immunosuppression via TGF-β activation, and the inner layer's mechanisms for regulating intraocular immune responses, though not fully understood, are critical. This intricate immunological structure highlights the need to understand these pathways for developing treatments for ocular damage from chronic or excessive inflammation, potentially leading to autoimmune reactions [82]. As an immune-privileged sensory organ, the eye is structured to resist immunogenic inflammation [83]. Nonetheless, it is susceptible to inflammatory and immune-mediated diseases impacting its anterior or posterior segments. Conditions including Sjögren's syndrome (SS) dry eye and graft-versus-host disease (GVHD)-associated dry eye, often resistant to standard treatments, can lead to serious complications.

Sjögren's syndrome, a chronic autoimmune disorder, is marked by lymphocytic infiltration in lacrimal and salivary glands, causing dryness of the eyes and mouth. This exocrinopathy may arise from primary Sjögren's syndrome (pSS) or as a secondary condition to other autoimmune disorders such as systemic lupus erythromatosus (SLE) or rheumatoid arthritis [84]. The pathogenesis of Sjögren's syndrome involves activating innate and adaptive immune pathways, such as B cell activating factor (BAFF)/BAFF receptor axis, interferon (IFN) signatures, and nuclear factor kappa B (NF-kB) signaling [85]. Prognostic markers for Sjögren's syndrome include anti-Ro/SSA, anti-La/SSB, and antinuclear autoantibodies (ANA) [84]. Salivary gland epithelial cells release these autoantigens via exosomes. Therefore, exosomes play a role in presenting intracellular autoantigens to the immune system, contributing to the progression of Sjögren's syndrome. Another study revealed that Epstein Barr Virus (EBV)-specific microRNA, namely EBV-miR-BART13-3p, is significantly elevated in the salivary glands of pSS patients [86]. Epstein Barr Virus (EBV) typically infects B cells, not salivary epithelial cells. Nonetheless, exosomes can transfer EBV-miR-BART13-3p from B cells to salivary epithelial cells. Subconjunctivally administered MSC-derived exosomes (MSC-Exos) have proven effective in alleviating induced autoimmune dacryoadenitis in rabbit models, closely resembling human Sjögren's syndrome. Research on mesenchymal stem cell (MSC)-based secretome and stem cell therapy indicates that subconjunctival administration of MSC-Exos significantly improves clinical outcomes and reduces inflammation in the lacrimal glands of affected rabbits. This improvement is attributed to the exosomes' capability to modulate lacrimal macrophage polarization and boost T-regulatory (Treg) and T-helper 2 (Th2) responses, specifically targeting the NF-kB signaling pathway [87].

Although the cornea is an immunologically privileged, avascular, transparent tissue, corneal grafts often face rejection due to an allogeneic immune response. The adaptive immune response begins with the recipient T cells recognizing donor MHC antigens [88]. Gonzalez-Nolasco et al. recently reviewed exosomes' roles in the recognition, rejection, and tolerance of allografts [89]. Another study demonstrates that short collagen-like peptides linked to polyethylene glycol (CLP-PEG) promote regeneration by stimulating EV production in host cells [90]. Additionally, exosomes derived from epithelial cells are reported to potentially aid in corneal wound healing and neovascularization. Han et al. have demonstrated that exosomes derived from epithelial cells facilitate communication among corneal epithelial cells, keratocytes, and vascular endothelial cells [68]. Moreover, exosomes from limbal stromal cells (LSCs) contribute to the proliferation and wound healing of limbal epithelial cells (LECs) [91]. Exosomes from human corneal mesenchymal stromal cells (cMSCs) are also reported to accelerate corneal epithelial wound healing.

Recent studies on the anti-inflammatory and reparative properties of MSC-Exos have yielded promising results. These exosomes can reprogram proinflammatory M1 macrophages into an immunosuppressive M2 phenotype, a process mediated by miR-204 targeting the IL-6/IL-6R/Stat3 pathway [92]. This method shows significant potential in reducing ocular surface inflammation and enhancing tear film integrity, providing new hope for patients who do not respond to conventional treatments.

4.2.2 Autoimmune Uveitis

Uveitis, often linked to autoimmune disorders, presents a significant risk for vision loss. Although corticosteroids are the standard first-line treatment, their efficacy can be limited, and they are associated with various side effects. In contrast, exosomes derived from MSCs and other immunosuppressive cells have emerged as a promising alternative. These exosomes have demonstrated potential in modulating immune responses, thereby protecting against inflammation and tissue damage associated with uveitis [93]. A notable study revealed that periocular administration of human MSC-Exo significantly reduced leucocyte infiltration, preserved retinal structure, and improved retinal function in rats with Experimental Autoimmune Uveitis (EAU) [94]. This finding highlights the therapeutic potential of MSC-Exo in managing and treating uveitis, offering a viable alternative to traditional corticosteroid therapies with the added benefit of reduced side effects.

4.3 *Chorioretinal Diseases*

4.3.1 Diabetic Retinopathy

Diabetic retinopathy, a major cause of blindness among working-age individuals, often exhibits limited responsiveness to standard treatments like anti-VEGF injections [95, 96]. In this context, MSC exosomes have emerged as a promising therapeutic option. Their potential in treating diabetic retinopathy lies in their ability to inhibit neovascularization and offer retinal protection [97]. Notably, exosomes from adipose tissue-derived mesenchymal stem cells (AT-MSC-Exos) have been found to carry miRNA-192 and miRNA-222. These miRNAs, when delivered to macrophages and endothelial cells (ECs) in the damaged retina, can suppress the production of inflammatory cytokines in macrophages and regulate neovascularization and blood flow, thus mitigating retinal injury [97]. Additionally, certain components within BMSC-Exos, such as miR-486-3p and miR-126, are known to counteract oxidative stress, inflammation, and apoptosis. They also promote the proliferation of high glucose-treated Muller cells, further contributing to the therapeutic efficacy of these exosomes in diabetic retinopathy [98].

4.3.2 Retinopathy of Prematurity (ROP)

ROP is a vasoproliferative disorder of the retina and a significant cause of visual impairment and childhood blindness worldwide [99]. This condition is uniquely associated with abnormal retinal vascular development in premature infants [100, 101]. In its early stages, oxygen-induced retinopathy (OIR) involves the activation of retinal microglia, which leads to the production of interleukin-1b (IL-1b) and subsequent microvascular injury [102]. While laser ablation of the peripheral avascular retina is the standard care for severe ROP [103], this treatment can lead to complications, including retinal detachment, macular ectopia, and significant myopia [104]. On the other hand, anti-vascular endothelial growth factor (VEGF) therapies present fewer ocular side effects compared to laser photocoagulation [105], but they may necessitate more frequent retreatments due to the possibility of late recurrences [106].

Recent research has focused on the intravitreal administration of exosomes derived from various cell types as potential innovative treatments for ROP [65, 107]. A notable study by Moisseiev et al. used a mouse model of oxygen-induced retinopathy (OIR) to demonstrate that exosomes from mesenchymal stem cells (MSCs) can preserve retinal vascular flow, reduce neovascularization, and mitigate retinal thinning, all without eliciting an immune response [108]. Given the established role of microglia in normal retinal vascular development [109, 110], these exosomes are thought to potentially exert protective effects against ROP. Experiments using these microglial exosomes in an oxygen-induced ROP mouse model have shown promising results, including a reduction in avascular areas, decreased expression of VEGF and TGF-b, and less neovascularization. Remarkably, injections of microglial exosomes were found to suppress photoreceptor apoptosis by inhibiting the IRE1-XBP1 cascade. This effect is partly attributed to elevated levels of microRNA-24-3p (miR-24-3p) in the exosomes, which is believed to play a crucial role in inhibiting hypoxia-induced IRE1a expression in photoreceptors.

4.3.3 Retinal Ischemia-Reperfusion Injury (IRI)

Ischemic Retinal Injury (IRI), characterized by the progressive degeneration of retinal ganglion cells (RGCs), plays a significant role in various vision-threatening diseases, including retinal vein occlusion and, notably, glaucoma [111]. In the case of glaucoma, transient spikes in intraocular hypertension can lead to the rapid occlusion of retinal blood flow. This initiates a detrimental cascade involving neuronal depolarization, calcium influx, and a compromise of the blood-retina barrier. Subsequently, during the reperfusion phase, the reintroduction of oxygen and glucose brings about a surge of free radicals and an exaggerated inflammatory response. Such a response often overwhelms the cellular antioxidant defenses, frequently resulting in cell death [112]. Despite its widespread occurrence in various ocular conditions, IRI currently lacks effective pharmacological treatment options.

In this context, MSC exosomes have shown promising neuroprotective and anti-inflammatory properties in models of IRI, aiding functional recovery and mitigating neuroinflammation [113]. Particularly, exosomes derived from human gingival MSCs (GMSCs) exhibited a high expression of miRNAs related to apoptosis and anti-inflammatory activity, such as miR-21-5p. Notably, this expression was significantly enhanced in GMSC exosomes stimulated with TNF-α. The neuroprotective effects of GMSC-derived exosomes were augmented through miR-21-5p upregulation post 48-hour TNF-α stimulation, which was found to be regulated by the lncRNA MEG3, as demonstrated through various gain- and loss-of-function experiments.

4.3.4 Choroidal Neovascularization (CNV)

Ocular neovascular disease, characterized by its progressive and degenerative nature, presents symptoms such as blurriness, darkness, and distortion in the central field of vision, leading to severe vision loss and impairing essential daily activities such as reading and driving. In the posterior segment of the eye, ocular neovascularization is typically classified as either CNV or retinal neovascularization (RNV). CNV is commonly seen in neovascular age-related macular degeneration (AMD), while RNV is frequently associated with diabetic retinopathy, both contributing significantly to the complexity of managing ocular neovascular diseases [114]. Angiogenesis and vasculogenesis, particularly in the form of CNV, are key pathobiological mechanisms in various chorioretinal diseases [115]. CNV involves the emergence of new blood vessels from the choroid into the subretinal pigment epithelium (subRPE) [116], the subretinal space, or both, necessitating often the use of anti-VEGF treatments for management.

Exosomes from various cell types, including MSCs and RPE cells, have been investigated as potential treatments, showing the ability to inhibit neovascularization and promote retinal health [117]. This study demonstrate that human umbilical cord blood mesenchymal stem cells (hUCMSCs)-derived exosomes can ameliorate retinal damage induced by blue light and laser, primarily through the downregulation of VEGF-A. This highlights the therapeutic potential of hUCMSC-derived exosomes in treating retinal diseases characterized by abnormal VEGF-A expression, offering a novel approach for managing conditions such as CNV.

4.4 *Other Ophthalmic Diseases*

4.4.1 Age-Related Macular Degeneration (AMD)

Age-Related Macular Degeneration (AMD) is a progressive condition leading to blindness in the elderly, with up to 90% of patients having no available treatment options [118]. This highlights an urgent need for innovative therapeutic strategies, especially for the untreatable dry form of AMD. Recent studies have been focusing on the use of exosomes as vehicles for delivering therapeutic agents to the retina to either treat or decelerate the progression of AMD. A key aspect of dry AMD pathogenesis is the oxidative stress-induced damage to retinal pigment epithelial cells. In addressing this, MSC exosomes have shown promising results, exhibiting anti-apoptotic and antioxidant properties in a dry AMD model. These exosomes have been found to protect retinal pigment epithelial cells from oxidative damage by regulating the Nrf2/Keap1 signaling pathway [119].

4.4.2 Glaucoma

Glaucoma, a primary cause of irreversible blindness worldwide, is characterized by progressive vision loss due to optic nerve damage. The current treatment strategies primarily aim at managing intraocular pressure (IOP) [120]. Research has shown that MSCs-Exos exhibit neuroprotective effects on retinal ganglionic cells (RGCs) in glaucoma and optic nerve crush (ONC) models. These effects include enhancing RGC survival through mechanisms such as the suppression of cis p-tau, modulation of miRNA, and secretion of neurotrophic factors [121, 122]. Exosomes from TNF-α primed MSCs provide increased neuroprotection. However, their effects tend to diminish over time, indicating a potential need for periodic administration.

Furthermore, these MSCs-Exos have shown promise in lowering IOP by addressing trabecular meshwork cell dysfunction, reducing oxidative stress and inflammation, and regulating matrix remodeling. The involvement of specific miRNAs, such as miR-21, miR-146a, and miR-17-92, underlines the dual role of MSCs-Exos in both neuroprotection and IOP management in glaucoma. This dual capacity positions MSCs-Exos as a potentially effective therapeutic approach for both halting the progression of optic nerve damage and managing the primary risk factor in glaucoma.

4.4.3 Posterior Capsule Opacification (PCO)

PCO is a common complication following cataract surgery, primarily resulting from the proliferation and transformation of residual LECs on the intraocular lens (IOL). To combat this issue, an innovative exosome-functionalized drug-loaded IOL has been developed, leveraging the homologous targeting capabilities and high biocompatibility of exosomes. These exosomes, derived from LECs, were loaded with the chemotherapeutic agent doxorubicin (Dox) and immobilized onto the IOL surface. This strategy enhances the direct delivery of the drug to LECs.

In vitro and in vivo studies have confirmed the efficacy of this approach. They showed significantly improved cellular uptake of Dox encapsulated in exosomes (Dox@Exos), leading to effective prevention of PCO and demonstrating excellent intraocular biocompatibility [123]. This novel strategy represents a significant advancement in the management of PCO, providing a targeted treatment method that minimizes side effects. It opens new avenues for clinical applications and research in the field of ophthalmology, especially in improving postoperative outcomes for cataract surgery patients.

5 Exosomes as Delivery Vectors

Since the 1970s, the development of drug delivery systems has garnered significant attention in biomedical research [124]. Innovative delivery platforms such as lipid nanoparticles, liposomes, and polymeric nanoparticles, collectively termed nanomedicines, have been instrumental in achieving targeted drug delivery to specific cells or organs [125]. Approximately 100 nanomedicines have received FDA approval for treating various diseases, including cancer and hepatitis C [126]. These platforms have been pivotal in the progress of nanomedicine, particularly in the administration of chemotherapy and other small molecular compounds. Despite their successes, ongoing safety and delivery efficiency challenges remain to be addressed.

Exosomes, serving as nanocarriers, share characteristics with cell membranes, notably their small size, negative charges, and ability to evade the immune system. These properties significantly reduce their immunogenicity and toxicity and enable them to cross biological barriers, including the blood–brain barrier [127]. This versatility makes exosomes highly effective for delivering a wide range of therapeutic agents, from chemicals and proteins to nucleic acids and gene therapies. Their potential as natural nanocarriers is rooted in their structural composition, multifaceted biological functions, and unique biological origin [98]. The review by Zhang et al. [128] outlines several key attributes that support the use of exosomes in drug delivery. Firstly, their abundance across various cellular environments suggests stability under different physiological and pathological conditions [129]. Secondly, compared to other drug delivery methods, exosomes offer intrinsic benefits such as low toxicity, minimal immunogenicity, ease of manipulation, high stability, biocompatibility, and efficient penetration [130, 131]. Thirdly, their unique biomolecular composition ensures targeted delivery to specific cells [132]. Finally, their ability to penetrate biological barriers like the blood–brain and blood–retina barriers addresses a significant challenge in systemic drug delivery, where many drugs and carriers fail to cross such barriers.

Exosomal therapies offer numerous potential applications in ophthalmology, particularly due to their ability to mediate diverse therapeutic effects such as anti-angiogenic, anti-inflammatory, neuroprotective, and regenerative actions. This is further illustrated by several studies. Samaeekia et al. [133] demonstrated that exosomes derived from human corneal mesenchymal stem cells not only enhanced in vitro cell migration and proliferation but also expedited wound healing in vivo, suggesting a novel approach for treating corneal wound healing disorders. Similarly, Mead and Tomarev [76] observed neuroprotective effects in a rat model of optic nerve crush following intravitreal injection of bone marrow-derived mesenchymal stem cell exosomes, leading to increased retinal ganglion cell survival and axonal regeneration. Yu et al. [134] reported that MSC-derived exosomes reduced damage, apoptosis, and inflammation in a mouse model of laser-induced retinopathy.

In the context of gene therapy, adeno-associated virus (AAV) vectors, though favored for their high gene transfer efficiency and safety profile, encounter limitations in capacity and potential immune responses. With their larger capacity and low immunogenicity, exosomes have emerged as innovative vectors for gene delivery. They have shown effectiveness in targeted gene therapy, notably in delivering the CRISPR-Cas9 system for tumor targeting and packaging viruses for immune evasion. This facilitates the repeated administration of large DNA quantities to specific tissues. Wassmer et al. [135] found that exosome-associated AAV2 enabled more effective gene delivery to the retina compared to conventional AAV2 vectors. The ability of exosomes containing AAV to improve gene transfer efficiency in the eye is significant. Studies indicate that exosome-based vectors can more effectively penetrate retinal layers and deliver AAV cargoes efficiently [136]. The synergistic combination of exosomal gene delivery with AAV capsid modifications has resulted in more efficient delivery systems at lower doses. Exo-AAV, therefore, represents a promising gene delivery technique for retinal applications, and its ease of manufacturing and isolation bodes well for extensive use in eye research.

Despite these advantages, the clinical translation of exosomal therapies faces challenges, particularly in isolation and enrichment. Additionally, the effects of exosomes in human subjects are not yet extensively studied, primarily due to ethical considerations and the reliance on animal models [128]. Nevertheless, when integrated with targeted drug delivery methods, exosomes show great promise as a therapy for a variety of ocular diseases. The heterogeneity of ocular diseases and their corresponding exosomes also point toward a future of personalized, highly effective treatments [137].

6 Corneal Stromal Stem Cell-Derived Exosomes (CSSC-Exos)

Limbal epithelial stem cells (LESCs), situated in the basal layer of the limbus, play a crucial role in preserving the integrity and functionality of the corneal epithelium [138]. Damage to LESCs, often due to factors such as chemical burns, extended contact lens usage, or chronic inflammation, can cause limbal stem cell deficiency (LSCD), leading to ocular irritation, impaired vision, neovascularization, and heightened risks of keratoplasty failure and immunologic rejection [139]. The standard treatment for LSCD, limbal stem cell transplantation, particularly involving in vitro-expanded LESCs, faces challenges in long-term corneal surface regeneration. The typical cultivation method using mouse 3T3 fibroblasts as feeder cells fails to fully mimic the natural limbal niche, underscoring the need to explore LESCs' molecular regulation by their niche [140].

In this context, corneal stromal stem cells (CSSCs) are of significant interest due to their crucial role in maintaining LESCs' stemness, primarily influencing LESCs through paracrine mechanisms rather than direct contact [141]. CSSCs, predominantly located in the anterior corneal stroma, are hypothesized to communicate with LESCs via exosomes [142]. These exosomes, adept at transferring bioactive molecules and altering the phenotype of target cells, offer a promising research path. Their noncellular composition, low immunogenicity, and ease of integrating them into mediums or topical eye drops make them potential candidates for treating LSCD and similar conditions. A focus on the role of exosomes in LESCs and CSSCs interaction could lead to breakthroughs in understanding the regulation of LESCs by niche factors and in developing more effective culture systems and treatment strategies that better replicate the limbal microenvironment [143].

Highlighting this, a study by Wang et al. [144] employed 2D and 3D coculture systems to investigate the paracrine impact of CSSCs on LESCs. Their findings underscored the vital role of CSSC-derived EVs (CSSC-EVs) in regulating LESCs, evidenced by an increased cell density and a higher proportion of p63α bright cells upon adding CSSC-EVs. Notably, miRNAs such as hsa-miR-663b, hsa-miR-16-5p, and hsa-miR-1290 within CSSC-EVs, which target the Notch pathway, were found in abundance. This underscores the pivotal role of CSSC-EVs in enhancing LESC proliferation and maintaining stemness by modulating the Notch signaling pathway through miRNAs, thereby offering new insights into LESC culture and potential treatments for corneal epithelial disorders.

7 Exosomes as Biomarkers for Eye Diseases

The burgeoning field of stem cell research in regenerative medicine has recently shifted its focus toward the utilization of exosomes in identifying biomarkers for ocular diseases [145]. Exosomes and extracellular vesicles (EVs) possess several characteristics that render them ideal for biomarker discovery: They have a protective lipid bilayer that safeguards the RNA, DNA, and proteins within from external nucleases and proteases, they contain cell-, tissue-, or disease-specific biomolecules, and they are robust enough to endure harsh conditions, facilitating their isolation and enrichment from various biological fluids, including plasma, serum, urine, saliva, semen, breast milk, aqueous humor, and cerebrospinal fluid. This potential is underscored by promising findings from cancer, cardiovascular, and diabetes research, where exosomes have been explored as diagnostic and prognostic biomarkers, and therapeutic vehicles [137, 146].

Exosomes are secreted by both healthy and pathological cells. In cancer patients, serum exosome concentrations have been instrumental in diagnosis and prognosis. Similarly, tear fluids offer a noninvasive medium for identifying exosome-based biomarkers. Despite the challenge of small sample volumes in tear fluid analysis, the expression levels of exosome markers such as CD9 and CD63 are significantly higher in tears than in serum, making them a promising diagnostic tool [147, 148]. While the literature on tear exosomes as disease biomarkers is still limited, tear-based proteomics has identified several exosome-associated proteins, indicating the potential for further research in this area [149–151].

Exosomes are also prevalent in other ocular fluids such as aqueous humor (AH) and vitreous humor, playing a role in both ocular surface health and disease progression [150, 152]. Protein profiling of AH and retinal pigment epithelium (RPE)-derived exosomes could yield novel diagnostic biomarkers for glaucoma, neovascular age-related macular degeneration (AMD), and other retinal diseases [152, 153]. The recent discovery that miR-146a from vitreous humor exosomes could serve as a biomarker for uveal melanoma further highlights this potential [154]. Although exosomes have been proposed as a treatment for ocular surface regeneration and various pathologies such as corneal fibrosis and dry eye diseases, there is a notable absence of literature on the diagnostic use of exosomes derived from pathological tears. However, with ongoing advancements in exosome analysis technologies, this remains a promising and largely unexplored area of research.

8 Potential Risks to Exosomes in

8.1 Interference with Neural Communication

While exosomes have shown promise in various therapeutic applications in ophthalmology, there are potential risks associated with their use, particularly in terms of neural communication [155]. This is a significant concern in the context of the optic nerve, an essential component of the visual system that cannot regenerate after injury. The apprehension stems from observations in other neurological conditions where exosomes have been linked to detrimental effects.

In Alzheimer's disease, for instance, exosomes have been found to transport amyloid-β peptide, a key factor in the disease's pathology, potentially exacerbating its progression [156, 157]. Similarly, in Parkinson's disease, the propagation of the α-synuclein protein, another neurodegenerative agent, has been observed to occur via exosomes. These exosomes can travel from enteric neurons to the brainstem and cortical centers, suggesting a mechanism for the disease's spread [158]. Such findings raise concerns that exosomes might inadvertently interfere with neural communication pathways. This is particularly relevant for ocular health, considering the crucial role of the optic nerve in visual processing. The potential for exosomes to affect the optic nerve negatively could have significant implications, highlighting the need for a cautious and thorough evaluation of exosomal therapies in ophthalmology.

8.2 *Risk of Intraocular Unexpected Proliferation*

The use of exosomes in intraocular therapies, while promising, raises concerns about the unintended proliferation of fibroblasts, as highlighted in studies [159]. Fibroblast proliferation is a double-edged sword; it can be beneficial for repairing damage caused by trauma or inflammation, but excessive proliferation may lead to complications. Overactive fibroblast activity can result in an accumulation of excess collagen, leading to the formation of scar tissue or angiogenesis, and ultimately, organ dysfunction [160].

For example, endothelial progenitor cell-derived exosomes have been shown to increase fibroblast proliferation and angiogenesis, which could complicate cardiac repair processes [161]. Similarly, exosomal microRNA-21 has been implicated in promoting the proliferation of keloid fibroblasts and enhancing collagen production, contributing to excessive scar formation [162]. These instances underscore the need for careful consideration and monitoring when using exosomes in intraocular treatments. The delicate balance required for effective tissue repair without triggering adverse proliferative responses is crucial, particularly in the eye, where such responses can significantly impair vision and ocular function. Thus, while exosomal therapies hold great promise, their potential risks, especially in delicate and complex systems like the eye, necessitate a cautious and thoroughly researched approach.

9 Future Challenges and Perspectives and Conclusion

Exosomes have emerged as a promising, cell-free, and adaptable method in ophthalmology, offering the potential to enhance patient outcomes significantly. The past decade has seen considerable advancements in understanding the biology of exosomes and their application in ophthalmic treatments. Researchers have gained deeper insights into their biophysical properties, origins, compositions, and methods for their isolation and characterization. Significant progress has been made in demonstrating their therapeutic potential in preclinical models for various ophthalmic diseases, including traumatic, autoimmune, and chorioretinal diseases.

Current research is focused on enhancing the therapeutic efficacy of exosomes. This includes improving their stability, targeting capabilities, and increasing their therapeutic load. The methods of administering exosomes in ophthalmology are in line with established clinical practices. For anterior segment diseases, direct applications to the ocular surface, such as eye drops and gels, are effective [87]. For treating the posterior segment, intravitreal injections are commonly used, while subconjunctival and periocular/retro-orbital injections provide alternative routes, minimizing infection and inflammation risks [3]. Surface modifications, like the addition of targeting peptides and proteins, are being explored to enhance exosome targeting.

Advancements in chemical, cellular, and genetic engineering are likely to yield innovative strategies for exosomal modifications to enhance therapeutic effects. An example is the use of macrophage tumor chimeric exosomes for antitumor therapy by Shuang Wang et al., combining the functions of two source cells for dual targeting and immune activation [163].

However, the field faces challenges, including understanding the biological mechanisms of exosomes, standardized large-scale production, and safety concerns. Future efforts should aim for a deeper understanding of exosomes from various sources, developing new enrichment and purification methods to meet pharmaceutical industry standards, and advancing clinical translation through guidelines for exosome separation and characterization. This encompasses optimizing manufacturing, culture conditions, storage, ensuring product purity, and quality control.

Before clinical approval, the safety and efficacy of exosomes must undergo rigorous testing through clinical trials. Regulatory agencies including the European Medicines Agency and the FDA provide guidelines for novel drug approval. While clinical trials involving exosome agents for ophthalmic therapeutics are currently limited, this number is expected to increase. In conclusion, exosomes present a promising avenue for ophthalmic therapies, but further research is essential to optimize their therapeutic applications, establish specific protocols, and understand their molecular actions for successful clinical translation.

Acknowledgements This study was financially supported by open funding from the Nankai University Eye Institute (NKYKD202203), the Research Project on Skin Injury & Repair (BKJ21J016, 2021XB008), and the Tianjin Natural Science Foundation (22JCZXJC00170, 21JCZDJC00070).

References

1. Messaoudi MD, Menelas BAJ, Mcheick H. Review of navigation assistive tools and technologies for the visually impaired. Sensors. 2022;22(20):7888.
2. Keel S, Müller A, Block S, Bourne R, Burton MJ, Chatterji S, He M, Lansingh VC, Mathenge W, Mariotti S, Muirhead D, Rabiu MM, Ravilla TD, Resnikoff S, Silva JC,

Tapply I, Vos T, Wang N, Cieza A. Keeping an eye on eye care: monitoring progress towards effective coverage. Lancet Global Health 2021;9(10):e1460–e1464. https://doi.org/10.1016/S2214-109X(21)00212-6. https://www.sciencedirect.com/science/article/pii/S2214109X21002126
3. Tian Y, Zhang T, Li J, Tao Y. Advances in development of exosomes for ophthalmic therapeutics. Adv Drug Deliv Rev. 2023;199:114899.
4. Mitry D, Awan MA, Borooah S, Siddiqui MAR, Brogan K, Fleck BW, Wright A, Campbell H, Singh J, Charteris DG, Yorston D. Surgical outcome and risk stratification for primary retinal detachment repair: results from the scottish retinal detachment study. Br J Ophthalmol. 2012;96(5):730–734. https://doi.org/10.1136/bjophthalmol-2011-300581. https://bjo.bmj.com/content/96/5/730. https://bjo.bmj.com/content/96/5/730.full.pdf
5. Li S, Jie Y. Cataract surgery and lens implantation. Curr Opin Ophthalmol. 2019;30(1):39–43.
6. Nichani P, Popovic MM, Schlenker MB, Park J, Ahmed IIK. Microinvasive glaucoma surgery: a review of 3476 eyes. Surv Ophthalmol. 2021;66(5):714–742. https://doi.org/10.1016/j.survophthal.2020.09.005. https://www.sciencedirect.com/science/article/pii/S0039625720301351
7. Storey PP, Ter-Zakarian A, Philander SA, De Koo LO, George M, Humayun MS, Rodger DC, Ameri H. Visual and anatomical outcomes after diabetic traction and traction-rhegmatogenous retinal detachment repair. Retina. 2018;38(10):1913–9.
8. Wubben TJ, Besirli CG, Zacks DN. Pharmacotherapies for retinal detachment. Ophthalmology. 2016;123(7):1553–62.
9. Efraim Y, Chen FYT, Stashko C, Cheong KN, Gaylord E, McNamara N, Knox SM. Alterations in corneal biomechanics underlie early stages of autoimmune-mediated dry eye disease. J Autoim. 2020;114: 102500.
10. Urtti A. Challenges and obstacles of ocular pharmacokinetics and drug delivery. Adv Drug Deliv Rev. 2006;58(11):1131–5.
11. Lamminsalo M, Urtti A, Ranta VP. Quantitative pharmacokinetic analyses of anterior and posterior elimination routes of intravitreal anti-vegf macromolecules using published human and rabbit data. Exp Eye Res. 2022;222: 109162.
12. Jiang Y, Lin S, Gao Y. Mesenchymal stromal cell-based therapy for dry eye: current status and future perspectives. Cell Transplant. 2022;31:09636897221133818.
13. Zhang X, Liu J, Yu B, Ma F, Ren X, Li X. Effects of mesenchymal stem cells and their exosomes on the healing of large and refractory macular holes. Graefe's Arch Clin Exp Ophthalmol. 2018;256:2041–52.
14. Singh MS, Park SS, Albini TA, Canto-Soler MV, Klassen H, MacLaren RE, Takahashi M, Nagiel A, Schwartz SD, Bharti K. Retinal stem cell transplantation: Balancing safety and potential. Prog Retinal Eye Res. 2020;75: 100779.
15. Park SS, Moisseiev E, Bauer G, Anderson JD, Grant MB, Zam A, Zawadzki RJ, Werner JS, Nolta JA. Advances in bone marrow stem cell therapy for retinal dysfunction. Prog Retinal Eye Res. 2017;56:148–65.
16. Ma J, Zhao Y, Sun L, Sun X, Zhao X, Sun X, Qian H, Xu W, Zhu W. Exosomes derived from akt-modified human umbilical cord mesenchymal stem cells improve cardiac regeneration and promote angiogenesis via activating platelet-derived growth factor d. Stem Cells Translat Med. 2017;6(1):51–9.
17. De Couto G, Gallet R, Cambier L, Jaghatspanyan E, Makkar N, Dawkins JF, Berman BP, Marbán E. Exosomal microrna transfer into macrophages mediates cellular postconditioning. Circulation. 2017;136(2):200–14.
18. Gray WD, French KM, Ghosh-Choudhary S, Maxwell JT, Brown ME, Platt MO, Searles CD, Davis ME. Identification of therapeutic covariant microrna clusters in hypoxia-treated cardiac progenitor cell exosomes using systems biology. Circ Res. 2015;116(2):255–63.
19. Ribeiro-Rodrigues TM, Laundos TL, Pereira-Carvalho R, Batista-Almeida D, Pereira R, Coelho-Santos V, Silva AP, Fernandes R, Zuzarte M, Enguita FJ, et al. Exosomes secreted

by cardiomyocytes subjected to ischaemia promote cardiac angiogenesis. Cardiovasc Res. 2017;113(11):1338–50.
20. Gilligan KE, Dwyer RM. Engineering exosomes for cancer therapy. Int J Molec Sci. 2017;18(6):1122.
21. Lin Z, Wu Y, Xu Y, Li G, Li Z, Liu T. Mesenchymal stem cell-derived exosomes in cancer therapy resistance: recent advances and therapeutic potential. Molec Cancer. 2022;21(1):179.
22. Ferreira D, Moreira JN, Rodrigues LR. New advances in exosome-based targeted drug delivery systems. Crit Rev Oncol/Hematol. 2022;172: 103628.
23. Nasir F. The promise of exosomes as drug delivery systems. Pharm Commun. 2023;2(01):01–2.
24. Moholkar DN, Kandimalla R, Gupta RC, Aqil F. Advances in lipid-based carriers for cancer therapeutics: liposomes, exosomes and hybrid exosomes. Cancer Lett. 2023;565: 216220.
25. Tenchov R, Sasso JM, Wang X, Liaw WS, Chen CA, Zhou QA. Exosomes-nature's lipid nanoparticles, a rising star in drug delivery and diagnostics. ACS Nano. 2022;16(11):17802–46.
26. Sadeghi S, Tehrani FR, Tahmasebi S, Shafiee A, Hashemi SM. Exosome engineering in cell therapy and drug delivery. Inflammopharmacology. 2023;31(1):145–69.
27. Zhao Y, Liu T, Zhou M. Immune-cell-derived exosomes for cancer therapy. Mol Pharm. 2022;19(9):3042–56.
28. Kimiz-Gebologlu I, Oncel SS. Exosomes: large-scale production, isolation, drug loading efficiency, and biodistribution and uptake. J. Controlled Release. 2022;347:533–43.
29. Basak M, Chaudhary DK, Takahashi Ru, Yamamoto Y, Tiwari S, Tahara H, Mittal A. Immunocyte derived exosomes: insight into the potential chemo-immunotherapeutic nanocarrier targeting the tumor microenvironment. ACS Biomat Sci Eng **9**(1), 20–39 (2022)
30. Joorabloo A, Liu T. Engineering exosome-based biomimetic nanovehicles for wound healing. J. Controlled Release. 2023;356:463–80.
31. Jiang Z, Wang F, Chen Z, Fan Y, Xu J, Yu M. Corneal biometric parameters and refractive properties in microcornea with normal axial length. Cornea. 2022;41(9):1074–9.
32. DelMonte DW, Kim T. Anatomy and physiology of the cornea. J Cataract Refractive Surg 2011;37(3):588–598. https://doi.org/10.1016/j.jcrs.2010.12.037. https://www.sciencedirect.com/science/article/pii/S0886335010019243
33. Eghrari AO, Riazuddin SA, Gottsch JD. Overview of the cornea: structure, function, and development. Progr Mol Biol Translat Sci. 2015;134:7–23.
34. Pflugfelder SC, Stern ME. Biological functions of tear film. Exp Eye Res. 2020;197: 108115.
35. Poli BM, Parisi G, Zampacavallo G, Mecatti M, Lupi P, Gualtieri M, Franci O. Quality outline of european sea bass (dicentrarchus labrax) reared in italy: shelf life, edible yield, nutritional and dietetic traits. Aquaculture. 2001;202(3–4):303–15.
36. Cholkar K, Dasari SR, Pal D, Mitra AK. Ocular transporters and receptors. Amsterdam: Elsevier; 2013. p. 1–36.
37. Rates ERD, Almeida CD, Costa EDPF, Farias RJDM, Santos-Oliveira R, Alencar LMR. Layer-by-layer investigation of ultrastructures and biomechanics of human cornea. Int J Mol Sci 2022;23(14):7833.
38. Lawman S, Mason S, Kaye SB, Shen YC, Zheng Y. Accurate in vivo bowman's thickness measurement using mirau ultrahigh axial resolution line field optical coherence tomography. Transl Vis Sci Technol. 2022;11(8):6–6.
39. Heidari Z, Mohammadpour M, Hajizadeh F, Fotouhi A, Hashemi H. Corneal layer thickness in keratoconus using optical coherence tomography. Clin Exp Optom 1–8 (2023)
40. Wilson SE, Sampaio LP, Shiju TM, Hilgert GS, de Oliveira RC. Corneal opacity: cell biological determinants of the transition from transparency to transient haze to scarring fibrosis, and resolution, after injury. Invest Ophthalmol Visual Sci. 2022;63(1):22–22.
41. Silver FH, Deshmukh T, Benedetto D, Gonzalez-Mercedes M, Mesica A. Measurement of the elastic modulus of cornea, sclera and limbus: The importance of the corneal-limbus-scleral biomechanical unit. Front Biosci-Schol. 2022;14(4):30.

42. Tang Q, Lu B, He J, Chen X, Fu Q, Han H, Luo C, Yin H, Qin Z, Lyu D, et al. Exosomes-loaded thermosensitive hydrogels for corneal epithelium and stroma regeneration. Biomaterials. 2022;280: 121320.
43. Hatami-Marbini H, Emu ME. The relationship between keratan sulfate glycosaminoglycan density and mechanical stiffening of cxl treatment. Exp Eye Res. 2023;234: 109570.
44. Petrela RB, Patel SP, The soil and the seed: the relationship between descemet's membrane and the corneal endothelium. Exp Eye Res. 2022;227:109376.
45. Vercammen H, Miron A, Oellerich S, Melles GR, Dhubhghaill SN, Koppen C, Van Den Bogerd B. Corneal endothelial wound healing: understanding the regenerative capacity of the innermost layer of the cornea. Transl Res. 2022;248:111–27.
46. Sekhon AS, He B, Iovieno A, Yeung SN. Pathophysiology of corneal endothelial cell loss in dry eye disease and other inflammatory ocular disorders. Ocul Immunol Inflammation. 2023;31(1):21–31.
47. Mohan RR, Kempuraj D, D'Souza S, Ghosh A. Corneal stromal repair and regeneration. Prog Retinal Eye Res. 2022;91: 101090.
48. Wu M, Hill LJ, Downie LE, Chinnery HR. Neuroimmune crosstalk in the cornea: the role of immune cells in corneal nerve maintenance during homeostasis and inflammation. Prog Retinal Eye Res. 2022;91:101105.
49. Pflugfelder SC, Cao A, Galor A, Nichols KK, Cohen NA, Dalton M. Nicotinic acetylcholine receptor stimulation: a new approach for stimulating tear secretion in dry eye disease. Ocul Surf. 2022;25:58–64.
50. Johnstone RM, Adam M, Hammond J, Orr L, Turbide C, Vesicle formation during reticulocyte maturation. Association of plasma membrane activities with released vesicles (exosomes). J Biol Chem **262**(19), 9412–9420 (1987)
51. Ferraris VA. Enhanced exosomes and spinal cord ischemia: excitement and a nobel prize but more to come. J Thoracic Cardiovas Surg. 2019;157(2):518–9.
52. Qi H, Liu C, Long L, Ren Y, Zhang S, Chang X, Qian X, Jia H, Zhao J, Sun J, et al. Blood exosomes endowed with magnetic and targeting properties for cancer therapy. ACS Nano. 2016;10(3):3323–33.
53. Pisitkun T, Shen RF, Knepper MA. Identification and proteomic profiling of exosomes in human urine. Proc Natl Acad Sci. 2004;101(36):13368–73.
54. Grigor'Eva A, Tamkovich S, Eremina A, Tupikin A, Kabilov M, Chernykh V, Vlassov V, Laktionov P, Ryabchikova E. Exosomes in tears of healthy individuals: isolation, identification, and characterization. Biochem (Moscow) Suppl Ser B Biomed Chem 2016;10:165–172
55. Calvo V, Izquierdo M. Inducible polarized secretion of exosomes in t and b lymphocytes. Int J Mol Sci. 2020;21(7):2631.
56. Ahmadi M, Rezaie J. Tumor cells derived-exosomes as angiogenenic agents: possible therapeutic implications. J Transl Med. 2020;18(1):1–17.
57. Huda MN, Nafiujjaman M, Deaguero IG, Okonkwo J, Hill ML, Kim T, Nurunnabi M. Potential use of exosomes as diagnostic biomarkers and in targeted drug delivery: progress in clinical and preclinical applications. ACS Biomat Sci Eng. 2021;7(6):2106–49.
58. Tanziela T, Shaikh S, Jiang H, Lu Z, Wang X. Efficient encapsulation of biocompatible nanoparticles in exosomes for cancer theranostics. Nano Today. 2020;35: 100964.
59. Meldolesi J. Exosomes and ectosomes in intercellular communication. Curr Biol. 2018;28(8):R435–44.
60. Egea-Jimenez AL, Zimmermann P. Lipids in exosome biology. In: Lipid Signaling in Human Diseases. Springer, Cham; 2020. pp. 309–336.
61. Lei Y, Fei X, Ding Y, Zhang J, Zhang G, Dong L, Song J, Zhuo Y, Xue W, Zhang P, et al. Simultaneous subset tracing and mirna profiling of tumor-derived exosomes via dual-surface-protein orthogonal barcoding. Sci Adv. 2023;9(40):eadi1556.
62. Chandrasekera D, Shah R, van Hout I, De Jonge W, Bunton R, Parry D, Davis P, Katare R. Combination of precipitation and size exclusion chromatography as an effective method

for exosome like extracellular vesicle isolation from pericardial fluids. Nanotheranostics. 2023;7(4):345.
63. Zhang M, Hu S, Liu L, Dang P, Liu Y, Sun Z, Qiao B, Wang C. Engineered exosomes from different sources for cancer-targeted therapy. Signal Transduction Targeted Ther. 2023;8(1):124.
64. Tidke SC, Tidake P. A review of corneal blindness: causes and management. Cureus 2022;14(10):e30097.
65. Wang E, Kong X, Wolle M, Gasquet N, Ssekasanvu J, Mariotti SP, Bourne R, Taylor H, Resnikoff S, West S. Global trends in blindness and vision impairment due to corneal opacity 1984-2020: a meta-analysis. Ophthalmology 2023;130(8):863–871.
66. Whitcher JP, Srinivasan M, Upadhyay MP. Corneal blindness: a global perspective. Bull World Health Organiz. 2001;79(3):214–21.
67. Tao H, Chen X, Cao H, Zheng L, Li Q, Zhang K, Han Z, Han ZC, Guo Z, Li Z, et al. Mesenchymal stem cell-derived extracellular vesicles for corneal wound repair. Stem Cells Int. 2019;2019:5738510.
68. Han KY, Tran JA, Chang JH, Azar DT, Zieske JD. Potential role of corneal epithelial cell-derived exosomes in corneal wound healing and neovascularization. Sci Rep. 2017;7(1):40548.
69. Shen T, Zheng Q, Luo H, Li X, Chen Z, Song Z, Zhou G, Hong C. Exosomal mir-19a from adipose-derived stem cells suppresses differentiation of corneal keratocytes into myofibroblasts. Aging. 2020;12(5):4093.
70. Shojaati G, Khandaker I, Funderburgh ML, Mann MM, Basu R, Stolz DB, Geary ML, Dos Santos A, Deng SX, Funderburgh JL. Mesenchymal stem cells reduce corneal fibrosis and inflammation via extracellular vesicle-mediated delivery of mirna. Stem Cells Transl Med. 2019;8(11):1192–201.
71. Nuzzi R, Buono L, Scalabrin S, De Iuliis M, Bussolati B, et al. Effect of stem cell-derived extracellular vesicles on damaged human corneal endothelial cells. Stem Cells Int. 2021;2021:6644463.
72. Cao H, Cheng Y, Gao H, Zhuang J, Zhang W, Bian Q, Wang F, Du Y, Li Z, Kong D, et al. In vivo tracking of mesenchymal stem cell-derived extracellular vesicles improving mitochondrial function in renal ischemia-reperfusion injury. ACS Nano. 2020;14(4):4014–26.
73. Butreddy A, Kommineni N, Dudhipala N. Exosomes as naturally occurring vehicles for delivery of biopharmaceuticals: insights from drug delivery to clinical perspectives. Nanomaterials. 2021;11(6):1481.
74. Nuzzi R, Caselgrandi P, Vercelli A, et al. Effect of mesenchymal stem cell-derived exosomes on retinal injury: a review of current findings. Stem cells Int. 2020;2020:8883616.
75. Cui Y, Liu C, Huang L, Chen J, Xu N. Protective effects of intravitreal administration of mesenchymal stem cell-derived exosomes in an experimental model of optic nerve injury. Exp Cell Res. 2021;407(1): 112792.
76. Mead B, Tomarev S. Bone marrow-derived mesenchymal stem cells-derived exosomes promote survival of retinal ganglion cells through mirna-dependent mechanisms. Stem Cells Transl Med. 2017;6(4):1273–85.
77. Kelly NE, Wendel RT. Vitreous surgery for idiopathic macular holes: results of a pilot study. Arch Ophthalmol. 1991;109(5):654–9.
78. Michalewska Z, Michalewski J, Cisiecki S, Adelman R, Nawrocki J. Correlation between foveal structure and visual outcome following macular hole surgery: a spectral optical coherence tomography study. Graefe's Arch Clin Exp Ophthalmol. 2008;246:823–30.
79. Ezra E, Gregor ZJ, ; Morfields Macular Hole Study Ggroup, et al. Surgery for idiopathic full-thickness macular hole: two-year results of a randomized clinical trial comparing naturalhistory, vitrectomy, and vitrectomy plus autologous serum: moorfields macularhole study group report no. 1. Arch Ophthalmol 2004;122(2):224–36.
80. Suda K, Hangai M, Yoshimura N. Axial length and outcomes of macular hole surgery assessed by spectral-domain optical coherence tomography. Am J Ophthalmol. 2011;151(1):118–27.

81. Ma M, Li B, Zhang M, Zhou L, Yang F, Ma F, Shao H, Li Q, Li X, Zhang X. Therapeutic effects of mesenchymal stem cell-derived exosomes on retinal detachment. Exp Eye Res. 2020;191:107899.
82. de Andrade FA, Fiorot SHS, Benchimol EI, Provenzano J, Martins VJ, Levy RA. The autoimmune diseases of the eyes. Autoimmuny Rev. 2016;15(3):258–71.
83. Keino H, Horie S, Sugita S, Immune privilege and eye-derived t-regulatory cells. J Immunol Res. 2018:2018:1679197.
84. Huang Y, Li R, Ye S, Lin S, Yin G, Xie Q. Recent advances in the use of exosomes in sjögren's syndrome. Front Immunol. 2020;11:1509.
85. Sandhya P, Theyilamannil Kurien B, Danda D, Hal Scofield R. Update on pathogenesis of sjogren's syndrome. Curr Rheumatol Rev. 2017;13(1):5–22.
86. Gallo A, Jang SI, Ong HL, Perez P, Tandon M, Ambudkar I, Illei G, Alevizos I. Targeting the Ca2+ sensor STIM1 by exosomal transfer of Ebv-miR-BART13-3p is associated with sjögren's syndrome. EBioMedicine. 2016;10:216–26.
87. Li N, Zhao L, Wei Y, Ea VL, Nian H, Wei R. Recent advances of exosomes in immune-mediated eye diseases. Stem Cell Res Therapy. 2019;10(1):1–10.
88. Marino J, Paster J, Benichou G. Allorecognition by t lymphocytes and allograft rejection. Front Immunol. 2016;7:582.
89. Gonzalez-Nolasco B, Wang M, Prunevieille A, Benichou G. Emerging role of exosomes in allorecognition and allograft rejection. Curr Opin Organ Transpl. 2018;23(1):22.
90. Jangamreddy JR, Haagdorens MK, Islam MM, Lewis P, Samanta A, Fagerholm P, Liszka A, Ljunggren MK, Buznyk O, Alarcon EI, et al. Short peptide analogs as alternatives to collagen in pro-regenerative corneal implants. Acta Biomater. 2018;69:120–30.
91. Leszczynska A, Kulkarni M, Ljubimov AV, Saghizadeh M. Exosomes from normal and diabetic human corneolimbal keratocytes differentially regulate migration, proliferation and marker expression of limbal epithelial cells. Sci Rep. 2018;8(1):15173.
92. Zhou T, He C, Lai P, Yang Z, Liu Y, Xu H, Lin X, Ni B, Ju R, Yi W, et al. MIR-204–containing exosomes ameliorate gvhd-associated dry eye disease. Sci Adv. 20222;8(2):eabj9617.
93. Rudraprasad D, Rawat A, Joseph J. Exosomes, extracellular vesicles and the eye. Exp Eye Res. 2022;214: 108892.
94. Bai L, Shao H, Wang H, Zhang Z, Su C, Dong L, Yu B, Chen X, Li X, Zhang X. Effects of mesenchymal stem cell-derived exosomes on experimental autoimmune uveitis. Sci Rep. 2017;7(1):4323.
95. Stewart MW. Treatment of diabetic retinopathy: recent advances and unresolved challenges. World J Diabetes. 2016;7(16):333.
96. Ajlan RS, Silva PS, Sun JK. Seminars in ophthalmology, vol. 31. Milton Park: Taylor & Francis; 2016. p. 40–48.
97. Harrell CR, Volarevic V, Djonov V, Volarevic A. Therapeutic potential of exosomes derived from adipose tissue-sourced mesenchymal stem cells in the treatment of neural and retinal diseases. Int J Mol Sci. 2022;23(9):4487.
98. Li W, Jin L, Cui Y, Nie A, Xie N, Liang G. Bone marrow mesenchymal stem cells-induced exosomal microrna-486-3p protects against diabetic retinopathy through tlr4/nf-κb axis repression. J Endocrinol Invest. 2021;44:1193–207.
99. Fevereiro-Martins M, Marques-Neves C, Guimarães H, Bicho M. Retinopathy of prematurity: a review of pathophysiology and signaling pathways. Surv Ophthalm 2023;68(2):175–210.
100. Solebo AL, Teoh L, Rahi J. Epidemiology of blindness in children. Arch Disease Childhood. 2017;102(9):853–7.
101. Hartnett ME, Penn JS. Mechanisms and management of retinopathy of prematurity. N Engl J Med. 2012;367(26):2515–26.
102. Rivera JC, Sitaras N, Noueihed B, Hamel D, Madaan A, Zhou T, Honoré JC, Quiniou C, Joyal JS, Hardy P, et al. Microglia and interleukin-1β in ischemic retinopathy elicit microvascular degeneration through neuronal semaphorin-3a. Arterioscler Thromb Vasc Biol. 2013;33(8):1881–91.

103. Hartnett ME. Advances in understanding and management of retinopathy of prematurity. Surv Ophthalmol. 2017;62(3):257–76.
104. Hwang CK, Hubbard GB, Hutchinson AK, Lambert SR. Outcomes after intravitreal bevacizumab versus laser photocoagulation for retinopathy of prematurity: a 5-year retrospective analysis. Ophthalmology. 2015;122(5):1008–15.
105. Seery CW, Betesh S, Guo S, Zarbin MA, Bhagat N, Wagner RS. Update on the use of anti-vegf drugs in the treatment of retinopathy of prematurity. J Pediatric Ophthalmol Strabismus. 2020;57(6):351–62.
106. Barnett JM, Hubbard GB. Complications of retinopathy of prematurity treatment. Curr Opin Ophthalmol. 2021;32(5):475.
107. Xu W, Wu Y, Hu Z, Sun L, Dou G, Zhang Z, Wang H, Guo C, Wang Y. Exosomes from microglia attenuate photoreceptor injury and neovascularization in an animal model of retinopathy of prematurity. Mol Therapy-Nucleic Acids. 2019;16:778–90.
108. Moisseiev E, Anderson JD, Oltjen S, Goswami M, Zawadzki RJ, Nolta JA, Park SS. Protective effect of intravitreal administration of exosomes derived from mesenchymal stem cells on retinal ischemia. Curr Eye Rese. 2017;42(10):1358–67.
109. Deliyanti D, Talia DM, Zhu T, Maxwell MJ, Agrotis A, Jerome JR, Hargreaves EM, Gerondakis S, Hibbs ML, Mackay F, et al. Foxp3+ tregs are recruited to the retina to repair pathological angiogenesis. Nat Commun. 2017;8(1):748.
110. Ebneter A, Kokona D, Schneider N, Zinkernagel MS. Microglia activation and recruitment of circulating macrophages during ischemic experimental branch retinal vein occlusion. Invest Ophthalmol Visual Sci. 2017;58(2):944–53.
111. Musayeva A, Unkrig JC, Zhutdieva MB, Manicam C, Ruan Y, Laspas P, Chronopoulos P, Göbel ML, Pfeiffer N, Brochhausen C, et al. Betulinic acid protects from ischemia-reperfusion injury in the mouse retina. Cells. 2021;10(9):2440.
112. Osborne NN, Casson RJ, Wood JP, Chidlow G, Graham M, Melena J. Retinal ischemia: mechanisms of damage and potential therapeutic strategies. Prog Retinal Eye Res. 2004;23(1):91–147.
113. Yu Z, Wen Y, Jiang N, Li Z, Guan J, Zhang Y, Deng C, Zhao L, Zheng SG, Zhu Y, et al. Tnf-α stimulation enhances the neuroprotective effects of gingival mscs derived exosomes in retinal ischemia-reperfusion injury via the meg3/mir-21a-5p axis. Biomaterials. 2022;284: 121484.
114. Tian Y, Zhang F, Qiu Y, Wang S, Li F, Zhao J, Pan C, Tao Y, Yu D, Wei W. Reduction of choroidal neovascularization via cleavable vegf antibodies conjugated to exosomes derived from regulatory t cells. Nat Biomed Eng. 2021;5(9):968–82.
115. Green WR, Wilson DJ. Choroidal neovascularization. Ophthalmology. 1986;93(9):1169–76.
116. Grossniklaus HE, Green WR. Choroidal neovascularization. Am J Ophthalmol. 2004;137(3):496–503.
117. He GH, Zhang W, Ma YX, Yang J, Chen L, Song J, Chen S. Mesenchymal stem cells-derived exosomes ameliorate blue light stimulation in retinal pigment epithelium cells and retinal laser injury by vegf-dependent mechanism. Int J Ophthalmol. 2018;11(4):559.
118. Ambati J, Fowler BJ. Mechanisms of age-related macular degeneration. Neuron. 2012;75(1):26–39.
119. Tang Y, Kang Y, Zhang X, Cheng C. Mesenchymal stem cell exosomes as nanotherapeutics for dry age-related macular degeneration. J Controlled Release. 2023;357:356–70.
120. Wu KY, Ahmad H, Lin G, Carbonneau M, Tran SD. Mesenchymal stem cell-derived exosomes in ophthalmology: a comprehensive review. Pharmaceutics. 2023;15(4):1167.
121. Mead B, Amaral J, Tomarev S. Mesenchymal stem cell-derived small extracellular vesicles promote neuroprotection in rodent models of glaucoma. Inves Ophthalmol Visual Sci. 2018;59(2):702–14.
122. Seyedrazizadeh SZ, Poosti S, Nazari A, Alikhani M, Shekari F, Pakdel F, Shahpasand K, Satarian L, Baharvand H. Extracellular vesicles derived from human ES-MSCs protect retinal ganglion cells and preserve retinal function in a rodent model of optic nerve injury. Stem Cell Res Therapy. 2020;11(1):1–13.

123. Zhu S, Huang H, Liu D, Wen S, Shen L, Lin Q. Augmented cellular uptake and homologous targeting of exosome-based drug loaded iol for posterior capsular opacification prevention and biosafety improvement. Bioact Mat. 2022;15:469–81.
124. Martín del Valle EM, Galan MA, Carbonell RG. Drug delivery technologies: the way forward in the new decade. Ind Eng Chem Res. 2009;48(5):2475–86.
125. Paolino D, Sinha P, Fresta M, Ferrari M. Drug delivery systems. In: Encyclopedia of medical devices and instrumentation. Hoboken: Wiley; 2006.
126. Farjadian F, Ghasemi A, Gohari O, Roointan A, Karimi M, Hamblin MR. Nanopharmaceuticals and nanomedicines currently on the market: challenges and opportunities. Nanomedicine. 2019;14(1):93–126.
127. Tran TH, Mattheolabakis G, Aldawsari H, Amiji M. Exosomes as nanocarriers for immunotherapy of cancer and inflammatory diseases. Clin Immunol. 2015;160(1):46–58.
128. Zhang Z, Mugisha A, Fransisca S, Liu Q, Xie P, Hu Z. Emerging role of exosomes in retinal diseases. Front Cell Develop Biol. 2021;9: 643680.
129. Kosaka N, Izumi H, Sekine K, Ochiya T. microrna as a new immune-regulatory agent in breast milk. Silence. 2010;1(1):1–8.
130. El Andaloussi S, Mäger I, Breakefield XO, Wood MJ. Extracellular vesicles: biology and emerging therapeutic opportunities. Nat Rev Drug Dis. 2013;12(5):347–57.
131. Ortega A, Martinez-Arroyo O, Forner MJ, Cortes R. Exosomes as drug delivery systems: endogenous nanovehicles for treatment of systemic lupus erythematosus. Pharmaceutics. 2021;13(1):3.
132. Alvarez-Erviti L, Seow Y, Yin H, Betts C, Lakhal S, Wood MJ. Delivery of sirna to the mouse brain by systemic injection of targeted exosomes. Nat Biotechnol. 2011;29(4):341–5.
133. Samaeekia R, Rabiee B, Putra I, Shen X, Park YJ, Hematti P, Eslani M, Djalilian AR. Effect of human corneal mesenchymal stromal cell-derived exosomes on corneal epithelial wound healing. Inves Ophthalmol Visual Sci. 2018;59(12):5194–200.
134. Yu B, Shao H, Su C, Jiang Y, Chen X, Bai L, Zhang Y, Li Q, Zhang X, Li X. Exosomes derived from mscs ameliorate retinal laser injury partially by inhibition of mcp-1. Sci Rep. 2016;6(1):34562.
135. Wassmer SJ, Carvalho LS, György B, Vandenberghe LH, Maguire CA. Exosome-associated AAV2 vector mediates robust gene delivery into the murine retina upon intravitreal injection. Sci Rep. 2017;7(1):45329.
136. Duan L, Xu L, Xu X, Qin Z, Zhou X, Xiao Y, Liang Y, Xia J. Exosome-mediated delivery of gene vectors for gene therapy. Nanoscale. 2021;13(3):1387–97.
137. Klingeborn M, Dismuke WM, Rickman CB, Stamer WD. Roles of exosomes in the normal and diseased eye. Prog Retinal Eye Res. 2017;59:158–77.
138. Ruan Y, Jiang S, Musayeva A, Pfeiffer N, Gericke A. Corneal epithelial stem cells-physiology, pathophysiology and therapeutic options. Cells. 2021;10(09):2302.
139. Dong Y, Peng H, Lavker RM, Emerging therapeutic strategies for limbal stem cell deficiency. J Ophthalmol. 2018;2018:7894647.
140. Matthyssen S, Van den Bogerd B, Ní Dhubhghaill S, Koppen C, Zakaria N. Stem cell applications in corneal regeneration and wound repair. In: Neurological regeneration. Cham: Springer; 2017. p. 213–255.
141. Tavakkoli F, Eleiwa TK, Elhusseiny AM, Damala M, Rai AK, Cheraqpour K, Ansari MH, Doroudian M, Keshel SH, Soleimani M, et al. Corneal stem cells niche and homeostasis impacts in regenerative medicine; concise review. Eur J Ophthalmol 2023;33(4):11206721221150065.
142. Nurković, JS, Vojinović R, Dolićanin Z. Corneal stem cells as a source of regenerative cell-based therapy. Stem Cells Int. 2020;2020:8813447.
143. Hancox Z, Keshel SH, Yousaf S, Saeinasab M, Shahbazi MA, Sefat F. The progress in corneal translational medicine. Biomat Sci. 2020;8(23):6469–504.
144. Wang L, Xu X, Chen Q, Wei Y, Wei Z, Jin ZB, Liang Q. Extracellular vesicle micrornas from corneal stromal stem cell enhance stemness of limbal epithelial stem cells by targeting the notch pathway. Inves Ophthalmol Visual Sci. 2023;64(12):42–42.

145. van der Merwe Y, Steketee MB. Extracellular vesicles: biomarkers, therapeutics, and vehicles in the visual system. Curr Ophthalmol Rep. 2017;5:276–82.
146. Lawson C, Vicencio JM, Yellon DM, Davidson SM. Microvesicles and exosomes: new players in metabolic and cardiovascular disease. J Endocrinol. 2016;228(2):R57–71.
147. Gonzalez E, Falcón-Pérez JM. Cell-derived extracellular vesicles as a platform to identify low-invasive disease biomarkers. Expert Rev Mol Diagnost. 2015;15(7):907–23.
148. Inubushi S, Kawaguchi H, Mizumoto S, Kunihisa T, Baba M, Kitayama Y, Takeuchi T, Hoffman RM, Sasaki R. Oncogenic mirnas identified in tear exosomes from metastatic breast cancer patients. Anticancer Re. 2020;40(6):3091–6.
149. Qin J, Xu Q. Functions and application of exosomes. Acta Pol Pharm. 2014;71(4):537–43.
150. Grigor'eva A, Tamkovich S, Eremina A, Tupikin A, Kabilov M, Chernykh V, Vlassov V, Laktionov P, Ryabchikova E. Characteristics of exosomes andmicroparticles discovered in human tears. Biomeditsinskaia Khimiia. 2016;62(1):99–106.
151. Liu C, Zhao J, Tian F, Cai L, Zhang W, Feng Q, Chang J, Wan F, Yang Y, Dai B, et al. Low-cost thermophoretic profiling of extracellular-vesicle surface proteins for the early detection and classification of cancers. Nat Biom Eng. 2019;3(3):183–93.
152. Dismuke WM, Challa P, Navarro I, Stamer WD, Liu Y. Human aqueous humor exosomes. Exp Eye Res.. 2015;132:73–7.
153. Kang GY, Bang JY, Choi AJ, Yoon J, Lee WC, Choi S, Yoon S, Kim HC, Baek JH, Park HS, et al. Exosomal proteins in the aqueous humor as novel biomarkers in patients with neovascular age-related macular degeneration. J Proteome Res. 2014;13(2):581–95.
154. Ragusa M, Barbagallo C, Statello L, Caltabiano R, Russo A, Puzzo L, Avitabile T, Longo A, Toro MD, Barbagallo D, et al. mirna profiling in vitreous humor, vitreal exosomes and serum from uveal melanoma patients: pathological and diagnostic implications. Cancer Biol Therapy. 2015;16(9):1387–96.
155. Caruso Bavisotto C, Scalia F, Marino Gammazza A, Carlisi D, Bucchieri F, Conway de Macario E, Macario AJ, Cappello F, Campanella C. Extracellular vesicle-mediated cell–cell communication in the nervous system: focus on neurological diseases. Int J Mol Sci 2019;20(2):434.
156. Rajendran L, Honsho M, Zahn TR, Keller P, Geiger KD, Verkade P, Simons K. Alzheimer's disease β-amyloid peptides are released in association with exosomes. Proc Natl Acad Sci. 2006;103(30):11172–7.
157. Sardar Sinha M, Ansell-Schultz A, Civitelli L, Hildesjö C, Larsson M, Lannfelt L, Ingelsson M, Hallbeck M. Alzheimer's disease pathology propagation by exosomes containing toxic amyloid-beta oligomers. Acta Neuropathol. 2018;136:41–56.
158. Guo M, Wang J, Zhao Y, Feng Y, Han S, Dong Q, Cui M, Tieu K. Microglial exosomes facilitate α-synuclein transmission in parkinson's disease. Brain. 2020;143(5):1476–97.
159. Shabbir A, Cox A, Rodriguez-Menocal L, Salgado M, Badiavas EV. Mesenchymal stem cell exosomes induce proliferation and migration of normal and chronic wound fibroblasts, and enhance angiogenesis in vitro. Stem Cells Develop. 2015;24(14):1635–47.
160. An Y, Lin S, Tan X, Zhu S, Nie F, Zhen Y, Gu L, Zhang C, Wang B, Wei W, et al. Exosomes from adipose-derived stem cells and application to skin wound healing. Cell Proliferat. 2021;54(3):e12993.
161. Ke X, Yang D, Liang J, Wang X, Wu S, Wang X, Hu C. Human endothelial progenitor cell-derived exosomes increase proliferation and angiogenesis in cardiac fibroblasts by promoting the mesenchymal-endothelial transition and reducing high mobility group box 1 protein b1 expression. DNA Cell Biol. 2017;36(11):1018–28.
162. Li Q, Fang L, Chen J, Zhou S, Zhou K, Cheng F, Cen Y, Qing Y, Wu J. Exosomal MicroRNA-21 promotes keloid fibroblast proliferation and collagen production by inhibiting Smad7. J Burn Care Res. 2021;42(6):1266–74.
163. Wang S, Li F, Ye T, Wang J, Lyu C, Qing S, Ding Z, Gao X, Jia R, Yu D, et al. Macrophage-tumor chimeric exosomes accumulate in lymph node and tumor to activate the immune response and the tumor microenvironment. Sci Transl Med 2021;13(615):eabb6981.

Chapter 11
The Role of Extracellular Vesicles in Inflammatory Bowel Diseases

Xinyue Wei and Xiaocang Cao

Abstract Despite extensive research, the precise pathogenesis of inflammatory bowel disease (IBD) remains incompletely understood, necessitating further investigations. Additionally, the treatment of IBD poses constant challenges, including adverse drug reactions, drug resistance, etc. In recent years, mounting evidence has demonstrated the potential of extracellular vesicles (EVs) for cell-free therapy in preclinical models of IBD. EVs have garnered significant research interest as an emerging therapeutic strategy. Consequently, an increasing number of studies have uncovered the role and mechanism of exosomes derived from various sources, which will greatly promote the research of IBD. This article aims to comprehensively review the therapeutic and pathogenic effects of EVs on IBD considering the diverse components of EVs originating from different sources.

Keywords Extracellular vesicles · Inflammatory bowel diseases · Gut microbiota · Intestinal barrier · Therapy

1 Introduction

Extracellular vesicles (EVs) are nanoscale, membrane-bound vesicles formed by the fusion of multivesicular bodies and the plasma membrane [1]. With the deepening exploration of EVs, numerous studies have shown that they possess diverse functions and originate from a variety of sources. Due to the lack of uniform nomenclature, the International Society for Extracellular Vesicles (ISEV) recommends using the term "extracellular vesicles" to refer to all types of membrane-derived vesicles [2]. All cellular organisms, including both prokaryotes and eukaryotes, release EVs [3]. There are two main types of EVs based on their biological origin: exosomes and ectosomes. Exosomes are small EVs with a diameter ranging from

X. Wei · X. Cao (✉)
Department of Hepato-Gastroenterology, Tianjin Medical University General Hospital, Tianjin Medical University, Tianjin, China

© The Author(s), under exclusive license to Springer Nature Singapore Pte Ltd. 2024

Z. Li (ed.), *Extracellular Vesicle: Biology and Translational Application*,
https://doi.org/10.1007/978-981-97-5536-3_11

approximately 40 to 160 nm (with a mean diameter of about 100 nm) [4], which are the primary types of EVs discussed in this paper. Therefore, in this review, the term "EV" will be used predominantly unless specific designations are used in some studies.

Actually, EVs not only play a therapeutic role but also cause disease [5]. EVs not only are involved in tissue repair and regeneration but also serve as biomarkers [6, 7], natural nanomedicine delivery systems, and effective gene delivery tools [8]. They can be derived from natural sources, including various cells in the human body, as well as nonhuman sources such as food (e.g., milk) and microorganisms (e.g., bacteria, fungi, and parasites) [9–11] (Fig. 11.1). Furthermore, despite their different origins, EVs share common components, and the composition of EVs is complex which includes various biological macromolecules such as proteins, microRNAs (miRNA), messenger RNAs (mRNA), and lipids [9] (Table 11.1).

Inflammatory bowel disease (IBD) is a chronic intestinal disorder that is typically categorized as one of two subtypes: Crohn's disease and ulcerative colitis [12]. Although the etiology and pathogenesis of IBD are not fully understood, studies have shown that IBD is related to the abnormal interaction of intestinal immune imbalances caused by environmental, genetic, and intestinal microecological factors [13, 14]. Over the past few decades, the global incidence of IBD has been increasing, and it has gradually become a common disorder of the digestive system [15, 16]. Although many first-line drugs, such as glucocorticoids, immunosuppressants, and biological agents, have been proven to be beneficial to some patients, they

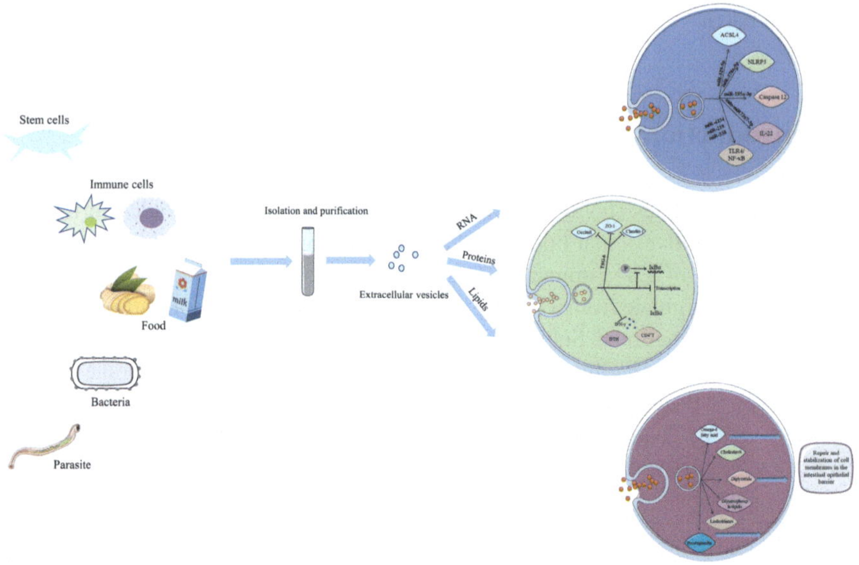

Fig. 11.1 Sources and composition of exosomes. Regardless of their source, exosomes share a similar structure and composition. The variety of molecules in exosomes are functionally characterized and grouped into exosomal RNAs, proteins, and lipids

Table 11.1 Exosome components and their effects in studies of IBD

Source	EV components	Effects	References
hUC-MSCs	miR-129-5p	Targets ACSL4 with miR-129-5p to inhibit lipid peroxidation and ferroptosis	[35]
hUC-MSCs	miR-378a-5p	Targets NLRP3 to hinder the assembly of inflammasomes and reduces pyroptosis	[38]
hUC-MSCs	TSG-6	Immunosuppression and intestinal barrier repair are mediated by TSG-6	[51]
BMSCs	Metallothionein-2	Suppresses NF-κB activation via MZF1 by enhancing IκBα transcription and inhibiting the phosphorylation of IκBα	[53]
Treg	miR-195a-3p	Promotes proliferation and inhibits apoptosis of YAMC cells by targeting caspase 12	[73]
G-MDSC	Arg-1	G-MDSC exo suppressed CD4 T cell proliferation, IFN-γ and inhibit the delayed-type hypersensitivity (DTH) response secretion in a dose-dependent manner, and this inhibitory effect correlated with Arg-1 activity	[74]
Porcine milk	miR-4334, miR-219, and miR-338	Prevents LPS-induced intestinal inflammation, apoptosis, and damage via inhibiting TLR4/NF-κB and p53 pathways	[78]
Ginger	mdo-miR7267-3p	Shapes gut microbiota	[81]
		Improves barrier function	
		Ameliorates colitis via IL-22-dependent mechanisms	
Hookworm	81 proteins including tetraspanin 14–3-3 protein, enolase, and heat shock proteins, together with 52 miRNA species	Protection against colitis by significantly suppressing IFN-γ, IL-6,IL-1β, and IL-17a and upregulating anti-inflammatory cytokine IL-10	[90]

also cause a series of adverse reactions when used long term [17–20]. As the need for effective and safe treatment options grows, more and more researchers are actively pursuing new methods with fewer adverse reactions and more precise treatment.

In recent years, a plethora of studies have unequivocally demonstrated that EVs are secreted in a paracrine manner, effectuating the regulation of the onset and progression of IBD through intercellular communication [21]. Owing to their low cytotoxicity and good biocompatibility, EVs can serve as carriers, facilitating the repair of intestinal mucosal injury and alleviating IBD [22]. Conversely, EVs from specific sources exert a stimulatory impact on inflammatory responses, thereby fostering the manifestation of IBD. Therefore, the therapeutic and pathogenic effects of EVs on IBD are reviewed in this article, and the relationship between EVs from different sources and IBD according to their composition is specifically introduced. It should be emphasized that this paper does not cover all EVs that have been identified but rather focuses on extracellular vesicles of some common origin. These researches on EVs from different sources will significantly aid in the exploration of novel therapeutic targets for IBD patients, consequently enhancing our comprehension of the EVs mechanism in IBD and assisting us in exploring novel therapeutic avenues.

2 Therapeutic Effects of EVs on IBD

More and more researchers have discussed that exosomes can effectively inhibit the pathological process of IBD and improve clinical symptoms through reference laboratory and animal model studies. Currently, exosome therapy is being gradually applied in the field of IBD, and several preclinical studies and preliminary clinical trials have been conducted to assess its safety, efficacy, and feasibility. Numerous clinical studies have indicated that exosomes can relieve IBD, promote tissue repair and regeneration, and ameliorate clinical symptoms. Furthermore, extensive molecular and cell biology investigations have found the mechanisms by which exosomes regulate the immune response. These mechanisms include inhibiting inflammatory responses, facilitating the repair and regeneration of intestinal tissue, modulating the immune response to maintain the balance of immune cell activity, and enhancing the barrier function of intestinal epithelial cells [23] (Fig. 11.2). Nonetheless, despite progress made, further research in this area is still in its nascent stage. The

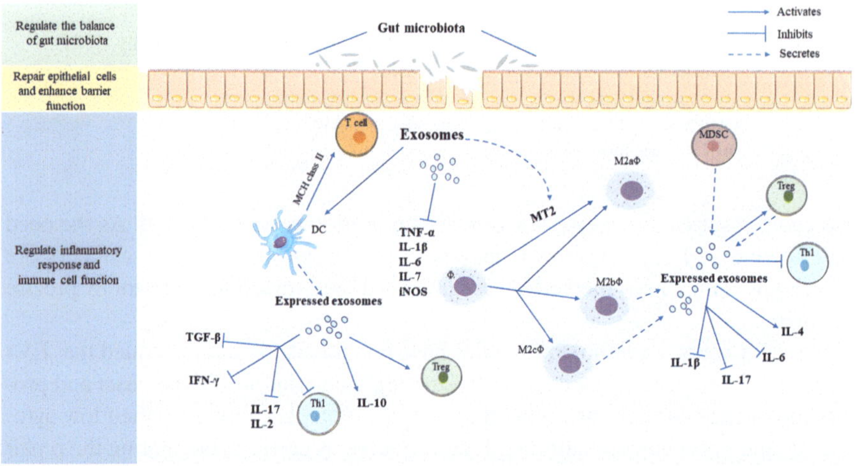

Fig. 11.2 Mechanisms of exosomes in IBD. *DC* dendritic cell, *IFN-γ* interferon-gamma, *IL* interleukin; *iNOS* inducible nitric oxide synthase, *M* macrophage, *MCH* major histocompatibility complex, *MDSC* myeloid-derived suppressor cell, *miR* microRNA, *MT2* melanotan 2, *TGF-β* transforming growth factor β, *Th* T helper, *TNF-α* tumor necrosis factor α, *Treg* regulatory T cells, *φ* macrophage

results of these preclinical studies and preliminary clinical trials provide useful clues and a basis for further research and development of exosome therapy in the field of IBD.

2.1 EVs from Natural Sources and IBD

Recent studies have emphasized the essential role of naturally derived extracellular vesicles in the development and advancement of IBD. In an inflammatory environment, intestinal cells can be stimulated to release a greater quantity of extracellular vesicles. These vesicles contain a diverse range of biomarkers and inflammatory mediators, such as cytokines and miRNAs. These molecules have the ability to influence the function of immune cells, thereby intensifying inflammation in the intestines. Conversely, specific constituents present in extracellular vesicles possess anti-inflammatory properties and the ability to facilitate the repair of intestinal tissue. As a result, they alleviate inflammatory responses and promote the regeneration of epithelial cells.

2.1.1 Stem Cell-Derived EVs

As common sources of EVs, stem cells play a crucial role in medical applications [24]. EVs can be derived from various types of stem cells, including perinatal stem cells and adult stem cells. Perinatal stem cells are isolated mainly from the umbilical cord, while adult stem cells are isolated mainly from the bone marrow and olfactory mucosa and so on [25]. Numerous studies have shown that exosomes derived from stem cells in different tissues can regulate inflammation-related pathways, target ferroptosis inhibition, or promote tissue repair [26–28]. It ameliorates the progression of colitis and contributes to the recovery of damaged tissues and organs [29]. The following paragraphs provide an overview of three types of stem cells, namely human umbilical cord mesenchymal stem cells (hUC-MSCs), bone marrow mesenchymal stem cells (BMSCs), and olfactory ectomesenchymal stem cells (OE-MSCs), and highlight the therapeutic effects of RNA, proteins, and lipids from different types of stem cells on IBD.

RNAs

The RNAs found in exosomes derived from stem cells include miRNAs, long noncoding RNAs (lncRNAs), and mRNAs [30]. The current research primarily concentrates on the regulation of inflammatory response and immune regulation as the main mechanisms of action for treating IBD. Furthermore, certain lncRNAs and mRNAs can facilitate the proliferation and repair of intestinal epithelial cells, thereby aiding in the restoration of intestinal barrier function [31].

Pyroptosis and ferroptosis are two distinct forms of cell death. Inhibiting iron and cell death factors, as well as the secretion of substances from outside the body, may affect the mechanisms related to cell death. Iron-related death, also known as ferroptosis, is a recently discovered form of cell death that is linked to disturbances in iron ion metabolism and oxidative stress [32]. The relationship between ferroptosis and the inflammatory response is highly intricate. On the one hand, ferroptosis can trigger an inflammatory response. On the other hand, the inflammatory response can influence the occurrence of ferroptosis by regulating iron metabolism and the expression of ferritin [33]. This creates a vicious cycle of mutual promotion. In the context of IBD, researchers have observed this phenomenon. Thus, inhibiting iron death has emerged as a potential target in the treatment of IBD [34].

Wei et al. conducted an experiment involving IBD mice injected with hUC-MSC-derived exosomes (hUC-MSCs-Exos). They assessed DAI scores, evaluated the condition of the spleen and colon, and analyzed inflammatory factors in colon tissues. The results indicated that hUC-MSCs-Exos exhibited anti-inflammatory and reparative effects on IBD mice at both the tissue and molecular levels [35]. To explore the potential mechanisms underlying these effects, researchers further examined markers of ferroptosis in mouse colon tissues. They discovered that treatment with hUC-MSCs-Exos resulted in the downregulation of the expression of acyl-CoA synthetase long-chain family member 4 (ACSL4), a critical enzyme involved in ferroptosis. According to the starBase genome database, miR-129-5p, which is abundantly expressed in hUC-MSCs-Exos, specifically targets ACSL4, inhibiting lipid peroxidation and attenuating iron-induced cell death. Moreover, miR-129-5p reduces intestinal inflammation, thereby alleviating the symptoms of IBD [35].

Pyroptosis is a controlled process of cell death that is usually associated with inflammatory reactions [36]. In certain circumstances, pyroptosis can be induced and regulated by inflammatory signals such as tumor necrosis factor-alpha (TNF-α). The release of intracellular contents such as DNA, RNA, and proteins during apoptosis can act as inflammatory stimuli, activating immune cells and triggering an inflammatory response. This may result in the release of inflammatory mediators and the aggregation of inflammatory cells, thereby exacerbating the inflammatory reaction [37].

To further elucidate the role of MSC-derived exosomes (MSCs-Exos) in mediating the biologically active ingredients responsible for their potentially beneficial effects in IBD, it is crucial to note that the evidence remains inconclusive. The research has suggested that hUC-MSCs-Exos effectively target NLRP3 (NOD-like receptor family, pyrin domain-containing 3) in a colitis mouse model through the use of miR-378a-5p. This targeting mechanism ultimately mitigates pyroptosis and the secretion of proinflammatory factors in vivo. This study, therefore, contributes significantly to our understanding of the potential of exosomes derived from hUC-MSCs and their exosomal RNA in the treatment of IBD [38]. Additionally, BMSCs are frequently employed in preclinical studies due to their anti-inflammatory and antifibrotic properties, making them a commonly used stem cell type in the context of autoimmune diseases [39–41]. The RNA within these exosomes plays a crucial role in regulating the inflammatory response, inhibiting ferroptosis and pyroptosis, and promoting the proliferation and repair of intestinal epithelial cells, thereby protecting the integrity of the intestinal barrier. These findings provide strong support for potential future applications of cell-free treatment for IBD [42, 43].

The RNA of exosomes not only plays a role in the treatment of IBD by regulating the inflammatory response but also can modulate immune cell function, impact immune homeostasis, mitigate excessive immune activation, and promote immune system equilibrium. For instance, a subtype of mesenchymal stem cells known as OE-MSCs, which are resident immunomodulatory stem cells in the nasal mucosa lamina propria, exhibits a high proliferation rate, self-renewal ability, and potential for multilineage differentiation [44].

Numerous studies have demonstrated that OE-MSCs mediate an anti-inflammatory balance between Treg cells and proinflammatory factors, thereby suppressing the inflammatory response. Exosomes derived from OE-MSCs (OE-MSC-Exos) exhibit similar and consistent biological effects both in vivo and in vitro. These exosomes also inhibit the proliferation of $CD4^+$ T cells, downregulate the expression of proinflammatory cytokines, and upregulate the expression of inhibitory cytokines. These findings indicate that OE-MSCs and their exosomes could be used as therapeutic agents for IBD by regulating the function and immune response of immune cells [45]. However, there is still a lack of studies examining the components of OE-MSC-Exos and how they exert their effects.

Stem cell-derived exosomes are safer and more effective than stem cell transplantation and have great potential in the treatment of IBD and other diseases [46]. Specifically, RNA is an important component of stem cell exosomes, and the diversity of exosome-enriched miRNAs resulted in exosomal miRNAs closely related to IBD progression. Although we see promising research results, many questions remain to be addressed. The precise mechanism underlying the anti-inflammatory efficacy of RNA encapsulated within stem cell-derived exosomes, the potential presence of additional small molecules exhibiting anti-inflammatory properties, and the extraction methodology for RNA from exosomes constitute challenges that necessitate further resolution in the forthcoming research endeavors. In addition, the wide range of origins of exosomes, coupled with the diversity of their enriched miRNAs, makes exosomal miRNAs closely related to the progression of IBD. The extensive array of exosome sources and the multifarious roles of miRNAs therein render these vesicles intimately associated with the advancement of intestinal inflammation. Currently, the involvement of exosomal DNA and other non-coding RNAs (e.g.circRNAs) in the modulation of IBD remains inadequately elucidated. Consequently, the exploration of these mechanisms constitutes a significant challenge and represents the focal point of our researchers' endeavors.

Proteins

The proteins and RNAs present in exosomes play crucial roles in the therapeutic potential of exosomes. While the secreted body RNAs hold significant value, it is equally important not to disregard the potential role of IBD in the secretion of body proteins. With advancements in technology and research on secreted foreign proteomics in the body, more than 900 different types of proteins have been identified in exosomes derived from MSCs (MSC-Exos) [47, 48]. This discovery represents a breakthrough in understanding the biological function and clinical application prospects of exosomes in IBD.

The proteins present in stem cell-derived exosomes include immunomodulatory molecules [49], such as cytokines, growth factors, and other functional proteins. These proteins primarily act through immune regulation and inhibition of inflammation, reducing the release of inflammatory cells and inflammatory mediators to exert anti-inflammatory effects. Previous studies have demonstrated that MSC-Exos can mitigate inflammatory damage in the gut, and the proteins and RNAs in MSC-Exos, both individually and in combination, play crucial roles in modulating inflammation. Additionally, Mardpour et al. conducted a proteomic analysis of MSC-Exos and identified certain cytokines and chemokines, including IL-10 and HGF, that either alleviate or promote the occurrence of an inflammatory response [50].

Although we have identified several inflammatory mediators that contribute to the initiation or resolution of inflammation, the upstream molecules and molecular pathways involved remain unclear. Several recent studies have provided new insights into this area. After different treatments, compared with those in the DSS + PBS group and control group, the expression of tumor necrosis factor-α-stimulated gene 6 (TSG-6) in MSC-Exos isolated from hUC-MSCs was significantly greater. TSG-6 is an immunomodulatory molecule secreted by MSC that has an anti-inflammatory effect. To further elucidate the role of TSG-6 in MSC-Exos therapy, colitis model mice with low TSG-6 expression were generated. These mice were administered MSC-Exos before and after the knockdown of TSG-6, and histological examination revealed significant differences in colon pathology between the two groups. These results suggest that TSG-6 may serve as an important mediator of the immunosuppressive and reparative effects of MSC-Exos in IBD, particularly in maintaining the integrity of the intestinal barrier. However, further research is needed to investigate whether other proteins have synergistic effects on the treatment of IBD [51].

The main component of EVs is likely protein. In addition to regulating inflammation-related signaling pathways and exerting anti-inflammatory effects such as cytokines and other functional proteins, exosome proteins can also modulate the function and immune response of immune cells to maintain immune system homeostasis. Macrophage polarization, as a crucial component of the immune system, plays a vital role in the treatment of IBD. Macrophages can modulate inflammatory and immune responses through various polarization states [52]. Therefore, regulating macrophage polarization may emerge as a therapeutic strategy for IBD. Through proteomic analysis, Liu et al. identified metallothionein-2 (MT2) in MSC-Exos as a key regulator and potential therapeutic target for inhibiting inflammatory responses in IBD. This regulation occurs by enhancing IκBα transcription and inhibiting the phosphorylation of IκBα, which ultimately inhibits NF-κB activation. Furthermore, researchers have investigated the impact of MSC-Exos on colonic macrophages and demonstrated that MSC-Exos alleviate colonic mucosal inflammation by promoting the polarization of M2b macrophages, underscoring the importance of macrophages in IBD [53]. Overall, this research, which is based on vesicle proteins from stem cell sources, provides a theoretical foundation for IBD treatment and offers a novel perspective on IBD without cell therapy.

Lipids

The lipids present in exosomes serve as essential constituents of the exosome membrane and also contribute to the repair and stability of the cellular membrane, thus facilitating the proper functioning of cells [54]. Similar to the cell membrane, lipids have both structural and functional roles in the formation and secretion of exosomes. Additionally, exosomes exhibit stronger resistance to washing than other EVs [55]. Through extensive investigations into extracellular vesicle lipids, we have gained a deeper understanding of the composition and function of exosomal lipids, as well as their pivotal role in vesicle biology. This knowledge contributes to the potential application of exosomal lipids in clinical therapy.

The key component responsible for the dynamic permeability barrier of the intestinal epithelium is the tight junction of intestinal epithelial cells, which is mainly mediated by tight junction proteins. These proteins are found primarily in cholesterol- and sphingomyelin-rich microdomains called lipid rafts [56]. The lipid composition of exosomes and lipid rafts exhibits similarities [57]. Researchers have collected intestinal biopsy samples of IBD to analyze lipid raft conditions. The authors observed that the protein concentration in UC mucosal homogenates was lower than that in control tissues. Additionally, the activity of lipase and alkaline phosphatase is reduced, indicating the presence of damaged lipid rafts in the intestinal mucosa of IBD patients. This evidence of lipid raft destruction suggests that this protein serves as an early warning sign of inflammation and subsequently promotes the destruction of the intestinal barrier in IBD patients [58]. Therefore, we propose that lipid components, such as cholesterol, within lipid rafts, play a vital role in maintaining the integrity of the intestinal epithelium. Furthermore, lipid components present in exosomes may contribute to the repair of lipid rafts and even the restoration of the intestinal epithelial barrier.

At present, there is a lack of relevant research on whether the lipid composition of EVs alone has a positive effect on IBD, but the importance of lipid composition is unclear. As a natural drug carrier, exosomes are nanoscale vesicles with stable phospholipid bilayer structures that show significant advantages in the application of intravenous nanocarriers. For example, EVs can pass through the blood–brain barrier and penetrate dense tissues, etc. [59]. In summary, the lipid composition of EVs endows them with great potential for drug delivery and treatment of IBD.

2.1.2 Immune Cell-Derived EVs

EVs are involved in intercellular communication in the immune system, including antigen presentation, immune cell activation (natural killer cells, T cells, etc.), immunosuppression, and other anti-inflammatory processes [60]. Many immune cells are involved in the pathogenesis of IBD, such as macrophages, dendritic cells (DC), and regulatory T cells (Treg), and these immune cell-derived exosomes also affect the progression of IBD in a variety of ways.

RNAs

Immune cell-derived exosomes contain miRNAs that are excellent regulators of immune responses. The source of immune cells can regulate T cells and the activity of immune cells, such as macrophages, to balance the immune response and inhibit excessive immune responses, thereby reducing inflammation [61].

A large number of studies have shown that regulating macrophage polarization may constitute a strategy for the treatment of IBD. Specifically, M1 macrophage polarization is a proinflammatory state, and M2 macrophage polarization is an anti-inflammatory and reparative state. Therefore, inhibiting the M1 macrophages and promoting M2 macrophages reduce inflammatory reactions and intestinal tissues [62, 63]. The M2a, M2b, and M2c subtypes of M2 macrophages play important roles in regulating immune activity and Th2 activation and are anti-inflammatory factors [64].

In an experimental colitis study in mice, both clinical disease scores and colonic pathology showed that M2b macrophage-derived EVs attenuated the damage caused by colonic inflammation in mice. In addition, in vitro experiments revealed that M2b-derived EVs can increase the frequency of Treg cells, thus preventing colitis. As shown by the in vivo and in vitro results, M2b macrophages and their sources for the treatment of colitis in extracellular vesicles are particularly promising. LncRNAs, which are key components of the secretory tract of the outer body, play important roles in this process. In a study of BMSC-mediated osteogenic adjustment, macrophages were secreted after extraction through high-throughput sequencing, bioinformatics analysis, and qRT–PCR to determine the importance of these lncRNAs. Moreover, the upregulated lncRNAs promoted the osteogenesis of BMSCs. These results suggest that RNAs may play a major role in the efficacy of macrophage-derived EVs in experimental colitis [65]. However, the effective ingredients of EVs in IBD have yet to be determined, and many researchers still need to explore and discover the role of EVs in IBD treatment.

In addition to macrophage-derived exosomes, dendritic cells (DCs) are the main members of the APC population of the immune system and are also key mediators that initiate and maintain T-cell-mediated responses. DC-derived exosomes (DC-Exos) can prevent the development of autoimmune diseases by regulating the activation of immune cells such as T cells. Immunomodulation targeting dendritic cells has been a research hotspot in the treatment of autoimmune diseases [66]. For example, DC-Exos have been reported to be an important mechanism for regulating inflammatory responses in IBD through the inhibition of T-cell proliferation [67, 68]. In addition, the lncRNAs expressed by exosomes derived from dendritic cells in different states may contribute to the different immunomodulatory functions of these exosomes in different recipient cells.

Various immune cells and immune cell-derived exosomes can not only alleviate the symptoms of IBD by regulating the activities of immune cells, such as T cells, but also regulate cell proliferation, differentiation, and apoptosis and accelerate the regeneration and repair of intestinal epithelial cells, thereby restoring the normal structure and function of the intestine [69]. Regulatory T cells (Tregs), a subset of T cells that control autoimmune responses in vivo, can be divided into naturally occurring natural regulatory T cells (nT-regs) and inducible adaptive regulatory T cells (aT-regs or iT-regs), which are closely related to the occurrence of autoimmune diseases [70]. Tregs may also play a therapeutic role in IBD by secreting exosomes (Treg-Exos). Mice treated with Treg-Exos exhibited increased colitis resistance and reduced intestinal injury with inflammatory cell infiltration via the inhibition of T helper-1 (Th1) cells proliferation and the promotion of Treg expansion [71]. Foxp3 is a master transcription factor of Treg cells, and these forkhead box P3 (Foxp3)-positive Treg cells protect against diseases such as IBD [72]. In addition, Treg regulates disease by inhibiting T cells through miRNA-containing exosomes via the mechanism of noncell autonomous gene silencing.

Another study showed that Treg-Exos promoted the proliferation and inhibited the apoptosis of colonic epithelial YAMC cells, thereby accelerating the repair of the intestinal mucosal barrier and restoring the normal structure and function of the intestine. When the expression of miR-195a-3p in Treg-Exos was inhibited, the above therapeutic effects of Treg-Exos were abolished. Therefore, miR-195-a-3p can promote the apoptosis of caspase 12 cells as a direct target, affect intestinal epithelial cell regeneration and repair, and play a role in treatment [73]. A large number of studies have shown that the outside source of immune cells secreted in the treatment of IBD plays a role, but additional research is needed to further confirm the specific mechanism involved.

Proteins

Current studies have shown that these proteins may include anti-inflammatory cytokines, proteins that inhibit inflammatory mediators, and proteins that regulate immune cell activity. Which are the main anti-inflammatory components of immune cell-derived exosomes.

Arginase-1 (Arg-1) is a key molecule of granulocyte-myeloid-derived suppressor cells (G-MDSCs) that exerts immunosuppressive effects. Myeloid-derived suppressor cells are derived from immature myeloid cells differentiated from hematopoietic stem cells. Myeloid-derived suppressor cells (MDSCs) can inhibit T-cell responses in inflammation and cancer. To further clarify the mechanism and key components of MDSCs in inhibiting inflammatory response, DSS-induced colitis mice were treated with G-MDSC-derived exosomes (G-MDSC-Exos). Histological examination revealed that the treated mice had more intact colon structures than the untreated mice. In vitro, treatment with G-MDSCs-Exos inhibited Th1-cell responses, enhanced Treg numbers, and inhibited $CD4^+$ T-cell proliferation and IFN-γ secretion, all of which contributed to the healing of colitis. This investigation demonstrated that G-MDSC-exos exerts its effects by specifically targeting Arg-1. The functional role of Arg-1 was substantiated to be intricately linked to the suppression of Th1 cell proliferation, Treg expansion, IFN-γ secretion, and the inhibition of delayed hypersensitivity responses in vitro [74].

In addition to their anti-inflammatory role in the treatment of IBD, immune cell-derived exosomes also regulate the activity of immune cells, such as dendritic cells, thereby inhibiting an excessive immune response and reducing IBD. Mature DC-Exos contain not only RNA but also a significant number of proteins, such as MHC-II, costimulatory molecules, and CC chemokine receptor 7 (CCR7), all of which contribute to their biological functions. Studies have indicated that the upregulation of CCR7 not only promotes the secretion of immature dendritic cells and stimulates more $CD4^+$ T cells in vitro but also leads to the accumulation of mature DC-Exos in the spleen, inducing inflammatory responses in vivo [75].

2.1.3 Food-Derived EVs

Exosomes can not only be produced by human cells but are also found in many foods that we eat. Food-derived exosomes are generally considered to be safe, such as milk, vegetables, fruits, etc. [76]. Some studies on edible food-derived exosomes have been conducted, and some plant-derived exosomes-like nanoparticles have been found to possess anti-inflammatory properties. For instance, recently reported findings suggest that exosomes derived from *Pueraria lobata* can induce M2 macrophage polarization, leading to the provision of anti-inflammatory effects and alleviation of colitis [77]. Moreover, plants serve as abundant and cost-effective biological resources, with a high growth rate, making them advantageous for large-scale production. The multifarious constituents of exosomes have the potential to ameliorate inflammatory bowel disease via diverse mechanisms. For instance, exosomes derived from foodstuffs, enriched with probiotics, are capable of modulating the pathophysiological processes in both animals and plants, augmenting anti-inflammatory responses, reinforcing intestinal barrier integrity, fine-tuning the immune system, and preserving the equilibrium of intestinal microbiota.

RNAs

The mechanism of RNA secretion from food sources holds great potential for the treatment of IBD. One mechanism involves exosomal miRNAs regulating gene expression. Both plant-derived and milk-derived exosomal miRNAs influence pathological processes, such as inflammation and tumor formation by modulating the expression of target genes. Porcine milk-derived exosomes reportedly contain components that upregulate three miRNAs, miR-4334, miR-219, and miR-338. These miRNAs could inhibit LPS-induced apoptosis and inflammation, reduce LPS-induced intestinal epithelial damage, and alleviate colitis [78].

Another mechanism involves enhancing immune suppression. Some edible plants, such as broccoli and ginger root, can strengthen immune suppression functions and thus alleviate IBD. Broccoli-derived exosome-like nanoparticles (BENs) have been found to relieve colitis-associated inflammation. The specific mechanisms by which the ELN achieves this goal have yet to be studied in depth [79]. ELNs sourced from ginger root inhibit NLRP3 activation, reducing the activity of inflammatory cells and subsequently suppressing downstream inflammation, such as the release of the cytokines IL-18 and IL-1β. These effects contribute to the reduction of IBD and pathological processes [80].

Additionally, food-derived RNA can affect gut microbial communities. Dietary-derived exosomes and their contents have been found to impact the diversity and composition of microbial communities in the mouse gut. Mdo-miR7267-3p, as one of the microRNAs of ginger ELNs (GERNs), can target *Lactobacillus rhamnosus* (LGG), leading to an increase in the expression of IL-22 and improvement of intestinal barrier function, thus helping to alleviate colitis [81]. These findings suggested that exosomes may influence the development and progression of IBD by regulating the balance of the intestinal flora. The specific mechanisms involved require further study, but these findings provide important clues for understanding the relationship between gut microbes and bowel disease.

The therapeutic mechanisms of exosomal RNA derived from food in IBD can be summarized as regulating gene expression, enhancing immunosuppressive function, and influencing the intestinal microbial community. These findings provide a valuable foundation for the development of novel therapeutic strategies and a deeper understanding of IBD pathogenesis. However, further studies are needed to explore and confirm these mechanisms and their application in clinical practice.

Proteins

Exosomes derived from milk play a crucial role in regulating intestinal microorganisms and the intestinal microecological environment in infants. The physiological function of protein components in the human intestinal immune system has become a research hotspot. Studies have shown that breast milk exosomes are rich in various bioactive components and microflora, and some of these proteins can regulate immune system stability and inhibit excessive immune function. Milk-derived exosomes also contribute to protein secretion outside the body and to inflammatory reactions, which are closely associated with immune responses [82].

Research has shown that milk-derived exosomal proteins play a significant role in treating IBD. Firstly, these proteins can reduce inflammation and damage to the intestinal epithelium, thereby protecting intestinal tissues from inflammation. Secondly, by adjusting the balance of gut microbes, the stability of the mucous membrane microenvironment can be maintained. This regulation can reduce the proliferation of harmful bacteria and promote the growth of beneficial bacteria, thereby improving the composition and function of the intestinal flora and positively influencing the development of IBD [83]. These findings offer new perspectives on the use of exosome proteins in regulating the intestinal barrier and inflammation-related immune responses.

2.1.4 Bacteria-Derived EVs

EVs are produced in almost all areas of life, including various prokaryotes and eukaryotes. Studies on EV generation from gram-positive bacteria and mycobacteria started in 2007. EVs were isolated from biofilms of *Mycobacterium ulcerans* and the fungal pathogen *Cryptococcus neoformans*. Symbiotic microbes present in the gut release EVs that help maintain the stability of the intestinal microecology [84]. In addition to enhancing intestinal barrier function, probiotics can alleviate IBD by inhibiting the expression of proinflammatory factors, providing a new target for exploring effective treatment strategies for IBD.

Similarly, the mechanisms of bacterial-derived exosomes in IBD treatment mainly involve enhancing the integrity of the intestinal epithelial barrier, regulating the intestinal immune response, remodeling the intestinal microbiota, and inhibiting inflammatory signaling pathways. The probiotic *E. coli* Nissle 1917 (EcN) releases outer membrane vesicles (OMVs) that promote the integrity of the intestinal epithelial barrier [85]. Studies have shown that OMVs released from EcN can upregulate the expression of ZO-1 and claudin-14 and downregulate claudin-2, thereby enhancing the integrity of the intestinal mucosa and the function of the intestinal epithelial barrier. This helps to reduce the penetration of harmful substances and bacterial invasion, preventing the progression of the inflammatory response [86]. EVs derived from *Akkermansia muciniphila* also have a protective effect against intestinal immunity in IBD patients. Experimental mice with colitis treated with *Akkermansia muciniphila*-derived EVs showed increased colon length, decreased DAI score, and improved histological and pathological trends, suggesting that *Akkermansia muciniphila*-derived EVs can alleviate IBD [87].

Secondly, bacterial-derived exosomes can reduce IBD, maintain immune system balance, promote immune tolerance, and inhibit excessive immune responses by regulating macrophage polarization and related inflammatory signaling pathways. *Clostridium butyricum*-derived exosomes can alleviate the symptoms of ulcerative colitis and colonic pathological manifestations by regulating macrophage polarization and altering the composition of the intestinal microbiota. Additionally, serum exosomes from *H. pylori*-positive gastritis patients promoted the expression of nucleotide-binding oligomerization domain-like receptor protein 12 (NLRP12) in intestinal epithelial cells (IECs). NLRP12 can reduce the expression of monocyte chemoattractant protein 1 (MCP1) and macrophage inflammatory protein 1α (MIP1α) by inhibiting the Notch signaling pathway, thereby improving the symptoms of colitis. These mechanisms work together to help alleviate IBD symptoms and promote intestinal recovery [88]. However, further research and verification are needed to elucidate the specific underlying mechanisms and effects involved.

2.1.5 Fungus-Derived EVs

Fungus-derived exosomes were initially discovered in the cell wall of *C. neoformans*, but their composition and function have not been fully elucidated. Nevertheless, with the isolation and detailed investigation of additional fungal EVs, their significant roles in cell metabolism, signal transduction, and other processes have been revealed. For instance, a study by Hu et al. demonstrated that downregulating let-7 miRNA promoted *Cryptosporidium parvum*-induced exosome secretion. These exosomes carry antimicrobial peptides that inhibit *Cryptosporidium* infection, revealing a novel pathway of mucosal immunity associated with antimicrobial defense. By further exploring the composition and function of fungal EVs, we can enhance our understanding of their impact on gut health and obtain additional clues for the development of relevant therapeutic strategies [89].

2.1.6 Parasite-Derived EVs

Gastrointestinal parasites, especially blood-feeding intestinal nematodes, have been shown to create an immunomodulatory environment within the host, playing a suppressive role in inflammation. Studies have identified various proteins and miRNAs in *Nippostrongylus brasiliensis* exosomes. Examples of these proteins include tetraspanin, 14–3-3 protein, enolase, and heat shock protein, which play crucial roles in the treatment of IBD. Exosomes significantly inhibit colitis-related cytokines such as IFN-γ, IL-6, IL-1β, and IL-17a, thereby reducing the occurrence of inflammatory response. Simultaneously, they upregulate the expression of the anti-inflammatory cytokine IL-10, further suppressing the inflammatory process [90]. These mechanisms collectively contribute to the regulation of colonic inflammation in mice by balancing cytokine levels. The proteins and miRNAs present in exosomes may be associated with immune regulation signaling pathways.

2.2 EVs from Unnatural Sources and IBD

The emergence of non-natural-derived extracellular vesicles, including preconditioned and engineered EVs, has garnered considerable interest in the field of IBD research. These artificially manipulated or modified vesicles present distinct possibilities for therapeutic intervention and disease modulation.

2.2.1 Preconditioned EVs

Natural exosomes not only possess a short half-life and poor targeting ability but are also prone to activity restrictions. To mitigate these limitations, studies have shown that EV donor cells or their microenvironments can be pretreated or modified using artificial methods such as hypoxia, cytokines, drugs, and other pretreatment strategies to enhance the functionality and therapeutic effects of natural exosomes on IBD.

Macrophage-derived EVs have been shown to have anti-inflammatory activity and specific immunomodulatory functions, while pretreated EVs also inhibit the progression of inflammatory responses. After lipopolysaccharide (LPS) stimulation, the cytokine content produced by macrophage-derived EV increased and mainly manifested as interleukin (IL) and TNF-α. In addition, researchers have used an injection of complete Freund's adjuvant (CFA) to treat inflammation model mice with LPS-stimulated EVs, and the swelling caused by inflammation was significantly reduced [91].

DC-derived exosomes regulate the immune response and alleviate the pathological damage in IBD. In addition, *S. japonicum* soluble egg antigen (SEA) also has curative effects on colitis mouse models, for example, by reducing the inflammatory infiltration of the colonic mucosa and other histological manifestations. Interestingly, the experiments demonstrated that DC-exos treated with SEA had stronger anti-inflammatory activity than DC-exos and SEA. Intraperitoneal injection of SEA-treated DC-exos into a mouse model of acute DSS-induced colitis significantly reduced clinical and histological scores and enhanced the expression of anti-inflammatory cytokines in colitis mice [92]. In addition, transforming growth factor-β1 (TGF-β1) gene-modified immature dendritic cells (imDCs; TGF-β1-DCs) may enhance the therapeutic effect of imDCs on IBD. TGF-β1 is a regulatory cytokine with multiple inhibitory effects, including the proliferation inhibition of T cells, B-cells, and other immune cell proliferation [93]. However, whether the protective effect of TGF-β1-DCs in mouse models of colitis is mediated by the secretion of EVs needs to be further investigated.

IL-10 is the most important cytokine that inhibits the proinflammatory response of the immune system and plays an important role in the development of mucosal immunity. When the researchers intraperitoneally injected IL-10 exosomes into a TNBS-induced colitis model rat, all the clinical and histopathological parameters were significantly reduced. Its therapeutic mechanism may be related to the downregulation of IL-2, IFN-γ, and TNF-α mRNA expression in colon tissue, as well as the significant upregulation of IL-10 mRNA expression in colon tissue and Tregs in the colon lamina propria [94]. Therefore, DC exosomes derived from IL-10 could lead to a new strategy for treating IBD.

Mycobacterium tuberculosis infection is extensively employed in scientific investigations as a pretreatment modality. Experimental evidence has shown that macrophage-derived exosomes infected with *Mycobacterium tuberculosis* can promote proinflammatory responses in immature macrophages. To determine the contributing factors, some studies have used *Mycobacterium tuberculosis* with the LspA gene knockout, which results in a reduction in the secretion of the 19-kDa lipoprotein. By comparing macrophages treated with LspA-deficient *Mycobacterium tuberculosis* and wild-type *Mycobacterium tuberculosis*, we found that the 19-kDa lipoprotein from exosomes released by macrophages infected with *Mycobacterium tuberculosis* is responsible for its proinflammatory activity [95].

2.2.2 Engineered EVs

After EVs enter systemic circulation, the RNA, proteins, and other transport substances they carry may be cleared by the immune system and excretory organs such as the liver and kidney. For EVs to be more targeted and effective, they need to be modified, engineered, or designed to improve their therapeutic efficacy and safety in treating diseases such as inflammation and cancer.

There is no doubt that MSCs, as the main members of the stem cell family, play a therapeutic role in clinical practice. MSC-EVs have also been confirmed to play an important role in immune regulation and tissue repair. Based on previous explorations of the mechanism and biological function of BM-MSCs in the treatment of IBD, researchers have used lentivirus transfection to introduce tyrosine-protein kinase receptor B2 (EphB2) into BM-MSCs to ultimately obtain engineered exosomes. These engineered exosomes can inhibit the RhoA/ROCK pathway to restore the intestinal barrier and maintain the immune balance of the intestinal microenvironment in patients with colitis, which is conducive to the remission and recovery of colitis. To date, engineered exosomes, which are produced through the modification of exosome surface molecules, have been infrequently employed in experimental studies pertaining to colitis [96]. With the continuous refinement and diversification of engineered exosome technology, there are expected to be more and more breakthrough experimental results for colitis in the future.

3 Pathogenic Effects of EVs on IBD

The understanding of the pathogenic effects of EVs on IBD is of paramount importance in the development of targeted therapeutic strategies. Through the elucidation of the mechanisms through which EVs contribute to the progression of the disease, researchers can identify potential therapeutic targets and develop interventions to modulate EV-mediated pathogenic processes.

3.1 EVs from Natural Sources and IBD

EVs derived from natural sources have been found to promote the development of IBD. These EVs, released by various cells in the body, play a significant role in the pathogenesis of IBD.

3.1.1 Immune Cell-Derived EVs

Mast cells originate from embryonic sources and adult bone marrow and are capable of releasing cytokines, such as IL-2, IL-10, and TGF-β1, to inhibit and terminate inflammation [97–99]. Additionally, mast cells can generate leukotrienes, prostaglandins, and other cytokines [100]. Certain mast cell products can promote local inflammation and enhance vascular permeability [101]. Similarly, EVs derived from inflamed mast cells may also have the same effect. EVs derived from HMCs-1 cells significantly upregulate intestinal epithelial permeability and disrupt intestinal barrier function through the mediation of functional miRNA transfer. They also downregulate the expression of tight junction-associated proteins. These vesicles are enriched with miR-223, which can inhibit the expression of CLDN8, consequently leading to the disruption of intestinal barrier function. The inhibitory effect of miR-223 on CLDN8 expression through HMC-1 derived EVs can be significantly reversed by a miR-223 inhibitor, further confirming the crucial role of miR-223. In conclusion, the EVs derived from HMC-1 cells, targeting the gene miR-223, may serve as new therapeutic targets for IBD [102].

3.1.2 Bacteria-Derived EVs

Pathogens deliver toxins to the host through outer membrane vesicles, inhibit the function of mitochondria, and activate NLRP3 bodies in the body to promote the progression of inflammation. *Mycobacteria*-derived exosomes can promote proinflammatory responses, such as TNF-α secretion, through MyD88, an adaptor molecule required for most Toll-like receptor signaling. The lipoprotein LpqH, present on hEVs released from Mtb-infected cells, is a TLR2 ligand and a major driver of this inflammatory response [95].

With the in-depth study of exosomal lipids, people have a deeper understanding of how EVs mediate the relationship between the host and microorganisms. It has been reported that EVs secreted by intestinal pathogens stimulate intestinal epithelial cells to produce EVs containing high levels of sphingosine-1-phosphate, affect the proliferation of Th17 cells through the MyD88-mediated pathway, and eventually promote the occurrence of colitis and colon cancer [103].

3.1.3 Fungus-Derived EVs

Fungal EVs have long been an interesting topic. Several researchers have shown that a common feature of fungi is the metabolic enzymes carried by fungal EVs [104]. *Candida auris*, a related emerging human pathogenic yeast, was the first fungus to be called a global public health threat by the World Health Organization [105]. EVs produced by fungi can exert a variety of biological effects, such as cell–cell communication and pathogen–host interactions. Fungi-derived exosomal RNAs include a variety of different classes of molecules, including small nucleolar RNAs (snoRNAs), small nuclear RNAs (snRNAs), and tRNA fragments (tfRNAs). These RNAs play important roles in the mechanisms of fungal pathogenesis, including interstrain virulence transfer in *Cryptococcus gattii* [106].

4 Prospective

EVs, which serve as a method of intercellular communication, have garnered significant attention in recent years. This approach offers advantages such as simplicity, safety, affordability, ease of storage and transportation. Additionally, EVs have a low risk of triggering adverse immune responses, are noncarcinogenic, and demonstrate high stability. These viruses are unable to replicate or infect viruses. Moreover, EVs play important roles in various physiological processes, including immune regulation, tissue repair, and regeneration. The application of EVs in nanotechnology is increasingly being recognized as a promising technique for the treatment of IBD. As nanometer-scale drug delivery systems, EVs can overcome the challenges posed by the inflammatory gut environment, such as epithelial rupture and mucus thickening. This allows for increased accumulation of delivered substances in the gut, enhancing efficacy while reducing systemic side effects. Thus, EVs show promise as nanocarriers for IBD treatment.

However, their clinical use is limited by factors such as rapid clearance and cytotoxicity by the mononuclear phagocyte system (MPS) or reticuloendothelial system (RES), which diminishes drug biodistribution in diseased tissues. Exogenous EVs pose significant challenges. Additionally, the clinical translation of engineered exosomes faces several issues. Different cell types can be utilized for exosome production, and variations in cell generations and culture conditions can affect exosome characteristics. Scaling up cell culture products is the most direct method for mass production of exosomes, but different extraction methods can impact exosome subset yield and homogeneity. Standardization of the cell culture scale, isolation, and extraction processes is necessary.

Currently, there are different views regarding the therapeutic effects of EVs on IBD. Vedolizumab, known for its approval for the treatment of IBD, has recently been found to be influenced by exosomes through its pharmacokinetics. This interference can affect the efficacy of drugs, especially for IBD patients who do not respond well to antitumor necrosis factor therapy. Vedolizumab selectively binds to the integrin $\alpha 4\beta 7$, hindering the trafficking of specific subsets of homing T lymphocytes to sites of gastrointestinal inflammation. When IBD patients receive anti-TNF treatment, the expression of $\alpha 4\beta 7$ integrin on the surface of serum exosomes increases, and this integrin competes with T cells and negatively affects the therapeutic efficacy of vedolizumab in IBD patients [107].

There is a certain relationship between viruses and exosomes. The process of exosome generation overlaps with various viral assembly and efflux pathways, suggesting that exosomes may play a role in viral infection. Viruses can exploit the exosomal release pathway to disseminate within cells, while exosomes can also contribute to antiviral immune responses. It is widely believed that the anti-inflammatory or proinflammatory effects of exosomes are dependent on their source. When the virus infects exosomes secreted by immune cells like T cells, macrophages, and DC cells, these exosomes contain immunomodulatory components such as IFN-α, IFN-β, and ILs, resulting in antiviral activity. Exosomes derived from cells are more likely to contain viral RNA and protein, facilitating the process of viral infection. These points underscore the potential clinical applications of exosomes while also highlighting the challenges and technical issues that need to be addressed. For instance, exosomes derived from virus-infected cells may possess viral infectivity. Thus, ensuring quality control and safety of exosomes is a crucial consideration for future practices.

EVs have been shown to participate in the regulation of immune cells, intestinal mucosal barrier integrity, and the microbiota through various mechanisms, thereby inhibiting inflammation, restoring the integrity of the intestinal barrier, and restoring the composition and diversity of the intestinal microbiome. As nanocarriers, EVs have potential clinical applications in the targeted therapy of IBD. Currently, there is a lack of relevant reports on the clinical application of EVs, and challenges remain in the isolation and purification of EVs from different sources. Therefore, factors such as the preparation method of EVs, the route of administration, and the dosage require further exploration. Basic research on EVs should be combined with emerging technologies to achieve successful clinical application, and the potential synergistic effects of EVs with existing clinical treatment methods should be investigated to foster breakthroughs in the treatment of IBD.

References

1. Raposo G, Stoorvogel W. Extracellular vesicles: exosomes, microvesicles, and friends. J Cell Biology. 2013;200(4):373–83. https://doi.org/10.1083/jcb.201211138.
2. Théry C, Witwer KW, Aikawa E, Alcaraz MJ, Anderson JD, Andriantsitohaina R, et al. Minimal information for studies of extracellular vesicles 2018 (MISEV2018): a position statement of the International Society for Extracellular Vesicles and update of the MISEV2014 guidelines. J Extracell Vesicles. 2018;7(1):1535750. https://doi.org/10.1080/20013078.2018.1535750.
3. Deatherage BL, Cookson BT. Membrane vesicle release in bacteria, eukaryotes, and archaea: a conserved yet underappreciated aspect of microbial life. Infect Immun. 2012;80(6):1948–57. https://doi.org/10.1128/iai.06014-11.
4. Cocucci E, Meldolesi J. Ectosomes and exosomes: shedding the confusion between extracellular vesicles. Trends Cell Biol. 2015;25(6):364–72. https://doi.org/10.1016/j.tcb.2015.01.004.
5. Mirzaei R, Zamani F, Hajibaba M, Rasouli-Saravani A, Noroozbeygi M, Gorgani M, et al. The pathogenic, therapeutic and diagnostic role of exosomal microRNA in the autoimmune diseases. J Neuroimmunology. 2021;358:577640. https://doi.org/10.1016/j.jneuroim.2021.577640.
6. Urabe F, Kosaka N, Ito K, Kimura T, Egawa S, Ochiya T. Extracellular vesicles as biomarkers and therapeutic targets for cancer. Am J Physiol Cell Physiol. 2020;318(1):C29–39. https://doi.org/10.1152/ajpcell.00280.2019.

7. Brennan M, Layrolle P, Mooney DJ. Biomaterials functionalized with MSC secreted extracellular vesicles and soluble factors for tissue regeneration. Adv Funct Mater. 2020;30(37):1909125. https://doi.org/10.1002/adfm.201909125.
8. Martinez E, de Castilla P, Tong L, Huang C, Sofias AM, Pastorin G, Chen X, et al. Extracellular vesicles as a drug delivery system: A systematic review of preclinical studies. Adv Drug Deliv Rev. 2021;175:113801. https://doi.org/10.1016/j.addr.2021.05.011.
9. Doyle LM, Wang MZ. Overview of extracellular vesicles, their origin, composition, purpose, and methods for exosome isolation and analysis. Cells. 2019;8(7):727. https://doi.org/10.3390/cells8070727.
10. Chen T, Wang F, Wei J, Feng L. Extracellular vesicles derived from different sources play various roles in diabetic retinopathy. Front Endocrinol. 2022;13:1064415. https://doi.org/10.3389/fendo.2022.1064415.
11. Blanco NG, Machado NM, Castro LL, Antunes MA, Takiya CM, Trugilho MRO, et al. Extracellular vesicles from different sources of Mesenchymal stromal cells have distinct effects on lung and distal organs in experimental sepsis. Int J Mol Sci. 2023;24(9):8234. https://doi.org/10.3390/ijms24098234.
12. Chang JT. Pathophysiology of inflammatory bowel diseases. New Engl J Med. 2020;383(27):2652–64. https://doi.org/10.1056/NEJMra2002697.
13. Baumgart DC, Carding SR. Inflammatory bowel disease: cause and immunobiology. Lancet. 2007;369(9573):1627–40. https://doi.org/10.1016/s0140-6736(07)60750-8.
14. Sanmarco LM, Chao CC, Wang YC, Kenison JE, Li Z, Rone JM, et al. Identification of environmental factors that promote intestinal inflammation. Nature. 2022;611(7937):801–9. https://doi.org/10.1038/s41586-022-05308-6.
15. Burisch J, Pedersen N, Čuković-Čavka S, Brinar M, Kaimakliotis I, Duricova D, et al. East-West gradient in the incidence of inflammatory bowel disease in Europe: the ECCO-EpiCom inception cohort. Gut. 2014;63(4):588–97. https://doi.org/10.1136/gutjnl-2013-304636.
16. Molodecky NA, Soon IS, Rabi DM, Ghali WA, Ferris M, Chernoff G, et al. Increasing incidence and prevalence of the inflammatory bowel diseases with time, based on systematic review. Gastroenterology. 2012;142(1):46–54. https://doi.org/10.1053/j.gastro.2011.10.001.
17. Abraham C, Cho JH. Inflammatory bowel disease. New Engl J Med. 2009;361(21):2066–78. https://doi.org/10.1056/NEJMra0804647.
18. Le Berre C, Honap S, Peyrin-Biroulet L. Ulcerative colitis. Lancet. 2023;402(10401):571–84. https://doi.org/10.1016/s0140-6736(23)00966-2.
19. Torres J, Mehandru S, Colombel JF, Peyrin-Biroulet L. Crohn's disease. Lancet. 2017;389(10080):1741–55. https://doi.org/10.1016/s0140-6736(16)31711-1.
20. Baumgart DC, Le Berre C. Newer biologic and small-molecule therapies for inflammatory bowel disease. New Engl J Med. 2021;385(14):1302–15. https://doi.org/10.1056/NEJMra1907607.
21. van Niel G, Carter DRF, Clayton A, Lambert DW, Raposo G, Vader P. Challenges and directions in studying cell-cell communication by extracellular vesicles. Nat Rev Mol Cell Biol. 2022;23(5):369–82. https://doi.org/10.1038/s41580-022-00460-3.
22. Banks WA, Sharma P, Bullock KM, Hansen KM, Ludwig N, Whiteside TL. Transport of extracellular vesicles across the blood-brain barrier: brain pharmacokinetics and effects of inflammation. Int J Mol Sci. 2020;21(12):4407. https://doi.org/10.3390/ijms21124407.
23. Ocansey DKW, Zhang L, Wang Y, Yan Y, Qian H, Zhang X, et al. Exosome-mediated effects and applications in inflammatory bowel disease. Biol Rev Cambridge Philos Soc. 2020;95(5):1287–307. https://doi.org/10.1111/brv.12608.
24. Keshtkar S, Azarpira N, Ghahremani MH. Mesenchymal stem cell-derived extracellular vesicles: novel frontiers in regenerative medicine. Stem Cell Res Ther. 2018;9(1):63. https://doi.org/10.1186/s13287-018-0791-7.
25. Matsuzaka Y, Yashiro R. Therapeutic strategy of Mesenchymal-stem-cell-derived extracellular vesicles as regenerative medicine. Int J Mol Sci. 2022;23(12):6480. https://doi.org/10.3390/ijms23126480.

26. Ocansey DKW, Yuan J, Wei Z, Mao F, Zhang Z. Role of ferroptosis in the pathogenesis and as a therapeutic target of inflammatory bowel disease (review). Int J Mol Med. 2023;51(6):53. https://doi.org/10.3892/ijmm.2023.5256.
27. Liang X, Li C, Song J, Liu A, Wang C, Wang W, et al. HucMSC-Exo promote mucosal healing in experimental colitis by accelerating intestinal stem cells and epithelium regeneration via Wnt signaling pathway. Int J Nanomed. 2023;18:2799–818. https://doi.org/10.2147/ijn.S402179.
28. Wang Y, Zhang Y, Lu B, Xi J, Ocansey DKW, Mao F, et al. hucMSC-Ex alleviates IBD-associated intestinal fibrosis by inhibiting ERK phosphorylation in intestinal fibroblasts. Stem Cells Int. 2023;2023:2828981. https://doi.org/10.1155/2023/2828981.
29. Duan L, Huang H, Zhao X, Zhou M, Chen S, Wang C, et al. Extracellular vesicles derived from human placental mesenchymal stem cells alleviate experimental colitis in mice by inhibiting inflammation and oxidative stress. International Journal of Molecular Medicine. 2020;46(4):1551–61. https://doi.org/10.3892/ijmm.2020.4679.
30. Valadi H, Ekström K, Bossios A, Sjöstrand M, Lee JJ, Lötvall JO. Exosome-mediated transfer of mRNAs and microRNAs is a novel mechanism of genetic exchange between cells. Nat Cell Biol. 2007;9(6):654–9. https://doi.org/10.1038/ncb1596.
31. Zou T, Jaladanki SK, Liu L, Xiao L, Chung HK, Wang JY, et al. H19 long noncoding RNA regulates intestinal epithelial barrier function via microRNA 675 by interacting with RNA-binding protein HuR. Mol Cell Biol. 2016;36(9):1332–41. https://doi.org/10.1128/mcb.01030-15.
32. Sun S, Shen J, Jiang J, Wang F, Min J. Targeting ferroptosis opens new avenues for the development of novel therapeutics. Signal Trans Target Ther. 2023;8(1):372. https://doi.org/10.1038/s41392-023-01606-1.
33. Chen Y, Fang ZM, Yi X, Wei X, Jiang DS. The interaction between ferroptosis and inflammatory signaling pathways. Cell Death Dis. 2023;14(3):205. https://doi.org/10.1038/s41419-023-05716-0.
34. Xu S, He Y, Lin L, Chen P, Chen M, Zhang S. The emerging role of ferroptosis in intestinal disease. Cell Death Dis. 2021;12(4):289. https://doi.org/10.1038/s41419-021-03559-1.
35. Wei Z, Hang S, Wiredu Ocansey DK, Zhang Z, Wang B, Zhang X, et al. Human umbilical cord mesenchymal stem cells derived exosome shuttling mir-129-5p attenuates inflammatory bowel disease by inhibiting ferroptosis. J Nanobiotechnol. 2023;21(1):188. https://doi.org/10.1186/s12951-023-01951-x.
36. Rao Z, Zhu Y, Yang P, Chen Z, Xia Y, Qiao C, et al. Pyroptosis in inflammatory diseases and cancer. Theranostics. 2022;12(9):4310–29. https://doi.org/10.7150/thno.71086.
37. You R, He X, Zeng Z, Zhan Y, Xiao Y, Xiao R. Pyroptosis and its role in autoimmune disease: A potential therapeutic target. Front Immunol. 2022;13:841732. https://doi.org/10.3389/fimmu.2022.841732.
38. Cai X, Zhang ZY, Yuan JT, Ocansey DKW, Tu Q, Zhang X, et al. hucMSC-derived exosomes attenuate colitis by regulating macrophage pyroptosis via the miR-378a-5p/NLRP3 axis. Stem Cell Res Ther. 2021;12(1):416. https://doi.org/10.1186/s13287-021-02492-6.
39. El-Jawhari JJ, El-Sherbiny YM, Jones EA, McGonagle D. Mesenchymal stem cells, autoimmunity and rheumatoid arthritis. QJM. 2014;107(7):505–14. https://doi.org/10.1093/qjmed/hcu033.
40. Fiorina P, Jurewicz M, Augello A, Vergani A, Dada S, La Rosa S, et al. Immunomodulatory function of bone marrow-derived mesenchymal stem cells in experimental autoimmune type 1 diabetes. J Immunol. 2009;183(2):993–1004. https://doi.org/10.4049/jimmunol.0900803.
41. Zhou K, Zhang H, Jin O, Feng X, Yao G, Hou Y, et al. Transplantation of human bone marrow mesenchymal stem cell ameliorates the autoimmune pathogenesis in MRL/lpr mice. Cell Mol Immunol. 2008;5(6):417–24. https://doi.org/10.1038/cmi.2008.52.
42. Zhang Z, Feng R, Niu L, Huang S, Deng W, Shi B, et al. Human umbilical cord mesenchymal stem cells inhibit T follicular helper cell expansion through the activation of iNOS in

lupus-prone B6.MRL-Fas(lpr) mice. Cell Transplant. 2017;26(6):1031–42. https://doi.org/10.3727/096368917x694660.
43. Xiang E, Han B, Zhang Q, Rao W, Wang Z, Chang C, et al. Human umbilical cord-derived mesenchymal stem cells prevent the progression of early diabetic nephropathy through inhibiting inflammation and fibrosis. Stem Cell Res Ther. 2020;11(1):336. https://doi.org/10.1186/s13287-020-01852-y.
44. Nivet E, Vignes M, Girard SD, Pierrisnard C, Baril N, Devèze A, et al. Engraftment of human nasal olfactory stem cells restores neuroplasticity in mice with hippocampal lesions. J Clin Invest. 2011;121(7):2808–20. https://doi.org/10.1172/jci44489.
45. Rui K, Hong Y, Zhu Q, Shi X, Xiao F, Fu H, et al. Olfactory ecto-mesenchymal stem cell-derived exosomes ameliorate murine Sjögren's syndrome by modulating the function of myeloid-derived suppressor cells. Cell Mol Immun. 2021;18(2):440–51. https://doi.org/10.1038/s41423-020-00587-3.
46. Harrell CR, Jovicic N, Djonov V, Arsenijevic N, Volarevic V. Mesenchymal stem cell-derived exosomes and other extracellular vesicles as new remedies in the therapy of inflammatory diseases. Cells. 2019;8(12):1605. https://doi.org/10.3390/cells8121605.
47. Kalra H, Simpson RJ, Ji H, Aikawa E, Altevogt P, Askenase P, et al. Vesiclepedia: a compendium for extracellular vesicles with continuous community annotation. PLoS Biol. 2012;10(12):e1001450. https://doi.org/10.1371/journal.pbio.1001450.
48. Keerthikumar S, Chisanga D, Ariyaratne D, Al Saffar H, Anand S, Zhao K, et al. ExoCarta: a web-based compendium of exosomal cargo. J Mol Biol. 2016;428(4):688–92. https://doi.org/10.1016/j.jmb.2015.09.019.
49. Jeppesen DK, Fenix AM, Franklin JL, Higginbotham JN, Zhang Q, Zimmerman LJ, et al. Reassessment of exosome composition. Cell. 2019;177(2):428–45.e18. https://doi.org/10.1016/j.cell.2019.02.029.
50. Mardpour S, Hamidieh AA, Taleahmad S, Sharifzad F, Taghikhani A, Baharvand H. Interaction between mesenchymal stromal cell-derived extracellular vesicles and immune cells by distinct protein content. J Cell Physiol. 2019;234(6):8249–58. https://doi.org/10.1002/jcp.27669.
51. Yang S, Liang X, Song J, Li C, Liu A, Luo Y, et al. A novel therapeutic approach for inflammatory bowel disease by exosomes derived from human umbilical cord mesenchymal stem cells to repair intestinal barrier via TSG-6. Stem cell Res Ther. 2021;12(1):315. https://doi.org/10.1186/s13287-021-02404-8.
52. Zhang K, Guo J, Yan W, Xu L. Macrophage polarization in inflammatory bowel disease. Cell Commun Signal. 2023;21(1):367. https://doi.org/10.1186/s12964-023-01386-9.
53. Liu H, Liang Z, Wang F, Zhou C, Zheng X, Hu T, et al. Exosomes from mesenchymal stromal cells reduce murine colonic inflammation via a macrophage-dependent mechanism. JCI Insight. 2019;4(24):e131273. https://doi.org/10.1172/jci.insight.131273.
54. Skotland T, Llorente A, Sandvig K. Lipids in extracellular vesicles: what can be learned about membrane structure and function? Cold Spring Harb Perspect Biol. 2023;15(8):a041415. https://doi.org/10.1101/cshperspect.a041415.
55. Zhang Q, Jeppesen DK, Higginbotham JN, Franklin JL, Coffey RJ. Comprehensive isolation of extracellular vesicles and nanoparticles. Nat Protoc. 2023;18(5):1462–87. https://doi.org/10.1038/s41596-023-00811-0.
56. Lingwood D, Simons K. Lipid rafts as a membrane-organizing principle. Science. 2010;327(5961):46–50. https://doi.org/10.1126/science.1174621.
57. Skotland T, Sandvig K, Llorente A. Lipids in exosomes: Current knowledge and the way forward. Prog Lipid Res. 2017;66:30–41. https://doi.org/10.1016/j.plipres.2017.03.001.
58. Bowie RV, Donatello S, Lyes C, Owens MB, Babina IS, Hudson L, et al. Lipid rafts are disrupted in mildly inflamed intestinal microenvironments without overt disruption of the epithelial barrier. Am J Physiol Gastrointest Liver Physiol. 2012;302(8):G781–93. https://doi.org/10.1152/ajpgi.00002.2011.

59. Rehman FU, Liu Y, Zheng M, Shi B. Exosomes based strategies for brain drug delivery. Biomaterials. 2023;293:121949. https://doi.org/10.1016/j.biomaterials.2022.121949.
60. Zhang L, Yu D. Exosomes in cancer development, metastasis, and immunity. Biochim Biophys Acta Rev Cancer. 2019;1871(2):455–68. https://doi.org/10.1016/j.bbcan.2019.04.004.
61. Hu Z, Chen G, Zhao Y, Gao H, Li L, Yin Y, et al. Exosome-derived circCCAR1 promotes CD8+T-cell dysfunction and anti-PD1 resistance in hepatocellular carcinoma. Mol Cancer. 2023;22(1):55. https://doi.org/10.1186/s12943-023-01759-1.
62. Shapouri-Moghaddam A, Mohammadian S, Vazini H, Taghadosi M, Esmaeili SA, Mardani F, et al. Macrophage plasticity, polarization, and function in health and disease. J Cell Physiol. 2018;233(9):6425–40. https://doi.org/10.1002/jcp.26429.
63. Murray PJ, Wynn TA. Obstacles and opportunities for understanding macrophage polarization. J Leukoc Biol. 2011;89(4):557–63. https://doi.org/10.1189/jlb.0710409.
64. Ferrante CJ, Leibovich SJ. Regulation of macrophage polarization and wound healing. Adv Wound Care. 2012;1(1):10–6. https://doi.org/10.1089/wound.2011.0307.
65. Yang R, Liao Y, Wang L, He P, Hu Y, Yuan D, et al. Exosomes derived from M2b macrophages attenuate DSS-induced colitis. Front Immunol. 2019;10:2346. https://doi.org/10.3389/fimmu.2019.02346.
66. Wu J, Zhao R, Dai J, Lai G, Khan AU, Yu X, et al. Analysis of differential expression of long non-coding RNAs in exosomes derived from mature and immature dendritic cells. Mol Med Rep. 2021;23(2):132. https://doi.org/10.3892/mmr.2020.11771.
67. Kim SH, Bianco NR, Shufesky WJ, Morelli AE, Robbins PD. Effective treatment of inflammatory disease models with exosomes derived from dendritic cells genetically modified to express IL-4. J Immunol. 2007;179(4):2242–9. https://doi.org/10.4049/jimmunol.179.4.2242.
68. Tkach M, Kowal J, Zucchetti AE, Enserink L, Jouve M, Lankar D, et al. Qualitative differences in T-cell activation by dendritic cell-derived extracellular vesicle subtypes. EMBO J. 2017;36(20):3012–28. https://doi.org/10.15252/embj.201696003.
69. Tavasolian F, Hosseini AZ, Rashidi M, Soudi S, Abdollahi E, Momtazi-Borojeni AA, et al. The impact of immune cell-derived exosomes on immune response initiation and immune system function. Curr Pharm Design. 2021;27(2):197–205. https://doi.org/10.2174/1381612826666201207221819.
70. Britton GJ, Contijoch EJ, Mogno I, Vennaro OH, Llewellyn SR, Ng R, et al. Microbiotas from humans with inflammatory bowel disease alter the balance of gut Th17 and RORγt(+) regulatory T cells and exacerbate colitis in mice. Immunity. 2019;50(1):212–24.e4. https://doi.org/10.1016/j.immuni.2018.12.015.
71. Ueno Y, Tanaka S, Sumii M, Miyake S, Tazuma S, Taniguchi M, et al. Single dose of OCH improves mucosal T helper type 1/T helper type 2 cytokine balance and prevents experimental colitis in the presence of valpha14 natural killer T cells in mice. Inflamm Bowel Dis. 2005;11(1):35–41. https://doi.org/10.1097/00054725-200501000-00005.
72. Dong Y, Yang C, Pan F. Post-translational regulations of Foxp3 in Treg cells and their therapeutic applications. Front Immun. 2021;12:626172. https://doi.org/10.3389/fimmu.2021.626172.
73. Liao F, Lu X, Dong W. Exosomes derived from T regulatory cells relieve inflammatory bowel disease by transferring miR-195a-3p. IUBMB Life. 2020;72(12):2591–600. https://doi.org/10.1002/iub.2385.
74. Wang Y, Tian J, Tang X, Rui K, Tian X, Ma J, et al. Exosomes released by granulocytic myeloid-derived suppressor cells attenuate DSS-induced colitis in mice. Oncotarget. 2016;7(13):15356–68. https://doi.org/10.18632/oncotarget.7324.
75. Wei G, Jie Y, Haibo L, Chaoneng W, Dong H, Jianbing Z, et al. Dendritic cells derived exosomes migration to spleen and induction of inflammation are regulated by CCR7. Sci Rep. 2017;7:42996. https://doi.org/10.1038/srep42996.
76. Munir J, Lee M, Ryu S. Exosomes in food: Health benefits and clinical relevance in diseases. Adv Nutr. 2020;11(3):687–96. https://doi.org/10.1093/advances/nmz123.
77. Lu Y, Xu J, Tang R, Zeng P, Li Z, You J, et al. Edible Pueraria lobata-derived exosome-like nanovesicles ameliorate dextran sulfate sodium-induced colitis associated lung inflammation

through modulating macrophage polarization. Biomed Pharmacother. 2024;170:116098. https://doi.org/10.1016/j.biopha.2023.116098.
78. Xie MY, Hou LJ, Sun JJ, Zeng B, Xi QY, Luo JY, et al. Porcine Milk exosome MiRNAs attenuate LPS-induced apoptosis through inhibiting TLR4/NF-κB and p53 pathways in intestinal epithelial cells. J Agric Food Chem. 2019;67(34):9477–91. https://doi.org/10.1021/acs.jafc.9b02925.
79. Duan T, Wang X, Dong X, Wang C, Wang L, Yang X, et al. Broccoli-derived exosome-like nanoparticles alleviate loperamide-induced constipation, in correlation with regulation on gut microbiota and tryptophan metabolism. J Agric Food Chem. 2023;71(44):16568–80. https://doi.org/10.1021/acs.jafc.3c04150.
80. Zhu H, He W. Ginger: A representative material of herb-derived exosome-like nanoparticles. Front Nutr. 2023;10:1223349. https://doi.org/10.3389/fnut.2023.1223349.
81. Teng Y, Ren Y, Sayed M, Hu X, Lei C, Kumar A, et al. Plant-derived exosomal microRNAs shape the gut microbiota. Cell Host Microbe. 2018;24(5):637–52.e8. https://doi.org/10.1016/j.chom.2018.10.001.
82. Kim KU, Han K, Kim J, Kwon DH, Ji YW, Yi DY, et al. The protective role of exosome-derived microRNAs and proteins from human breast milk against infectious agents. Metabolites. 2023;13(5):635. https://doi.org/10.3390/metabo13050635.
83. He S, Liu G, Zhu X. Human breast milk-derived exosomes may help maintain intestinal epithelial barrier integrity. Pediatr Res. 2021;90(2):366–72. https://doi.org/10.1038/s41390-021-01449-y.
84. Brown L, Wolf JM, Prados-Rosales R, Casadevall A. Through the wall: extracellular vesicles in Gram-positive bacteria, mycobacteria and fungi. Nat Rev Microbiol. 2015;13(10):620–30. https://doi.org/10.1038/nrmicro3480.
85. Hu R, Lin H, Li J, Zhao Y, Wang M, Sun X, et al. Probiotic Escherichia coli Nissle 1917-derived outer membrane vesicles enhance immunomodulation and antimicrobial activity in RAW264.7 macrophages. BMC Microbiol. 2020;20:268. https://doi.org/10.1186/s12866-020-01953-x.
86. Alvarez CS, Badia J, Bosch M, Giménez R, Baldomà L. Outer membrane vesicles and soluble factors released by probiotic Escherichia coli Nissle 1917 and commensal ECOR63 enhance barrier function by regulating expression of tight junction proteins in intestinal epithelial cells. Front Microbiol. 2016;7:1981. https://doi.org/10.3389/fmicb.2016.01981.
87. Kang CS, Ban M, Choi EJ, Moon HG, Jeon JS, Kim DK, et al. Extracellular vesicles derived from gut microbiota, especially Akkermansia muciniphila, protect the progression of dextran sulfate sodium-induced colitis. PLoS One. 2013;8(10):e76520. https://doi.org/10.1371/journal.pone.0076520.
88. Liang L, Yang C, Liu L, Mai G, Li H, Wu L, et al. Commensal bacteria-derived extracellular vesicles suppress ulcerative colitis through regulating the macrophages polarization and remodeling the gut microbiota. Microb Cell Fact. 2022;21(1):88. https://doi.org/10.1186/s12934-022-01812-6.
89. Hu G, Gong AY, Roth AL, Huang BQ, Ward HD, Zhu G, et al. Release of luminal exosomes contributes to TLR4-mediated epithelial antimicrobial defense. PLoS Pathog. 2013;9(4):e1003261. https://doi.org/10.1371/journal.ppat.1003261.
90. Jacobs BA, Prince S, Smith KA. Gastrointestinal nematode-derived antigens alter colorectal cancer cell proliferation and migration through regulation of cell cycle and epithelial-mesenchymal transition proteins. Int J Mol Sci. 2020;21(21):7845. https://doi.org/10.3390/ijms21217845.
91. McDonald MK, Tian Y, Qureshi RA, Gormley M, Ertel A, Gao R, et al. Functional significance of macrophage-derived exosomes in inflammation and pain. Pain. 2014;155(8):1527–39. https://doi.org/10.1016/j.pain.2014.04.029.
92. Wang L, Yu Z, Wan S, Wu F, Chen W, Zhang B, et al. Exosomes derived from dendritic cells treated with Schistosoma japonicum soluble egg antigen attenuate DSS-induced colitis. Front Pharmacol. 2017;8:651. https://doi.org/10.3389/fphar.2017.00651.

93. Cai Z, Zhang W, Li M, Yue Y, Yang F, Yu L, et al. TGF-beta1 gene-modified, immature dendritic cells delay the development of inflammatory bowel disease by inducing CD4(+) Foxp3(+) regulatory T cells. Cell Mol Immunol. 2010;7(1):35–43. https://doi.org/10.1038/cmi.2009.107.
94. Kim SH, Lechman ER, Bianco N, Menon R, Keravala A, Nash J, et al. Exosomes derived from IL-10-treated dendritic cells can suppress inflammation and collagen-induced arthritis. J Immunol. 2005;174(10):6440–8. https://doi.org/10.4049/jimmunol.174.10.6440.
95. Schorey JS, Cheng Y, McManus WR. Bacteria- and host-derived extracellular vesicles - two sides of the same coin? J Cell Sci. 2021;134:jcs256628. https://doi.org/10.1242/jcs.256628.
96. Chu S, Yu T, Wang W, Wu H, Zhu F, Wei C, et al. Exosomes derived from EphB2-overexpressing bone marrow mesenchymal stem cells regulate immune balance and repair barrier function. Biotechnol Lett. 2023;45(5–6):601–17. https://doi.org/10.1007/s10529-023-03358-y.
97. Enerbäck L. Mucosal mast cells in the rat and in man. Int Arch Allergy Appl Immunol. 1987;82(3–4):249–55. https://doi.org/10.1159/000234199.
98. Rothenberg ME, Austen KF. Influence of the fibroblast environment on the structure of mast cell proteoglycans. Ann N Y Acad Sci. 1989;556:233–44. https://doi.org/10.1111/j.1749-6632.1989.tb22506.x.
99. Shanahan F, Denburg JA, Bienenstock J, Befus AD. Mast cell heterogeneity. Can J Physiol Pharmacol. 1984;62(6):734–7. https://doi.org/10.1139/y84-121.
100. Mukai K, Tsai M, Saito H, Galli SJ. Mast cells as sources of cytokines, chemokines, and growth factors. Immunol Rev. 2018;282(1):121–50. https://doi.org/10.1111/imr.12634.
101. Boesiger J, Tsai M, Maurer M, Yamaguchi M, Brown LF, Claffey KP, et al. Mast cells can secrete vascular permeability factor/vascular endothelial cell growth factor and exhibit enhanced release after immunoglobulin E-dependent upregulation of fc epsilon receptor I expression. J Exp Med. 1998;188(6):1135–45. https://doi.org/10.1084/jem.188.6.1135.
102. Li M, Zhao J, Cao M, Liu R, Chen G, Li S, et al. Mast cells-derived MiR-223 destroys intestinal barrier function by inhibition of CLDN8 expression in intestinal epithelial cells. Biol Res. 2020;53(1):12. https://doi.org/10.1186/s40659-020-00279-2.
103. Chen T, Gu K, Lin R, Liu Y, Shan Y. The function of Sphingosine-1-phosphate receptor 2 (S1PR2) in maintaining intestinal barrier and inducing ulcerative colitis. Bioengineered. 2022;13(5):13703–17. https://doi.org/10.1080/21655979.2022.2076500.
104. Zamith-Miranda D, Nimrichter L, Rodrigues ML, Nosanchuk JD. Fungal extracellular vesicles: modulating host-pathogen interactions by both the fungus and the host. Microbes Infect. 2018;20(9–10):501–4. https://doi.org/10.1016/j.micinf.2018.01.011.
105. Jeffery-Smith A, Taori SK, Schelenz S, Jeffery K, Johnson EM, Borman A, et al. Candida auris: A review of the literature. Clin Microbiol Rev. 2018;31(1):e00029–17. https://doi.org/10.1128/cmr.00029-17.
106. Chen SC, Meyer W, Sorrell TC. Cryptococcus gattii infections. Clin Microbiol Rev. 2014;27(4):980–1024. https://doi.org/10.1128/cmr.00126-13.
107. Domenis R, Marino M, Cifù A, Scardino G, Curcio F, Fabris M. Circulating exosomes express α4β7 integrin and compete with CD4+ T cells for the binding to Vedolizumab. PLoS One. 2020;15(11):e0242342. https://doi.org/10.1371/journal.pone.0242342.

Chapter 12
The Applications of Exosomes in the Treatment of Autoimmune-Mediated Eye Diseases

Huan Li and Xiaomin Zhang

Abstract In recent years, exosomes possess therapeutic potential for ocular disorders. They have similar cellular functions as their mother cell, especially mesenchymal stem cells (MSCs). In addition, exosomes are ideal carriers for drug delivery because of their unique structure, including nanoparticle size, biological barrier penetration, and low immunogenicity. Exosomes have already been successfully tested for loading several drugs in the treatment of ocular diseases. At present, some studies suggest that exosomes derived from diverse cellular sources can be used as a biomarker for the predictive diagnosis of disease. In this chapter, we summarize the application of exosomes in autoimmune-mediated eye diseases including dry eye or Sjögren's Syndrome, corneal transplant rejection, autoimmune uveitis, age-related macular degeneration (AMD), and thyroid-associated ophthalmopathy (TAO).

Keywords Exosomes · Mesenchymal stem cells · Uveitis · Experimental autoimmune uveitis

1 Introduction

Exosomes are nano-sized membrane vesicles secreted by most cell types with a diameter of 40–150 nm. These vesicles contain a wide range of components including RNA, proteins, and lipids which play essential roles in modulating cell-to-cell signaling. Exosomes are widely distributed in body fluids, such as aqueous humor, blood, and urine. Almost all types of cells can secrete exosomes. Exosomes secreted by pathological cells are involved in the development and progression of disease.

H. Li · X. Zhang (✉)
Tianjin Key Laboratory of Retinal Functions and Diseases, Tianjin Branch of National Clinical Research Center for Ocular Disease, Eye Institute and School of Optometry, Tianjin Medical University Eye Hospital, Tianjin, China
e-mail: xzhang08@tmu.edu.cn

© The Author(s), under exclusive license to Springer Nature Singapore Pte Ltd. 2024

Z. Li (ed.), *Extracellular Vesicle: Biology and Translational Application*,
https://doi.org/10.1007/978-981-97-5536-3_12

The characteristics of exosomes are greatly affected by the parental cell. MSCs can generate exosomes that can exert the same therapeutic effects as parental cells, including promoting tissue regeneration, inhibiting autoimmune responses, and protecting neuron function. In addition, exosomes are ideal carriers for drug delivery and biomarkers for diagnosis because of their distinctive advantages, including small particle size, high biocompatibility, biological barrier penetration, and low immunogenicity.

Autoimmune-mediated eye diseases are one of the major diseases that have affected ocular health, including dry eye or Sjögren's syndrome, corneal transplant rejection, autoimmune uveitis, AMD, and TAO. The pathophysiological mechanism of this group of diseases is so complex that the treatment becomes extremely intractable. The mainstream treatment methods including glucocorticoids, immunosuppressants and biological agents often come with significant side effects, and this disease has a tendency to recurrence. Therefore, there is an urgent need to discover new therapeutic modalities to treat this disease.

In this chapter, we summarize the research progress of exosomes in the treatment of autoimmune-mediated eye diseases and discuss the potential to be used as biomarker and drug delivery systems in the eye (Fig. 12.1).

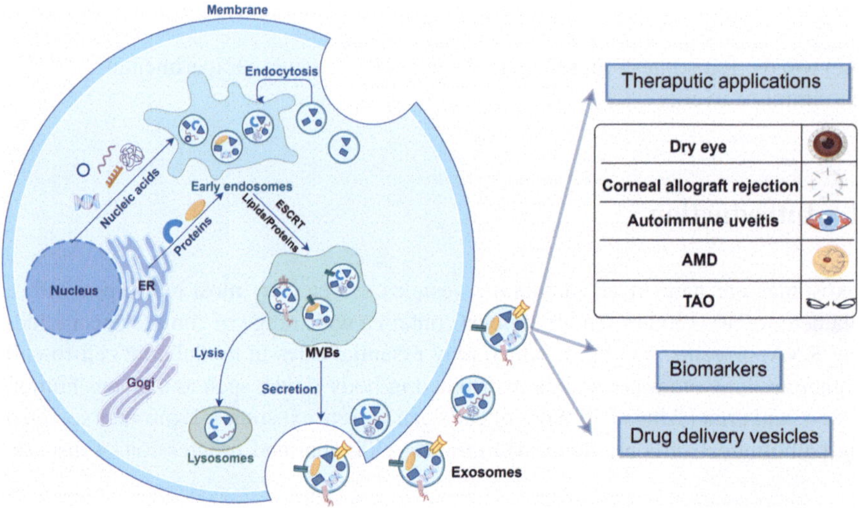

Fig. 12.1 The biogenesis and applications of exosomes (by figdraw). The plasma membrane invades to form early endosomes. Then, these early endosomes encapsulate cellular components, such as nucleic acids, proteins, and lipids. Further invagination leads to the formation of intraluminal vesicles (ILV), which gradually assemble into the multivesicular body (MVB). The MVB then undergoes two distinct fusion events: one with the lysosome for degradation of its contents, and another with cell membranes for the release of exosomes. These exosomes play crucial roles in the treatment, identification of biomarkers, and drug delivery for immune-mediated eye diseases

2 Basic Characteristics and Biological Functions of Exosomes

Exosome was first reported in the 1980s to describe endosome-derived nanovesicles released during the maturation of sheep [1–3]. Initially, the exosome was considered as a bin to remove unwanted cell components [3, 4]. Afterward, increasing evidence suggested that exosome is a key medium in intercellular communication [5]. Exosome biogenesis is finely regulated by complex mechanisms. During this biogenesis process, the plasma membrane invades to form early endosomes. Then, the early endosomes encapsulate cellular components including nucleic acids, proteins, and lipids, and subsequently, these vesicles further invaginate to form intraluminal vesicles (ILV) which gradually form the multivesicular body (MVB). One part fuses with the lysosome where their contents are degraded and another fuses with cell membranes to release exosomes [6]. The center of exosome biogenesis is MVB formation. Overall, the mechanism of MVB biogenesis contains the endosomal sorting complex required for transport (ESCRT)-dependent mechanisms and ESCRT-independent mechanisms. ESCRT involves ESCRT-0, ESCRT-I, ESCRT-II, ESCRT-III, and ATPase VPS4. There are several alternative mechanisms also called non-typical ESCRT-dependent pathways containing Alix and HD-PTP-dependent pathways. ESCRT-independent pathway largely relies on lipids, syndecan, and syntenin [7, 8]. Of these, the neutral sphingomyelinase 2 (nSMase2)-ceramide pathway is the most studied mechanism. However, despite the increasing interest in this field, the processes involved in the formation and release of exosomes are still not exhaustive.

It was only in 1996 that the biological function of exosomes was just beginning to be discovered [9]. Signal transduction is one of the most basic and also one of the most important functions of exosomes. The surface molecule of exosomes aids in their attachment to target cells and activates cellular signaling systems via ligand–receptor interactions [5]. The immune regulation role of exosomes has been extensively investigated. Although the studies on cancer disease are relatively numerous, the immunomodulatory effect of exosomes in ocular diseases has emerged in recent years [10, 11]. Furthermore, exosomes have been reported to play important roles in exerting reparation and regeneration roles in multiple diseases and disease models [12, 13]. In addition to being diagnostic biomarkers, exosomes possess the potential to be therapeutic application systems for delivering specific cargo [14, 15].

3 Exosomes in Immune-Mediated Eye Diseases

Exosomes can exist stably in ocular fluids such as tears, aqueous humor, vitreous humor, and blood. They play important roles in regulating ocular cell migration, proliferation, apoptosis, immune response, and angiogenesis. Exosomes have a tight relationship with immune-mediated eye diseases including Sjögren's syndrome dry eye, corneal allograft rejection, autoimmune uveitis, AMD, and TAO.

3.1 Ocular Surface Diseases

3.1.1 Dry Eye

Sjögren's Syndrome (SS) Dry Eye

Sjögren's syndrome (SS) is a chronic autoimmune condition characterized by diminished exocrine gland function. It can be classified into primary Sjogren's syndrome (pSS) and secondary Sjogren's syndrome (sSS) as part of another connective tissue disease such as systemic lupus erythematosus (SLE) or rheumatoid arthritis [16]. The etiology of pSS is unclear, and the prevalence of pSS is 0.3–0.7% in China. As the most common autoimmune connective tissue disease, its clinical manifestations are complex and diverse including dry eyes, nose, and mouth. The main pathological feature of pSS manifests as focal infiltration of lymphocytes and secretion of multiple inflammatory cytokines including IFN-γ, IL-1β, and TNF-α, which results in tissue impairment and functional loss of exocrine glands [17]. Abnormal proliferation, differentiation, and apoptosis of $CD4^+$ T cells are key contributors to pSS disturbances [18, 19]. The management of pSS is chronic and complicated. Current treatments mainly include non-steroidal anti-inflammatory drugs (NSAIDs), glucocorticoids, and immunosuppression which can produce side effects. To date, with the lack of effective treatments for pSS, novel therapeutic approaches are actively being explored.

Exosomes are involved in the development and progression of pSS by the presentation of self-antigens to the immune system. Several prognostic biomarkers of pSS, such as anti-Ro/SSA, anti-La/SSB, and antinuclear autoantibodies (ANA), are released by salivary gland epithelial cells via exosomes [16]. BART13-3p, a microRNA encoded by Epstein–Barr virus (EBV), was significantly increased in salivary glands (SGs) of pSS patients. Exosomes could transfer miR-BART13-3p from B lymphocytes into salivary gland epithelial cells (SGEC) [20]. Another study has shown that miR-1423p contained exosomes derived from T cells that were involved in the pathogenesis of pSS by affecting intracellular Ca^{2+} signaling [21]. Laboratory examinations of pSS patients, such as ANA, RF, anti-Ro/SSA, and anti-La/SSB, are incapable of achieving both high sensitivity and specificity simultaneously [22]. Thus, more accurate diagnosis methods are required. MiRNA levels in exosomes isolated from parotid saliva of pSS patients could serve as diagnostic markers [23]. Potential biomarkers of extracellular vesicles in the tear and saliva of pSS patients were detected by liquid chromatography-mass spectrometry (LC-MS) [24]. Recently, the therapeutic effects of exosomes in pSS have gradually attracted great interest. Exosomes from different sources have shown therapeutic potential. Li et al. found that exosomes derived from MSC (MSC-exo) could effectively alleviate the induced autoimmune dacryoadenitis in rabbit models by subconjunctival injection [10]. It has been demonstrated that umbilical mesenchymal stem cell-derived exosomes (UCMSC-exo) could reduce the autophagy level in peripheral $CD4^+$ T cells [25]. In addition, labial gland-derived MSCs and their exosomes have been well-established to attenuate murine Sjögren's syndrome by regulating the balance of Treg and Th17 cells [26]. More interestingly, some studies developed novel nanoparticle delivery systems on the basis of MSC-exo to serve as treatment of SS [27, 28].

Graft-Versus-Host Disease (GVHD) Dry Eye

Graft-versus-host disease (GVHD) is a life-threatening complication for patients who have undergone allogeneic hemopoietic stem cell transplantation (HSCT) [29, 30]. Dry eye syndrome is more prone to occur as part of GVHD. The clinical manifestations of GVHD can be classified into acute and chronic presentations. Acute GVHD usually involves the liver, skin, and gastrointestinal tract. However, chronic GVHD affects all organ systems of an individual [31]. Ocular GVHD is a common manifestation of chronic GVHD and most frequently presents with dry eye and its sequelae. The manifestations are similar to SS dry eye. However, unlike SS dry eye, fibrosis might occur in the ocular surface of GVHD patients. During the progression of GVHD-associated dry eye disease, the donor $CD8^+$ and $CD4^+$ T cells are stimulated by the interaction between donor T cells and host antigen-presenting cells, and hematopoietic tissue or ocular epithelial tissues are targeted, such as the cornea, conjunctiva, and lacrimal gland [32, 33]. This process leads to significant inflammatory damage. The condition is distinguished by symptoms of irritation, pain, and persistent visual impairment, significantly impacting the patient's quality of life. However, the effectiveness of existing clinical medications, such as tear supplements, steroids, and others, remains limited [34]. Thus, new therapeutic strategies are needed.

Exosomes become a potential cell-free alternative because of their immunomodulatory effects [35]. Clinical trials have proved that MSC-exo had a favorable therapeutic effect in GVHD patients, and no local or systemic side effects were reported within 24 months [36]. In addition, the reprogramming of proinflammatory M1 macrophages into immunosuppressive M2 macrophages by MSC-exo was achieved through the targeting of the IL-6/IL-6R/Stat3 pathway mediated by miR-204. The effects of MSC-exo were nullified when miR-204 was blocked, whereas the introduction of miR-204 into L929-exo significantly reduced the severity of dry eye. Consequently, MSC-exo exhibits effectiveness in the treatment of dry eye disease associated with GVHD and emphasizes the potential of miR-204 as a therapeutic agent [37].

3.2 Corneal Allograft Rejection

Keratopathy is one of the global blinding eye diseases. In the world, corneal transplants are one of the most frequent transplant surgeries performed to restore vision to those suffering from corneal blindness. The cornea is a transparent avascular tissue, which makes it experience immune privilege. Despite this, it is considered to be one of the major causes of corneal transplant failure due to rejection of the graft by the immune system [38], mainly allograft rejection. The problem of graft rejection is mainly attributed to the recognition of donor major histocompatibility complex (MHC) molecules by recipient T cells [39]. At present, anti-rejection therapy consists of systemic corticosteroids or immunosuppressive agents which may yield life-threatening complications [40]. New therapies are needed to ensure the viability of the corneal transplant.

Exosomes could be used as cell-free therapies to increase immune tolerance in corneal transplantation [41, 42]. Han et al. found that exosomes derived from epithelium could promote corneal wound healing and neovascularization [43]. It was demonstrated that exosomes derived from limbal keratocytes activated AKT signaling and promoted limbal epithelial proliferation during wound healing [44]. Again, exosomes derived from mesenchymal stromal cells of the corneal were proven to promote corneal wound healing [45]. Furthermore, MSC-exo could cross biological barriers and prolong graft survival [46]. Therefore, exosomes could be used as potential therapeutic targets for corneal transplantation. The shortage of corneal donors is an enormous challenge across the world [47]. New biomaterials were introduced to address this question including collagen gels, synthetic polymers, and tissue-engineered scaffolds [48]. Jangamreddy et al. found that short collagen-like peptides facilitated corneal regeneration by stimulating the secretion of extracellular vesicles from growing corneal epithelium cells for generating matrix components [49]. In conclusion, exosomes provide new therapeutic strategies for corneal allograft rejection and corneal injuries.

3.3 Autoimmune Uveitis

Uveitis refers to an inflammation affecting the uveal and retinal layers of the eye. Based on the anatomical localization of inflammation, uveitis may be classified as anterior, posterior, and panuveitis. Depending on the etiology, uveitis is classified clinically by the International Uveitis Study Group as infectious uveitis (IU), non-infectious uveitis (NIU), masquerade syndrome (malignant), or other (idiopathic). According to the clinical pathological classification, uveitis can be classified as granulomatous and non-granulomatous uveitis [50, 51]. Autoimmune uveitis is one of the leading causes of blindness in developed countries [52]. Uveitis progresses most commonly to relapsing chronic uveitis, which causes permanent vision loss in some cases. Glucocorticoids and immunosuppressants are traditional medications.

However, some side effects may occur at a high dose or for a prolonged course including hypertension, hyperlipidemia, cataracts, and ocular hypertension [53–55].

In recent years, exosomes have exhibited enormous potential in the treatment of uveitis. Previous research indicated that MSCs can ameliorate experimental autoimmune uveitis (EAU) in rats and mice [56–58]. Exosomes derived from MSCs have similar biological functions to blast cells, which are expected to become an alternative cell-free therapy [59]. Bai et al. reported that MSC-exo inhibited the chemoattractive effects of CCL2 and CCL21 on inflammatory cells via peribulbar injection. This study suggested that MSC-Exo can inhibit the migration of inflammatory cells (including T cells, NK cells, and macrophages) to ameliorate EAU [60]. It has been suggested that vaccination of naïve rats with circulating exosomes derived from EAU rats reduced the incidence of recurrent uveitis. Autologous circulating exosomes may become a vaccine for the suppression of antigen-specific inflammation in patients with recurrent AU [61]. Furthermore, engineered exosomes have generated an enormous amount of interest in recent years. Li et al. found that exosomes derived from IL-10-overexpressing MSCs (MSC-exo-IL-10) efficiently inhibited the development of EAU in mice by a single intravenous injection [62]. MSC-exo-IL-10 can suppress Th1 and Th17 cell proliferation while increasing regulatory T cell subsets in vivo and in vitro [62]. In addition, interleukin 35 (IL-35) and interleukin 27 (IL-27)-producing exosomes derived from IL-35 and IL-27-producing Breg population have been shown to suppress neuroinflammation and EAU [63, 64]. Exosomes are considered to be natural carriers because of their lipid bilayer and nanostructure. The findings from our previous study have shown that MSC-Exo can enhance the penetration of rapamycin. Following subconjunctival injection, MSC-exo loaded with rapamycin was able to reach the retinal foci. In comparison to the administration of exo or rapamycin alone, the combined use of MSC-exo and rapamycin resulted in a more pronounced therapeutic effect and a reduction in the infiltration of inflammatory cells in the eye [14]. Some proteins of exosomes can be used as biomarkers for active inflammation in uveitis including carbonic anhydrase 2 and Ras-related protein Rap-1b [65].

3.4 Age-Related Macular Degeneration (AMD)

AMD is a complex eye disorder and one of the leading causes of blindness in the elderly [66]. Depending on clinical manifestations and pathology changes, AMD is classified as dry AMD (atrophic) or wet AMD (exudative). Wet AMD (wAMD) is characterized by the presence of choroidal neovascularization (CNV) and retinal hemorrhage, which cause irreversible vision loss and even blindness. Retinal pigment epithelium (RPE) forms the outer blood-retinal barrier of the eye which plays an important role in AMD pathogenesis. The pathogenesis of AMD is not yet fully comprehended, such as RPE cell senescence, oxidative stress, lipid metabolism, inflammation and immunity, and neovascularization [67]. Recently, the pathogenesis of "inflammation and immunity" has received particular interest [68]. Ample evidence has proven that the complement system plays a major role in AMD pathogenesis [68, 69]. It was shown that the abnormal function of complement factor H (CFH), a regulator for the alternative complement pathway, might promote drusen formation [70, 71]. In addition, the absence of complement C3 increased the risk of AMD [72]. Dysfunction of CFH might contribute to aggression against exosomes from RPE cells coated with C3 by the invading leukocytes in the aging retina. This might induce instability of the exosome membrane and result in the release of intracellular proteins and thus aggravate AMD [71]. Many different types of retinal cells including RPE cells and retinal astroglial cells (RACs) can generate and release exosomes which might participate in the pathophysiology of AMD. It is shown that exosomes derived from RACs contain anti-angiogenic components that could suppress the formation of CNV [73]. This study might provide new ideas for the future treatment of AMD. However, RPE cells in stressed states could release a large number of exosomes containing vascular endothelial growth factor (VEGF) 2 to promote the formation of NV. In addition, MSC-exos are capable of regulating the polarization of macrophages and downregulating VEGF expression [74, 75]. Some content of exosomes can be used as a diagnostic biomarker of AMD, especially microRNA and protein [76, 77]. The potential therapeutic applications of exosomes in AMD warrant further investigations.

3.5 Thyroid-Associated Ophthalmopathy (TAO)

TAO, also called Graves' ophthalmopathy (GO), is a common organ-specific autoimmune disorder that threatens vision and damages appearance. The most common extrathyroidal manifestation of Graves' disease (GD) is GO [78]. GO develops in approximately 30–50% of patients with GD [79]. It may occur in euthyroid or hypothyroid patients. Immune tolerance is in a dysfunctional state due to shared autoantigens between thyroid tissue and the orbit. GO patients could experience photophobia, diplopia, conjunctival hyperemia, exposure-related dry eye, eyelid retraction, exophthalmos, swelling of periorbital tissues, and may even develop exposure keratopathy and compressive neuropathy [80]. The pathogenesis of GD is not yet fully understood, and there is no cure for the disease. Intravenous GCs (ivGCs) are the first-line treatment for moderate-to-severe and active GO, which is effective in 70–80% of patients and might contribute to serious complications. Therefore, there is an urgent need to discover novel therapeutic strategies for GO.

Currently, direct research evidence of exosomes on the occurrence and development of GD is still missing. However, some studies have demonstrated that exosomes are highly associated with the development of TAO. Han et al. have found that specific proteins, such as vitamin D-binding (VDB) protein, C-reactive protein (CRP), chitinase 3-like 1 (CHI3L1), matrix metalloproteinase-9 (MMP-9), vascular adhesion molecule-1 (VCAM-1), caspase-3, and complement C4A, displayed higher expression in tear-derived exosomes from TAO patients [81, 82]. In addition, some cytokines, including IL-1 and IL-18, were significantly increased in the tear exosomes and serum of GO patients [82]. Furthermore, miRNA from the plasma exosome was transferred into recipient cells and bound to the target cells via exosome. Plasma exosomes deliver miR-885-3p and inhibit the AKT/NF-κB signaling pathway to improve the glucocorticoid sensitivity of orbital fibroblasts [83]. Exosome miR-885-3p can be used as a biomarker of GC sensitivity in GO patients.

4 Exosome Biomarkers for Eye Diseases

Exosomes are attractive targets in the search for novel biomarkers because of their unique features. Exosomes contain various sets of nucleotide acids and proteins which could act as stimulatory signals influencing pathway expression [84]. During the progression of the disease, exosomes carry multiple bioactive molecules for signal transduction via intercellular communication. Thus, exosomes derived from various cell sources are considered to be excellent biomarkers, such as cancer screening [85], cardiovascular disease [86], urinary system [87], and autoimmune diseases [88]. For example, as the composition of exosomes differs between healthy individuals and cancer patients, exosomes are thought to have great potential for early clinical diagnosis of cancer [89, 90].

Recently, exosomes as biomarkers have attracted great interest in the diagnosis and treatment of immune-mediated eye diseases. Exosomes are widely present in body fluids that are closely related to the eye, such as tears, vitreous, aqueous humor, and blood. To evaluate patients with pSS, noninvasive and more accurate diagnostic techniques are urgently needed. LC-MS was used to develop proteomic biomarker profiles of exosomes in salivary and tears from 27 patients with pSS and 32 healthy controls [24]. In this study, results showed that the upregulation of proteins such as LCN2, CALM, GRN, CALML5, and CPNE1 was explored in the saliva and tears of pSS patients [24]. It may also apply to staging and monitoring pSS, providing novel disease biomarkers. In addition, Xiao et al. found that expression levels of TNFAIP8 in plasma exosomes might be considered a potential biomarker for diabetic retinopathy (DR) based on proteomic analysis [91]. Han et al. isolated exosomes from tear fluids (TF) of TAO patients and analyzed the proteins. It was found that some kind of protein expression level in patients was higher than that in healthy controls. Increased exosome count in TF and expression of a unique protein in exosome might reflect the pathological process and disease severity of TAO [81]. Furthermore, RPE-derived exosomes could attenuate an inflammatory response by killing monocytes in an inflammatory environment [92]. Exosome proteomic findings from RPEs may provide potential biomarkers for prognostication, diagnosis, and treatment of retinal diseases [93]. Consistent with our recent study, we found a significant negative correlation between plasma endothelin-converting enzyme 1 level and mean retinal thickness by comprehensive profiling of extracellular vesicles in uveitis and scleritis [94]. Our study proposed potential therapeutic drugs and identified valuable biomarker candidates. In conclusion, exosomes as biomarkers have great potential and large room in the diagnosis and treatment of ocular diseases.

5 Exosomes as Drug Delivery Vesicles

Exosomes are potential natural nanocarriers for drug delivery because of their structure, multiple biological functions, and unique biogenesis. The lipid bilayers of the exosome membrane are considered to protect loaded drugs from degradation. Due to their nanostructures, exosomes could traverse all kinds of biological barriers to transport cargo [73], such as the blood–ocular barrier and blood–brain barrier. Exosomes have extensive sources and are secreted by a wide spectrum of cell types. Moreover, exosomes could maintain their original morphology and other characteristics after repeated freezing and thawing. As cell-free therapies, exosomes have a low risk of immunogenicity and tumorigenicity. Approaches of engineering exosomes for cargo delivery include passive loading and active loading [95]. Passive loading mainly refers to the incubation of exosomes or donor cells with free drugs. Active loading includes sonication, extrusion, freeze/thaw, electroporation, incubation with saponin, click chemistry, and antibody binding.

Traditionally, eye drops have been the primary treatment option for many ocular disorders, especially for anterior ocular segments. However, there are some limitations, such as frequent dosing and low bioavailability. Exosomes have been demonstrated to serve as potential nanocarriers to carry functional RNA, DNA, peptides, and drugs with poor bioavailability [96]. Dry eye is associated with inflammation, corneal injuries, and oxidative stress. Gold nanoparticles (AuNPs) reduced by ascorbic acid (AA) have been deposited onto the phospholipid membrane of MSC-derived exosomes (mExo@AA) to create a multi-functional eye drop [27]. The research team also created another nanoparticle by growing cerium oxide (Ce) nanocrystals in situ on MSC-exos, called MSCExo-Ce [28]. The two nanoparticles were all used for the treatment of dry eye and effectively inhibit inflammation. They might offer new therapeutic options in the future. In addition, exosomes derived from 293 T cells that could generate adeno-associated virus type 2 (AAV-2) showed higher transduction efficiency than conventional AAV-2 when administered via intravitreal injection [97]. Exosome-associated AAV2 was considered a robust gene delivery vehicle into the retina upon intravitreal injection [97]. MSC-exo overexpressed miR-126 could effectively inhibit the signaling pathway associated with inflammation response of DR [98]. Furthermore, we loaded MSC-exo with rapamycin which has poor bioavailability by subconjunctival injection. Our study found that MSC-exo loaded with rapamycin could be detected in the retina and showed improved therapeutic efficacy [14]. Exosomes as drug delivery systems provide new ideas for clinical treatment.

6 Conclusions and Prospects

The existing body of evidence indicates that exosomes possess a diverse array of proteins and genetic material, which significantly contribute to their extensive range of biological functions encompassing signal transduction, immune regulation, and tissue repair. On the one hand, exosomes derived from disease cells mediate the incidence and development of autoimmune-mediated eye diseases including inflammation, neuronal degeneration, oxidative stress, and neovascularization, and, on the other hand, they have an excellent application prospect in the treatment of ocular disorders. Exosomes are considered to be potential biomarkers and natural drug delivery systems due to their unique structure.

However, based on the complexity and diversity of exosomes, the applications mentioned above require further investigation. At present, there are still many dilemmas that need to be urgently addressed, such as the most optimal cell source of exosomes, isolation method, quality control, and ethical issues. In addition, the toxicity of exosomes to the eye is a highly concerning issue that needs to be addressed. To date, relatively few such studies have been conducted. It has been demonstrated that IgG-loaded exosomes could activate the complement pathway via the classic pathway to aggravate DR progression [99]. Furthermore, exosomes released by RPE cells in the state of oxidative stress could induce inflammation and apoptosis of normal RPE cells through the Apaf1/caspase-9 axis [100]. Thus, more related studies about the toxicity of exosomes are needed to be initiated. The research on exosomes for ocular diseases is expected to result in a breakthrough with unceasing efforts. Thus, they will provide new insight into how the disorders develop and offer new therapeutic options.Conflict of Interest StatementThe authors declare that they have no competing interests.

Author Contributions Xiaomin Zhang conceived the article, reviewed, and finalized the manuscript. Huan Li wrote the chapter. All authors reviewed and approved the final manuscript.

Funding This work was supported by grants from the National Natural Science Foundation of China (82171042) and Tianjin Key Medical Discipline (Specialty) Construction Project (TJYXZDXK-037A).

References

1. Pan BT, Johnstone RM. Fate of the transferrin receptor during maturation of sheep reticulocytes in vitro: selective externalization of the receptor. Cell. 1983;33(3):967–78. https://doi.org/10.1016/0092-8674(83)90040-5.
2. Pan BT, Teng K, Wu C, Adam M, Johnstone RM. Electron microscopic evidence for externalization of the transferrin receptor in vesicular form in sheep reticulocytes. J Cell Biol. 1985;101(3):942–8. https://doi.org/10.1083/jcb.101.3.942.

3. Johnstone RM, Adam M, Hammond JR, Orr L, Turbide C. Vesicle formation during reticulocyte maturation. Association of plasma membrane activities with released vesicles (exosomes). J Biol Chem. 1987;262(19):9412–20.
4. Kalluri R. The biology and function of exosomes in cancer. J Clin Invest. 2016;126(4):1208–15. https://doi.org/10.1172/jci81135.
5. Tkach M, Théry C. Communication by extracellular vesicles: Where we are and where we need to go. Cell. 2016;164(6):1226–32. https://doi.org/10.1016/j.cell.2016.01.043.
6. Boriachek K, Islam MN, Möller A, Salomon C, Nguyen NT, Hossain MSA, et al. Biological functions and current advances in isolation and detection strategies for exosome nanovesicles. Small. 2018;14(6):201702153. https://doi.org/10.1002/smll.201702153.
7. Mathieu M, Martin-Jaular L, Lavieu G, Théry C. Specificities of secretion and uptake of exosomes and other extracellular vesicles for cell-to-cell communication. Nat Cell Biol. 2019;21(1):9–17. https://doi.org/10.1038/s41556-018-0250-9.
8. van Niel G, D'Angelo G, Raposo G. Shedding light on the cell biology of extracellular vesicles. Nat Rev Mol Cell Biol. 2018;19(4):213–28. https://doi.org/10.1038/nrm.2017.125.
9. Raposo G, Nijman HW, Stoorvogel W, Liejendekker R, Harding CV, Melief CJ, et al. B lymphocytes secrete antigen-presenting vesicles. J Exp Med. 1996;183(3):1161–72. https://doi.org/10.1084/jem.183.3.1161.
10. Li N, Zhao L, Wei Y, Ea VL, Nian H, Wei R. Recent advances of exosomes in immune-mediated eye diseases. Stem Cell Res Ther. 2019;10(1):278. https://doi.org/10.1186/s13287-019-1372-0.
11. Shen Z, Huang W, Liu J, Tian J, Wang S, Rui K. Effects of mesenchymal stem cell-derived Exosomes on autoimmune diseases. Front Immunol. 2021;12:749192. https://doi.org/10.3389/fimmu.2021.749192.
12. Chen B, Li Q, Zhao B, Wang Y. Stem cell-derived extracellular vesicles as a novel potential therapeutic tool for tissue repair. Stem Cells Transl Med. 2017;6(9):1753–8. https://doi.org/10.1002/sctm.16-0477.
13. Atienzar-Aroca S, Flores-Bellver M, Serrano-Heras G, Martinez-Gil N, Barcia JM, Aparicio S, et al. Oxidative stress in retinal pigment epithelium cells increases exosome secretion and promotes angiogenesis in endothelial cells. J Cell Mol Med. 2016;20(8):1457–66. https://doi.org/10.1111/jcmm.12834.
14. Li H, Zhang Z, Li Y, Su L, Duan Y, Zhang H, et al. Therapeutic effect of rapamycin-loaded small extracellular vesicles derived from mesenchymal stem cells on experimental autoimmune uveitis. Front Immunol. 2022;13:864956. https://doi.org/10.3389/fimmu.2022.864956.
15. Yu B, Xiao M, Yang F, Xiao J, Zhang H, Su L, et al. MicroRNA-431-5p encapsulated in serum extracellular vesicles as a biomarker for proliferative diabetic retinopathy. Int J Biochem Cell Biol. 2021;135:105975. https://doi.org/10.1016/j.biocel.2021.105975.
16. Huang Y, Li R, Ye S, Lin S, Yin G, Xie Q. Recent advances in the use of exosomes in Sjögren's syndrome. Front Immunol. 2020;11:1509. https://doi.org/10.3389/fimmu.2020.01509.
17. Christodoulou MI, Kapsogeorgou EK, Moutsopoulos HM. Characteristics of the minor salivary gland infiltrates in Sjögren's syndrome. J Autoimmun. 2010;34(4):400–7. https://doi.org/10.1016/j.jaut.2009.10.004.
18. An Q, Zhao J, Zhu X, Yang B, Wu Z, Su Y, et al. Exploiting the role of T cells in the pathogenesis of Sjögren's syndrome for therapeutic treatment. Front Immunol. 2022;13:995895. https://doi.org/10.3389/fimmu.2022.995895.
19. Ramos-Casals M, Brito-Zerón P, Bombardieri S, Bootsma H, De Vita S, Dörner T, et al. EULAR recommendations for the management of Sjögren's syndrome with topical and systemic therapies. Ann Rheum Dis. 2020;79(1):3–18. https://doi.org/10.1136/annrheumdis-2019-216114.
20. Gallo A, Jang SI, Ong HL, Perez P, Tandon M, Ambudkar I, et al. Targeting the Ca(2+) sensor STIM1 by exosomal transfer of Ebv-miR-BART13-3p is associated with Sjögren's syndrome. EBioMedicine. 2016;10:216–26. https://doi.org/10.1016/j.ebiom.2016.06.041.

21. Cortes-Troncoso J, Jang SI, Perez P, Hidalgo J, Ikeuchi T, Greenwell-Wild T, et al. T cell exosome-derived miR-142-3p impairs glandular cell function in Sjögren's syndrome. JCI Insight. 2020;5(9):e133497. https://doi.org/10.1172/jci.insight.133497.
22. Tan EM, Feltkamp TE, Smolen JS, Butcher B, Dawkins R, Fritzler MJ, et al. Range of antinuclear antibodies in "healthy" individuals. Arthritis Rheum. 1997;40(9):1601–11. https://doi.org/10.1002/art.1780400909.
23. Michael A, Bajracharya SD, Yuen PS, Zhou H, Star RA, Illei GG, et al. Exosomes from human saliva as a source of microRNA biomarkers. Oral Dis. 2010;16(1):34–8. https://doi.org/10.1111/j.1601-0825.2009.01604.x.
24. Aqrawi LA, Galtung HK, Vestad B, Øvstebø R, Thiede B, Rusthen S, et al. Identification of potential saliva and tear biomarkers in primary Sjögren's syndrome, utilising the extraction of extracellular vesicles and proteomics analysis. Arthritis Res Ther. 2017;19(1):14. https://doi.org/10.1186/s13075-017-1228-x.
25. Ma D, Wu Z, Zhao X, Zhu X, An Q, Wang Y, et al. Immunomodulatory effects of umbilical mesenchymal stem cell-derived exosomes on CD4(+) T cells in patients with primary Sjögren's syndrome. Inflammopharmacology. 2023;31(4):1823–38. https://doi.org/10.1007/s10787-023-01189-x.
26. Li B, Xing Y, Gan Y, He J, Hua H. Labial gland-derived mesenchymal stem cells and their exosomes ameliorate murine Sjögren's syndrome by modulating the balance of Treg and Th17 cells. Stem Cell Res Ther. 2021;12(1):478. https://doi.org/10.1186/s13287-021-02541-0.
27. Ma F, Feng J, Liu X, Tian Y, Wang WJ, Luan FX, et al. A synergistic therapeutic nano-eyedrop for dry eye disease based on ascorbic acid-coupled exosomes. Nanoscale. 2023;15(4):1890–9. https://doi.org/10.1039/d2nr05178h.
28. Tian Y, Zhang Y, Zhao J, Luan F, Wang Y, Lai F, et al. Combining MSC exosomes and cerium oxide nanocrystals for enhanced dry eye syndrome therapy. Pharmaceutics. 2023;15(9):2301. https://doi.org/10.3390/pharmaceutics15092301.
29. Wolff D, Radojcic V, Lafyatis R, Cinar R, Rosenstein RK, Cowen EW, et al. National Institutes of Health consensus development project on criteria for clinical trials in chronic graft-versus-host disease: IV. The 2020 highly morbid forms report. Transplant Cell Ther. 2021;27(10):817–35. https://doi.org/10.1016/j.jtct.2021.06.001.
30. Zeiser R, Blazar BR. Pathophysiology of chronic graft-versus-host disease and therapeutic targets. N Engl J Med. 2017;377(26):2565–79. https://doi.org/10.1056/NEJMra1703472.
31. Nair S, Vanathi M, Mukhija R, Tandon R, Jain S, Ogawa Y. Update on ocular graft-versus-host disease. Indian J Ophthalmol. 2021;69(5):1038–50. https://doi.org/10.4103/ijo.IJO_2016_20.
32. Herretes S, Ross DB, Duffort S, Barreras H, Yaohong T, Saeed AM, et al. Recruitment of donor T cells to the eyes during ocular GVHD in recipients of MHC-matched allogeneic hematopoietic stem cell transplants. Invest Ophthalmol Vis Sci. 2015;56(4):2348–57. https://doi.org/10.1167/iovs.14-15630.
33. Royer DJ, Echegaray-Mendez J, Lin L, Gmyrek GB, Mathew R, Saban DR, et al. Complement and CD4(+) T cells drive context-specific corneal sensory neuropathy. eLife. 2019;8:e48378. https://doi.org/10.7554/eLife.48378.
34. Balaram M, Rashid S, Dana R. Chronic ocular surface disease after allogeneic bone marrow transplantation. Ocul Surf. 2005;3(4):203–11. https://doi.org/10.1016/s1542-0124(12)70207-0.
35. Yin K, Wang S, Zhao RC. Exosomes from mesenchymal stem/stromal cells: a new therapeutic paradigm. Biomark Res. 2019;7:8. https://doi.org/10.1186/s40364-019-0159-x.
36. Kalluri R, LeBleu VS. The biology, function, and biomedical applications of exosomes. Science. 2020;367(6478):eaau6977. https://doi.org/10.1126/science.aau6977.
37. Zhou T, He C, Lai P, Yang Z, Liu Y, Xu H, et al. miR-204-containing exosomes ameliorate GVHD-associated dry eye disease. Sci Adv. 2022;8(2):eabj9617. https://doi.org/10.1126/sciadv.abj9617.

38. Niederkorn JY. High-risk corneal allografts and why they lose their immune privilege. Curr Opin Allergy Clin Immunol. 2010;10(5):493–7. https://doi.org/10.1097/ACI.0b013e32833dfa11.
39. Marino J, Paster J, Benichou G. Allorecognition by T lymphocytes and allograft rejection. Front Immunol. 2016;7:582. https://doi.org/10.3389/fimmu.2016.00582.
40. Abud TB, Di Zazzo A, Kheirkhah A, Dana R. Systemic immunomodulatory strategies in high-risk corneal transplantation. J Ophthalmic Vis Res. 2017;12(1):81–92. https://doi.org/10.4103/2008-322x.200156.
41. Wen D, Peng Y, Liu D, Weizmann Y, Mahato RI. Mesenchymal stem cell and derived exosome as small RNA carrier and Immunomodulator to improve islet transplantation. J Control Release. 2016;238:166–75. https://doi.org/10.1016/j.jconrel.2016.07.044.
42. Aiello S, Rocchetta F, Longaretti L, Faravelli S, Todeschini M, Cassis L, et al. Extracellular vesicles derived from T regulatory cells suppress T cell proliferation and prolong allograft survival. Sci Rep. 2017;7(1):11518. https://doi.org/10.1038/s41598-017-08617-3.
43. Han KY, Tran JA, Chang JH, Azar DT, Zieske JD. Potential role of corneal epithelial cell-derived exosomes in corneal wound healing and neovascularization. Sci Rep. 2017;7:40548. https://doi.org/10.1038/srep40548.
44. Ghafouri-Fard S, Glassy MC, Abak A, Hussen BM, Niazi V, Taheri M. The interaction between miRNAs/lncRNAs and Notch pathway in human disorders. Biomed Pharmacother. 2021;138:111496. https://doi.org/10.1016/j.biopha.2021.111496.
45. Samaeekia R, Rabiee B, Putra I, Shen X, Park YJ, Hematti P, et al. Effect of human corneal mesenchymal stromal cell-derived exosomes on corneal epithelial wound healing. Invest Ophthalmol Vis Sci. 2018;59(12):5194–200. https://doi.org/10.1167/iovs.18-24803.
46. Jia Z, Lv Y, Zhang W, Zhang X, Li F, Lu X, et al. Mesenchymal stem cell derived exosomes-based immunological signature in a rat model of corneal allograft rejection therapy. Front Biosci. 2022;27(3):86. https://doi.org/10.31083/j.fbl2703086.
47. Gain P, Jullienne R, He Z, Aldossary M, Acquart S, Cognasse F, et al. Global survey of corneal transplantation and eye banking. JAMA Ophthalmol. 2016;134(2):167–73. https://doi.org/10.1001/jamaophthalmol.2015.4776.
48. Williams R, Lace R, Kennedy S, Doherty K, Levis H. Biomaterials for regenerative medicine approaches for the anterior segment of the eye. Adv Healthc Mater. 2018;7(10):e1701328. https://doi.org/10.1002/adhm.201701328.
49. Jangamreddy JR, Haagdorens MKC, Mirazul Islam M, Lewis P, Samanta A, Fagerholm P, et al. Short peptide analogs as alternatives to collagen in pro-regenerative corneal implants. Acta Biomater. 2018;69:120–30. https://doi.org/10.1016/j.actbio.2018.01.011.
50. Jabs DA, Nussenblatt RB, Rosenbaum JT, Standardization of Uveitis Nomenclature Working G. Standardization of uveitis nomenclature for reporting clinical data. Results of the First International Workshop. Am J Ophthalmol. 2005;140(3):509–16. https://doi.org/10.1016/j.ajo.2005.03.057.
51. Barry RJ, Nguyen QD, Lee RW, Murray PI, Denniston AK. Pharmacotherapy for uveitis: current management and emerging therapy. Clin Ophthalmol. 2014;8:1891–911. https://doi.org/10.2147/OPTH.S47778.
52. Horai R, Caspi RR. Microbiome autoimmune uveitis. Front Immunol. 2019;10:232. https://doi.org/10.3389/fimmu.2019.00232.
53. Jabs DA, Rosenbaum JT, Foster CS, Holland GN, Jaffe GJ, Louie JS, et al. Guidelines for the use of immunosuppressive drugs in patients with ocular inflammatory disorders: recommendations of an expert panel. Am J Ophthalmol. 2000;130(4):492–513. https://doi.org/10.1016/s0002-9394(00)00659-0.
54. Heo J, Sepah Y, Yohannan J, Renner M, Akhtar A, Gregory A, et al. The role of biologic agents in the management of non-infectious uveitis. Expert Opin Biol Ther. 2012;12(8):995–1008. https://doi.org/10.1517/14712598.2012.688021.

55. Sen H, Vitale S, Gangaputra S, Nussenblatt R, Liesegang T, Levy-Clarke G, et al. Periocular corticosteroid injections in uveitis: effects and complications. Ophthalmology. 2014;121(11):2275–86. https://doi.org/10.1016/j.ophtha.2014.05.021.
56. Oh JY, Kim TW, Jeong HJ, Lee HJ, Ryu JS, Wee WR, et al. Intraperitoneal infusion of mesenchymal stem/stromal cells prevents experimental autoimmune uveitis in mice. Mediators Inflamm. 2014;2014:624640. https://doi.org/10.1155/2014/624640.
57. Zhang L, Zheng H, Shao H, Nian H, Zhang Y, Bai L, et al. Long-term therapeutic effects of mesenchymal stem cells compared to dexamethasone on recurrent experimental autoimmune uveitis of rats. Invest Ophthalmol Vis Sci. 2014;55(9):5561–71. https://doi.org/10.1167/iovs.14-14788.
58. Zhang X, Ren X, Li G, Jiao C, Zhang L, Zhao S, et al. Mesenchymal stem cells ameliorate experimental autoimmune uveoretinitis by comprehensive modulation of systemic autoimmunity. Invest Ophthalmol Vis Sci. 2011;52(6):3143–52. https://doi.org/10.1167/iovs.10-6334.
59. Ren K. Exosomes in perspective: a potential surrogate for stem cell therapy. Odontology. 2019;107(3):271–84. https://doi.org/10.1007/s10266-018-0395-9.
60. Bai L, Shao H, Wang H, Zhang Z, Su C, Dong L, et al. Effects of mesenchymal stem cell-derived exosomes on experimental autoimmune uveitis. Sci Rep. 2017;7(1):4323. https://doi.org/10.1038/s41598-017-04559-y.
61. Jiang G, Yun J, Kaplan HJ, Zhao Y, Sun D, Shao H. Vaccination with circulating exosomes in autoimmune uveitis prevents recurrent intraocular inflammation. Clin Exp Ophthalmol. 2021;49(9):1069–77. https://doi.org/10.1111/ceo.13990.
62. Li Y, Ren X, Zhang Z, Duan Y, Li H, Chen S, et al. Effect of small extracellular vesicles derived from IL-10-overexpressing mesenchymal stem cells on experimental autoimmune uveitis. Stem Cell Res Ther. 2022;13(1):100. https://doi.org/10.1186/s13287-022-02780-9.
63. Kang M, Yadav MK, Mbanefo EC, Yu CR, Egwuagu CE. IL-27-containing exosomes secreted by innate B-1a cells suppress and ameliorate uveitis. Front Immunol. 2023;14:1071162. https://doi.org/10.3389/fimmu.2023.1071162.
64. Kang M, Choi JK, Jittayasothorn Y, Egwuagu CE. Interleukin 35-producing exosomes suppress neuroinflammation and autoimmune uveitis. Front Immunol. 2020;11:1051. https://doi.org/10.3389/fimmu.2020.01051.
65. Zheng H, Yang F, Ea V, Zhou L, Wu L, Zhao G, et al. Proteomics profiling of plasma exosomes in VKH patients. Curr Mol Med. 2021;21(8):675–89. https://doi.org/10.2174/1566524020666200719021653.
66. Wang Y, Zhong Y, Zhang L, Wu Q, Tham Y, Rim TH, et al. Global incidence, progression, and risk factors of age-related macular degeneration and projection of disease statistics in 30 years: A modeling study. Gerontology. 2022;68(7):721–35. https://doi.org/10.1159/000518822.
67. Deng Y, Qiao L, Du M, Qu C, Wan L, Li J, et al. Age-related macular degeneration: epidemiology, genetics, pathophysiology, diagnosis, and targeted therapy. Genes Dis. 2022;9(1):62–79. https://doi.org/10.1016/j.gendis.2021.02.009.
68. Perez VL, Caspi RR. Immune mechanisms in inflammatory and degenerative eye disease. Trends Immunol. 2015;36(6):354–63. https://doi.org/10.1016/j.it.2015.04.003.
69. Bora NS, Matta B, Lyzogubov VV, Bora PS. Relationship between the complement system, risk factors and prediction models in age-related macular degeneration. Mol Immunol. 2015;63(2):176–83. https://doi.org/10.1016/j.molimm.2014.07.012.
70. Wang AL, Lukas TJ, Yuan M, Du N, Tso MO, Neufeld AH. Autophagy and exosomes in the aged retinal pigment epithelium: possible relevance to drusen formation and age-related macular degeneration. PLoS One. 2009;4(1):e4160. https://doi.org/10.1371/journal.pone.0004160.
71. Liszewski MK, Java A, Schramm EC, Atkinson JP. Complement dysregulation and disease: insights from contemporary genetics. Annu Rev Pathol. 2017;12:25–52. https://doi.org/10.1146/annurev-pathol-012615-044145.

72. Toomey CB, Johnson LV, Bowes RC. Complement factor H in AMD: bridging genetic associations and pathobiology. Prog Retin Eye Res. 2018;62:38–57. https://doi.org/10.1016/j.preteyeres.2017.09.001.
73. Hajrasouliha AR, Jiang G, Lu Q, Lu H, Kaplan HJ, Zhang HG, et al. Exosomes from retinal astrocytes contain antiangiogenic components that inhibit laser-induced choroidal neovascularization. J Biol Chem. 2013;288(39):28058–67. https://doi.org/10.1074/jbc.M113.470765.
74. Li J, Xue H, Li T, Chu X, Xin D, Xiong Y, et al. Exosomes derived from mesenchymal stem cells attenuate the progression of atherosclerosis in ApoE(−/−) mice via miR-let7 mediated infiltration and polarization of M2 macrophage. Biochem Biophys Res Commun. 2019;510(4):565–72. https://doi.org/10.1016/j.bbrc.2019.02.005.
75. Pakravan K, Babashah S, Sadeghizadeh M, Mowla SJ, Mossahebi-Mohammadi M, Ataei F, et al. MicroRNA-100 shuttled by mesenchymal stem cell-derived exosomes suppresses in vitro angiogenesis through modulating the mTOR/HIF-1α/VEGF signaling axis in breast cancer cells. Cell Oncol. 2017;40(5):457–70. https://doi.org/10.1007/s13402-017-0335-7.
76. Kang GY, Bang JY, Choi AJ, Yoon J, Lee WC, Choi S, et al. Exosomal proteins in the aqueous humor as novel biomarkers in patients with neovascular age-related macular degeneration. J Proteome Res. 2014;13(2):581–95. https://doi.org/10.1021/pr400751k.
77. Elbay A, Ercan Ç, Akbaş F, Bulut H, Ozdemir H. Three new circulating microRNAs may be associated with wet age-related macular degeneration. Scand J Clin Lab Invest. 2019;79(6):388–94. https://doi.org/10.1080/00365513.2019.1637931.
78. Prabhakar BS, Bahn RS, Smith TJ. Current perspective on the pathogenesis of Graves' disease and ophthalmopathy. Endocr Rev. 2003;24(6):802–35. https://doi.org/10.1210/er.2002-0020.
79. Elia G, Fallahi P, Ragusa F, Paparo SR, Mazzi V, Benvenga S, et al. Precision medicine in Graves' disease and ophthalmopathy. Front Pharmacol. 2021;12:754386. https://doi.org/10.3389/fphar.2021.754386.
80. Bahn RS. Graves' ophthalmopathy. N Engl J Med. 2010;362(8):726–38. https://doi.org/10.1056/NEJMra0905750.
81. Han JS, Kim SE, Jin JQ, Park NR, Lee JY, Kim HL, et al. Tear-derived exosome proteins are increased in patients with thyroid eye disease. Int J Mol Sci. 2021;22(3):1115. https://doi.org/10.3390/ijms22031115.
82. Shi TT, Zhao RX, Xin Z, Hou ZJ, Wang H, Xie RR, et al. Tear-derived exosomal biomarkers of Graves' ophthalmopathy. Front Immunol. 2022;13:1088606. https://doi.org/10.3389/fimmu.2022.1088606.
83. Sun J, Wei J, Zhang Y, Li J, Li J, Yan J, et al. Plasma exosomes transfer miR-885-3p targeting the AKT/NFκB signaling pathway to improve the sensitivity of intravenous glucocorticoid therapy against graves ophthalmopathy. Front Immunol. 2022;13:819680. https://doi.org/10.3389/fimmu.2022.819680.
84. Zhang L, Yu D. Exosomes in cancer development, metastasis, and immunity. Biochim Biophys Acta Rev Cancer. 2019;1871(2):455–68. https://doi.org/10.1016/j.bbcan.2019.04.004.
85. González E, Falcón-Pérez JM. Cell-derived extracellular vesicles as a platform to identify low-invasive disease biomarkers. Expert Rev Mol Diagn. 2015;15(7):907–23. https://doi.org/10.1586/14737159.2015.1043272.
86. Neves KB, Rios FJ, Sevilla-Montero J, Montezano AC, Touyz RM. Exosomes and the cardiovascular system: role in cardiovascular health and disease. J Physiol. 2022:4923–36. https://doi.org/10.1113/jp282054.
87. Choi JY, Kim S, Kwak HB, Park DH, Park JH, Ryu JS, et al. Extracellular vesicles as a source of urological biomarkers: lessons learned from advances and challenges in clinical applications to major diseases. Int Neurourol J. 2017;21(2):83–96. https://doi.org/10.5213/inj.1734961.458.
88. Xu H, Jia S, Xu H. Potential therapeutic applications of exosomes in different autoimmune diseases. Clin Immunol. 2019;205:116–24. https://doi.org/10.1016/j.clim.2019.06.006.
89. Logozzi M, Mizzoni D, Di Raimo R, Fais S. Exosomes: a source for new and old biomarkers in cancer. Cancers. 2020;12(9):2566. https://doi.org/10.3390/cancers12092566.

90. Rahbarghazi R, Jabbari N, Sani NA, Asghari R, Salimi L, Kalashani SA, et al. Tumor-derived extracellular vesicles: reliable tools for cancer diagnosis and clinical applications. Cell Commun Signal. 2019;17(1):73. https://doi.org/10.1186/s12964-019-0390-y.
91. Xiao J, Zhang H, Yang F, Xiao M, Zhou L, Yu R, et al. Proteomic analysis of plasma sEVs reveals that TNFAIP8 is a new biomarker of cell proliferation in diabetic retinopathy. J Proteome Res. 2021;20(3):1770–82. https://doi.org/10.1021/acs.jproteome.0c01048.
92. Knickelbein JE, Liu B, Arakelyan A, Zicari S, Hannes S, Chen P, et al. Modulation of immune responses by extracellular vesicles from retinal pigment epithelium. Invest Ophthalmol Vis Sci. 2016;57(10):4101–7. https://doi.org/10.1167/iovs.15-18353.
93. Klingeborn M, Dismuke WM, Skiba NP, Kelly U, Stamer WD, Bowes RC. Directional exosome proteomes reflect polarity-specific functions in retinal pigmented epithelium monolayers. Sci Rep. 2017;7(1):4901. https://doi.org/10.1038/s41598-017-05102-9.
94. Wu L, Zhou L, An J, Shao X, Zhang H, Wang C, et al. Comprehensive profiling of extracellular vesicles in uveitis and scleritis enables biomarker discovery and mechanism exploration. J Transl Med. 2023;21(1):388. https://doi.org/10.1186/s12967-023-04228-x.
95. Luan X, Sansanaphongpricha K, Myers I, Chen H, Yuan H, Sun D. Engineering exosomes as refined biological nanoplatforms for drug delivery. Acta Pharmacol Sin. 2017;38(6):754–63. https://doi.org/10.1038/aps.2017.12.
96. Barile L, Vassalli G. Exosomes: therapy delivery tools and biomarkers of diseases. Pharmacol Ther. 2017;174:63–78. https://doi.org/10.1016/j.pharmthera.2017.02.020.
97. Wassmer SJ, Carvalho LS, György B, Vandenberghe LH, Maguire CA. Exosome-associated AAV2 vector mediates robust gene delivery into the murine retina upon intravitreal injection. Sci Rep. 2017;7:45329. https://doi.org/10.1038/srep45329.
98. Zhang W, Wang Y, Kong Y. Exosomes derived from mesenchymal stem cells modulate miR-126 to ameliorate hyperglycemia-induced retinal inflammation via targeting HMGB1. Invest Ophthalmol Vis Sci. 2019;60(1):294–303. https://doi.org/10.1167/iovs.18-25617.
99. Huang C, Fisher KP, Hammer SS, Navitskaya S, Blanchard GJ, Busik JV. Plasma exosomes contribute to microvascular damage in diabetic retinopathy by activating the classical complement pathway. Diabetes. 2018;67(8):1639–49. https://doi.org/10.2337/db17-1587.
100. Ke Y, Fan X, Rui H, Xinjun R, Dejia W, Chuanzhen Z, et al. Exosomes derived from RPE cells under oxidative stress mediate inflammation and apoptosis of normal RPE cells through Apaf1/caspase-9 axis. J Cell Biochem. 2020;121(12):4849–61. https://doi.org/10.1002/jcb.29713.

Chapter 13
Extracellular Vesicles in Pregnancy: Functional Insights, Diagnostic Potential for Maternal–Fetal Disorders, and Therapeutic Implications

Yixin Wang, Ying Chang, and Zongjin Li

Abstract Extracellular vesicles (EVs) are receiving increasing amounts of attention during pregnancy. The unique properties of EVs, especially their potential for cargo protection and tissue-specific delivery, make them key players in regulating various physiological processes at the fetal–maternal interface. In particular, EVs have great potential as noninvasive biomarkers for monitoring maternal development and fetal health. The ability of these devices to encapsulate specific molecules associated with pathological conditions provides a new avenue for the early detection and accurate assessment of pregnancy-related complications. Additionally, new evidence suggests the promise of using EVs for therapeutic purposes to ameliorate pregnancy-associated disorders, demonstrating their potential to advance individualized medicine. Taken together, the results of this review lead to a better understanding of the multifaceted functions of EVs throughout pregnancy, explore their

Y. Wang
School of Medicine, Nankai University, Tianjin, China

Tianjin Key Laboratory of Human Development and Reproductive Regulation, Tianjin Central Hospital of Gynecology Obstetrics, Nankai University Affiliated Hospital of Obstetrics and Gynecology, Tianjin, China

Y. Chang
Tianjin Key Laboratory of Human Development and Reproductive Regulation, Tianjin Central Hospital of Gynecology Obstetrics, Nankai University Affiliated Hospital of Obstetrics and Gynecology, Tianjin, China

Z. Li (✉)
School of Medicine, Nankai University, Tianjin, China

Tianjin Key Laboratory of Human Development and Reproductive Regulation, Tianjin Central Hospital of Gynecology Obstetrics, Nankai University Affiliated Hospital of Obstetrics and Gynecology, Tianjin, China

National Key Laboratory of Kidney Diseases, Chinese PLA General Hospital, Beijing, China

Henan Key Laboratory of Cardiac Remodeling and Transplantation, Zhengzhou Seventh People's Hospital, Zhengzhou, China
e-mail: zongjinli@nankai.edu.cn

© The Author(s), under exclusive license to Springer Nature Singapore Pte Ltd. 2024
Z. Li (ed.), *Extracellular Vesicle: Biology and Translational Application*, https://doi.org/10.1007/978-981-97-5536-3_13

diagnostic potential for fetal–maternal disorders, and provide insights into their role as therapeutic agents in the treatment of pregnancy-related diseases. As the field continues to evolve, understanding the complex contributions of EVs to pregnancy-related processes and fully utilizing the diagnostic and therapeutic potential of EVs will hold great promise for improving maternal and fetal health outcomes.

Keywords Pregnancy · Extracellular vesicles · Exosomes · Pregnancy-related complications · Diagnostic biomarkers · Therapeutic implications

1 Introduction

Physiological adaptations necessary to uphold a salubrious gestation demand enduring interaction between the developing fetus and the maternal figure. Fetal–maternal communication also holds paramount importance in accurately timing the onset of parturition [1]. Multiple mediators serve this purpose, encompassing endocrine factors [2], immune cells [3], cytokines [4], and mechanical agents [5]. Recently, extracellular vesicles (EVs) have emerged as key players in facilitating the communicative functions of pregnancy [6]. The latest research has elucidated the crucial involvement of extracellular vesicles in various pivotal physiological events during pregnancy, including embryo implantation, the establishment of immune tolerance, placental formation, the maintenance of pregnancy, fetal development, the initiation of labor, and postpartum tissue and organ repair. These findings underscore the essential role of extracellular vesicles in modulating molecular and cellular interactions associated with pregnancy, providing compelling impetus for further in-depth investigations.

An advantage of EVs over non-vesicular transport methods lies in the protective nature of their phospholipid bilayer, allowing potential tissue-specific delivery of their cargo [7]. In addition to transporting mRNAs, EVs effectively transport various other cargoes, such as proteins, lipids, and non-coding RNAs, between cells located in disparate regions of the body. Consequently, they assume pivotal roles as regulators of cellular communication [8–10]. Recent studies have revealed that distinct subtypes of EVs can manifest discernible regulatory functions, which are mediated both by their surface receptors and their contents.

Apart from their fundamental roles in mediating cell-to-cell communication, the stability of EVs and their capacity to safeguard specific miRNAs from degradation [11] have aroused considerable research interest in exploring their potential as diagnostic biomarkers [12, 13], indicators of disease progression [14, 15], and even as targeted therapies [16]. Analyzing extracellular vesicles during pregnancy and characterizing their contents will facilitate the development of novel noninvasive diagnostic tools for identifying maternal or fetal abnormalities. Therefore, this review aims to shed light on the functions of pregnancy-associated extracellular vesicles, outline the diagnostic potential of EVs in pregnancy-related maternal–fetal diseases, and summarize the current research on the therapeutic effects of extracellular vesicles in pregnancy-related diseases.

2 Sources of EVs During Pregnancy

Extracellular vesicles (EVs), which are organelles enclosed within a membrane, are released by all kinds of cells and carry distinct cargos, such as lipids, proteins, RNAs, microRNAs, and DNA fragments. Additionally, larger EVs may encapsulate entire organelle space, such as mitochondria. Moreover, it has been confirmed that EVs possess an extensive array of enzymes and are capable of mediating a wide range of biological activities, even following their separation from their parent cells [17]. The comprehensive classification of EVs encompasses three primary subtypes. First, exosomes (50–150 nm) are derived from the endocytic pathway and are enriched in endosomal proteins. Second, microvesicles/microparticles/exosomes (ranging in size from a few micrometers to the scale of exosomes) can form through outward protrusion of the plasma membrane. Finally, apoptotic bodies (greater than 1 μm) are released from apoptotic or dying cells [18–22]. EVs derived from specific cell types or encountered in certain pathological conditions are designated as distinct appellations, resulting in multiple terms being used to describe the same EV subtype under varying circumstances (Fig. 13.1). Throughout this paper, we refer to EVs collectively unless researchers are able to confirm their size and cellular origin.

Many studies provide evidence showing the escalation of maternal blood EV concentrations during pregnancy. However, the exact source of these EVs has yet to be determined. The potential origins of EVs include maternal tissue, where EVs are secreted directly into the maternal bloodstream, or fetal tissue, wherein EVs are secreted into fetal blood and subsequently traverse the placenta, where they enter the maternal circulation. Additionally, it is plausible that the placenta itself produces these EVs. To discern between these possibilities, it becomes imperative to employ specific endogenous markers that are indicative of the placenta or other fetal tissues. These markers facilitate the identification of fetal-origin EVs within maternal blood (Fig. 13.2). Fortunately, fortuitous circumstances arise as several proteins and miRNAs exhibit placenta-specific expression patterns, thereby serving as valuable indicators of EVs released by the placenta. Many of these markers are exclusively found in humans and other primates, rendering them unsuitable for rodent studies. Notably, these markers include placental alkaline phosphatase (PLAP), syncytin, human leukocyte antigen-G (HLA-G), and others [23].

Although the multifaceted role of EVs as biomarkers and functional mediators of intercellular communication in various physiological and pathological contexts, including pregnancy, has only recently been recognized, the generation of EVs from maternal, fetal, and gestational structures has long been recognized. These extracellular vesicles have been discerned in a wide array of biological fluids, encompassing blood, urine, milk, tears, saliva, amniotic fluid, semen, and vaginal secretions. Notably, EVs originating from the placenta assume a pivotal function during

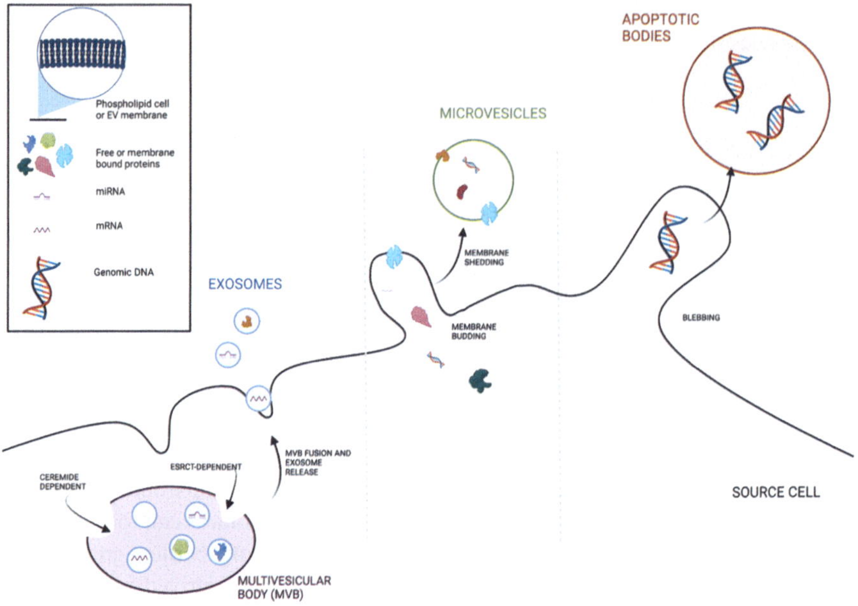

Fig. 13.1 Biogenesis of exosomes, microvesicles, and apoptotic bodies. The biogenesis of exosomes, microvesicles, and apoptotic bodies represents a multifaceted process. Exosomes, with diameters spanning from 50 to 150 nm, emerge via release from the multivesicular body through fusion with the cellular membrane. Remarkably, these cells exhibit identical membrane orientations to those of the originating cells. While a similar phenomenon holds for microvesicles, which range in size from a few micrometers to the scale of exosomes, their genesis involves a heterogeneous course. This encompasses the budding of a cellular membrane encompassing the intended contents, followed by the subsequent shedding of this membrane. In the realm of apoptotic bodies, their dimensions surpass 1 μm in diameter. Their formation is the outcome of a process whereby the plasma of the source cell becomes separated from the cytoskeleton. This reaction is triggered in response to an increase in hydrostatic pressure following cellular contraction. (Reprinted with permission from *J Physiol 2023;601:4973–4988.* https://doi.org/10.1113/JP282849)

pregnancy. Simultaneously, EVs derived from diverse sources, such as follicular fluid, the endometrium, the embryo, and trophoblasts, may also influence fertility, implantation, and the early stages of pregnancy [24].

3 EV Transport During Pregnancy

The placenta releases EVs into the maternal and fetal circulations, takes up maternal EVs, and allows the transit of maternal and exogenous EVs to the fetus [25]. The capacity of placental EVs (pEVs) to enter maternal cells has been demonstrated by the uptake of fluorescently labeled EVs by cultured placenta [26]. These EVs are localized to the maternal lungs, liver, and kidneys following venous administration in pregnant mice [27]. Approximately 12–25% of all maternal circulating EVs are

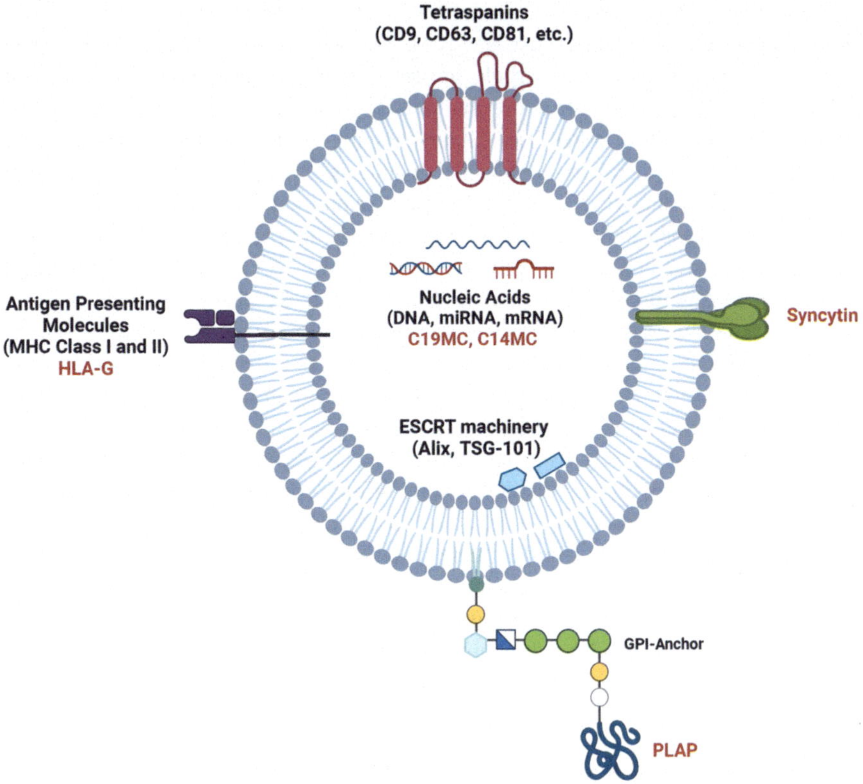

Fig. 13.2 Endogenous markers of placental EVs. EVs express various proteins, glycoproteins, and lipid-associated molecules at or near the membrane. These include tetraspanins (CD9, CD63, and CD81), antigen-presenting molecules (MHC class I and class II), and ESCRT complex proteins. EVs also carry cargo, including nucleic acids (DNA, miRNA, mRNA). Placental-specific EV markers are shown in red and include HLA-G, syncytin, the lipid-anchored phosphatase PLAP, and miRNAs expressed from the C19MC and C14MC clusters. (Reprinted with permission from *J Mol Med (Berl) 2023;101(1–2):65–81*. https://doi.org/10.1007/s00109-022-02278-0)

of placental origin [28]. Throughout the course of pregnancy, the concentration of placenta-derived EVs in maternal plasma progressively increases, as evidenced by the exosomal PLAP concentration. Bidirectional trafficking of placenta-specific miRNAs from chromosome 14 and chromosome 19 has been demonstrated both in vivo in mice and in placental biopsies from matched patients in maternal and fetal compartments [29, 30]. The role of EVs in promoting bidirectional communication between maternal tissues and the placenta was confirmed by visualizing the internalization of fluorescently labeled maternal macrophage-derived EVs by the placenta. Furthermore, they discovered that maternal macrophage-derived EVs, which modulate placental cytokine production, are actively transported into the placenta via clathrin-mediated endocytosis [31].

Fetal blood likely contains EVs derived from developing organs, which could yield insights into congenital diseases [25]. The mechanisms underlying EV passage across the placenta have largely not been explored. The syncytiotrophoblast (STB) is a multinucleated structure that acts as an impermeable barrier devoid of intercellular junctions or pores. Thus, it appears most plausible that EVs traverse the syncytium itself, being taken up on one side and subsequently released on the other. To the best of our knowledge, no studies have examined the process by which the placenta itself uptakes EVs, leaving the precise trafficking of fetal EVs across the placenta uncertain. It is conceivable that these fetal EVs can be used as placental biomarkers during this intricate process. The finding that placental-specific proteins are expressed on the surface of EVs has facilitated antibody-based or immune-based identification and isolation of placental-origin EVs as a distinct subset within the maternal circulation. As the placenta is a transient organ during pregnancy, these biomarkers are not expressed in adult tissues. However, the definitive identification of EVs originating directly from the fetus within maternal circulation poses challenges due to the lack of specific biomarkers. While certain organ-specific biomarkers can be detected, most of them are also expressed in adult tissues. One approach to circumvent this issue is to investigate discrepancies in pregnancies affected by congenital diseases. Such comparisons provide indirect evidence for the transfer of fetal EVs. Nevertheless, to draw accurate conclusions from these studies, one must assume that there are no alterations in the mother's physiology in such cases [23].

4 The Importance of EV-Mediated Communication in Normal Pregnancy

4.1 EVs in the Establishment of Maternal Immune Tolerance

The induction of immune tolerance during pregnancy is crucial for averting the risk of rejection or abortion while simultaneously safeguarding the body against external pathogens, and EVs play pivotal roles in modulating maternal immunity for the successful progression of pregnancy. A reciprocal interaction between the placenta and the maternal immune system is established through EVs. Placental EVs are assimilated by neighboring and distant maternal immune cells, while EVs generated by maternal immune cells modulate placental responses. This represents a multifaceted scenario of immunomodulation aimed at maintaining the efficacy of pathogen elimination without jeopardizing the well-being of the fetus [32, 33].

Throughout pregnancy, the human placenta releases escalating concentrations of EVs, which disseminate to various organs and cells. The primary producer of placental EVs is the STB, which encompasses the maternal villi and is in direct contact with the maternal bloodstream. The specific function of placental EVs has not been fully elucidated, but numerous studies have reported the involvement of placental EVs in immune responses associated with the induction of immune tolerance through diverse mechanisms. Many of the immunoregulatory surface molecules found on the STB have been detected on its EVs, where they are presumed to fulfill similar functions. The following examples have been investigated: trophoblast-derived EVs containing FasL may eliminate activated maternal T cells, thus conferring immunotolerance to the fetus [34, 35]. Human placenta explants release exosomes harboring ligands (MHC class I chain-related (MIC) proteins A and B, UL-16 binding proteins) for the NKG2D receptor on NK cells, cytotoxic T cells, and γδ T cells. The binding of these ligands to NKG2D leads to diminished in vitro cytotoxicity without affecting the perforin-mediated lytic pathway [36]. Furthermore, placental exosomes suppress the expression of the CD3-zeta chain and Janus kinase 3 (JAK3) while inducing a suppressor of cytokine signaling 2 (SOCS2) in T cells [37]. This may favor the expansion of lymphocytes with suppressive phenotypes, such as Treg cells. Another aspect of EVs in feto-maternal tolerance is their involvement in antigen presentation. This can occur directly through MHC–peptide complexes on the EV surface or indirectly via MHC cross-dressing of antigen-presenting cells (APCs). The latter refers to the acquisition of intact MHC molecules preloaded with antigen-derived peptides by leukocytes, particularly APCs [38]. These mechanisms are of great interest in the establishment and maintenance of maternal immune tolerance (Fig. 13.3).

Research has demonstrated that the transmission of certain viruses (such as CMV, HSV, and coxsackie B3) to endothelial and fibroblast compartments is facilitated through the action of microRNAs (miRNAs) from the chromosome 19 miRNA cluster (C19MC). These miRNAs induce autophagy in recipient cells, thereby limiting infection. C19MC is an imprinted cluster of primate-specific miRNAs exclusively expressed in the human placenta [39]. Notably, placenta-derived small EVs carry specific miRNAs, including miR-512-3p and miR-517-3p. In addition,

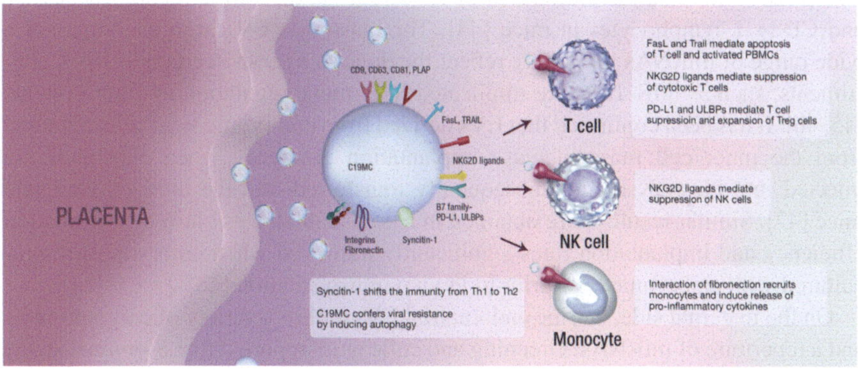

Fig. 13.3 Schematic representation of the immunomodulatory effects of placental exosomes and STB-MVs (STBMs) on the maternal immune system. Immune-suppressive ligands such as FasL, TRAIL, NKG2D, and PDL-1 are present on the exosomal membrane and regulate the apoptosis and cytotoxicity of maternal leucocytes. Exosomes mediate proinflammatory effects by recruiting and activating monocytes. (Reprinted with permission from *Semin Immunopathol 2018;40:425–437*. https://doi.org/10.1007/s00281-018-0680-2)

miR-141, which is transported by small EVs derived from the STB, suppresses T-cell proliferation and plays a role in regulating the activation of the nitric oxide/cGMP signaling pathway. Furthermore, activated endothelial nitric oxide synthase (eNOS), the primary isoform responsible for nitric oxide transport in STB-derived EVs, is expressed in all types of EVs derived from the STB. During normal pregnancy, there is a greater abundance of small EVs than medium/large EVs in terms of quantity and weight. Small EVs are generally considered to have anti-inflammatory properties, while medium/large EVs are more likely to exhibit proinflammatory characteristics [40].

4.2 EVs in Embryo Implantation

Embryo implantation, the initial event in the sequence of processes crucial for a successful pregnancy, represents a pivotal stage requiring a cascade of molecular operations that ultimately facilitates the adhesion of the trophectoderm to endometrial epithelial cells. While mediators such as adhesion molecules, growth factors, hormones, and cytokines are recognized as vital for endometrial receptivity, extracellular vesicles (EVs) have recently emerged as contributors to the implantation process [29, 41]. The communication mediated by exosomes, which support the establishment of pregnancy, initiates prior to implantation and involves the exchange of information between the maternal endometrium and the preimplantation embryo. Notably, the cargo of blastocyst-derived EVs has been shown to include HLA-G, which serves as an inhibitory molecule for NK cells and potentially macrophages, T cells, and other immune cells within the endometrium [42, 43]. Blastocyst-derived EVs that bind to annexin V also contain progesterone-induced blocking factor, which may stimulate the production of IL-10 and modulate the function of CD4+ and CD8+ T lymphocytes in mice [44]. The blastocyst EV cargo encompasses a wide range of miRNAs that likely reflect the donor's age, and certain miRNA constituents, such as miR-155i, are implicated in immune regulation in other systems [45, 46]. It has been confirmed that EVs derived from embryonic stem cells obtained from the inner cell mass increase implantation efficiency when exosomes are injected into blastocysts and subsequently transferred into the uteri of surrogate mice [47]. Similar results were obtained in a bovine model, as blastocyst formation efficiency and implantation rates significantly improved when embryos were precultured with embryonic EVs prior to in vivo transplantation [48].

On the maternal side, uterine endometrial EVs harbor a variety of cargo proteins and a repertoire of miRNAs. Greening and colleagues reported that hormonal status influences the packaging of cargo within exosomes released by endometrial epithelial cells. Specifically, exosomes released during the receptive phase, which is dependent on hormonally regulated molecular processes, exhibit enrichment of proteins related to adhesion, migration, invasion, and extracellular matrix remodeling. Importantly, these exosomes possess the ability to modulate gene expression through their miRNA content. In this context, Ng YH and colleagues analyzed a panel of 227 endometrial exosomal miRNAs and demonstrated that numerous target genes of these miRNAs play crucial roles in implantation. These target genes regulate essential pathways, such as the VEGF pathway, the Toll-like receptor pathway, and the Jak-STAT pathway, as well as processes related to extracellular matrix (ECM)-receptor interactions and adherens junctions [49]. Similarly, it has been shown that endometrium exosome-derived hsa-miR-30d, when taken up by trophoblasts, enhances the gene expression of proteins including Integrin Subunit Alpha 7 (Itg7), Integrin Subunit Beta-3 (Itgb3), and Cadherin 5 (Cdh5), all of which are required for blastocyst implantation [50]. Subsequent studies by Greening and colleagues demonstrated that when endometrial EVs are internalized by the trophectoderm, they enhance adhesiveness through activation of the focal adhesion kinase (FAK) signaling pathway [51].

4.3 EVs in Spiral Artery Remodeling and Angiogenesis

During pregnancy, EVs can either stimulate or hinder angiogenesis through their expression and expression, which is contingent upon the cell of origin and pathophysiological state. Extra villus trophoblast (EVT) cells play a crucial role in the restructuring of uterine spiral arteries and are essential for effective placentation [52]. Remodeling of the uterine spiral arteries (SpA) is necessary during placentation to facilitate adequate exchange of gases and nutrients from the maternal to the fetal side [53]. Abnormalities in vascular restructuring result in the formation of a narrow SpA, which impacts placental blood flow [54]. This, in turn, leads to tissue inflammation, ischemia, endothelial injury, thrombosis, and other complications [55].

During SpA remodeling, exosomes derived from EVT cells promote the migration of endothelial cells, leading to the formation of vascular tubes [56]. Within this stage, the migration of vascular smooth muscle cells (VSMCs) is assumed to be important. EVs derived from EVT cell lines (JEG-3 and HTR-8/sVneo cells) promote the migration of VSMCs from vessel walls, thereby facilitating SpA remodeling [57], and placental EVs of different sizes can interact with endothelial cells in various organs in vivo, potentially affecting endothelial function [58]. Placental EVs play pivotal roles in both angiogenesis and endothelial dysfunction. MiR-520c-3p in placental chorionic villous trophoblast (CVT)-derived EVs leads to increased invasion of EVTs and promotes angiogenesis by targeting CD44 (Fig. 13.4). Another significant regulator of endothelial function is nitric oxide, a potent vasodilator. Evidence suggests that STB-EVs harbor endothelial nitric oxide synthase (eNOS), which enables the production of nitric oxide [40, 59]. Furthermore, placental EVs have been found to contain miR-155, which is known to inhibit eNOS, consequently contributing to endothelial dysfunction [43], and the capacity of human umbilical vein-derived endothelial cells (HUVECs) to remodel SpA can be modulated by exosomes derived from both maternal and placental serum [60].

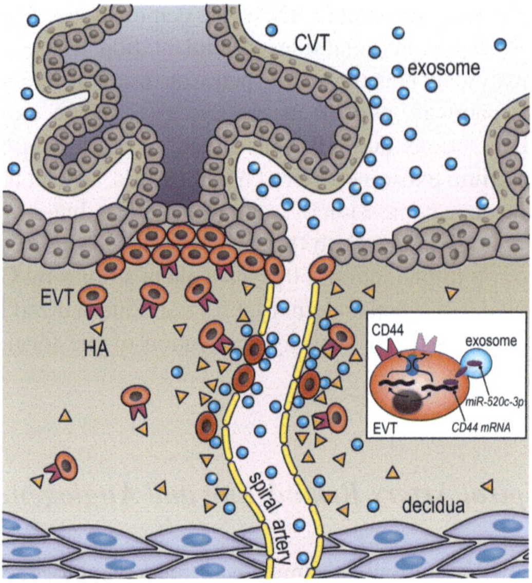

Fig. 13.4 Proposed model for chorionic villous trophoblast (CVT)-EVT communication via exosomes during early pregnancy. CVT highly expresses C19MC miR-520c-3p but not CD44. During the CVT-EVT, the expression of the CD44-targeting gene miR-520c-3p is downregulated, while that of the CD44 standard form (CD44s) is simultaneously upregulated. A remarkable gradient of hyaluronic acid (HA) is present toward the periarterial area in the decidua. Thus, CD44s-expressing EVTs are attracted to the periarterial area to trigger the remodeling of the uterine spiral artery. Exosomal CVT-derived miR-520c-3p is delivered to the periarterial area and transferred into miR-520c-3p-downregulated and CD44s-upregulated EVTs, resulting in the inhibition of cell invasion by upregulation of miR-520c-3p in the recipient EVTs. For spiral artery remodeling in the decidua, CD44-mediated EVT invasion should be increased toward the spiral artery by the interaction of CD44-HA, and CD44-mediated EVT invasion should also be positioned within or around the appropriate site by the modulation of CD44 expression by miR-520c-3p. (Reproduced with permission from *Placenta 2017;50:25–31*. https://doi.org/10.1016/j.placenta.2016.12.016)

EVs carry proangiogenic factors that support normal gestational development and maintain endothelial homeostasis [61]. The concentration of exosomes in maternal plasma increases more than twofold toward the end of pregnancy [27]. The ability of exosomes to modulate endothelial cell migration diminishes over time. Exosomes can originate from maternal circulation and cord blood and promote endothelial cell proliferation, migration, and tube formation [62]. The activity of exosomes derived from maternal serum is greater than that of exosomes from cord blood [63], likely due to differential expression of miRNAs involved in regulating cell migration, such as miRNA-550a-5p, miRNA-122-5p, miRNA-210-3p, miR376c-3p, miRNA-151a-5p, and miRNA-296-5p [60]. Another source of exosomes is human umbilical cord mesenchymal stem cells (UC-MSCs), which possess proangiogenic activity [64]. Various exosomal miRNAs, including miR486-1-5p and miR486-2-5p, participate in migration, placental development, and angiogenesis [65].

The success of pregnancy hinges upon SpA remodeling. Insufficient remodeling of the SpA leads to several pregnancy-related complications, such as preeclampsia and intrauterine growth restriction (IUGR), which will be discussed further in subsequent sections.

5 EVs as Effectors and Biomarkers in Maternal–Fetal Complications

In light of the mounting evidence concerning the significance of extracellular vesicles (EVs) in maintaining a healthy pregnancy, it is no surprise that there has been a corresponding surge of interest in exploring the roles of EVs in obstetric diseases. This exploration highlights the potential role of EVs as mediators of pregnancy complications and their promising utility as reservoirs of protein and RNA biomarkers for disease prediction and diagnosis. An expanding body of research indicates that both the concentration and content of EVs hold substantial diagnostic value for identifying women at risk of pregnancy-related diseases.

5.1 EVs as Potential Biomarkers in Obstetric Diseases

5.1.1 EVs in Gestational Hypertension and Preeclampsia (PE)

Preeclampsia, characterized by the onset of hypertension and proteinuria, is a highly severe complication that occurs in approximately 5% of pregnant women worldwide. However, delivery remains the most efficacious treatment for preeclampsia. For many years, its pathogenesis has been attributed predominantly to (1) compromised invasion of trophoblasts, (2) dysregulated immune tolerance in the mother, and (3) dysfunction of endothelial cells [66]. The etiology of preeclampsia is still under investigation, considering the prevailing notion that preeclampsia primarily stems from placental dysfunction, certain studies have focused on identifying discrepancies in placenta-derived EVs associated with preeclampsia. The concentration of placenta-derived EVs varies among normal pregnancies, early-stage preeclampsia, and late-stage preeclampsia. Placental alkaline phosphatase (PLAP), an integral membrane protein produced by the STB, has been utilized as a specific marker for the placenta [37]. During preeclampsia, the levels of proteins and miRNAs in placental EVs vary and can serve as diagnostic markers for preeclampsia.

Galectin 13 (LGALS13), also known as placental protein 13 (PP13), is typically expressed in the placenta. STBs generally synthesize this protein and release it into the circulation. Throughout gestation, the level of PP13 gradually increased. As preeclampsia progresses, the concentration of LGALS13 in placenta-derived EVs decreases [67]. Neprilysin (also called NEP or CD10), a widely expressed membrane-bound metalloprotease, binds to and cleaves various types of peptides, including vasodilators, natriuretics, and diuretics. NEP expression is typically upregulated in preeclampsia, leading to hypertension [68]. Studies have demonstrated that the upregulation of sFlt-1, sEng, and neprilysin in small EVs from preeclampsia patients elicits a proinflammatory response [69]. Furthermore, serum-derived EVs from preeclampsia patients exhibit reduced expression of syncytin-2, an indicator of placental pathology specific to humans that can contribute to maternal and perinatal morbidity [70]. Furthermore, studies have shown that STB-derived microvesicles stimulate peripheral blood mononuclear cells in in vitro culture conditions, resulting in the secretion of a proinflammatory response [71].

Promising biomarkers of preeclampsia are miRNAs encapsulated within EVs. The elevated expression of exosomal miRNAs in pregnant women with preeclampsia could serve as an indicator for early detection of this condition. For example, statistically significant differences in the expression of exosomal miRNAs, such as hsa-miR-423-5p, hsa-miR-451a, and hsa-miR-107 [72], have been found in pregnant women with preeclampsia compared to healthy individuals. The miRNAs miR-376c and miR-520 g, which appear to play a significant role in the migration and proliferation of trophoblast cells, have been identified as differentially expressed between pregnant women with preeclampsia and healthy women [73]. Li and colleagues reported a noteworthy twofold increase in the abundance of miRNA cargos, specifically miR-153-3p and miR-325-3p, within the context of preeclamptic pregnancies compared to their healthy counterparts (Fig. 13.5). In addition, miR-153, apart from its participation in this context, has also been demonstrated to bind to the 3′ untranslated region of hypoxia-inducible factor 1 (HIF1). This binding event induces a suppression of expression [74]. Consequently, such regulatory activity is linked to the attenuation of tube formation within the endothelial cells of the primary human umbilical vein. Moreover, it prompts a reduction in vascular endothelial growth factor A (VEGFA) expression, thereby impeding angiogenesis.

5.1.2 EVs in Gestational Diabetes Mellitus (GDM)

Gestational diabetes mellitus (GDM) is a prevalent metabolic disorder that can occur during pregnancy. Its occurrence affects approximately 13.2% of expectant mothers in developed nations. When left untreated, GDM may lead to adverse consequences, such as preterm birth, fetal mortality, and other pregnancy-related complications caused by impaired placentation induced by hyperglycemia [75]. Despite the potential for prevention and effective management, infants born to mothers with GDM face an elevated risk of developing heart disease, obesity, or type 2 diabetes [9].

Evidence suggests an increase in placental EVs within maternal plasma in patients with gestational diabetes [76]. These vesicles exhibit varied expression of proteins [77] and miRNAs [78] associated with alterations in insulin sensitivity. Notably, in GDM patients, EVs exhibit a distinctive cargo composition. Dipeptidyl peptidase IV (DPPIV) plays a significant role in type 2 diabetes by degrading glucagon-like peptide 1 (GLP-1), which in turn regulates glucose-dependent insulin

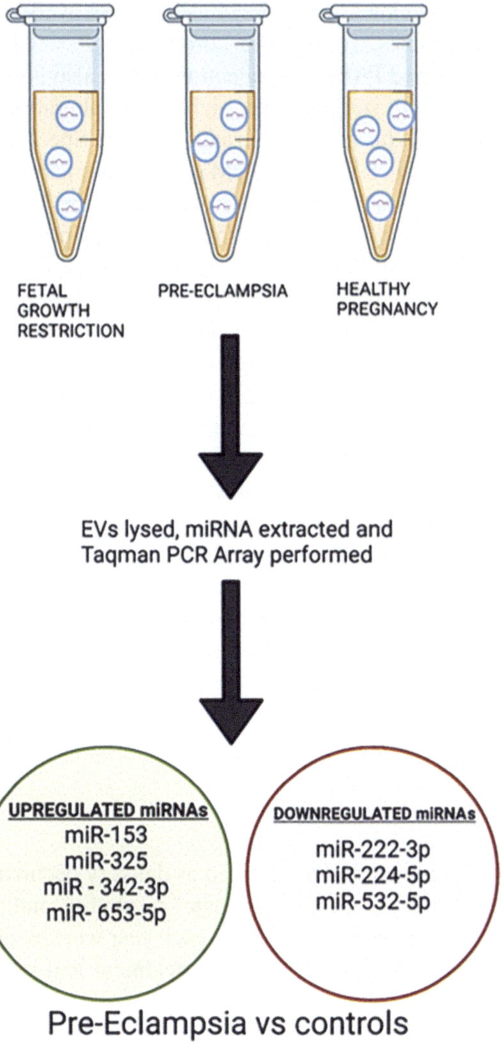

Fig. 13.5 Exosomes extracted from the plasma of women with preeclampsia exhibit a unique miRNA profile. EVs were extracted from the maternal plasma of 60 mothers who had either a healthy pregnancy or suffered from FGR and preeclampsia. Further experiments explored whether miRNA cargo was altered between each control group and found a collection was up- and downregulated in pre-eclampsia compared to the two control groups. (Reprinted with permission from *J Physiol 2023;601:4973–4988.* https://doi.org/10.1113/JP282849)

secretion. Remarkably, perfused placentae from GDM pregnancies demonstrated increased dipeptidyl peptidase-4 (DPPIV) activity within STB-EVs, and the circulation of DPPIV-bound STB-EVs increased eightfold among women with GDM [79]. Compared with those in EVs from healthy women, proteins related to metabolic processes and biological regulation in EVs from GDM patients were significantly different. Spectrin alpha erythrocytic protein (SPTA)-1, CAMK2β, PAPP-A, perilipin 4, fatty acid binding protein (FABP) 4, and hexokinase-3 exhibited notable abundance within EVs from GDM patients. Interestingly, pappalisin-1 (PAPP-A) and protein kinase II beta-calcium/calmodulin-dependent protein kinase 2β (CAMK2β), two proteins implicated in insulin sensitivity regulation, were downregulated and upregulated, respectively, in small EVs isolated from GDM pregnancies. A negative correlation ($P = 0.0135$) between maternal BMI and EV-carried PAPP-A concentration was also demonstrated [77].

Several studies have highlighted alterations in EV characteristics in GDM patients who involve carried miRNAs. Specifically, upregulation of miR-326 was observed in EVs from GDM patients compared to those from healthy women, while miR-36 was negatively correlated with the levels of adiponectin, its target. Several other miRNAs (miR-122-5p, miR-132-3p, miR-1323, miR–136-5p, miR-182-3p, miR-210-3p, miR-29a-3p, miR-29b-3p, miR-342-3p, and miR-520h) were significantly upregulated in GDM patients compared to healthy controls. In the context of pregnancy, these miRNAs have been implicated in processes such as proliferation, trophoblast differentiation, insulin secretion and regulation, and glucose transport [78]. Moreover, it has been demonstrated that hypertrophic adipose tissue in GDM patients exhibits a differential expression of small EV-carried miRNAs, which contributes to the systemic inflammatory state and insulin resistance (IR) and impacts placental metabolism, nutrient uptake, and the dysregulation of placental nutrient signaling pathways in obese GDM patients [80].

Given the relevant role of placental hormones in the multifactorial pathogenesis of GDM, certain studies have delved into the involvement of placental EVs in this disease, highlighting their contribution to insulin secretion mechanisms and the generation of a systemic inflammatory state. The alteration of EV cargo strongly correlates with the pathogenesis of GDM, necessitating further investigation to elucidate the precise mechanisms through which EVs are involved in this disease.

5.1.3 EVs in Preterm Birth

Preterm birth (PTB), defined as delivery occurring before 37 weeks of gestation, is the primary cause of neonatal morbidity and mortality. Approximately 11% of infants are born prematurely each year worldwide. Despite its high prevalence, PTB poses challenges in terms of treatment and prediction due to its association with various underlying conditions, presenting a significant clinical dilemma [81, 82]. Hence, the prediction of PTB is imperative, and the utilization of blood-based EV testing holds promise as a noninvasive and practical approach for early detection and intervention of this complication.

Numerous reports have highlighted miRNAs, such as hsa-miR-381, hsa-miR-154, hsa-miR-377, and hsa-miR-150-5p, packaged within EVs as potential biomarkers for predicting PTB [83–85]. Comprehensive analysis of miRNAs in plasma EVs isolated from pregnant women revealed significant alterations in the amounts and profiles of EV miRNAs between full-term patients and PTB patients, particularly those miRNAs associated with TGF-β, p53, and glucocorticoid receptor signaling, which are known to be involved in cytotrophoblast (CTB) proliferation and fetal membrane apoptosis [85]. Notably, exosomal proteins from plasma also exhibit potential as reliable diagnostic tools.

Further studies, including the investigation by Menon and colleagues, have examined groups of patients, including those who were not in labor, those who were in term labor, those who had preterm premature rupture of membranes, and those with PTB, comparing the proteomes of maternal plasma exosomes. The study identified 72 proteins that exhibited significant changes among these four groups, leading to the hypothesis that hematological dysfunctions, rather than inflammation, serve as the primary instigators of PTB (Fig. 13.6). Inflammation, however, occurs as a consequence of functional changes, such as complement activation, acute phase signaling, and liver X receptor (LXR)/retinoid X receptor (RXR) activation [86].

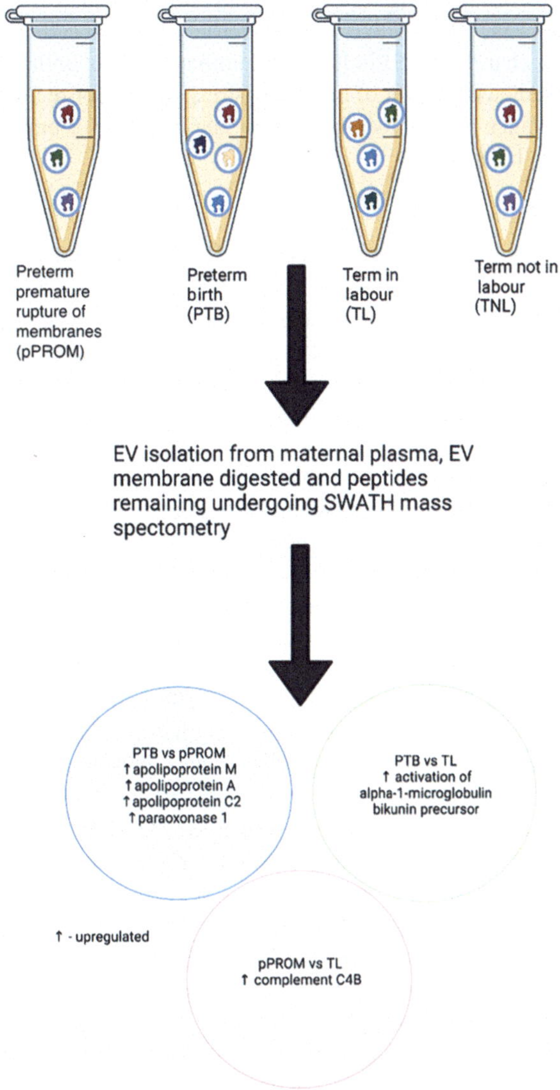

Fig. 13.6 Differences in maternal plasma exosome proteomics contents in term and preterm births. EVs isolated from maternal plasma were collected from four groups: term not in labor (TNIL, $n = 13$), term in labor (TL, $n = 11$), preterm premature rupture of membranes (pPROM, $n = 8$), and preterm birth (PTB, $n = 13$). SWATH mass spectrometry (MS) proteomics was performed and revealed notable differences among the four groups. (Reprinted with permission from *J Physiol 2023;601:4973–4988*. https://doi.org/10.1113/JP282849)

Another study reported that a panel of proteins in plasma EVs (F13A, FBLN1, IC1, etc.) at 10–12 weeks of gestation could serve as a valuable biomarker for predicting PTB occurring before 35 weeks of gestation, with an area under the receiver operating characteristic (ROC) curve of 0.74 [87]. Multiple reports indicate that PTB signatures can be detected early in pregnancy, well before they can occur; these findings not only indicate that PTB can be used as a clinical tool for prediction but also prove highly intriguing [83, 85, 87].

The distinct exosomal signatures found in patients who subsequently develop PTB support the involvement of exosomes in the etiology of PTB [88]. One study reported that placental exosome-derived bta-miR-499, which is enriched in plasma exosomes during early pregnancy in mice and cows, inhibits NF-κB activation via the Lin28B/let-7 axis in endometrial epithelial cells in vitro in response to lipopolysaccharide (LPS)-induced inflammation. This finding suggested that the exosome-mediated transfer of miRNAs is responsible for regulating the uterine inflammatory balance to establish the appropriate window for maintaining pregnancy and initiating labor [89].

5.1.4 EVs in Viral Infections During Pregnancy

It has been demonstrated that viral infections occurring during pregnancy can elicit a heightened risk of pregnancy loss through spontaneous abortion or fetal infection, resulting in subsequent congenital viral syndromes. Presently, the field lacks a genuine standard for prenatal care regarding viral infections during pregnancy, aside from the limited spectrum of pathogens encompassed by the TORCH acronym, including *Toxoplasma gondii*, rubella virus, cytomegalovirus (CMV), and herpes simplex virus (HSV). While these guidelines enable early diagnosis of an infection, preventing adverse pregnancy effects remains uncertain [90]. Patients with viral infections exhibit a notable predisposition to preterm labor or delivery. Consequently, comprehending the augmented risk to which pregnant women are exposed is pivotal for formulating an appropriate treatment or prevention plan [91].

EVs fulfill a role in immune responses by serving as modulators, with EVs derived from antigen-presenting cells either stimulating or modulating immune responses. Furthermore, EVs originating from nonpathological syncytiotrophoblasts in the placenta appear to impact resistance pathways against pathogen infection, and akin to those from epithelial cells and tumors, they can possess either tolerogenic or suppressive properties [92]. EVs derived from embryos encapsulate nucleic acids, including DNA fragments, microRNAs, long non-coding RNAs, messenger RNAs, and an array of proteins, which can be internalized by cells in the maternal vascular and immune systems, subsequently influencing pregnancy-related changes in maternal physiology [93]. Several studies have revealed that the transmission of certain viruses to fibroblast and endothelial compartments is mediated by the miRNA cluster of chromosome 19, a cluster exclusively expressed in the human placenta, effectively limiting infection through induced autophagy in recipient cells

[39]. Similarly, recent findings have implicated trophoblast-derived exosomes in hindering viral replication by upregulating key proteins involved in the autophagy process, such as light chain 3 (LC3), ultraviolet irradiation resistance-associated gene (UVRAG), and autophagy-related 4C cysteine peptidase (ATG4C) [94]. Furthermore, it has been speculated that EVs derived from trophoblast cells release IFNλ1, thereby safeguarding the fetal compartment against infection by pathogens such as the Zika virus [95]. Extracellular vesicles also contribute to the antiviral response by facilitating the transfer of nucleic acids. In this manner, cells can receive signals of viral infection even when they are not directly infected by the virus, as they may lack specific receptors [96]. Conversely, exosomal trafficking can also serve the pathogen's interests by facilitating immune evasion through enclosing the virus within a protective vesicle, ultimately promoting viral spread and overall infectivity [97].

5.2 EVs as Potential Biomarkers in Fetal Disease

5.2.1 EVs in Fetal Growth Restriction

Research has revealed that EVs assume specific functions in the interaction between the endometrium and the embryo, concurrently fostering the release of proinflammatory cytokines that play a significant role in modulating inflammatory responses throughout pregnancy by dampening the activation of natural killer cells and macrophages [25]. In a collective investigation in which women who delivered small fetuses were targeted, the concentrations of both total and placental EVs in the circulatory system were discerned through the use of CD63 and PLAP labeling. The ratio of PLAP+ CD63+ exosomes to PLAP− CD63+ exosomes was used to determine the contribution of the relative number of placental exosomes to the overall count. By correlating this ratio with the percentile of birth weight, a discernible trend consistent with the severity of the condition could be deduced (Fig. 13.7). Consequently, the relative concentration of placental EVs could serve as a marker for fetal growth [98]. The overall concentration of EVs in maternal plasma was notably lower in pregnancies with fetal growth restriction (FGR), defined as birth weight below the tenth centile and exhibiting absent or reversed umbilical artery end-diastolic flow, than in normal pregnancies. Strikingly, the EV concentrations in maternal plasma of FGR pregnancies resembled those found in non-pregnant individuals. Moreover, the concentration of Fas ligand (FasL) within EVs, known for promoting immune tolerance by suppressing the nuclear factor κB subunit p65, was also found to be lower in EVs from FGR pregnancies than in those from appropriately grown pregnancies (AGA). This observation potentially indicates the pathophysiological implication of EVs in the context of fetal growth restriction (FGR) [99]. In another investigation, increased expression levels of miR-942-5p, miR-223-5p, miR-20b-5p, miR-324-3p, and miR127-3p were reported in patients in

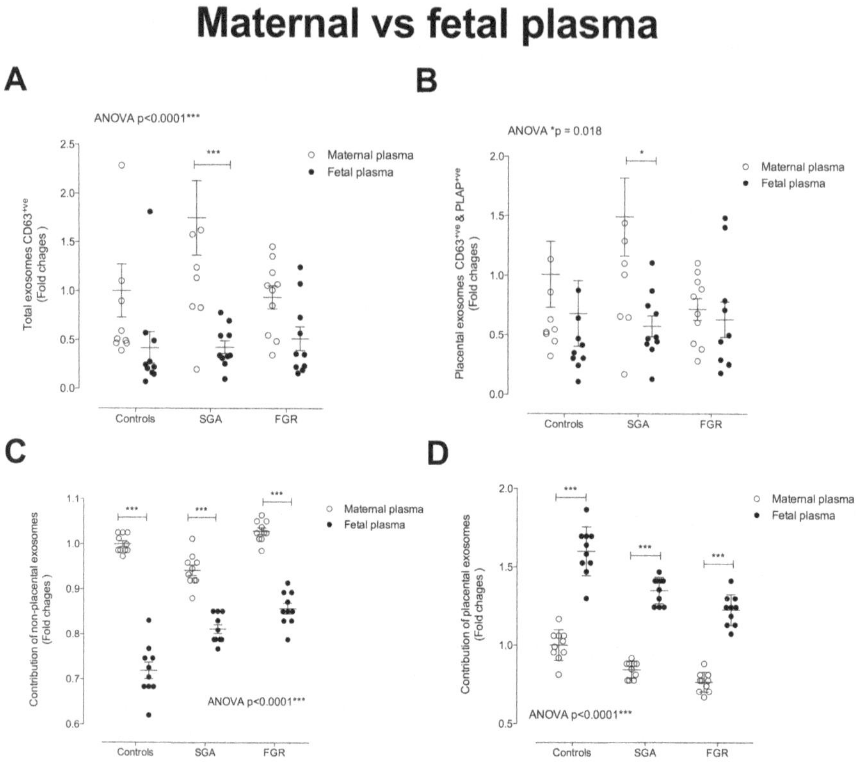

Fig. 13.7 The maternal PLAP^{+ve} ratio is a marker of fetal growth and placental function. The contribution of placental exosomes (expressed as the percentage of exosomes positive for PLAP^{+ve} versus total exosomes CD63^{+ve}) to maternal circulation was <14% lower in small for gestational age (SGA) patients and <23% lower in patients with FGR; moreover, placental exosomes may be used as a marker for placental insufficiency and fetal growth. The fetal blood PLAP^{+ve} ratio behaved similarly to the maternal PLAP^{+ve} ratio. (Reprinted with permission from *Placenta 2018;64:34–43*. https://doi.org/10.1016/j.placenta.2018.02.006)

the second trimester who subsequently gave birth to small-for-gestational-age (SGA) infants. Notably, miR-127-3p was also found to be associated with abnormal fetal growth [100, 101].

5.2.2 EVs in Down Syndrome

Down syndrome (DS) is a chromosomal anomaly that occurs in approximately 1 in 800 births worldwide and is characterized by trisomy of human chromosome 21 (Hsa21). It is primarily associated with intellectual disability and neurodevelopmental abnormalities that manifest early in embryonic life [102]. Extensive investigations into the etiology of this common chromosomal disorder have been conducted, revealing a correlation between clinical variations and the involvement of epigenetic factors [103]. In approximately 95% of cases, the underlying genetic mechanism involves meiotic nondisjunction, resulting in the presence of 47 chromosomes. Other implicated mechanisms include the Robertsonian translocation of an additional chromosome 21 to chromosomes 13, 14, 15, or 22, as well as the duplication of a delimited segment of chromosome 21 [104].

To gain a deeper understanding of the underlying processes involved in DS, numerous studies have been undertaken, suggesting the involvement of both neural-derived and non-neuronal extracellular vesicles (EVs) in its pathogenesis. This ability is attributed to the capacity of these cells to facilitate intercellular communication within the central nervous system (CNS) [105]. Given the diverse compositions of the CNS, which comprises various cell types with distinct morphologies and functions, EVs may contain distinct cargos depending on their cellular origin. Through the delivery of their cargo, EVs can modulate signal transduction pathways, thereby influencing neurogenesis, gliogenesis, synaptogenesis, network circuit formation, myelination, and synaptic pruning [106].

Several investigations have demonstrated that in the context of CNS disorders, including DS, the composition of EVs varies depending on biogenetic pathways and parental cells [107]. In addition to the commonly accepted exosomal markers, such as endosomal sorting complexes required for transport (ESCRT), apoptosis-linked gene 2-interacting protein X (Alix), and tumor susceptibility gene 101 (TSG101) proteins, various membrane, cytosolic, and cytoskeletal proteins have been consistently identified in exosomes derived from neuronal cells [106].

In the pursuit of facilitating noninvasive prenatal testing, miRNAs originating from trophoblast-derived EVs have been examined in pregnant women with fetuses affected by DS. Their findings revealed significantly greater levels of miR-99a and miR-3156 in pregnant patients than in healthy fetuses [108]. Similarly, a proof-of-concept study utilizing high-throughput quantitative PCR (HT-qPCR) on samples from DS pregnancies and control subjects successfully identified 36 mature miRNAs exhibiting notable differential expression in affected pregnancies, thereby highlighting their potential as stable biomarkers [109]. Furthermore, Lim and colleagues employed microarray-based genome-wide expression profiling to analyze placental samples from euploid and DS fetuses and detected elevated expression levels of miR-3196 and miR-1973 in placentas from affected fetuses [110].

5.2.3 EVs in Congenital Heart Disease

Congenital heart defects (CHDs) are the most prevalent birth defects and the primary cause of infant mortality resulting from such defects. The incidence of NAFLD is 8 per 1000 newborns. Approximately 20% of these circulatory system abnormalities can be attributed to genetic syndromes, including trisomy 13, 18, and 21, and Turner syndrome, as well as submicroscopic deletions or duplications [111]. Despite extensive research, the exact genetic etiology of these abnormalities has not been fully elucidated. However, the development of CHD is multifactorial and arises from a combination of non-coding genetic factors, epigenetic factors, and environmental influences [112].

Comprehensive population-based studies conducted in recent decades revealed that pregnancies affected by CHDs are at increased risk of developing placenta-associated pathologies. This correlation may arise from the concurrent development of the heart and placenta as vascular organs, potentially sharing common developmental pathways [113]. Nonetheless, the literature on cardiovascular diseases (CVDs) and the involvement of extracellular vesicles (EVs) in their etiology do not provide a definitive understanding of the molecular mechanisms and communication pathways involved in these syndromes. EVs have been implicated in both physiological and pathological processes associated with the placenta, such as inflammation and vascular calcification. Additionally, they play regulatory roles in the cardiovascular system, influencing thrombosis and vasoactive responses [112].

Furthermore, a noteworthy disparity in the expression levels of more than 100 circulating miRNAs, including miR-1275, miR-142-5p, miR-34a, miR-4666a-3p, and miR-3664-3p, has been documented in the peripheral blood of pregnant women carrying fetuses with aberrant heart growth compared to women with uncomplicated pregnancies [114]. These findings substantiate the regulatory role of circulatory exosomes or circulating miRNAs in influencing fetal heart growth, thus yielding novel perspectives on the prevention, diagnosis, and treatment of CHD [115]. Conversely, the transfer of exosomal miRNAs assumes a pivotal role in safeguarding against myocardial ischemia and mitigating myocyte death by suppressing genes associated with cell death [116]. For example, it has been demonstrated that the exosomal expression of maternal serum miR-146a-5p is correlated with fetal ventricular septal defects (VSDs), a prominent type of CHD accounting for 20–30% of CHD cases [117]. Maternal and umbilical exosomes contribute to facilitating angiogenesis by promoting the proliferation and migration of human umbilical vein endothelial cells (HUVECs) and the formation of the heart tube, thus facilitating cardiovascular development. They identified a group of upregulated miRNAs present in both types of EVs that were associated with the modulation of cell migration and proliferation [60]. Similarly, miRNA expression in CHD pregnancies was investigated using microarray analysis followed by RT-PCR. Based on the high level of statistical significance, they identified a consistent panel of four dysregulated miRNAs in CHD pregnancies, indicating the potential of exosome-derived miRNAs as noninvasive biomarkers for CHD [118].

5.2.4 EVs in Neural Tube Defects

Neural tube defects (NTDs) represent the second most prevalent category of congenital anomalies involving the central nervous system (CNS) and affect approximately 1 in 1000 newborns worldwide. These defects primarily arise due to the incomplete closure of the neural tube during the process of neurulation in early embryonic development. The manifestations of this anomaly vary depending on the location, ranging from cranial to more caudal regions, and encompass conditions such as exencephaly progressing to anencephaly, encephalocele, spina bifida, and myelomeningocele, among others [119].

Although the exact pathogenesis of NTDs is incompletely understood, it is known to be multifactorial and involves a combination of genetic and epigenetic factors [120]. Successful closure of the neural tube relies on numerous signaling pathways and genes that facilitate intercellular communication. These include the wingless/integrated (Wnt) pathway, the bone morphogenetic protein (BMP) signaling pathway, and the Sonic hedgehog (Shh) signaling pathway [121], as well as genes associated with neural patterning, growth, and cell intercalation, such as axis inhibition protein 2 (AXIN2), lymphoid enhancer-binding factor 1 (LEF1), and paired box gene 3 (PAX3) [122]. Epigenetic modifications of histones, DNA methylation, abnormal folate metabolism, and related enzymes have also been implicated [123].

Several miRNAs, including CBS, DNMT3A, FPGS, and MTHFD2, have been identified as potential targets involved in folate methylation or homeostasis. Thus, misrepresentation of miRNAs may help NTDs by modulating folate methylation or metabolism [124]. Although the search for highly sensitive and specific biomarkers in maternal serological analysis for the early detection of NTDs is ongoing, the differential expression of certain proteins within the serum exosomes of pregnant women between the 14th and 20th weeks of gestation holds promise for earlier clinical screening and diagnosis of NTDs. For instance, Wang and colleagues demonstrated a significant reduction in the expression of the DNM2 and CORO1A proteins in pregnant women carrying fetuses with NTDs compared to those in the normal control group [125]. In a recent study employing a fetal lamb model of spina bifida, the secretome of placenta-derived mesenchymal stromal cells (PMSCs) was analyzed, and exosomes derived from these cells were found to potentially alleviate limb paralysis. These exosomes were found to be enriched with RNAs and proteins associated with neuronal survival, such as galectin 1, thereby highlighting their neuroprotective function [126]. Similarly, studies have investigated the placental miRNA profiles in pregnancies affected by NTDs and identified six upregulated miRNAs unique to these cases after analyzing a panel of 887 human miRNAs [124]. These findings suggest that EVs could serve as a promising avenue for early prenatal diagnosis of fetal neurological diseases.

5.2.5 EVs in Prenatal Diagnosis

Prenatal diagnosis encompasses a range of conditions, including chromosomal abnormalities, sexually transmitted genetic diseases, genetic metabolic disorders, and congenital structural anomalies. Circulating EVs have been shown to be vital contributors to both prenatal conditions and neurodegenerative diseases [127]. Research has indicated that EV miRNAs can serve as valuable biomarkers for the prenatal diagnosis of congenital hydronephrosis [128]. Specifically, a reduction in the expression of EV-transmitted miR-300 and miR-299-5p in amniotic fluid from patients with congenital hydronephrosis has been associated with predictive indications of renal fibrosis. Notably, investigations have revealed that EVs derived from the fetal central nervous system can be isolated from maternal plasma [129]. This phenomenon correlates with the abnormal proliferation and differentiation of neural stem cells. Consequently, EVs hold potential as a means for early prenatal diagnosis of fetal neurological diseases [130]. Despite the association of EVs and the transfer of bioactive factors with certain prenatal and neurodegenerative conditions, further studies are needed to elucidate their precise role in prenatal diagnosis.

6 The Therapeutic Potential of EVs in Pregnancy-Related Diseases

The placenta is assumed to play a vital role as the crucial maternal–fetal interface, facilitating the transfer of nutrients and oxygen from the mother to the developing fetus, thereby ensuring normal fetal growth [131]. Numerous perinatal disorders are intrinsically linked to the health of the placenta. Conditions such as preeclampsia (PE) and chorioamnionitis can have significant implications for fetal well-being throughout pregnancy, encompassing fetal growth restriction (FGR), preterm birth, and, in severe cases, stillbirth. Furthermore, these placental ailments can also impart long-lasting effects on neonatal health, a realm that is only beginning to be comprehended. The exceptional capacity of MSCs to ameliorate diverse disease processes has been extensively characterized. These cells, endowed with pluripotent capabilities, predominantly contribute to the maintenance of tissue homeostasis. Consequently, these studies unveiled novel therapeutic prospects for treating a wide array of conditions. Despite the ability of MSCs to migrate to injured tissues, their long-term engraftment and expansion capacities remain limited. As a result, recent years have witnessed the emergence of acellular approaches in therapy, capitalizing on the knowledge gained regarding the paracrine actions of MSCs through extracellular vesicles and their pivotal biological role within the organism [132]. Extracellular vesicles can promote trophoblast invasion, safeguard the vascular endothelium, mitigate inflammatory injury, enhance cell proliferation, and suppress apoptosis [133]. Therapeutic interventions based on MSC-derived extracellular vesicles present a highly promising avenue, with the potential to enhance maternal–fetal outcomes and offer protective and regenerative benefits.

Several studies have demonstrated the protective influence of mesenchymal stem cell-derived extracellular vesicles (MSC-EVs) on preeclampsia incidence, preeclampsia-associated intrauterine growth restriction (IUGR), and lung outcomes [134, 135]. In a preclinical model of preeclampsia utilizing heme oxygenase (Hmox1)-null mice, intravenous antenatal MSC-EV therapy effectively prevented core preeclamptic features, significantly improved fetal survival, and mitigated intrauterine growth restriction [134]. Newborn pups from preeclamptic mothers displayed noteworthy alveolar simplification, altered bronchial epithelial morphology, and modifications in lung development genes, thus reaffirming the adverse impact of prenatal conditions on lung development [135]. Remarkably, weekly systemic administration of MSC-EVs to pregnant preeclamptic mothers effectively prevented the aforementioned deleterious effects on neonatal lung development. Intriguingly, the therapeutic effects of MSC-EVs might be indirectly mediated, as direct application of MSC-EVs to lung explants had no discernible impact. MSC-EV therapy substantially altered the cytokine and proteomic profiles of preeclamptic amniotic fluid (AF), which clearly served as a mediator of the therapeutic effects of MSC-EVs on lung development. These changes may be associated with the immunomodulation of uteroplacental leukocytes, as mass cytometry analysis indicated that a single injection of MSC-EVs altered the abundance, surface marker repertoire, and cytokine profile of various immune cell populations in the uteroplacental environment [134, 135].

Similarly, the effects of intrauterine MSC-EV treatment on the placenta and neonatal lung were evaluated using a rat model of endotoxin (ETX)-induced chorioamnionitis. The placentas from the ETX group exhibited elevated levels of inflammatory markers (NLRP-3 and IL-1β) and discernible alterations in spiral artery morphology. Assessment of the neonatal lungs in the ETX group revealed reduced alveolarization, decreased pulmonary vessel density, increased right ventricular hypertrophy, and impaired lung mechanics compared to those of the healthy controls. These findings further substantiate the influence of the antenatal environment on postnatal lung health. Intriguingly, intrauterine administration of MSC-EVs mitigated placental inflammatory cytokines and restored the architecture of spiral arteries. Moreover, offspring from the MSC-EV group demonstrated preserved distal lung growth and improved lung mechanics. Furthermore, treatment of fetal lung explants with MSC-EVs in vitro resulted in enhanced distal lung branching and increased expression of VEGF and surfactant protein C genes, surpassing the effects of ETX exposure [136]. Consequently, previous studies have shed light on the detrimental impact of dysregulated intrauterine environments on postnatal lung development. These findings also underscore the remarkable potential of MSC-EVs to modulate uteroplacental equilibrium and restore neonatal lung development, even within the antenatal context.

Despite the burgeoning interest and research in MSC-EVs, the domain is still nascent, and several challenges need to be addressed to facilitate a seamless transition to clinical applications. The specific mechanisms underlying the therapeutic effects of extracellular vesicles in perinatal disease treatment have not been elucidated and are currently limited to the study of individual components, neglecting the comprehensive functions of these intricate entities. Furthermore, studies elucidating the underlying actions of the abundant proteins, lipids, and metabolites within extracellular vesicles are still lacking. Standardization and optimization of isolation and purification techniques, administration modalities, and effective dosing regimens for extracellular vesicles have yet to be achieved. Additionally, exploring the application of engineered extracellular vesicles to enhance drug delivery efficiency or target-specific organs holds promise and warrants further investigation. While extracellular vesicle therapy, as a cell-free approach, has advantages such as low toxicity, minimal immunogenicity, and facile preservation, it allows for the circumvention of potential side effects associated with cell-based therapies. However, large-scale clinical trials assessing both the efficacy and safety of these treatments are lacking. Additionally, its application in the treatment of perinatal diseases has not been studied, necessitating a comprehensive evaluation of its safety for both mothers and fetuses.

7 Conclusion

All the above observations confirm the possible role of EVs in the complex process of normal and pathological pregnancy and associated perinatal complications and treatment. It is now firmly established that the intercellular exchange of EVs at the maternal–fetal interface serves as an evolving avenue of communication. A comprehensive understanding of the molecular pathways and mechanisms through which EVs participate in feto-maternal cross-talk will unveil their precise impact as either beneficial or detrimental factors in maintaining a healthy pregnancy. Advancing our foundational comprehension of EV structure and biogenesis and their intricate roles in intercellular communication, coupled with an in-depth exploration of the processes governing the recognition and internalization of EV cargo, holds the potential to refine methods for precise and efficacious manipulation and delivery of EVs. Nevertheless, as exemplified earlier, a host of pivotal challenges necessitate attention. These include optimizing dosage and administration routes to achieve desired therapeutic outcomes, devising strategies to track EVs within the body, and discerning any potential long-term side effects that may arise from this therapeutic modality. Furthermore, further investigations of maternal plasma EV concentrations and molecular content could enrich the understanding of perinatal disease etiology and complications affecting both mothers and children. The acquired insights can then be translated into the development of diagnostic and predictive biomarkers, fostering their clinical application (Fig. 13.8). Upon attaining full mastery of harnessing the favorable attributes of EVs, thereby revealing their potential for drug delivery and targeted gene expression correction in specific cells and tissues, these nanoscale vesicles may emerge as potent and sophisticated tools within the burgeoning domain of nanomedicine.

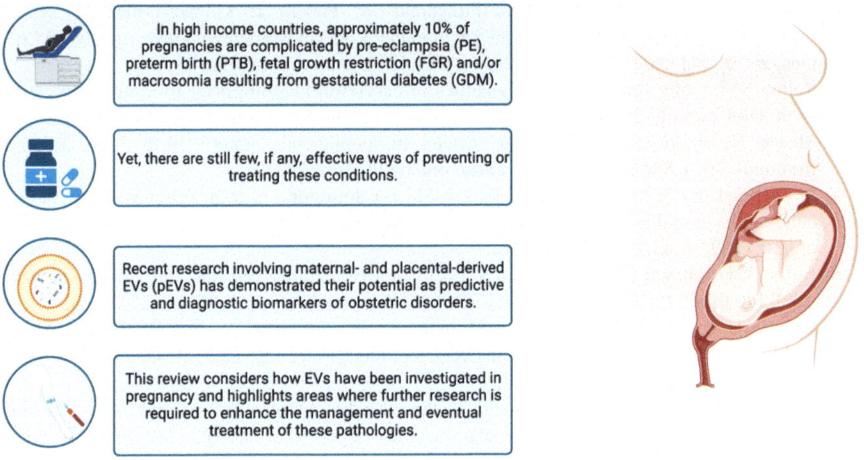

Fig. 13.8 Extracellular vesicles are involved in the pathophysiology of pregnancy complications and have potential for biomarkers and therapy. In high-income countries, approximately 10% of pregnancies are associated with adverse complications such as preeclampsia, preterm birth, fetal growth restriction, or gestational diabetes. Despite the burden of disease this places on pregnant people and their newborns, there are still few, if any, effective ways of preventing or treating these conditions. There are also gaps in our understanding of the underlying pathophysiologies and our ability to predict which mothers will be affected. The placenta plays a crucial role in pregnancy, and alterations in placental structure and function have been implicated in all of these conditions. As extracellular vesicles (EVs) have emerged as important molecules in cell-to-cell communication in health and disease, recent research involving maternal- and placental-derived EVs has demonstrated their potential as predictive and diagnostic biomarkers of obstetric disorders. This review will consider how EVs have been investigated in maternal–fetal complications and aims to highlight areas where further research is required to enhance the management and eventual treatment of these pathologies. (Reprinted with permission from *J Physiol 2023;601:4973–4988.* https://doi.org/10.1113/JP282849)

Acknowledgments This study was financially supported by open funding from the Nankai University Eye Institute (NKYKD202203), the Research Project on Skin Injury and Repair (BKJ21J016, 2021XB008), and the Tianjin Natural Science Foundation (22JCZXJC00170, 21JCZDJC00070).

References

1. Kaur S, Teoh AN, Shukri NHM, Shafie SR, Bustami NA, Takahashi M, et al. Circadian rhythm and its association with birth and infant outcomes: research protocol of a prospective cohort study. BMC Pregnancy Childbirth. 2020;20(1):96. https://doi.org/10.1186/s12884-020-2797-2.
2. Smith R, Nicholson RC. Corticotrophin releasing hormone and the timing of birth. Front Biosci. 2007;12:912–8. https://doi.org/10.2741/2113.
3. Golightly E, Jabbour HN, Norman JE. Endocrine immune interactions in human parturition. Mol Cell Endocrinol. 2011;335(1):52–9. https://doi.org/10.1016/j.mce.2010.08.005.

4. Bhatti G, Romero R, Rice GE, Fitzgerald W, Pacora P, Gomez-Lopez N, et al. Compartmentalized profiling of amniotic fluid cytokines in women with preterm labor. PLoS One. 2020;15(1):e0227881. https://doi.org/10.1371/journal.pone.0227881.
5. Yellon SM. Contributions to the dynamics of cervix remodeling prior to term and preterm birth. Biol Reprod. 2017;96(1):13–23. https://doi.org/10.1095/biolreprod.116.142844.
6. Menon R, Shahin H. Extracellular vesicles in spontaneous preterm birth. Am J Reprod Immunol. 2021;85(2):e13353. https://doi.org/10.1111/aji.13353.
7. Manier S, Liu CJ, Avet-Loiseau H, Park J, Shi J, Campigotto F, et al. Prognostic role of circulating exosomal miRNAs in multiple myeloma. Blood. 2017;129(17):2429–36. https://doi.org/10.1182/blood-2016-09-742296.
8. Kalra H, Drummen GP, Mathivanan S. Focus on extracellular vesicles: introducing the next small big thing. Int J Mol Sci. 2016;17(2):170. https://doi.org/10.3390/ijms17020170.
9. Saez T, de Vos P, Sobrevia L, Faas MM. Is there a role for exosomes in foetoplacental endothelial dysfunction in gestational diabetes mellitus? Placenta. 2018;61:48–54. https://doi.org/10.1016/j.placenta.2017.11.007.
10. McVey MJ, Kuebler WM. Extracellular vesicles: biomarkers and regulators of vascular function during extracorporeal circulation. Oncotarget. 2018;9(98):37229–51. https://doi.org/10.18632/oncotarget.26433.
11. Ge Q, Zhou Y, Lu J, Bai Y, Xie X, Lu Z. miRNA in plasma exosome is stable under different storage conditions. Molecules. 2014;19(2):1568–75. https://doi.org/10.3390/molecules19021568.
12. Simeone P, Bologna G, Lanuti P, Pierdomenico L, Guagnano MT, Pieragostino D, et al. Extracellular vesicles as signaling mediators and disease biomarkers across biological barriers. Int J Mol Sci. 2020;21(7):2514. https://doi.org/10.3390/ijms21072514.
13. Zhao Y, Yang G. Potential of extracellular vesicles in the Parkinson's disease—pathological mediators and biomarkers. Neurochem Int. 2021;144:104974. https://doi.org/10.1016/j.neuint.2021.104974.
14. Lee YT, Tran BV, Wang JJ, Liang IY, You S, Zhu Y, et al. The role of extracellular vesicles in disease progression and detection of hepatocellular carcinoma. Cancers. 2021;13(12) https://doi.org/10.3390/cancers13123076.
15. Simionescu N, Zonda R, Petrovici AR, Georgescu A. The multifaceted role of extracellular vesicles in glioblastoma: microRNA nanocarriers for disease progression and gene therapy. Pharmaceutics. 2021;13(7):988. https://doi.org/10.3390/pharmaceutics13070988.
16. Pezzana C, Agnely F, Bochot A, Siepmann J, Menasche P. Extracellular vesicles and biomaterial design: new therapies for cardiac repair. Trends Mol Med. 2021;27(3):231–47. https://doi.org/10.1016/j.molmed.2020.10.006.
17. Pieragostino D, Cicalini I, Lanuti P, Ercolino E, di Ioia M, Zucchelli M, et al. Enhanced release of acid sphingomyelinase-enriched exosomes generates a lipidomics signature in CSF of multiple sclerosis patients. Sci Rep. 2018;8(1):3071. https://doi.org/10.1038/s41598-018-21497-5.
18. Stein JM, Luzio JP. Ectocytosis caused by sublytic autologous complement attack on human neutrophils. The sorting of endogenous plasma-membrane proteins and lipids into shed vesicles. Biochem J. 1991;274(Pt 2):381–6. https://doi.org/10.1042/bj2740381.
19. Tkach M, Kowal J, Thery C. Why the need and how to approach the functional diversity of extracellular vesicles. Philos Trans R Soc Lond B Biol Sci. 2018;373(1737):20160479. https://doi.org/10.1098/rstb.2016.0479.
20. Cocucci E, Meldolesi J. Ectosomes and exosomes: shedding the confusion between extracellular vesicles. Trends Cell Biol. 2015;25(6):364–72. https://doi.org/10.1016/j.tcb.2015.01.004.
21. Gasser O, Hess C, Miot S, Deon C, Sanchez JC, Schifferli JA. Characterisation and properties of ectosomes released by human polymorphonuclear neutrophils. Exp Cell Res. 2003;285(2):243–57. https://doi.org/10.1016/s0014-4827(03)00055-7.
22. Caruso S, Poon IKH. Apoptotic cell-derived extracellular vesicles: more than just debris. Front Immunol. 2018;9:1486. https://doi.org/10.3389/fimmu.2018.01486.

23. Adamova P, Lotto RR, Powell AK, Dykes IM. Are there foetal extracellular vesicles in maternal blood? Prospects for diagnostic biomarker discovery. J Mol Med. 2023;101(1–2):65–81. https://doi.org/10.1007/s00109-022-02278-0.
24. Ghafourian M, Mahdavi R, Akbari Jonoush Z, Sadeghi M, Ghadiri N, Farzaneh M, et al. The implications of exosomes in pregnancy: emerging as new diagnostic markers and therapeutics targets. Cell Commun Signal. 2022;20(1):51. https://doi.org/10.1186/s12964-022-00853-z.
25. Buca D, Bologna G, D'Amico A, Cugini S, Musca F, Febbo M, et al. Extracellular vesicles in feto-maternal crosstalk and pregnancy disorders. Int J Mol Sci. 2020;21(6):2120. https://doi.org/10.3390/ijms21062120.
26. Tong M, Stanley JL, Chen Q, James JL, Stone PR, Chamley LW. Placental nano-vesicles target to specific organs and modulate vascular tone in vivo. Hum Reprod. 2017;32(11):2188–98. https://doi.org/10.1093/humrep/dex310.
27. Salomon C, Torres MJ, Kobayashi M, Scholz-Romero K, Sobrevia L, Dobierzewska A, et al. A gestational profile of placental exosomes in maternal plasma and their effects on endothelial cell migration. PLoS One. 2014;9(6):e98667. https://doi.org/10.1371/journal.pone.0098667.
28. Elfeky O, Longo S, Lai A, Rice GE, Salomon C. Influence of maternal BMI on the exosomal profile during gestation and their role on maternal systemic inflammation. Placenta. 2017;50:60–9. https://doi.org/10.1016/j.placenta.2016.12.020.
29. Chang G, Mouillet JF, Mishima T, Chu T, Sadovsky E, Coyne CB, et al. Expression and trafficking of placental microRNAs at the feto-maternal interface. FASEB J. 2017;31(7):2760–70. https://doi.org/10.1096/fj.201601146R.
30. Paquette AG, Chu T, Wu X, Wang K, Price ND, Sadovsky Y. Distinct communication patterns of trophoblastic miRNA among the maternal-placental-fetal compartments. Placenta. 2018;72-73:28–35. https://doi.org/10.1016/j.placenta.2018.10.004.
31. Holder B, Jones T, Sancho Shimizu V, Rice TF, Donaldson B, Bouqueau M, et al. Macrophage exosomes induce placental inflammatory cytokines: a novel mode of maternal-placental messaging. Traffic. 2016;17(2):168–78. https://doi.org/10.1111/tra.12352.
32. Nair S, Salomon C. Extracellular vesicles and their immunomodulatory functions in pregnancy. Semin Immunopathol. 2018;40(5):425–37. https://doi.org/10.1007/s00281-018-0680-2.
33. Mor G, Cardenas I. The immune system in pregnancy: a unique complexity. Am J Reprod Immunol. 2010;63(6):425–33. https://doi.org/10.1111/j.1600-0897.2010.00836.x.
34. Frangsmyr L, Baranov V, Nagaeva O, Stendahl U, Kjellberg L, Mincheva-Nilsson L. Cytoplasmic microvesicular form of Fas ligand in human early placenta: switching the tissue immune privilege hypothesis from cellular to vesicular level. Mol Hum Reprod. 2005;11(1):35–41. https://doi.org/10.1093/molehr/gah129.
35. Abrahams VM, Straszewski-Chavez SL, Guller S, Mor G. First trimester trophoblast cells secrete Fas ligand which induces immune cell apoptosis. Mol Hum Reprod. 2004;10(1):55–63. https://doi.org/10.1093/molehr/gah006.
36. Hedlund M, Stenqvist AC, Nagaeva O, Kjellberg L, Wulff M, Baranov V, et al. Human placenta expresses and secretes NKG2D ligands via exosomes that down-modulate the cognate receptor expression: evidence for immunosuppressive function. J Immunol. 2009;183(1):340–51. https://doi.org/10.4049/jimmunol.0803477.
37. Sabapatha A, Gercel-Taylor C, Taylor DD. Specific isolation of placenta-derived exosomes from the circulation of pregnant women and their immunoregulatory consequences. Am J Reprod Immunol. 2006;56(5–6):345–55. https://doi.org/10.1111/j.1600-0897.2006.00435.x.
38. Zeng F, Morelli AE. Extracellular vesicle-mediated MHC cross-dressing in immune homeostasis, transplantation, infectious diseases, and cancer. Semin Immunopathol. 2018;40(5):477–90. https://doi.org/10.1007/s00281-018-0679-8.
39. Donker RB, Mouillet JF, Chu T, Hubel CA, Stolz DB, Morelli AE, et al. The expression profile of C19MC microRNAs in primary human trophoblast cells and exosomes. Mol Hum Reprod. 2012;18(8):417–24. https://doi.org/10.1093/molehr/gas013.
40. Motta-Mejia C, Kandzija N, Zhang W, Mhlomi V, Cerdeira AS, Burdujan A, et al. Placental vesicles carry active endothelial nitric oxide synthase and their activity is

reduced in preeclampsia. Hypertension. 2017;70(2):372–81. https://doi.org/10.1161/HYPERTENSIONAHA.117.09321.
41. Hemmatzadeh M, Shomali N, Yousefzadeh Y, Mohammadi H, Ghasemzadeh A, Yousefi M. MicroRNAs: small molecules with a large impact on pre-eclampsia. J Cell Physiol. 2020;235(4):3235–48. https://doi.org/10.1002/jcp.29286.
42. Ferreira LMR, Meissner TB, Tilburgs T, Strominger JL. HLA-G: at the interface of maternal-fetal tolerance. Trends Immunol. 2017;38(4):272–86. https://doi.org/10.1016/j.it.2017.01.009.
43. Giacomini E, Vago R, Sanchez AM, Podini P, Zarovni N, Murdica V, et al. Secretome of in vitro cultured human embryos contains extracellular vesicles that are uptaken by the maternal side. Sci Rep. 2017;7(1):5210. https://doi.org/10.1038/s41598-017-05549-w.
44. Pallinger E, Bognar Z, Bogdan A, Csabai T, Abraham H, Szekeres-Bartho J. PIBF+ extracellular vesicles from mouse embryos affect IL-10 production by CD8+ cells. Sci Rep. 2018;8(1):4662. https://doi.org/10.1038/s41598-018-23112-z.
45. Ceppi M, Pereira PM, Dunand-Sauthier I, Barras E, Reith W, Santos MA, et al. MicroRNA-155 modulates the interleukin-1 signaling pathway in activated human monocyte-derived dendritic cells. Proc Natl Acad Sci U S A. 2009;106(8):2735–40. https://doi.org/10.1073/pnas.0811073106.
46. Parks JC, McCallie BR, Patton AL, Al-Safi ZA, Polotsky AJ, Griffin DK, et al. The impact of infertility diagnosis on embryo-endometrial dialogue. Reproduction. 2018;155(6):543–52. https://doi.org/10.1530/REP-17-0566.
47. Desrochers LM, Bordeleau F, Reinhart-King CA, Cerione RA, Antonyak MA. Microvesicles provide a mechanism for intercellular communication by embryonic stem cells during embryo implantation. Nat Commun. 2016;7:11958. https://doi.org/10.1038/ncomms11958.
48. Qu P, Qing S, Liu R, Qin H, Wang W, Qiao F, et al. Effects of embryo-derived exosomes on the development of bovine cloned embryos. PLoS One. 2017;12(3):e0174535. https://doi.org/10.1371/journal.pone.0174535.
49. Ng YH, Rome S, Jalabert A, Forterre A, Singh H, Hincks CL, et al. Endometrial exosomes/microvesicles in the uterine microenvironment: a new paradigm for embryo-endometrial cross talk at implantation. PLoS One. 2013;8(3):e58502. https://doi.org/10.1371/journal.pone.0058502.
50. Vilella F, Moreno-Moya JM, Balaguer N, Grasso A, Herrero M, Martinez S, et al. Hsa-miR-30d, secreted by the human endometrium, is taken up by the pre-implantation embryo and might modify its transcriptome. Development. 2015;142(18):3210–21. https://doi.org/10.1242/dev.124289.
51. Greening DW, Nguyen HP, Elgass K, Simpson RJ, Salamonsen LA. Human endometrial exosomes contain hormone-specific cargo modulating trophoblast adhesive capacity: insights into endometrial-embryo interactions. Biol Reprod. 2016;94(2):38. https://doi.org/10.1095/biolreprod.115.134890.
52. Lyall F. Priming and remodelling of human placental bed spiral arteries during pregnancy—a review. Placenta. 2005;26(Suppl A):S31–6. https://doi.org/10.1016/j.placenta.2005.02.010.
53. Harris LK. IFPA Gabor Than Award lecture: transformation of the spiral arteries in human pregnancy: key events in the remodelling timeline. Placenta. 2011;32(Suppl 2):S154–8. https://doi.org/10.1016/j.placenta.2010.11.018.
54. Lee SM, Romero R, Lee YJ, Park IS, Park CW, Yoon BH. Systemic inflammatory stimulation by microparticles derived from hypoxic trophoblast as a model for inflammatory response in preeclampsia. Am J Obstet Gynecol. 2012;207(4):337 e1–8. https://doi.org/10.1016/j.ajog.2012.06.047.
55. Roland CS, Hu J, Ren CE, Chen H, Li J, Varvoutis MS, et al. Morphological changes of placental syncytium and their implications for the pathogenesis of preeclampsia. Cell Mol Life Sci. 2016;73(2):365–76. https://doi.org/10.1007/s00018-015-2069-x.
56. Zhang Y, Fei M, Xue G, Zhou Q, Jia Y, Li L, et al. Elevated levels of hypoxia-inducible microRNA-210 in pre-eclampsia: new insights into molecular mechanisms for the disease. J Cell Mol Med. 2012;16(2):249–59. https://doi.org/10.1111/j.1582-4934.2011.01291.x.

57. Salomon C, Yee S, Scholz-Romero K, Kobayashi M, Vaswani K, Kvaskoff D, et al. Extravillous trophoblast cells-derived exosomes promote vascular smooth muscle cell migration. Front Pharmacol. 2014;5:175. https://doi.org/10.3389/fphar.2014.00175.
58. Tong M, Kleffmann T, Pradhan S, Johansson CL, DeSousa J, Stone PR, et al. Proteomic characterization of macro-, micro- and nano-extracellular vesicles derived from the same first trimester placenta: relevance for feto-maternal communication. Hum Reprod. 2016;31(4):687–99. https://doi.org/10.1093/humrep/dew004.
59. Takahashi H, Ohkuchi A, Kuwata T, Usui R, Baba Y, Suzuki H, et al. Endogenous and exogenous miR-520c-3p modulates CD44-mediated extravillous trophoblast invasion. Placenta. 2017;50:25–31. https://doi.org/10.1016/j.placenta.2016.12.016.
60. Jia L, Zhou X, Huang X, Xu X, Jia Y, Wu Y, et al. Maternal and umbilical cord serum-derived exosomes enhance endothelial cell proliferation and migration. FASEB J. 2018;32(8):4534–43. https://doi.org/10.1096/fj.201701337RR.
61. Ahmadi M, Rezaie J. Tumor cells derived-exosomes as angiogenenic agents: possible therapeutic implications. J Transl Med. 2020;18(1):249. https://doi.org/10.1186/s12967-020-02426-5.
62. Zhong XQ, Yan Q, Chen ZG, Jia CH, Li XH, Liang ZY, et al. Umbilical cord blood-derived exosomes from very preterm infants with bronchopulmonary dysplasia impaired endothelial angiogenesis: roles of exosomal microRNAs. Front Cell Dev Biol. 2021;9:637248. https://doi.org/10.3389/fcell.2021.637248.
63. Zhang N, Zhu J, Ma Q, Zhao Y, Wang Y, Hu X, et al. Exosomes derived from human umbilical cord MSCs rejuvenate aged MSCs and enhance their functions for myocardial repair. Stem Cell Res Ther. 2020;11(1):273. https://doi.org/10.1186/s13287-020-01782-9.
64. Muralikumar M, Manoj Jain S, Ganesan H, Duttaroy AK, Pathak S, Banerjee A. Current understanding of the mesenchymal stem cell-derived exosomes in cancer and aging. Biotechnol Rep. 2021;31:e00658. https://doi.org/10.1016/j.btre.2021.e00658.
65. Ma R, Liang Z, Shi X, Xu L, Li X, Wu J, et al. Exosomal miR-486-5p derived from human placental microvascular endothelial cells regulates proliferation and invasion of trophoblasts via targeting IGF1. Human Cell. 2021;34(5):1310–23. https://doi.org/10.1007/s13577-021-00543-x.
66. Phipps EA, Thadhani R, Benzing T, Karumanchi SA. Pre-eclampsia: pathogenesis, novel diagnostics and therapies. Nat Rev Nephrol. 2019;15(5):275–89. https://doi.org/10.1038/s41581-019-0119-6.
67. Than NG, Balogh A, Romero R, Karpati E, Erez O, Szilagyi A, et al. Placental protein 13 (PP13)—a placental immunoregulatory galectin protecting pregnancy. Front Immunol. 2014;5:348. https://doi.org/10.3389/fimmu.2014.00348.
68. Morgan TK. Cell- and size-specific analysis of placental extracellular vesicles in maternal plasma and pre-eclampsia. Transl Res. 2018;201:40–8. https://doi.org/10.1016/j.trsl.2018.08.004.
69. Chang X, Yao J, He Q, Liu M, Duan T, Wang K. Exosomes from women with preeclampsia induced vascular dysfunction by delivering sFlt (soluble Fms-like tyrosine kinase)-1 and sEng (soluble endoglin) to endothelial cells. Hypertension. 2018;72(6):1381–90. https://doi.org/10.1161/HYPERTENSIONAHA.118.11706.
70. Vargas A, Zhou S, Ethier-Chiasson M, Flipo D, Lafond J, Gilbert C, et al. Syncytin proteins incorporated in placenta exosomes are important for cell uptake and show variation in abundance in serum exosomes from patients with preeclampsia. FASEB J. 2014;28(8):3703–19. https://doi.org/10.1096/fj.13-239053.
71. Southcombe J, Tannetta D, Redman C, Sargent I. The immunomodulatory role of syncytiotrophoblast microvesicles. PLoS One. 2011;6(5):e20245. https://doi.org/10.1371/journal.pone.0020245.
72. Salomon C, Guanzon D, Scholz-Romero K, Longo S, Correa P, Illanes SE, et al. Placental exosomes as early biomarker of preeclampsia: potential role of exosomal microRNAs across gestation. J Clin Endocrinol Metab. 2017;102(9):3182–94. https://doi.org/10.1210/jc.2017-00672.

73. Jiang L, Long A, Tan L, Hong M, Wu J, Cai L, et al. Elevated microRNA-520g in preeclampsia inhibits migration and invasion of trophoblasts. Placenta. 2017;51:70–5. https://doi.org/10.1016/j.placenta.2017.02.001.
74. Li H, Ouyang Y, Sadovsky E, Parks WT, Chu T, Sadovsky Y. Unique microRNA signals in plasma exosomes from pregnancies complicated by preeclampsia. Hypertension. 2020;75(3):762–71. https://doi.org/10.1161/HYPERTENSIONAHA.119.14081.
75. Melchior H, Kurch-Bek D, Mund M. The prevalence of gestational diabetes. Dtsch Arztebl Int. 2017;114(24):412–8. https://doi.org/10.3238/arztebl.2017.0412.
76. Salomon C, Scholz-Romero K, Sarker S, Sweeney E, Kobayashi M, Correa P, et al. Gestational diabetes mellitus is associated with changes in the concentration and bioactivity of placenta-derived exosomes in maternal circulation across gestation. Diabetes. 2016;65(3):598–609. https://doi.org/10.2337/db15-0966.
77. Jayabalan N, Lai A, Nair S, Guanzon D, Scholz-Romero K, Palma C, et al. Quantitative proteomics by SWATH-MS suggest an association between circulating exosomes and maternal metabolic changes in gestational diabetes mellitus. Proteomics. 2019;19(1–2):e1800164. https://doi.org/10.1002/pmic.201800164.
78. Gillet V, Ouellet A, Stepanov Y, Rodosthenous RS, Croft EK, Brennan K, et al. miRNA profiles in extracellular vesicles from serum early in pregnancies complicated by gestational diabetes mellitus. J Clin Endocrinol Metab. 2019;104(11):5157–69. https://doi.org/10.1210/jc.2018-02693.
79. Kandzija N, Zhang W, Motta-Mejia C, Mhlomi V, McGowan-Downey J, James T, et al. Placental extracellular vesicles express active dipeptidyl peptidase IV; levels are increased in gestational diabetes mellitus. J Extracell Vesicles. 2019;8(1):1617000. https://doi.org/10.1080/20013078.2019.1617000.
80. Jayabalan N, Nair S, Nuzhat Z, Rice GE, Zuniga FA, Sobrevia L, et al. Cross talk between adipose tissue and placenta in obese and gestational diabetes mellitus pregnancies via exosomes. Front Endocrinol. 2017;8:239. https://doi.org/10.3389/fendo.2017.00239.
81. Blencowe H, Cousens S, Oestergaard MZ, Chou D, Moller AB, Narwal R, et al. National, regional, and worldwide estimates of preterm birth rates in the year 2010 with time trends since 1990 for selected countries: a systematic analysis and implications. Lancet. 2012;379(9832):2162–72. https://doi.org/10.1016/S0140-6736(12)60820-4.
82. Goldenberg RL, Culhane JF, Iams JD, Romero R. Epidemiology and causes of preterm birth. Lancet. 2008;371(9606):75–84. https://doi.org/10.1016/S0140-6736(08)60074-4.
83. Cook J, Bennett PR, Kim SH, Teoh TG, Sykes L, Kindinger LM, et al. First trimester circulating microRNA biomarkers predictive of subsequent preterm delivery and cervical shortening. Sci Rep. 2019;9(1):5861. https://doi.org/10.1038/s41598-019-42166-1.
84. Fallen S, Baxter D, Wu X, Kim TK, Shynlova O, Lee MY, et al. Extracellular vesicle RNAs reflect placenta dysfunction and are a biomarker source for preterm labour. J Cell Mol Med. 2018;22(5):2760–73. https://doi.org/10.1111/jcmm.13570.
85. Menon R, Debnath C, Lai A, Guanzon D, Bhatnagar S, Kshetrapal PK, et al. Circulating exosomal miRNA profile during term and preterm birth pregnancies: a longitudinal study. Endocrinology. 2019;160(2):249–75. https://doi.org/10.1210/en.2018-00836.
86. Menon R, Dixon CL, Sheller-Miller S, Fortunato SJ, Saade GR, Palma C, et al. Quantitative proteomics by SWATH-MS of maternal plasma exosomes determine pathways associated with term and preterm birth. Endocrinology. 2019;160(3):639–50. https://doi.org/10.1210/en.2018-00820.
87. McElrath TF, Cantonwine DE, Jeyabalan A, Doss RC, Page G, Roberts JM, et al. Circulating microparticle proteins obtained in the late first trimester predict spontaneous preterm birth at less than 35 weeks' gestation: a panel validation with specific characterization by parity. Am J Obstet Gynecol. 2019;220(5):488. e1–e11. https://doi.org/10.1016/j.ajog.2019.01.220.
88. Sheller-Miller S, Trivedi J, Yellon SM, Menon R. Exosomes cause preterm birth in mice: evidence for paracrine signaling in pregnancy. Sci Rep. 2019;9(1):608. https://doi.org/10.1038/s41598-018-37002-x.

89. Zhao G, Yang C, Yang J, Liu P, Jiang K, Shaukat A, et al. Placental exosome-mediated Bta-miR-499-Lin28B/let-7 axis regulates inflammatory bias during early pregnancy. Cell Death Dis. 2018;9(6):704. https://doi.org/10.1038/s41419-018-0713-8.
90. Silasi M, Cardenas I, Kwon JY, Racicot K, Aldo P, Mor G. Viral infections during pregnancy. Am J Reprod Immunol. 2015;73(3):199–213. https://doi.org/10.1111/aji.12355.
91. Chernyshov VP, Slukvin II, Bondarenko GI. Phenotypic characterization of CD7+, CD3+, and CD8+ lymphocytes from first trimester human decidua using two-color flow cytometry. Am J Reprod Immunol. 1993;29(1):5–16. https://doi.org/10.1111/j.1600-0897.1993.tb00831.x.
92. Kaminski VL, Ellwanger JH, Chies JAB. Extracellular vesicles in host-pathogen interactions and immune regulation—exosomes as emerging actors in the immunological theater of pregnancy. Heliyon. 2019;5(8):e02355. https://doi.org/10.1016/j.heliyon.2019.e02355.
93. Floriano JF, Willis G, Catapano F, Lima PR, Reis F, Barbosa AMP, et al. Exosomes could offer new options to combat the Long-term complications inflicted by gestational diabetes mellitus. Cells. 2020;9(3):675. https://doi.org/10.3390/cells9030675.
94. Delorme-Axford E, Donker RB, Mouillet JF, Chu T, Bayer A, Ouyang Y, et al. Human placental trophoblasts confer viral resistance to recipient cells. Proc Natl Acad Sci U S A. 2013;110(29):12048–53. https://doi.org/10.1073/pnas.1304718110.
95. Bayer A, Lennemann NJ, Ouyang Y, Bramley JC, Morosky S, Marques ET Jr, et al. Type III interferons produced by human placental trophoblasts confer protection against Zika virus infection. Cell Host Microbe. 2016;19(5):705–12. https://doi.org/10.1016/j.chom.2016.03.008.
96. Urbanelli L, Buratta S, Tancini B, Sagini K, Delo F, Porcellati S, et al. The role of extracellular vesicles in viral infection and transmission. Vaccines. 2019;7(3):102. https://doi.org/10.3390/vaccines7030102.
97. Raab-Traub N, Dittmer DP. Viral effects on the content and function of extracellular vesicles. Nat Rev Microbiol. 2017;15(9):559–72. https://doi.org/10.1038/nrmicro.2017.60.
98. Miranda J, Paules C, Nair S, Lai A, Palma C, Scholz-Romero K, et al. Placental exosomes profile in maternal and fetal circulation in intrauterine growth restriction—liquid biopsies to monitoring fetal growth. Placenta. 2018;64:34–43. https://doi.org/10.1016/j.placenta.2018.02.006.
99. Ariyakumar G, Morris JM, McKelvey KJ, Ashton AW, McCracken SA. NF-kappaB regulation in maternal immunity during normal and IUGR pregnancies. Sci Rep. 2021;11(1):20971. https://doi.org/10.1038/s41598-021-00430-3.
100. Zhang J, Li H, Fan B, Xu W, Zhang X. Extracellular vesicles in normal pregnancy and pregnancy-related diseases. J Cell Mol Med. 2020;24(8):4377–88. https://doi.org/10.1111/jcmm.15144.
101. Rodosthenous RS, Burris HH, Sanders AP, Just AC, Dereix AE, Svensson K, et al. Second trimester extracellular microRNAs in maternal blood and fetal growth: an exploratory study. Epigenetics. 2017;12(9):804–10. https://doi.org/10.1080/15592294.2017.1358345.
102. Bull MJ. Down syndrome. N Engl J Med. 2020;382(24):2344–52. https://doi.org/10.1056/NEJMra1706537.
103. El Hajj N, Dittrich M, Bock J, Kraus TF, Nanda I, Muller T, et al. Epigenetic dysregulation in the developing Down syndrome cortex. Epigenetics. 2016;11(8):563–78. https://doi.org/10.1080/15592294.2016.1192736.
104. Zhao WW, Wu M, Chen F, Jiang S, Su H, Liang J, et al. Robertsonian translocations: an overview of 872 Robertsonian translocations identified in a diagnostic laboratory in China. PLoS One. 2015;10(5):e0122647. https://doi.org/10.1371/journal.pone.0122647.
105. Gauthier SA, Perez-Gonzalez R, Sharma A, Huang FK, Alldred MJ, Pawlik M, et al. Enhanced exosome secretion in Down syndrome brain—a protective mechanism to alleviate neuronal endosomal abnormalities. Acta Neuropathol Commun. 2017;5(1):65. https://doi.org/10.1186/s40478-017-0466-0.

106. Gomes AR, Sangani NB, Fernandes TG, Diogo MM, Curfs LMG, Reutelingsperger CP. Extracellular vesicles in CNS developmental disorders. Int J Mol Sci. 2020;21(24):9428. https://doi.org/10.3390/ijms21249428.
107. Abdel-Haq H. The potential of liquid biopsy of the brain using blood extracellular vesicles: the first step toward effective neuroprotection against neurodegenerative diseases. Mol Diagn Ther. 2020;24(6):703–13. https://doi.org/10.1007/s40291-020-00493-4.
108. Erturk B, Karaca E, Aykut A, Durmaz B, Guler A, Buke B, et al. Prenatal evaluation of microRNA expressions in pregnancies with Down syndrome. Biomed Res Int. 2016;2016:5312674. https://doi.org/10.1155/2016/5312674.
109. Kamhieh-Milz J, Moftah RF, Bal G, Futschik M, Sterzer V, Khorramshahi O, et al. Differentially expressed microRNAs in maternal plasma for the noninvasive prenatal diagnosis of Down syndrome (trisomy 21). Biomed Res Int. 2014;2014:402475. https://doi.org/10.1155/2014/402475.
110. Lim JH, Lee DE, Kim SY, Kim HJ, Kim KS, Han YJ, et al. MicroRNAs as potential biomarkers for noninvasive detection of fetal trisomy 21. J Assist Reprod Genet. 2015;32(5):827–37. https://doi.org/10.1007/s10815-015-0429-y.
111. van Nisselrooij AEL, Lugthart MA, Clur SA, Linskens IH, Pajkrt E, Rammeloo LA, et al. The prevalence of genetic diagnoses in fetuses with severe congenital heart defects. Genet Med. 2020;22(7):1206–14. https://doi.org/10.1038/s41436-020-0791-8.
112. Courtney JA, Cnota JF, Jones HN. The role of abnormal placentation in congenital heart disease; cause, correlate, or consequence? Front Physiol. 2018;9:1045. https://doi.org/10.3389/fphys.2018.01045.
113. Hu P, Qiao F, Wang Y, Meng L, Ji X, Luo C, et al. Clinical application of targeted next-generation sequencing in fetuses with congenital heart defect. Ultrasound Obstet Gynecol. 2018;52(2):205–11. https://doi.org/10.1002/uog.19042.
114. Wu KH, Xiao QR, Yang Y, Xu JL, Zhang F, Liu CM, et al. MicroRNA-34a modulates the Notch signaling pathway in mice with congenital heart disease and its role in heart development. J Mol Cell Cardiol. 2018;114:300–8. https://doi.org/10.1016/j.yjmcc.2017.11.015.
115. Yang H, Ma Q, Wang Y, Tang Z. Clinical application of exosomes and circulating microRNAs in the diagnosis of pregnancy complications and foetal abnormalities. J Transl Med. 2020;18(1):32. https://doi.org/10.1186/s12967-020-02227-w.
116. Milano G, Biemmi V, Lazzarini E, Balbi C, Ciullo A, Bolis S, et al. Intravenous administration of cardiac progenitor cell-derived exosomes protects against doxorubicin/trastuzumab-induced cardiac toxicity. Cardiovasc Res. 2020;116(2):383–92. https://doi.org/10.1093/cvr/cvz108.
117. Ying X, Zhu Y, Jin X, Chang X. Umbilical cord plasma-derived exosomes from preeclamptic women induce vascular dysfunction by targeting HMGCS1 in endothelial cells. Placenta. 2021;103:86–93. https://doi.org/10.1016/j.placenta.2020.10.022.
118. Gu H, Chen L, Xue J, Huang T, Wei X, Liu D, et al. Expression profile of maternal circulating microRNAs as non-invasive biomarkers for prenatal diagnosis of congenital heart defects. Biomed Pharmacother. 2019;109:823–30. https://doi.org/10.1016/j.biopha.2018.10.110.
119. Salih MA, Murshid WR, Seidahmed MZ. Epidemiology, prenatal management, and prevention of neural tube defects. Saudi Med J. 2014;35(Suppl 1):S15–28.
120. Wilde JJ, Petersen JR, Niswander L. Genetic, epigenetic, and environmental contributions to neural tube closure. Annu Rev Genet. 2014;48:583–611. https://doi.org/10.1146/annurev-genet-120213-092208.
121. Anderson MJ, Schimmang T, Lewandoski M. An FGF3-BMP signaling axis regulates caudal neural tube closure, neural crest specification and anterior-posterior Axis extension. PLoS Genet. 2016;12(5):e1006018. https://doi.org/10.1371/journal.pgen.1006018.
122. Au KS, Ashley-Koch A, Northrup H. Epidemiologic and genetic aspects of spina bifida and other neural tube defects. Dev Disabil Res Rev. 2010;16(1):6–15. https://doi.org/10.1002/ddrr.93.

123. Crider KS, Yang TP, Berry RJ, Bailey LB. Folate and DNA methylation: a review of molecular mechanisms and the evidence for folate's role. Adv Nutr. 2012;3(1):21–38. https://doi.org/10.3945/an.111.000992.
124. Gu H, Li H, Zhang L, Luan H, Huang T, Wang L, et al. Diagnostic role of microRNA expression profile in the serum of pregnant women with fetuses with neural tube defects. J Neurochem. 2012;122(3):641–9. https://doi.org/10.1111/j.1471-4159.2012.07812.x.
125. Wang Y, Ma L, Jia S, Liu D, Gu H, Wei X, et al. Serum exosomal coronin 1A and dynamin 2 as neural tube defect biomarkers. J Mol Med. 2022;100(9):1307–19. https://doi.org/10.1007/s00109-022-02236-w.
126. Condrat CE, Varlas VN, Duica F, Antoniadis P, Danila CA, Cretoiu D, et al. Pregnancy-related extracellular vesicles revisited. Int J Mol Sci. 2021;22(8):3904. https://doi.org/10.3390/ijms22083904.
127. Cai X, Janku F, Zhan Q, Fan JB. Accessing genetic information with liquid biopsies. Trends Genet. 2015;31(10):564–75. https://doi.org/10.1016/j.tig.2015.06.001.
128. Xie J, Zhou Y, Gao W, Li Z, Xu Z, Zhou L. The relationship between amniotic fluid miRNAs and congenital obstructive nephropathy. Am J Transl Res. 2017;9(4):1754–63.
129. Goetzl L, Darbinian N, Merabova N. Noninvasive assessment of fetal central nervous system insult: potential application to prenatal diagnosis. Prenat Diagn. 2019;39(8):609–15. https://doi.org/10.1002/pd.5474.
130. Hamlett ED, LaRosa A, Mufson EJ, Fortea J, Ledreux A, Granholm AC. Exosome release and cargo in Down syndrome. Dev Neurobiol. 2019;79(7):639–55. https://doi.org/10.1002/dneu.22712.
131. Sun C, Groom KM, Oyston C, Chamley LW, Clark AR, James JL. The placenta in fetal growth restriction: what is going wrong? Placenta. 2020;96:10–8. https://doi.org/10.1016/j.placenta.2020.05.003.
132. Sdrimas K, Kourembanas S. MSC microvesicles for the treatment of lung disease: a new paradigm for cell-free therapy. Antioxid Redox Signal. 2014;21(13):1905–15. https://doi.org/10.1089/ars.2013.5784.
133. Zhaoer Y, Mingming G, Wei Z, Dan Y, Yating Q, Ruizhe J. Extracellular vesicles for the treatment of preeclampsia. Tissue Cell. 2022;77:101860. https://doi.org/10.1016/j.tice.2022.101860.
134. Taglauer ES, Fernandez-Gonzalez A, Willis GR, Reis M, Yeung V, Liu X, et al. Mesenchymal stromal cell-derived extracellular vesicle therapy prevents preeclamptic physiology through intrauterine immunomodulation dagger. Biol Reprod. 2021;104(2):457–67. https://doi.org/10.1093/biolre/ioaa198.
135. Taglauer ES, Fernandez-Gonzalez A, Willis GR, Reis M, Yeung V, Liu X, et al. Antenatal mesenchymal stromal cell extracellular vesicle therapy prevents preeclamptic lung injury in mice. Am J Respir Cell Mol Biol. 2022;66(1):86–95. https://doi.org/10.1165/rcmb.2021-0307OC.
136. Abele AN, Taglauer ES, Almeda M, Wilson N, Abikoye A, Seedorf GJ, et al. Antenatal mesenchymal stromal cell extracellular vesicle treatment preserves lung development in a model of bronchopulmonary dysplasia due to chorioamnionitis. Am J Physiol Lung Cell Mol Physiol. 2022;322(2):L179–L90. https://doi.org/10.1152/ajplung.00329.2021.

Chapter 14
Extracellular Vesicles in Aging and Age-Related Diseases

Jinfeng Liu, Huaxin Yu, Yunpeng Tian, and Na Liu

Abstract Aging is characterized by a decline in organ function and a loss of tissue repair ability. Consequently, the incidence of chronic degenerative diseases has increased, accompanied by an elevated risk of neurodegenerative diseases, cardiovascular disease, diabetes, osteoarthritis, cancer, and osteoporosis. Aging and age-related diseases impose a tremendous burden on the healthcare system. Therefore, developing strategies for preventing, delaying, or alleviating multiple age-related diseases is imperative. Cellular senescence is specifically attributed to the senescence-associated secretory phenotype (SASP), which is the primary factor of the aging microenvironment. Emerging evidence indicates that extracellular vesicles (EVs) play crucial roles in mediating senescent cells and their microenvironment. Their communication among cells and organs is closely associated with the transmission of aging signals in the context of age-related diseases. In addition, the ability of EVs from young plasma to promote tissue regeneration suggested that stem cell-derived EVs could be used therapeutically for rejuvenation. Here, we focus on the characteristics of stem cells and EVs, as well as the potential applications of stem cell- and EV-based therapies in aging and age-related chronic diseases.

Keywords Aging · Rejuvenation · Stem cells · Extracellular vesicles · Pluripotent stem cells

J. Liu · H. Yu · N. Liu (✉)
Nankai University School of Medicine, Tianjin, China
e-mail: liuna@nankai.edu.cn

Y. Tian
Department of Cardiology, Tianjin First Central Hospital, Tianjin, China

© The Author(s), under exclusive license to Springer Nature Singapore Pte Ltd. 2024

Z. Li (ed.), *Extracellular Vesicle: Biology and Translational Application*,
https://doi.org/10.1007/978-981-97-5536-3_14

1 Introduction

Aging is a complex degenerative state with many factors affecting the process and longevity. These factors include telomere shortening, nutrient sensing, mitochondrial dysfunction, oxidative stress, DNA repair deterioration, accumulation of DNA damage, changes in protein homeostasis leading to the accumulation and aggregation of misfolded proteins, and changes in epigenetic regulation [1]. Cell senescence, proposed more than 60 years ago, refers to cells with constrained division capacity. Telomerase decreasing-induced cell proliferation halt is a key hallmark of aged cells [2, 3]. The continuous accumulation of senescent cells can lead to tissue dysfunction and age-related diseases, such as neurodegenerative diseases, cardiovascular illnesses, diabetes, cancer, musculoskeletal disorders, and fibrotic diseases [4]. Therefore, the exploration of new and effective therapeutic strategies for healthy aging is ongoing.

Most cells can release extracellular vesicles, which can be divided into three subtypes: exosomes, microvesicles, and apoptotic bodies [5, 6]. Exosomes are lipid bilayer vesicles measuring 40–100 nm in diameter. Microvesicles, with a diameter ranging from 150 to 1000 nm, represent a more diverse population than exosomes. Both exosomes and microvesicles are smaller than other types of EVs known as apoptotic bodies; these vesicles are vesicular bodies with a diameter of 1000–5000 nm [5]. Among the three types of EVs, exosomes have the smallest average diameter, the highest uniformity, the most intricate composition, and the most diverse functionality. Consequently, they hold the highest theoretical and practical value, leading to in-depth research and widespread application. Under healthy and pathological conditions, almost all mammalian cells can release extracellular vesicles, which are widely dispersed through bodily fluids that are simple to access, such as blood, saliva, breast milk or urine [7, 8]. Extracellular vesicles carry bioactive substances such as nucleic acids, proteins, lipids, and cytokines. They have various biomedical applications, including medication delivery [9], cell-free vaccine development, and regenerative medicine [8].

Stem cells are a unique class of cells that can self-renew and differentiate into various cell types and play a crucial role in maintaining the function of tissues and organs in the human body. The ability of stem cells to self-renew and undergo multiple differentiations makes them an ideal focus for the treatment of age-related diseases [10]. However, immune rejection, safety, and ethical issues need to be addressed before stem cells can be successfully used in the clinic [11]. Moreover, extracellular vesicles derived from stem cells contain bioactive factors and successfully circumvent certain limitations. Therefore, stem cell-derived extracellular vesicles play a unique role in the treatment of aging-related diseases. Here, we primarily review the crucial roles of EVs derived from adult stem cells (ASCs) and pluripotent stem cells (PSCs) in rejuvenation.

2 Exploring the Potential of Stem Cells to Reverse Aging

Stem cells can be divided into two main types: pluripotent stem cells and adult stem cells. PSCs include embryonic stem cells (ESCs) and induced pluripotent stem cells (iPSCs). ASCs, also known as undifferentiated cells in tissues and organs, can self-renew and develop into specific cell types [12–14]. Given the association between the decline in regenerative potential and aging, many experimental and clinical studies have indicated the effectiveness of stem cell therapy for preventing or slowing the progression of age-related diseases. Next, we will provide a concise overview of the role that different types of stem cells play in addressing the process of aging.

3 Adult Stem Cells Offer Promising Solutions for Age-Related Diseases

Adult stem cells are present in numerous organs and tissues within the human body and possess multipotent characteristics that are crucial for maintaining homeostasis. Adult stem cell treatment can restore certain aging-related functions. The best-known type of ASCs is mesenchymal stem cells (MSCs), which have become one of the most interesting cell types in clinical research due to their ease of isolation and culture and low immunogenicity [15].

Mesenchymal stem cells are derived from the mesoderm and possess potent multipotency and regenerative properties [16]. They have been utilized in numerous clinical studies worldwide for the treatment of various diseases, including bone and cartilage pathologies [17], diabetes, cardiovascular disease, immune-related and neurological problems, and lung, kidney, and liver diseases [18, 19]. MSCs treatment can restore certain aging-related functions. Neurodegenerative diseases encompass a diverse group of brain disorders, such as Alzheimer's disease (AD), Parkinson's disease (PD), and Huntington's disease (HD) [20]. Preclinical studies indicate that MSCs infusion can ameliorate cognitive impairments in AD patients by reducing Aβ deposition, promoting its clearance in AD animal models, and leading to the rescue of memory deficits [21]. Furthermore, human MSCs have been found to decrease Aβ levels by enhancing the autophagy of pathological neurons in AD mouse models [22]. In PD models, MSCs have been demonstrated to improve α-synuclein clearance and regulate autophagy-lysosomal activity [23]. MSCs may activate autophagy signaling through the upregulation of Beclin-1, a key positive regulator of mammalian autophagy [22]. Cardiovascular diseases (CVDs) and aging frailty often coexist. Several preclinical studies have demonstrated that MSCs can exert cardioprotective effects and promote cardiac functions through different mechanisms. Treatment with MSCs significantly improves cerebrovascular disease (CVD) symptoms by promoting angiogenesis, repairing damaged areas, and enhancing myocardial function. The use of MSCs for CVD treatment is a promising treatment strategy [24]. The left ventricular ejection fraction (LVEF) is a significant parameter for evaluating cardiac function. Several clinical trials have demonstrated that transplantation of MSCs could significantly attenuate adverse ventricular remodeling and improve LVEF in patients with heart failure [25, 26]. Umbilical cord mesenchymal stem cells (UC-MSCs) reduce insulin resistance in type 2 diabetic rats by inducing an anti-inflammatory phenotype in adipose tissue macrophages and inhibiting NLRP3-mediated inflammation. An increasing number of clinical trials support the efficacy and safety of MSCs in the treatment of type 2 diabetes mellitus (T2DM) [27]. MSCs can regulate immune cell activity and reduce adverse effects that occur outside of the tumor site [28]. MSCs attenuate sarcopenia by increasing skeletal muscle weight and myofiber cross-sectional area in animal models of sarcopenia [29]. MSCs hold great promise in the treatment of a wide range of inflammatory diseases [30]. They have the capacity to modulate the immune system and mitigate tissue damage resulting from inflammation. The immunomodulatory function of MSCs can help immune cells reduce proinflammatory marker levels and restore the T-cell pool, suggesting that this approach is promising for rejuvenating the aging immune system [31]. These findings suggest that adult stem cell therapy holds promise for conferring a beneficial impact on patients with antiaging and age-related diseases.

4 Perspectives on Rejuvenation Through Pluripotent Stem Cells

Pluripotent stem cells, such as ESCs and iPSCs, are characterized by their ability to indefinitely self-renew and differentiate into all three germ layer cell types and thus have numerous potential applications in human developmental biology, regenerative medicine, and drug screening [32, 33]. Many other features of PSCs, such as high telomerase activity, mitochondrial status and epigenetic state, reflect features of young cells, suggesting that they have potential roles in preventing aging.

Many studies suggest that ESCs have the potential to treat Alzheimer's and Parkinson's diseases. In mouse models of brain damage, cognitive function can be restored by transplanting ESCs. The cholinergic neurons generated by ESCs can enhance spatial memory function, as they are functionally incorporated into hippocampal neural circuits [34]. Transplanting human ESC-derived dopaminergic neurons can enhance motor function in a rat model of Parkinson's disease [35]. Additional preclinical studies have shown that ESC-derived cardiomyocytes are capable of electromagnetically integrating into the host myocardium, thereby positively influencing the cardiac remodeling process, diminishing scar formation, and improving heart function [36–38].

Induced pluripotent stem cells, produced from adult somatic cells through a reprogramming process, have properties similar to those of ESCs [39, 40]. iPSC-based therapies hold significant promise in the treatment of a diverse array of diseases, including but not limited to cardiovascular diseases, gastrointestinal disorders, immune-related conditions, and metabolic disorders [41]. iPSC-derived cardiomyocytes significantly improved the contractile function, spontaneous beating, and ion channel expression in the heart after infarction [42]. Transplanted iPSC-derived cardiomyocytes can enhance the function of cardiomyocytes in damaged heart tissue, which could mechanically improve cardiac function [43]. iPSCs are directed to differentiate into functionally mature β-like cells that secrete insulin and can be used to treat diabetes [44]. iPSCs also play an important role in the treatment of AD. Transplanting Neprilysin-2-expressing iPSC-derived macrophages into a mouse model of AD significantly reduced the expression of extracellular amyloid-β [45]. iPSC-derived natural killer (NK) cells have a similar phenotype and function to NK cells isolated from peripheral blood and can kill a variety of tumor cells both in vitro and in vivo [46]. iPSCs also serve as a suitable source of liver cells for the treatment of liver disease, gradually generating hepatocyte-like cells (HLCs) with homologous phenotypic and physiological functions [41]. iPSCs hold significant promise across a range of medical applications, including drug discovery, disease modeling for research, and regenerative medicine for potential treatments [47].

Therefore, both ESCs and iPSCs have good prospects and great potential for the treatment of aging-related diseases.

5 Rejuvenation by Remodeling the Aging Microenvironment

Stem cells have great potential for cell therapy due to their pluripotency/multipotency, but ethical issues and security concerns remain. The primary safety risk is tumorigenicity, particularly for pluripotent stem cell-based therapy. Tumorigenicity concerns could arise in vivo as a result of the heterogeneity of cell products created by PSCs [11]. Adult stem cells are considerably safer than PSCs. However, certain factors, such as PDGF and VEGF secreted from MSC-derived cells, may facilitate tumor invasion and angiogenesis [48]. Additionally, embryonic stem cells raise ethical concerns due to their origination from embryos, complicating the development and approval of ESC-based cell therapies. It is essential to exploit the therapeutic potential of these cells and overcome their limitations to improve their application in treating age-related diseases [49].

Senescent cells release a variety of SASP factors, including interleukin-1 (IL-1), interleukin-6 (IL-6), and insulin-Like Growth Factor Binding Protein 7 (IGFBP7). These SASP factors, in combination with senescent cells, extracellular matrix components, and other elements, collectively contribute to the aging microenvironment. Senescent cells can transmit aging signals via the SASP to amplify aging among organs. Moreover, the aging microenvironment can be transformative and reprogrammed. Reprogramming the microenvironment could serve as a new strategy for rejuvenation. Recent studies suggest that the microenvironment of younger organisms may rejuvenate aging. The lifetime of older mice increased by 10% after 3 months of conjoined symbiosis between young and old mice, and the epigenetic aging of the blood and liver was dramatically reduced [50]. This is due to the effect of factors secreted by younger organisms, which can alter the microenvironment of older organisms. Additionally, extracellular vesicles secreted by stem cells can also contribute to modifying the microenvironment. EVs are pivotal in reversing aging, offering numerous advantages. They facilitate the transport of various biomolecules, including proteins, nucleic acids, and metabolites, between cells. Ongoing progress in this field has brought new hope for enhancing quality of life and extending longevity. In the following sections, we will provide a brief overview of various types of EVs involved in antisenescence.

6 Achieving Rejuvenation via Young Blood Extracellular Vesicles

Recent evidence suggests that young blood can reverse aging. Treating old rats with EVs extracted from the plasma of young pigs has been shown to have regenerative effects on various tissues, improving organ function, such as the blood, heart, and liver. Additionally, cognitive abilities, including learning and memory, improved in the treated rats, and the epigenetic age of their brain samples was younger than that of untreated aged rats. EVs in young blood play a crucial role in regulating aging skeletal muscle regeneration. Younger serum has the potential to enhance both skeletal muscle regeneration and functional recovery while also improving the mitochondrial function of aging progenitor cells [51].

Serum-derived extracellular vesicles have demonstrated effectiveness in enhancing vascular remodeling and preventing muscle damage in a mouse model of acute hindlimb ischemia. Proteomic and transcriptomic analyses of serum-derived extracellular vesicles have revealed that their therapeutic benefits are closely linked to the activation of matrix metalloproteinases and the modulation of the extracellular matrix, cytokine and chemokine signaling pathways [52]. Young blood has the capacity to ameliorate the vascular aging phenotype by reducing oxidative stress, stimulating mitochondrial regeneration, enhancing endothelial function, and correcting imbalanced vascular gene expression in aging mouse models. These effects collectively contribute to the preservation of vascular system functionality and the deceleration of the aging process [53]. Young blood has been shown to have a favorable impact on hippocampal neurogenesis in mice, concurrently reducing neuroinflammation. These findings suggest that young blood may contain certain components or signaling molecules capable of stimulating nerve cell generation and mitigating neuroinflammation, which are vital for maintaining brain health. Furthermore, this effect has the potential to enhance the cognitive function of aging mice [54, 55].

The rejuvenating potential of young blood in aging mice implies that specific components in the blood can facilitate the formation of nerve cells, indicating a pivotal role in regulating neurogenesis, intricately linked with cellular aging [56]. The youthful blood environment may contribute to the downregulation of angiotensin, PAI-1, and associated signaling pathways, thereby mitigating renal fibrosis and elevating autophagy levels within elderly kidneys, resulting in reduced cell apoptosis and diminished inflammation [57]. The administration of young plasma to aging mice has the potential to enhance synaptic plasticity and cognitive function in the hippocampus as well as positively impact organ functionality and stimulate stem cell activity [58]. Young plasma therapy has emerged as a natural and minimally invasive intervention that holds the promise of enhancing the health of elderly individuals through epigenetic regeneration and an extended, healthier lifespan. The potential clinical applications of this method suggest that it could be employed to prolong lifespan and enhance the overall quality of life in elderly individuals [59]. In summary, young blood has the potential to reverse aging, reduce epigenetic age, and reduce the expression of markers of chronic inflammation and oxidative stress.

7 Roles of Extracellular Vesicles in the Aging Microenvironment

The aging microenvironment plays an important role in the onset and progression of aging. Extracellular vesicles, key components of the microenvironment, also contribute significantly to the aging process and rejuvenation. Senescent cells secrete cytokines that form the senescent microenvironment, facilitating the transmission of senescence-related signals to other cells in a scale-up manner. This contributes to the spread of aging-associated traits within tissues. Injection of EVs isolated from AD or PD patients into mice reportedly leads to neurotoxic effects and the observation of aging-related symptoms [60, 61].

Conversely, EVs derived from young cells have been shown to significantly ameliorate the senescent microenvironment and reverse senescence. EVs serve as cell-free therapeutic tools by carrying bioactive molecules, such as microRNAs and proteins, that regulate the behavior and functionality of recipient cells [62]. Notably, treatment with EVs has been found to reduce inflammatory cytokines associated with aging, such as IL-6, interleukin-1β (IL-1β), and tumor necrosis factor-alpha (TNF-α) [63]. Moreover, increasing evidence suggests that stem cell-derived EVs have beneficial effects on various aging-related diseases, such as skin aging and osteoarthritis. For instance, stem cell-derived EVs have been shown to enhance the proliferation and migration of skin fibroblasts and inhibit the aging process [64, 65]. In osteoarthritis models, EVs enhanced cartilage anabolism and reduced inflammation in vitro [66–68] and had protective effects against cartilage degradation and osteoarthritis progression in mice [69, 70]. Additionally, EVs have been investigated for their potential in the treatment of T2DM, with research indicating their ability to improve metabolic balance and insulin resistance in animal models [71]. Overall, EVs play pivotal roles in cellular senescence and age-related diseases, and they hold promise as targets for identifying and therapeutically addressing senescent cells.

8 Antisenescence Effects of Extracellular Vesicles from Adult Stem Cells

Mesenchymal stem cell-derived extracellular vesicles (MSC-EVs) are currently receiving significant attention in the field of antiaging research due to their possession of the MSC active factor and ability to cross the blood–brain barrier (BBB) [72]. MSCs-EVs reversed the loss of hippocampal neurons in AD mice by reducing Aβ42-induced synaptic damage [73, 74].

Osteoarthritis (OA) is characterized by articular cartilage degradation, alterations in bone structure, synovial inflammation, and pain. By decreasing the quantity of plasma cells and increasing the release of IL-10 Bregs, MSC-EVs have the potential to alleviate inflammatory arthritis [75]. They have shown beneficial effects on extracellular matrix regeneration [69], promotion of chondrocyte proliferation and survival [76, 77], and immunomodulation [78]. MSC-EVs have demonstrated potential as therapeutics in preclinical models of this disease [77] and can effectively improve healthy longevity by reducing oxidative stress and inflammation, slowing the progression of aging, and promoting survival signaling [79].

MSC-EVs demonstrate potential therapeutic effects in the treatment of myocardial infarction by improving cardiac function, reducing damage, promoting neovascular formation, and diminishing inflammation and apoptosis through various mechanisms. One of these mechanisms involves the activation of the PI3K/Akt signaling pathway, which contributes to the reduction in the myocardial infarction area [80]. Treatment with MSC-EVs improves angiogenesis, migration and proliferative capability, SASP, mitochondrial dysfunction, and ROS levels while shielding it against aging-induced biomarkers and malfunction [81]. When treating renal illness, MSC-EVs have been shown to reduce immunogenicity and tumorigenicity compared to mesenchymal cells. Via antiapoptotic, anti-inflammatory, antifibrotic, antioxidant, and other mechanisms, MSC-EVs can treat kidney disease [82]. MSC-EVs show promise as potential treatments for promoting kidney regeneration and repair. These compounds seem to prevent the apoptosis of damaged tubular cells and aid in maintaining the integrity of kidney tissue. Additionally, they can potentially enhance kidney function by reducing the inflammatory response, thereby mitigating kidney damage [80]. Previous studies have shown that EVs can inhibit apoptosis in renal tubular epithelial cells, reduce the overexpression of Caspase-3, and promote the proliferation of glomerular endothelial cells. Furthermore, EVs contain a variety of growth factors, including transforming growth factor-β1 (TGF-β1), angiopoietin, and bone morphogenetic protein 7, which contribute to the maintenance of kidney cell survival [83]. MSC-EVs have been shown to enhance the regeneration and repair of damaged tubular cells [84].

T2DM is a relevant disease associated with aging and contributes to the development of many age-related conditions [85]. EVs from human mesenchymal stem cells could alleviate peripheral insulin resistance and ameliorate β-cell destruction, providing an alternative approach for treating T2DM [86]. In general, extracellular vesicles from adult stem cells play a role in combating the aging process through the regulation of cellular signaling, gene expression and overall cellular function.

9 Role of Extracellular Vesicles from Pluripotent Stem Cells in Rejuvenation

Pluripotent stem cells, such as ESCs and iPSCs, can differentiate into various cell types through their pluripotency. Consequently, these materials can be used to treat age-related diseases. Research has demonstrated that EVs derived from ESCs (ESC-EVs) can rejuvenate aging endothelial cells and restore their tube formation, migration, and proliferation capabilities. Activation of Nrf2 by ESC-EVs has been found to accelerate the wound healing process in elderly mice [87]. Furthermore, treatment with human ESC-EVs has been shown to significantly ameliorate mesenchymal stem cell senescence and prevent age-related bone loss [88]. In addition, extensive research on EVs has revealed their potential role in tissue damage repair, chronic wound healing, inhibition of wound scar formation, and antiaging [89, 90]. Specifically, ESC-EVs play a crucial role in promoting the structural integrity and proliferation of renal tubular epithelial cells, stimulating renal angiogenesis, impeding the progression of renal fibrosis, and facilitating functional recovery of injured kidneys. Endogenous progenitor cells or adult stem cells play critical roles in organ damage recovery. Studies have shown that ES-EVs can rejuvenate aged MSCs in vitro and restore their physiological repair function [91]. In vivo, ESC-EVs have been found to significantly promote the recovery of acute kidney injury (AKI) partially by activating $Sox9^+$ kidney progenitor cells [92]. Additionally, the research group demonstrated that ESC-EV treatment can rejuvenate senescence both in vitro and in vivo, suggesting the therapeutic potential of ESC-EVs as a novel cell-free alternative to ESCs for age-related diseases [93].

In addition to the direct application of PSC-derived EVs, EVs generated from PSC-derived cells have also been investigated for various clinical therapeutic applications. In mouse models of myocardial infarction (MI), the injection of EVs from iPSC-derived cardiac cells has several beneficial effects, including proangiogenic and cytoprotective effects. Furthermore, EVs from iPSC-derived mesenchymal stem cells can enhance endothelial cell migration and the formation of vascular tubes, exhibiting robust therapeutic effects [94].

EVs from PSCs or PSC-derived cells offer exciting avenues for addressing the aging process and age-related diseases. The multifaceted effects of these compounds on tissue regeneration, immune modulation, antioxidation, and cellular senescence reversal suggest their potential as novel antiaging agents. However, further research is needed to fully understand the underlying mechanisms involved and to ensure their safety and efficacy in clinical applications.

10 Conclusions and Perspectives

Cellular senescence has been implicated in various diseases, not only because of its association with organ/systemic aging but also because of its inherent properties. The rejuvenation of aging represents a significant area of interest. Both stem cells and extracellular vesicles have been investigated for their potential to reverse or decelerate cellular and tissue aging. Despite the therapeutic potential of stem cells in reversing aging and treating age-related diseases, there are also certain drawbacks and challenges associated with these cells. First, safety is a major concern in the field of stem cell therapy, as uncontrolled cell proliferation poses the risk of tumor formation or abnormal tissue growth. Second, exogenous stem cells may be recognized and rejected by the immune system of recipients, necessitating the use of immunosuppressive drugs. This highlights the potential immunological challenges associated with the use of stem cells in transplantation therapies. Third, embryonic stem cells have raised ethical concerns and controversies, leading to legal restrictions in some regions. In summary, while stem cell therapy offers great potential for reversing the aging process, it is accompanied by notable drawbacks, including safety concerns, potential for immune rejection, ethical issues, high cost, and technical challenges.

EVs originate as intraluminal vesicles and can be precisely localized and internalized into target cells by transporting specific proteins, nucleic acids, and biomolecules from their source cells. This unique feature renders them highly effective as drug delivery vehicles, enabling the precise delivery of therapeutic drugs or genetic materials to specific tissues or cells [84]. For instance, they can facilitate communication and coordination among immune cells, transport genetic material between neurons in the nervous system, and even promote or suppress cancer progression by delivering specific molecules to neighboring or distant cells. While further research is needed, EVs represent a promising therapeutic approach that can help slow or reverse the biological changes and diseases associated with aging. EVs rejuvenate aging through their cellular communication, anti-inflammatory and antioxidant effects, activation of stem cells, genetic regulation, tissue repair and regulation of cellular metabolism. For instance, growth factors and cell signaling molecules within extracellular vesicles promote the repair and regeneration of damaged tissues and are essential for restoring the youthful state and functionality of tissues. Furthermore, EVs contain antioxidant enzymes and molecules that contribute to the clearance of free radicals and attenuate oxidative stress in senescent cells.

Compared to stem cells, extracellular vesicles offer several advantages, such as antisenescence effects. First, EVs have lower immunogenicity than stem cells, thereby eliminating the need for immunosuppressive medications. Second, EVs do not pose the risk of uncontrolled cell growth or tumor formation, as they do not contain living cells. Third, the utilization of EVs bypasses ethical concerns. Finally, compared with stem cells, EVs can be freeze-dried and stored for extended periods without losing their biological activity, making them more practical for storage and transportation.

In conclusion, EVs show promise as candidates for reversing aging processes and are increasingly attractive in the field of rejuvenation. However, further research is required to improve the application of these materials in clinical practice. Additionally, stringent standards must be established to regulate the preparation and application of EVs in the clinical treatment of aging.

Acknowledgments This work was supported by grants from the National Natural Science Foundation of China (32070860, 32371217), the National Key R&D Plan (2021YFA1101002), the Tianjin Natural Science Foundation (22JCYBJC01220), and the National Students' Platform for Innovation and Entrepreneurship Training Program (202410055676, 202410055320, 20231005540 7, 202310055893).

Conflict of Interest Statement The authors have no competing interests to declare.

References

1. López-Otín C, Blasco MA, Partridge L, Serrano M, Kroemer G. Hallmarks of aging: an expanding universe. Cell. 2023;186(2):243–78. https://doi.org/10.1016/j.cell.2022.11.001.
2. Chakravarti D, LaBella KA, DePinho RA. Telomeres: history, health, and hallmarks of aging. Cell. 2021;184(2):306–22. https://doi.org/10.1016/j.cell.2020.12.028.
3. Campisi J, Kapahi P, Lithgow GJ, Melov S, Newman JC, Verdin E. From discoveries in ageing research to therapeutics for healthy ageing. Nature. 2019;571(7764):183–92. https://doi.org/10.1038/s41586-019-1365-2.
4. Calcinotto A, Kohli J, Zagato E, Pellegrini L, Demaria M, Alimonti A. Cellular senescence: aging, cancer, and injury. Physiol Rev. 2019;99(2):1047–78. https://doi.org/10.1152/physrev.00020.2018.
5. Kalluri R, LeBleu VS. The biology, function, and biomedical applications of exosomes. Science. 2020;367(6478):640. https://doi.org/10.1126/science.aau6977.
6. Gupta D, Zickler AM, El Andaloussi S. Dosing extracellular vesicles. Adv Drug Deliv Rev. 2021;178:113961. https://doi.org/10.1016/j.addr.2021.113961.
7. Zhang Y, Bi JY, Huang JY, Tang YN, Du SY, Li PY. Exosome: a review of its classification, isolation techniques, storage, diagnostic and targeted therapy applications. Int J Nanomed. 2020;15:6917–34. https://doi.org/10.2147/Ijn.S264498.
8. Yang D, Zhang W, Zhang H, Zhang F, Chen L, Ma L, et al. Progress, opportunity, and perspective on exosome isolation—efforts for efficient exosome-based theranostics. Theranostics. 2020;10(8):3684–707. https://doi.org/10.7150/thno.41580.
9. Cully M. Exosome-based candidates move into the clinic. Nat Rev Drug Dis. 2020;20(1):6–7. https://doi.org/10.1038/d41573-020-00220-y.
10. Laplane L, Solary E. Towards a classification of stem cells. eLife. 2019;8:e46563. https://doi.org/10.7554/eLife.46563.
11. Yamanaka S. Pluripotent stem cell-based cell therapy- promise and challenges. Cell Stem Cell. 2020;27(4):523–31. https://doi.org/10.1016/j.stem.2020.09.014.
12. Godoy-Parejo C, Deng C, Zhang Y, Liu W, Chen G. Roles of vitamins in stem cells. Cell Mol Life Sci. 2019;77(9):1771–91. https://doi.org/10.1007/s00018-019-03352-6.

13. Zakrzewski W, Dobrzyński M, Szymonowicz M, Rybak Z. Stem cells: past, present, and future. Stem Cell Res Ther. 2019;10(1):68. https://doi.org/10.1186/s13287-019-1165-5.
14. Hashemzadeh MR, Taghavizadeh Yazdi ME, Amiri MS, Mousavi SH. Stem cell therapy in the heart: biomaterials as a key route. Tissue Cell. 2021;71:101504. https://doi.org/10.1016/j.tice.2021.101504.
15. Wang ZB, Wang ZT, Sun Y, Tan L, Yu JT. The future of stem cell therapies of Alzheimer's disease. Ageing Res Rev. 2022;80:101655. https://doi.org/10.1016/j.arr.2022.101655.
16. Baryawno N, Przybylski D, Kowalczyk MS, Kfoury Y, Severe N, Gustafsson K, et al. A cellular taxonomy of the bone marrow stroma in homeostasis and leukemia. Cell. 2019;177(7):1915–32.e16. https://doi.org/10.1016/j.cell.2019.04.040.
17. Malekpour K, Hazrati A, Zahar M, Markov A, Zekiy AO, Navashenaq JG, et al. The potential use of mesenchymal stem cells and their derived exosomes for orthopedic diseases treatment. Stem Cell Rev Rep. 2022;18(3):933–51. https://doi.org/10.1007/s12015-021-10185-z.
18. Wang R, Ji Q, Meng C, Liu H, Fan C, Lipkind S, et al. Role of gingival mesenchymal stem cell exosomes in macrophage polarization under inflammatory conditions. Int Immunopharmacol. 2020;81:106030. https://doi.org/10.1016/j.intimp.2019.106030.
19. Squillaro T, Peluso G, Galderisi U. Clinical trials with mesenchymal stem cells: an update. Cell Transplant. 2016;25(5):829–48. https://doi.org/10.3727/096368915x689622.
20. Culig L, Chu X, Bohr VA. Neurogenesis in aging and age-related neurodegenerative diseases. Ageing Res Rev. 2022;78:101636. https://doi.org/10.1016/j.arr.2022.101636.
21. Harach T, Jammes F, Muller C, Duthilleul N, Cheatham V, Zufferey V, et al. Administrations of human adult ischemia-tolerant mesenchymal stem cells and factors reduce amyloid beta pathology in a mouse model of Alzheimer's disease. Neurobiol Aging. 2017;51:83–96. https://doi.org/10.1016/j.neurobiolaging.2016.11.009.
22. Shin JY, Park HJ, Kim HN, Oh SH, Bae JS, Ha HJ, et al. Mesenchymal stem cells enhance autophagy and increase β-amyloid clearance in Alzheimer disease models. Autophagy. 2014;10(1):32–44. https://doi.org/10.4161/auto.26508.
23. Park HJ, Shin JY, Kim HN, Oh SH, Lee PH. Neuroprotective effects of mesenchymal stem cells through autophagy modulation in a parkinsonian model. Neurobiol Aging. 2014;35(8):1920–8. https://doi.org/10.1016/j.neurobiolaging.2014.01.028.
24. Yun CW, Lee SH. Enhancement of functionality and therapeutic efficacy of cell-based therapy using mesenchymal stem cells for cardiovascular disease. Int J Mol Sci. 2019;20(4):982.
25. Bartolucci J, Verdugo FJ, González PL, Larrea RE, Abarzua E, Goset C, et al. Safety and efficacy of the intravenous infusion of umbilical cord mesenchymal stem cells in patients with heart failure: a phase 1/2 randomized controlled trial (RIMECARD trial [Randomized clinical trial of intravenous infusion umbilical cord mesenchymal stem cells on cardiopathy]). Circ Res. 2017;121(10):1192–204. https://doi.org/10.1161/circresaha.117.310712.
26. Lalu MM, Mazzarello S, Zlepnig J, Dong YYR, Montroy J, McIntyre L, et al. Safety and efficacy of adult stem cell therapy for acute myocardial infarction and ischemic heart failure (SafeCell Heart): a systematic review and meta-analysis. Stem Cells Transl Med. 2018;7(12):857–66. https://doi.org/10.1002/sctm.18-0120.
27. Zang L, Li Y, Hao H, Liu J, Cheng Y, Li B, et al. Efficacy and safety of umbilical cord-derived mesenchymal stem cells in Chinese adults with type 2 diabetes: a single-center, double-blinded, randomized, placebo-controlled phase II trial. Stem Cell Res Ther. 2022;13(1):180. https://doi.org/10.1186/s13287-022-02848-6.
28. Eom YW, Akter R, Li W, Lee S, Hwang S, Kim J, et al. M1 macrophages promote TRAIL expression in adipose tissue-derived stem cells, which suppresses colitis-associated colon cancer by increasing apoptosis of CD133+ cancer stem cells and decreasing M2 macrophage population. Int J Mol Sci. 2020;21(11):3887.
29. Wang QQ, Jing XM, Bi YZ, Cao XF, Wang YZ, Li YX, et al. Human umbilical cord Wharton's jelly derived mesenchymal stromal cells may attenuate sarcopenia in aged mice induced by hindlimb suspension. Med Sci Monit. 2018;24:9272–81. https://doi.org/10.12659/msm.913362.

30. Wang J, Chen Z, Sun M, Xu H, Gao Y, Liu J, et al. Characterization and therapeutic applications of mesenchymal stem cells for regenerative medicine. Tissue Cell. 2020;64:101330. https://doi.org/10.1016/j.tice.2020.101330.
31. Yeo GEC, Ng MH, Nordin FB, Law JX. Potential of mesenchymal stem cells in the rejuvenation of the aging immune system. Int J Mol Sci. 2021;22(11):5749.
32. Liu G, David BT, Trawczynski M, Fessler RG. Advances in pluripotent stem cells: history, mechanisms, technologies, and applications. Stem Cell Rev Rep. 2019;16(1):3–32. https://doi.org/10.1007/s12015-019-09935-x.
33. Xu J, Yu L, Guo J, Xiang J, Zheng Z, Gao D, et al. Generation of pig induced pluripotent stem cells using an extended pluripotent stem cell culture system. Stem Cell Res Ther. 2019;10(1):193. https://doi.org/10.1186/s13287-019-1303-0.
34. Duncan T, Valenzuela M. Alzheimer's disease, dementia, and stem cell therapy. Stem Cell Res Ther. 2017;8(1):111. https://doi.org/10.1186/s13287-017-0567-5.
35. Doi D, Morizane A, Kikuchi T, Onoe H, Hayashi T, Kawasaki T, et al. Prolonged maturation culture favors a reduction in the tumorigenicity and the dopaminergic function of human ESC-derived neural cells in a primate model of Parkinson's disease. Stem Cells. 2012;30(5):935–45. https://doi.org/10.1002/stem.1060.
36. Shiba Y, Fernandes S, Zhu WZ, Filice D, Muskheli V, Kim J, et al. Human ES-cell-derived cardiomyocytes electrically couple and suppress arrhythmias in injured hearts. Nature. 2012;489(7415):322–5. https://doi.org/10.1038/nature11317.
37. Yeghiazarians Y, Gaur M, Zhang Y, Sievers RE, Ritner C, Prasad M, et al. Myocardial improvement with human embryonic stem cell-derived cardiomyocytes enriched by p38MAPK inhibition. Cytotherapy. 2012;14(2):223–31. https://doi.org/10.3109/14653249.2011.623690.
38. Chong JJ, Yang X, Don CW, Minami E, Liu YW, Weyers JJ, et al. Human embryonic-stem-cell-derived cardiomyocytes regenerate non-human primate hearts. Nature. 2014;510(7504):273–7. https://doi.org/10.1038/nature13233.
39. Takahashi K, Tanabe K, Ohnuki M, Narita M, Ichisaka T, Tomoda K, et al. Induction of pluripotent stem cells from adult human fibroblasts by defined factors. Cell. 2007;131(5):861–72. https://doi.org/10.1016/j.cell.2007.11.019.
40. Yoshida Y, Yamanaka S. Induced pluripotent stem cells 10 years later: for cardiac applications. Circ Res. 2017;120(12):1958–68. https://doi.org/10.1161/circresaha.117.311080.
41. Li Y, Yang X, Plummer R, Hayashi Y, Deng X-S, Nie Y-Z, et al. Human pluripotent stem cell-derived hepatocyte-like cells and organoids for liver disease and therapy. Int J Mol Sci. 2021;22(19):10471.
42. Caldas HC, Lojudice FH, Dias C, Fernandes-Charpiot IMM, Baptista M, Kawasaki-Oyama RS, et al. Induced pluripotent stem cells reduce progression of experimental chronic kidney disease but develop Wilms' tumors. Stem Cells Int. 2017;2017:7428316. https://doi.org/10.1155/2017/7428316.
43. Kawamura M, Miyagawa S, Fukushima S, Saito A, Miki K, Ito E, et al. Enhanced survival of transplanted human induced pluripotent stem cell-derived cardiomyocytes by the combination of cell sheets with the pedicled omental flap technique in a porcine heart. Circulation. 2013;128(11 Suppl 1):S87–94. https://doi.org/10.1161/circulationaha.112.000366.
44. Kondo Y, Toyoda T, Inagaki N, Osafune K. iPSC technology-based regenerative therapy for diabetes. J Diabetes Investig. 2018;9(2):234–43. https://doi.org/10.1111/jdi.12702.
45. Takamatsu K, Ikeda T, Haruta M, Matsumura K, Ogi Y, Nakagata N, et al. Degradation of amyloid beta by human induced pluripotent stem cell-derived macrophages expressing Neprilysin-2. Stem Cell Res. 2014;13(3):442–53. https://doi.org/10.1016/j.scr.2014.10.001.
46. Woll PS, Grzywacz B, Tian X, Marcus RK, Knorr DA, Verneris MR, et al. Human embryonic stem cells differentiate into a homogeneous population of natural killer cells with potent in vivo antitumor activity. Blood. 2009;113(24):6094–101. https://doi.org/10.1182/blood-2008-06-165225.

47. Karagiannis P, Takahashi K, Saito M, Yoshida Y, Okita K, Watanabe A, et al. Induced pluripotent stem cells and their use in human models of disease and development. Physiol Rev. 2019;99(1):79–114. https://doi.org/10.1152/physrev.00039.2017.
48. Papaccio F, Paino F, Regad T, Papaccio G, Desiderio V, Tirino V. Concise review: Cancer cells, cancer stem cells, and mesenchymal stem cells: influence in cancer development. Stem Cell Transl Med. 2017;6(12):2115–25. https://doi.org/10.1002/sctm.17-0138.
49. Karimian M, Nouri N, Ghasemi LV, Mohammadi AH, Behjati M. Administration of stem cells against cardiovascular diseases with a focus on molecular mechanisms: current knowledge and prospects. Tissue Cell. 2023;81:102030. https://doi.org/10.1016/j.tice.2023.102030.
50. Zhang B, Lee DE, Trapp A, Tyshkovskiy A, Lu AT, Bareja A, et al. Multi-omic rejuvenation and life span extension on exposure to youthful circulation. Nat Aging. 2023;3(8):948–64. https://doi.org/10.1038/s43587-023-00451-9.
51. Sahu A, Clemens ZJ, Shinde SN, Sivakumar S, Pius A, Bhatia A, et al. Regulation of aged skeletal muscle regeneration by circulating extracellular vesicles. Nat Aging. 2021;1(12):1148–61. https://doi.org/10.1038/s43587-021-00143-2.
52. Cavallari C, Ranghino A, Tapparo M, Cedrino M, Figliolini F, Grange C, et al. Serum-derived extracellular vesicles (EVs) impact on vascular remodeling and prevent muscle damage in acute hind limb ischemia. Sci Rep. 2017;7(1):8180. https://doi.org/10.1038/s41598-017-08250-0.
53. Kiss T, Tarantini S, Csipo T, Balasubramanian P, Nyúl-Tóth Á, Yabluchanskiy A, et al. Circulating anti-geronic factors from heterochonic parabionts promote vascular rejuvenation in aged mice: transcriptional footprint of mitochondrial protection, attenuation of oxidative stress, and rescue of endothelial function by young blood. GeroScience. 2020;42(2):727–48. https://doi.org/10.1007/s11357-020-00180-6.
54. Huffman DM, Csiszar A, Ungvari Z. Heterochronic blood exchange attenuates age-related neuroinflammation and confers cognitive benefits: do microvascular protective effects play a role? GeroScience. 2021;43(1):111–3. https://doi.org/10.1007/s11357-021-00324-2.
55. Mehdipour M, Mehdipour T, Skinner CM, Wong N, Liu C, Chen C-C, et al. Plasma dilution improves cognition and attenuates neuroinflammation in old mice. GeroScience. 2020;43(1):1–18. https://doi.org/10.1007/s11357-020-00297-8.
56. Rojas-Vázquez S, Blasco-Chamarro L, López-Fabuel I, Martínez-Máñez R, Fariñas I. Vascular senescence: a potential bridge between physiological aging and neurogenic decline. Front Neurosci. 2021;15:666881. https://doi.org/10.3389/fnins.2021.666881.
57. Huang Q, Ning Y, Liu D, Zhang Y, Li D, Zhang Y, et al. A young blood environment decreases aging of senile mice kidneys. J Gerontol A Biol Sci Med Sci. 2018;73(4):421–8. https://doi.org/10.1093/gerona/glx183.
58. Villeda SA, Plambeck KE, Middeldorp J, Castellano JM, Mosher KI, Luo J, et al. Young blood reverses age-related impairments in cognitive function and synaptic plasticity in mice. Nat Med. 2014;20(6):659–63. https://doi.org/10.1038/nm.3569.
59. Chiavellini P, Mallat MC, Lehmann M, Zoller JA, Gordevicius J, Gallardo MD, et al. Young plasma rejuvenates blood dna methylation profile, prolongs mean lifespan and improves health in old rats. bioRxiv. 2022;2022:518747. https://doi.org/10.1101/2022.12.01.518747.
60. Goldring SR, Goldring MB. Changes in the osteochondral unit during osteoarthritis: structure, function and cartilage-bone crosstalk. Nat Rev Rheumatol. 2016;12(11):632–44. https://doi.org/10.1038/nrrheum.2016.148.
61. Wang X, Han Z, Yu Y, Xu Z, Cai B, Yuan Y. Potential applications of induced pluripotent stem cells for cardiovascular diseases. Curr Drug Targets. 2019;20(7):763–74. https://doi.org/10.2174/1389450120666181211164147.
62. Yin Y, Chen H, Wang Y, Zhang L, Wang X. Roles of extracellular vesicles in the aging microenvironment and age-related diseases. J Extracell Vesicles. 2021;10(12):e12154. https://doi.org/10.1002/jev2.12154.
63. Sanz-Ros J, Mas-Bargues C, Romero-García N, Huete-Acevedo J, Dromant M, Borrás C. Therapeutic potential of extracellular vesicles in aging and age-related diseases. Int J Mol Sci. 2022;23(23):14632.

64. Wu J-Y, Wu S-N, Zhang L-P, Zhao X-S, Li Y, Yang Q-Y, et al. Stem cell-derived exosomes: a new method for reversing skin aging. Tissue Eng Regen Med. 2022;19(5):961–8. https://doi.org/10.1007/s13770-022-00461-5.
65. Deng M, Yu TZ, Li D, Wang X, Zhou G, Liu W, et al. Human umbilical cord mesenchymal stem cell-derived and dermal fibroblast-derived extracellular vesicles protect dermal fibroblasts from ultraviolet radiation-induced photoaging in vitro. Photochem Photobiol Sci. 2020;19(3):406–14. https://doi.org/10.1039/c9pp00421a.
66. Cosenza S, Ruiz M, Toupet K, Jorgensen C, Noël D. Mesenchymal stem cells derived exosomes and microparticles protect cartilage and bone from degradation in osteoarthritis. Sci Rep. 2017;7(1):16214. https://doi.org/10.1038/s41598-017-15376-8.
67. Mao G, Zhang Z, Hu S, Zhang Z, Chang Z, Huang Z, et al. Exosomes derived from miR-92a-3p-overexpressing human mesenchymal stem cells enhance chondrogenesis and suppress cartilage degradation via targeting WNT5A. Stem Cell Res Ther. 2018;9(1):247. https://doi.org/10.1186/s13287-018-1004-0.
68. Vonk LA, van Dooremalen SFJ, Liv N, Klumperman J, Coffer PJ, Saris DBF, et al. Mesenchymal stromal/stem cell-derived extracellular vesicles promote human cartilage regeneration in vitro. Theranostics. 2018;8(4):906–20. https://doi.org/10.7150/thno.20746.
69. Wang Y, Yu D, Liu Z, Zhou F, Dai J, Wu B, et al. Exosomes from embryonic mesenchymal stem cells alleviate osteoarthritis through balancing synthesis and degradation of cartilage extracellular matrix. Stem Cell Res Ther. 2017;8(1):189. https://doi.org/10.1186/s13287-017-0632-0.
70. Zhu Y, Wang Y, Zhao B, Niu X, Hu B, Li Q, et al. Comparison of exosomes secreted by induced pluripotent stem cell-derived mesenchymal stem cells and synovial membrane-derived mesenchymal stem cells for the treatment of osteoarthritis. Stem Cell Res Ther. 2017;8(1):64. https://doi.org/10.1186/s13287-017-0510-9.
71. Zhao H, Shang Q, Pan Z, Bai Y, Li Z, Zhang H, et al. Exosomes from adipose-derived stem cells attenuate adipose inflammation and obesity through polarizing M2 macrophages and beiging in white adipose tissue. Diabetes. 2018;67(2):235–47. https://doi.org/10.2337/db17-0356.
72. Nakano M, Fujimiya M. Potential effects of mesenchymal stem cell derived extracellular vesicles and exosomal miRNAs in neurological disorders. Neural Regen Res. 2021;16(12):2359–66. https://doi.org/10.4103/1673-5374.313026.
73. Díaz-Moreno M, Armenteros T, Gradari S, Hortigüela R, García-Corzo L, Fontán-Lozano Á, et al. Noggin rescues age-related stem cell loss in the brain of senescent mice with neurodegenerative pathology. Proc Natl Acad Sci U S A. 2018;115(45):11625–30. https://doi.org/10.1073/pnas.1813205115.
74. Yang L, Zhai Y, Hao Y, Zhu Z, Cheng G. The regulatory functionality of exosomes derived from hUMSCs in 3D culture for Alzheimer's disease therapy. Small. 2020;16(3):e1906273. https://doi.org/10.1002/smll.201906273.
75. Cosenza S, Ruiz M, Maumus M, Jorgensen C, Noël D. Pathogenic or therapeutic extracellular vesicles in rheumatic diseases: role of mesenchymal stem cell-derived vesicles. Int J Mol Sci. 2017;18(4):889. https://doi.org/10.3390/ijms18040889.
76. Li S, Liu J, Liu S, Jiao W, Wang X. Mesenchymal stem cell-derived extracellular vesicles prevent the development of osteoarthritis via the circHIPK3/miR-124-3p/MYH9 axis. J Nanobiotechnol. 2021;19(1):194. https://doi.org/10.1186/s12951-021-00940-2.
77. Woo CH, Kim HK, Jung GY, Jung YJ, Lee KS, Yun YE, et al. Small extracellular vesicles from human adipose-derived stem cells attenuate cartilage degeneration. J Extracellular Vesicles. 2020;9(1):1735249. https://doi.org/10.1080/20013078.2020.1735249.
78. Toh WS, Lai RC, Hui JHP, Lim SK. MSC exosome as a cell-free MSC therapy for cartilage regeneration: Implications for osteoarthritis treatment. Semin Cell Dev Biol. 2017;67:56–64. https://doi.org/10.1016/j.semcdb.2016.11.008.
79. Dorronsoro A, Santiago FE, Grassi D, Zhang T, Lai RC, McGowan SJ, et al. Mesenchymal stem cell-derived extracellular vesicles reduce senescence and extend health span in mouse models of aging. Aging Cell. 2021;20(4):e13337. https://doi.org/10.1111/acel.13337.

80. Tsiapalis D, O'Driscoll L. Mesenchymal stem cell derived extracellular vesicles for tissue engineering and regenerative medicine applications. Cells. 2020;9(4):991.
81. Xiao X, Xu M, Yu H, Wang L, Li X, Rak J, et al. Mesenchymal stem cell-derived small extracellular vesicles mitigate oxidative stress-induced senescence in endothelial cells via regulation of miR-146a/Src. Signal Transduct Target Ther. 2021;6(1):354. https://doi.org/10.1038/s41392-021-00765-3.
82. Huang Y, Yang L. Mesenchymal stem cells and extracellular vesicles in therapy against kidney diseases. Stem Cell Res Ther. 2021;12(1):219. https://doi.org/10.1186/s13287-021-02289-7.
83. Mao J, Li C, Wu F, She Z, Luo S, Chen X, et al. MSC-EVs transferring mitochondria and related components: a new hope for the treatment of kidney disease. Front Immunol. 2022;13:978571. https://doi.org/10.3389/fimmu.2022.978571.
84. Grange C, Skovronova R, Marabese F, Bussolati B. Stem cell-derived extracellular vesicles and kidney regeneration. Cells. 2019;8(10):1240.
85. Longo M, Bellastella G, Maiorino MI, Meier JJ, Esposito K, Giugliano D. Diabetes and aging: from treatment goals to pharmacologic therapy. Front Endocrinol. 2019;10:45. https://doi.org/10.3389/fendo.2019.00045.
86. Sun Y, Shi H, Yin S, Ji C, Zhang X, Zhang B, et al. Human mesenchymal stem cell derived exosomes alleviate type 2 diabetes mellitus by reversing peripheral insulin resistance and relieving β-cell destruction. ACS Nano. 2018;12(8):7613–28. https://doi.org/10.1021/acsnano.7b07643.
87. Chen B, Sun Y, Zhang J, Zhu Q, Yang Y, Niu X, et al. Human embryonic stem cell-derived exosomes promote pressure ulcer healing in aged mice by rejuvenating senescent endothelial cells. Stem Cell Res Ther. 2019;10(1):142. https://doi.org/10.1186/s13287-019-1253-6.
88. Gong L, Chen B, Zhang J, Sun Y, Yuan J, Niu X, et al. Human ESC-sEVs alleviate age-related bone loss by rejuvenating senescent bone marrow-derived mesenchymal stem cells. J Extracellular Vesicles. 2020;9(1):1800971. https://doi.org/10.1080/20013078.2020.1800971.
89. Shabbir A, Cox A, Rodriguez-Menocal L, Salgado M, Van Badiavas E. Mesenchymal stem cell exosomes induce proliferation and migration of normal and chronic wound fibroblasts, and enhance angiogenesis in vitro. Stem Cells Develop. 2015;24(14):1635–47. https://doi.org/10.1089/scd.2014.0316.
90. Zhang Y, Kim MS, Jia B, Yan J, Zuniga-Hertz JP, Han C, et al. Hypothalamic stem cells control ageing speed partly through exosomal miRNAs. Nature. 2017;548(7665):52–7. https://doi.org/10.1038/nature23282.
91. Zhang Y, Xu J, Liu S, Lim M, Zhao S, Cui K, et al. Embryonic stem cell-derived extracellular vesicles enhance the therapeutic effect of mesenchymal stem cells. Theranostics. 2019;9(23):6976–90. https://doi.org/10.7150/thno.35305.
92. Yu L, Liu S, Wang C, Zhang C, Wen Y, Zhang K, et al. Embryonic stem cell-derived extracellular vesicles promote the recovery of kidney injury. Stem Cell Res Ther. 2021;12(1):379. https://doi.org/10.1186/s13287-021-02460-0.
93. Yu L, Wen H, Liu C, Wang C, Yu H, Zhang K, et al. Embryonic stem cell-derived extracellular vesicles rejuvenate senescent cells and antagonize aging in mice. Bioact Mater. 2023;29:85–97. https://doi.org/10.1016/j.bioactmat.2023.06.011.
94. Zhang J, Guan J, Niu X, Hu G, Guo S, Li Q, et al. Exosomes released from human induced pluripotent stem cells-derived MSCs facilitate cutaneous wound healing by promoting collagen synthesis and angiogenesis. J Transl Med. 2015;13(1):49. https://doi.org/10.1186/s12967-015-0417-0.

Chapter 15
Extracellular Vesicle Imaging in Theranostic Applications

Kaiyue Zhang, Huifang Li, and Ke Cheng

Abstract Extracellular vesicles (EVs) are lipid bilayer vesicles with a diameter ranging from 30 to 150 nm, playing a crucial role in mediating cell-to-cell communication under both physiological and pathophysiological conditions. However, the clinical utilization of EVs is currently hindered by the challenge of monitoring them effectively in vitro and in vivo, primarily due to their small size. In response to the limitations, scientists worldwide have dedicated significant efforts to advancing research aimed at addressing this issue and enhancing the visualization of EVs for optimized theranostic applications. Here, we comprehensively discussed existing strategies for labeling and tracking EVs in both in vitro and in vivo settings. Furthermore, we explored their potential applications in the fields of cancer theranostics, neurodegenerative diseases, and tissue injury.

Keywords Extracellular vesicle · Imaging · Cancer theranostics · Neurodegenerative disease · Tissue injury

1 Introduction

Extracellular vesicles (EVs), which are composed of the plasma membrane with transmembrane proteins and the inclusive contents, play a critical role in cell-to-cell communication under both physiological and pathophysiological conditions [1, 2]. EVs can usually be divided into two types, exosomes (30–100 nm) and macrovesicles (MVs, 100–1000 nm), according to their size and subcellular origin [3]. Both exosomes and MVs have been reported to carry a multitude of bioactive molecules, including proteins, peptides, genetic materials (mRNAs and microRNAs), and

K. Zhang · K. Cheng (✉)
Columbia University, New York City, NY, USA
e-mail: kc3727@columbia.edu

H. Li
School of Medicine, Nankai University, Tianjin, China

© The Author(s), under exclusive license to Springer Nature Singapore Pte Ltd. 2024

Z. Li (ed.), *Extracellular Vesicle: Biology and Translational Application*,
https://doi.org/10.1007/978-981-97-5536-3_15

lipids, which can regulate recipient cells by shuttling information in an endocrine or paracrine manner [1, 4]. Recently, increasing research has investigated the diagnostic and therapeutic applications of EVs in cancer and regeneration medicine [5]. For instance, EVs derived from mesenchymal stem cells (MSCs) have shown promising application prospects for delivering antitumor drugs and promoting the recovery of injured tissues [6–9]. However, the current clinical application of EVs is restricted by their inability to monitor EVs both in vitro and in vivo at a reliable resolution. Intense efforts from scientists around the world have been devoted to advancing studies to fill these gaps in visualizing EVs to maximize their theranostic applications [10]. In this chapter, we will discuss the current strategies that label and track EVs both in vitro and in vivo for theranostic applications.

2 Current Strategies for EV Imaging

EV imaging involves various techniques that allow for the visualization and tracking of EVs. There are several commonly used imaging methods for EV studies and applications.

2.1 Label-Free Strategies

2.1.1 Electron Microscopy

Electron microscopy (EM) is a classic imaging technique for capturing the natural shape of EVs and allows for detailed visualization of EVs at high resolution [11]. There are three primary types of EM used for EV research, transmission electron microscopy (TEM), scanning electron microscopy (SEM), and cryogenic electron microscopy (cryo-EM) (Fig. 15.1).

TEM

As the most commonly used method that characterizes EVs, TEM is based on the use of a beam of electrons passing through a thin EV layer to obtain detailed information about the EVs via electron microscopy. In this method, EV samples are typically loaded on a carbon film-coated copper grid and negatively stained with electron-dense stains, such as uranyl acetate or phosphotungstic acid, to enhance contrast [15]. TEM images can reveal the morphology, size, and lipid bilayer of EVs at a resolution down to a few nanometers, which makes TEM the gold standard for the morphological characterization of EVs. Additionally, TEM plays a crucial role in the field of EV biogenesis. By imaging cell slices through TEM, researchers can evaluate the process of EV biogenesis and secretion in donor cells.

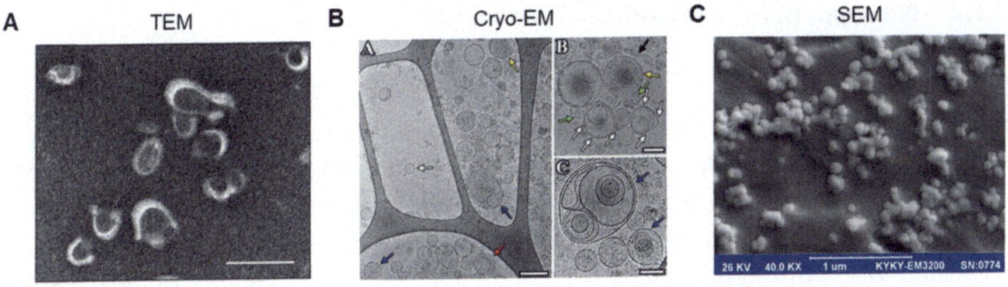

Fig. 15.1 Electron microscopy images of EVs (**a**) TEM images of EVs isolated from cell culture media. Scale bar = 100 nm [12] (**b**) Cryo-EM images of EVs from cerebrospinal fluid [13]. Scale bars are 200 nm for A, B and 100 nm for C in this panel. (**c**) SEM image of purified EVs from the cell culture supernatant of K562 cells [14]

Cryo-EM

Cryo-EM, an imaging technique that has been widely used in structural biology, was recently used in the study of EVs. In this method, EVs are rapidly frozen to a thin layer of vitreous ice on the grid in their native state without the need for fixation or staining [13]. This sample preparation process helps preserve the native structure of EVs. Thus, in addition to revealing the shape and size of EVs, cryo-EM could also be used to visualize the 3D structure of EVs and gain insights into their cargo. Cryo-EM avoids the need for heavy metal stains, which are commonly used in traditional electron microscopy. This approach is important for studying EV samples, as heavy metal stains can obscure details and may affect the native structure of EVs. Therefore, EVs observed by TEM usually exhibit a cup-shaped morphology, whereas cryo-EM images of EVs usually reveal a perfect circle. In addition, Cryo-EM can provide near-atomic resolution information. This advantage allows visualization of the molecular details of the protein and lipid composition of EVs as well as their overall morphology.

SEM

In contrast to TEM and Cryo-EM, which can characterize the internal structure of EVs, SEM usually provides surface imaging of EVs by scanning EV samples with a focused beam of electrons [16]. EV samples for SEM usually need to be fixed, dehydrated, and subjected to gold coating for enhanced conductivity. The gold on the surface of EVs will interact with electrons and emit secondary electrons to construct SEM images. SEM images can provide information about the size, shape, and surface characteristics of EVs. This imaging technique is widely used in EV studies, including surface morphology and topography studies of EVs, evaluation of EV release from donor cells and uptake by recipient cells, and studies of EV-modified materials or devices.

2.1.2 Nanoparticle Tracking Analysis (NTA)

NTA is also widely used to trace EVs in liquid buffer. The commonly used equipment is NanoSight from Malvern Panalytical. On this basis, several new devices, such as ZetaView, have been developed to specifically detect EVs. In these methods, a laser beam is directed through the samples, EVs in the buffer scatter the laser beam, and the Brownian motion of individual EVs is captured and recorded as a real-time video through a microscope and camera system. Then, specialized software will be used to analyze the video to track the movement of all individual EVs and determine their size distribution and concentration [17].

NTA was developed for tracing nanoparticles and has become a widely used technique for analyzing the size distribution and concentration of EVs. NTA allows for the measurement of the concentration and size distribution of EVs in various types of biological samples, including cell culture medium, serum, urine, and many other bodily fluids. This information helps to quantify EVs and understand the heterogeneity of EV populations in samples [18]. Furthermore, NTA can provide real-time, high-resolution analysis of individual EVs in samples, which is critical for investigating dynamic changes in EV pollution over time [17]. Moreover, real-time video of EVs provides insights into their morphology and behavior. All these advantages mean that NTA analysis is always required for both EV research and EV industrialization. However, NTA still has several limitations, such as NTA cannot provide information about the composition or cargo of EVs and cannot distinguish the individual EVs and aggregated EVs (Fig. 15.2A).

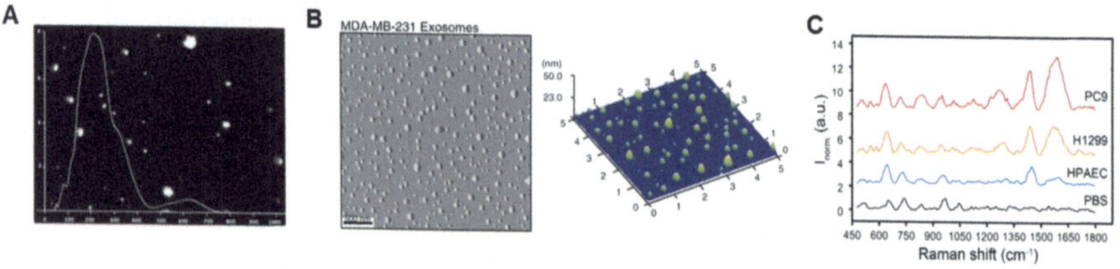

Fig. 15.2 NTA, AFM, and Raman spectroscopy images of EVs. (**a**) Screenshot from video of light scatter of placental vesicles overlaid with a graph of vesicle size and concentration as determined by nanoparticle tracking analysis [19]. (**b**) 2D and 3D AFM images of MDA-MB-231 derived EVs showed individual, regular, fairly rounded structures [20]. (Reprinted from The Lancet, Volume 26, Issue 5, 10 November 2014, Pages 707–721, Copyright (2014), with permission from Elsevier). (**c**). SERS spectra of EVs derived from HPAEC (normal) and H1299, PC9 (lung cancer) cell lines, with PBS as a control [21]. (Reprinted (adapted) with permission from "Correlation between Cancerous Exosomes and Protein Markers Based on Surface-Enhanced Raman Spectroscopy (SERS) and Principal Component Analysis (PCA)." Copyright 2018 American Chemical Society)

2.1.3 Atomic Force Microscopy (AFM)

AFM is a high-resolution imaging technique used for visualizing the surfaces of a wide variety of materials at the nanoscale, including EVs. AFM is a type of scanning probe microscopy in which the image is gathered by touching the surface of samples with a mechanical probe. The piezoelectric element on the probe that facilitates tiny but accurate and precise movement responding to electronic changes enables precise scanning [22]. For AFM imaging, EV samples do not need to be fixed and are typically imaged in their native state. EVs are deposited onto a mica surface, a probe composed of a flexible cantilever and a sharp tip interacts with the EV surface, and the deflection of the cantilever is measured to create an image [23]. The probe scans the sample surface line by line and records the height and other properties of the surface features at the same time [24]. Thus, AFM allows researchers to observe the size, shape, topography, and surface characteristics of EVs. In addition, AFM can also be used for force spectroscopy studies, providing information about the mechanical properties of EVs (Fig. 15.2B).

2.1.4 Raman Spectroscopy

Raman spectroscopy is a spectroscopic technique typically used to determine the vibrational modes of molecules and this technique relies upon inelastic scattering of photons, known as Raman scattering. A source of monochromatic light interacts with molecular vibrations, phonons, and other excitations in the system. The energy of the monochromatic light photons is shifted up and down, which provides information about the vibration modes in the system [25]. It can provide a structural fingerprint by which molecules can be identified in both chemistry and biology. Regarding EV imaging, Raman spectroscopy can identify detailed information and map the spatial distribution of the molecular composition of EV samples [26, 27]. Benefiting from the ability of Raman imaging to distinguish the spatial map molecular components within EVs, Raman imaging can be used to investigate the heterogeneity of EV populations and monitor changes in EV cargos under different conditions, such as disease progression or responses to stimuli [28] (Fig. 15.2C).

2.1.5 Microfluid-Based Imaging

Microfluid-based imaging integrates microfluidics devices and imaging technologies. Microfluidic devices consist of channels and chambers for precisely manipulating small-scale fluid [29]. When combined with imaging technologies, the whole system can visualize and analyze the behavior of fluids, cells, or particles at the nanoscale [29]. Due to these properties, microfluidics-based systems have gained significant attention in the study of EV. Distinguished to other imaging techniques, microfluidics-based systems can also be designed to isolate EVs from complex samples based on their size, charge, surface molecules, and other characteristics by combining them with other isolation methods such as filtration, immunoaffinity capture, or acoustic-based separation [30–32]. Moreover, microfluid-based imaging offers many advantages for imaging EVs, including enhanced sensitivity, minimal sample consumption, integration with multiple analysis techniques, and high-throughput analysis.

Microfluidic devices are usually integrated with imaging systems, such as microscopy or Raman spectroscopy, to enable real-time observation of individual EVs [30]. Moreover, this system can be combined with various analytical techniques, including mass spectrometry and nucleic acid analysis [33, 34]. The integrated system provides comprehensive information about the composition of every individual EV in the population at high throughput. It is particularly valuable to understand the heterogeneity in cargo content and biomarker expression within the EV population. In addition, this imaging system can also be used to capture the dynamic interaction between live cells and EVs and it is a very good tool for investigating EV secretion, uptake and interactions with recipient cells. For example, the imaging flow cytometer Cytek® Amnis® CellStream® from CYTEK can be used to image and analyze single EVs as well as their internalization [35, 36] (Fig. 15.3).

2.2 *Label-Based Imaging*

In addition to the aforementioned label-free imaging methods, EV imaging technology also includes a variety of label-dependent imaging methods, such as fluorescence imaging, bioluminescence imaging, computed tomography scan, magnetic resonance imaging, and radionuclide imaging.

2.2.1 Fluorescence Imaging

As the most common EV imaging method, fluorescence imaging uses a variety of dyes, probes, or other labels to visualize EVs and their characteristics, distribution, and interactions under fluorescence microscopy during different physiological and pathological processes. The commonly used fluorescent probes include fluorescent proteins (e.g., GFP and RFP), lipophilic dyes (e.g., DiI and DiD), membrane permeable dyes (e.g., PKH26), and other regular fluorescent dyes (e.g., cyanine and rhodamine) [37–40]. Recently, an increasing number of fluorescent molecules (e.g., quantum dots and aggregation-induced emission molecules) have been reported that can be used to label EVs [41]. The different types of fluorescent dyes plus different colors allow researchers to label different EV components, whether lipids, proteins, or nucleic acids, in one sample simultaneously.

To modify EVs with these dyes, researchers have developed various strategies. Initially, biologists first proposed inserting fluorescent proteins into the EV marker protein-encoding gene frame of donor cells by using genetic engineering methods to specifically label EVs. The classic scaffold proteins are CD63, Lamp-2b,

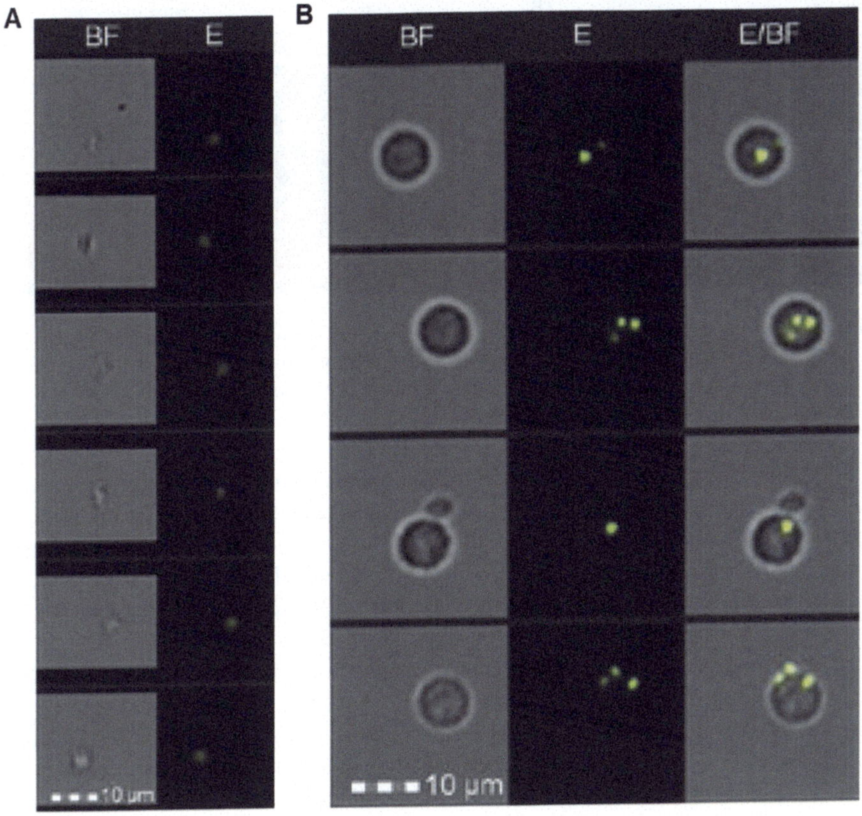

Fig. 15.3 Imaging flow cytometry images of isolated and cell-bound fluorescently labeled EVs. (**a**) EVs labeled with PKH67 were analyzed with the Amnis ImageStream®X cytometer. BF and FITC-fluorescence images are shown. (**b**) PKH67-prelabeled EVs were internalized by peripheral blood mononuclear cells [36]. (The copyright information was obtained from Springer Nature and the relevant information is as follows: "Clark, R. Imaging flow cytometry enhances particle detection sensitivity for extracellular vesicle analysis. Nat Methods 12, i–ii (2015), Springer Nature.")

syntenin, and MFGE8 [42–44]. However, this method faces several challenges, including variable transfection efficacy and time consumption [56]. Therefore, scientists have developed physical and chemical modification methods to overcome these deficiencies. For example, simple incubation of isolated EVs with the lipophilic dye DiI at room temperature for 10 min could label EVs with strong red fluorescent signals [39]. Similarly, using a click reaction allows for the modification of specific molecules in EVs with Cyanine 5. An increasing number of dyes and labeling methods are available, allowing researchers to label an EV sample with multiple fluorescent probes or even use different fluorescent probes to label different proteins in each EV (Fig. 15.4).

On these bases, EV fluorescence imaging relies on a series of fluorescent imaging equipment. An IVIS® Spectrum In Vivo Imaging System is often used to detect

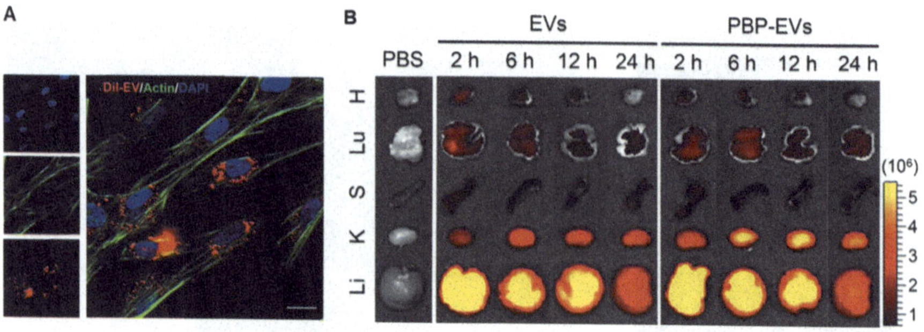

Fig. 15.4 Fluorescent imaging of EVs in vitro and vivo. (**a**) The uptake of DiI-labeled (red) EVs derived from MSCs [40]. Scale bars, 20 μm. (**b**) Organ distributions of EVs and PBP-EVs in renal IRI mice were detected by Cy5.5 radiant efficiency. *H* heart, *Lu* lung, *S* spleen, *K* kidney, *Li* liver [43]

the organ biodistribution of EVs in living animals via fluorescent imaging [43, 45]. Confocal laser scanning microscopy can observe the internalization of fluorescent dye labeled EVs by recipient cells [12]. To visualize a single EV by fluorescence imaging, fluorescent labeling needs to be combined with NTA, nanoflow cytometry, super-resolution microscopy, and other methods [42]. After fluorescent labeling, a subpopulation of EV samples can be distinguished by the above detection techniques. Moreover, super-resolution microscopy is now enabling scientists to assign the localization of various EV markers as being either present on the surface or within the EV as cargo [42, 46]. The Nanoimager from the ONI company offers super-resolution mode, which can resolve and visualize structures down to 20 nm in the XY plane. After the image is captured, the Nanoimager can quantify the individual molecules and resolve the structures via the dSTORM reconstruction process.

2.2.2 Bioluminescence Imaging

Bioluminescence imaging (BLI) is a technique used to visualize cells of interest, vesicles, and biological processes in living organisms [47]. Unlike fluorescence imaging, BLI relies on the detection of light produced by the expression of the bioluminescent protein "luciferase" in cells or tissues. Therefore, the methods for modifying luciferase on EVs are very limited. Initially, scientists directly expressed free luciferase in donor cells, and EVs carried luciferase to achieve BLI through a passive loading process [48]. Then, researchers started to fuse the luciferase protein with EV marker proteins to actively sort luciferase into EVs, which increased the labeling efficiency. For example, a successful technique named "surface display" to generate modified EVs is made possible through cloning the Gaussia luciferase sequence to be "displayed" on the EV surface as a translational fusion to the C1C2 domain of the lactadherin protein, which could be used to trace EVs in vivo by BLI [49]. The detection of EVs in cells or animals by BLI requires the use of an IVIS® Spectrum In Vivo Imaging System or Spectral Instruments Imaging systems. After

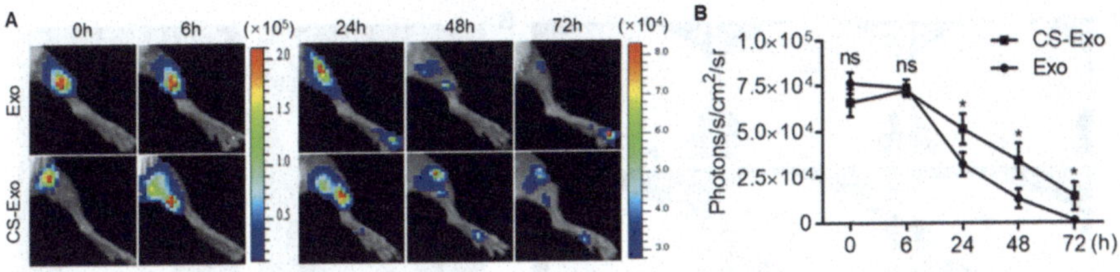

Fig. 15.5 Bioluminescent imaging of EVs in vivo. (**a**). In vivo monitoring of the retention of transplanted chitosan hydrogel loaded EVs or EVs through tracking Gluc signals by BLI analysis. (**b**) Quantitative analysis of Gluc signals indicated retention of EVs. The signal activity was expressed as photons/s/cm²/steradian [52]. (Reprinted (adapted) with permission from "Enhanced Therapeutic Effects of Mesenchymal Stem Cell-Derived Exosomes with an Injectable Hydrogel for Hindlimb Ischemia Treatment." Copyright 2018 American Chemical Society)

incubation or injection of the substrate, luciferase labeled EVs can be visualized by a spectral imaging system both in vitro and in vivo.

Compared with fluorescence imaging, BLI of EVs has distinct advantages and limitations. BLIs do not need external light to excite fluorophores, making them noninvasive and leading to low phototoxicity. This property allows real-time monitoring and repeated measurement of EVs over time in the same animals. Another distinguishing advantage of BLI is that bioluminescence signals are very sensitive and have a very low background density. This is because BLI is based on the photons emitted from a specific enzymatic reaction with a higher signal-to-background ratio [50]. This feature makes BLI labeled EVs much more accurate at very low concentrations, preventing false positive signals. In addition, the intensity of bioluminescent signals can be quantified to estimate the number of EVs. This quantitative approach is often simpler and more accurate than fluorescence imaging because fluorophore concentration and photobleaching may complicate the measurements [51]. However, the limited tissue penetration of bioluminescence signals makes it very difficult to observe labeled EVs in deep tissues or large animals (Fig. 15.5).

2.2.3 Computed Tomography

Computed tomography (CT) scanning is a commonly used medical imaging procedure in which specialized X-ray equipment is used to obtain detailed cross-sectional images of the body. These cross-sectional images with high spatial and temporal resolution, which are reconstructed based on the data collected during the scan, can provide more detailed information than regular X-rays and are commonly used to diagnose and monitor various medical conditions [53]. These properties make CT hopeful for tracking EVs in vivo. However, EVs cannot be detected by CT scanning directly, and they must be labeled with CT contrast agents for visualization [53, 54]. Thus, researchers established a protocol to monitor EVs by using CT scanning through labeling EVs with gold nanoparticles [53, 55, 56]. Gold nanoparticle-labeled EVs can be monitored in vivo by CT to investigate their biodistribution and

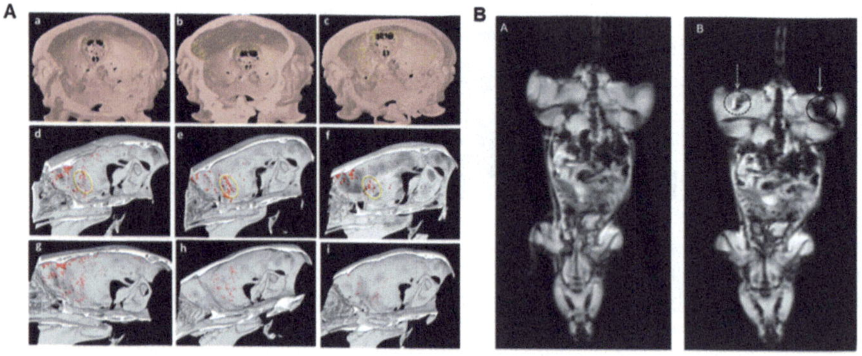

Fig. 15.6 CT and MRI scanning of EVs in vivo. (**a**) In vivo CT imaging of gold nanoparticle-labeled EVs after acute striatal stroke in a mouse model. EVs can be seen to migrate and accumulate at the ischemic brain region at 3 h. (The ischemic region is demarcated in yellow circle.) [54] (Reprinted (adapted) with permission from "In Vivo Neuroimaging of Exosomes Using Gold Nanoparticles." Copyright 2017 American chemical Society). (**b**) In vivo MR images acquired pre-intramuscular (A) and post-intramuscular (B) injections of exosome–MSC/FTH1 (circle and arrow) and exosome–MSCs (dotted circle and arrow) [58]. (Reprinted from Magnetic Resonance Imaging, Vol 68, Tianqi Liu, Yurong Zhu, Ruiting Zhao, Xinhua Wei, Xuegang Xin, Visualization of exosomes from mesenchymal stem cells in vivo by magnetic resonance imaging, Pages 75–82, Copyright (2020), with permission from Elsevier)

migration [57]. However, CT scanning is still in the early stages of development, and further studies are needed to develop additional CT contrast agents and more efficient labeling strategies (Fig. 15.6A).

2.2.4 Magnetic Resonance Imaging

Magnetic resonance imaging (MRI) is an ionizing radiation-free imaging technique in which powerful magnets, radio waves, and computer processing are used to visualize the internal structure of the body at a very high resolution [59]. On an MRI instrument, the hydrogen atoms in the patient's body align with the strong magnetic field of the machine. Then, radiofrequency pulses are applied to temporarily disrupt the alignment of the hydrogen atoms. The radiofrequency signals emitted from the hydrogen atoms then return to their original alignment would be detected and processed by a computer to generate detailed cross-sectional images [60]. Since MRI relies only on hydrogen atoms in the body, EVs cannot be distinguished from other tissues or body fluids by MRI without labeling [61]. To trace EVs via MRI, researchers usually use superparamagnetic iron oxide nanoparticles (SPIONs) or ferritin heavy chain (FTH1) as contrast agents to label EVs [58, 60, 61]. After injection of contrast agent-labeled EVs, the contrast agent causes a signal void in the MR images where the labeled exosomes are present, which can track their distribution, accumulation, and clearance in specific tissues or organs [62]. Because of the high-resolution properties of MRI, using MRI to track EVs is valuable for understanding the pharmacokinetics and biodistribution of EVs in preclinical and clinical applications. However, the stability of contrast agent labeling still needs to be improved to ensure the reliability and reproducibility of EV MRI research (Fig. 15.6B).

2.2.5 Radionuclide Imaging

Radionuclide imaging is an imaging technique in which radionuclides are used to label cells or molecules of interest to diagnose and treat diseases. Unlike X-rays and CT scans, radionuclide imaging does not rely on external radiation sources but requires the introduction of a radiotracer into the body, which can provide information about the structure and function of organs and tissues of interest through the emission of gamma rays or positrons [63]. In the context of EV imaging, radionuclide imaging can be employed to track and visualize the distribution of EVs in preclinical studies. For example, EVs labeled by the common radiotracers ^{99m}Tc and ^{64}Cu can be detected by single-photon emission computed tomography (SPECT) and positron emission tomography (PET), respectively [64–67]. Compared with conventional fluorescent protein or bioluminescent protein labeling, radiotracer labeling is more suitable for evaluating the in vivo distribution of EVs. On the one hand, the radiotracer is much more stable than proteins that can still be detected after EVs are degraded; on the other hand, the tissue penetration depth of the radiotracer is also better than that of proteins [63]. However, because of the radioactivity of the radiotracer, it is crucial to carefully control and monitor the use of radioactive materials to ensure patient safety. In addition, the amount of radiation exposure should be kept as low as reasonably achievable while still capturing the necessary information (Fig. 15.7A).

2.2.6 Photoacoustic Imaging

Photoacoustic imaging (PAI) is a novel noninvasive imaging technique that combines laser-induced ultrasound and optical imaging. The ultrasonic wave is generated by illuminating the target of interest with a pulsed laser and is reconstructed as an image of light energy absorption in tissues, which can be used to visualize structures and functions in biological tissues at high resolution [69]. When PAI is used in EV research, the process typically involves the following steps. First, EVs are labeled with contrast agents that can absorb laser light. The contrast agents could be organic dyes, gold particles, carbon nanostructures, semiconducting nanoparticles, or fluorescent proteins [70–72]. Second, the labeled EVs are irradiated by laser light. When the labeling contrast agents absorb the laser energy agent absorbs the laser energy and rapidly convert it to localized heat, it generates a brief but intense ultrasound wave. Third, the generated ultrasound waves are captured by ultrasound detectors and reconstructed into images that represent the distribution of the labeled EVs in the tissues [73].

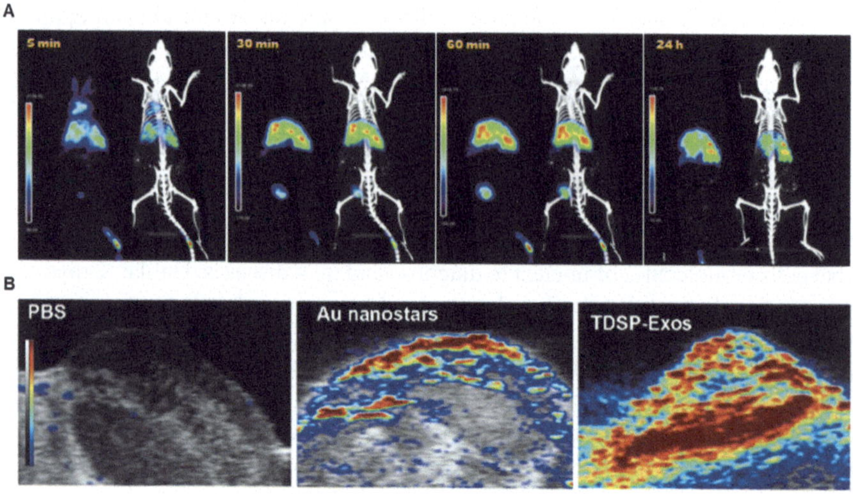

Fig. 15.7 SPECT and PAI images of EVs in vivo. (**a**) In vivo SPECT and CT imaging of [^{99m}Tc]-EVs 5 min, 30 min, 60 min, and 24 h post I.V. injection [66]. (**b**) Representative PAI images of murine tumors at 24 h post injection of PBS, Au nanostars, or tumor cell-derived stellate plasmonic EVs (TDSP-EVs) [68]. (Reprinted (adapted) with permission from "Stellate Plasmonic Exosomes for Penetrative Targeting Tumor NIR-II Thermo-Radiotherapy." Copyright 2020 American Chemical Society)

This emerging imaging technique allows for visualizing EVs in tissues with high sensitivity and spatial resolution. Moreover, PAI can selectively image molecules with specific optical absorption properties, which allows high-contrast imaging at the molecular level. Additionally, compared to traditional optical imaging methods, PAI is more suitable for imaging EVs at greater depths because it can penetrate deeper into tissues [74]. Unlike X-ray-based imaging, PAI does not require ionizing radiation, which means that PAI is a much safer imaging modality. Moreover, PAI can provide complementary information when combined with other imaging technologies, such as conventional ultrasound imaging and MRI [75]. Although the PAI of EVs shows promise, further research is needed to optimize the technology and expand its applications [10]. Researchers are working on developing novel contrast agents to enhance their capabilities and exploring ways to improve imaging depth and resolution (Fig. 15.7B).

3 Applications of EV Imaging in Theranostics

EV imaging is a rapidly evolving field with diverse applications in both research and clinical settings. The imaging techniques we discussed above provide valuable tools for investigating the biogenesis, release, and uptake of EVs under natural conditions and during many disease processes. There are several applications of EV imaging in theranostics.

3.1 Cancer Theranostics

As for cancer theranostics, EV imaging seamlessly integrates cancer early diagnosis, targeted drug delivery, and real-time treatment monitoring into a single framework. In the realm of cancer early diagnosis, EV imaging has emerged as a groundbreaking tool. Circulating tumor cell-derived EVs carry specific markers that can act as diagnostic biomarkers, enabling the identification of tumors with enhanced sensitivity [76]. By leveraging sophisticated imaging techniques, such as super-resolution fluorescent imaging or microfluid-based imaging, researchers can detect the unique biomolecular signatures carried by tumor cell-derived EVs, providing a noninvasive strategy for diagnosing cancer at nascent stages [75–78]. EV-based cancer screening has made some progress, for example, several companies have captured and detected EVs from blood to help potential prostate cancer patients decide whether they should undergo biopsy [79]. This early diagnostic capability of EV imaging not only facilitates timely intervention but also lays the foundation for personalized treatment tailored to the molecular profile of the detected cancer.

In addition to early diagnosis, EV imaging also serves as an invaluable companion of modern therapeutics in the field of monitoring treatment response. Researchers can assess the dynamics of treatment-induced changes by real-time visualization and analysis of EVs posttreatment [80]. Combining EV imaging with other detection technologies, such as nanoflow cytometry or high-throughput sequencing, allows researchers to monitor alterations in the EV profile, including changes in protein expression or nucleic acid content, providing a dynamic and comprehensive overview of treatment response and allowing real-time adjustments. Using EV imaging in cancer theranostics not only enhances the precision of personalized medicine but also contributes to our understanding of the molecular landscape of treatment response.

Furthermore, EV imaging plays a critical role in the development of targeted drug delivery strategies during cancer therapeutics. Benefiting from their natural ability to carry and transfer biomolecules from donor cells to recipient cells, EVs can be used as effective carriers to load and deliver therapeutic agents directly to tumor cells. To achieve targeted drug delivery, minimize off-target effects and maximize therapeutic efficacy, researchers have engineered EVs to carry therapeutic agents, such as specific drugs or proteins and RNA molecules [78]. By incorporating imaging reporters into EV cargos, the release kinetics of therapeutic agents can be tracked, offering insights into the effectiveness of treatment [81, 82]. Moreover, imaging techniques assist in validating the accuracy of targeted drug delivery, enabling the visualization of interactions between EVs and tumor cells [83]. The multiple roles of EV imaging in early diagnosis, treatment response monitoring and targeted drug delivery tracing underscore its significance in the pursuit of more precise, personalized, and effective cancer theranostics.

3.2 Neurodegenerative Disease

In neurodegenerative disease research, EV imaging has emerged as a powerful tool for identifying biomarkers and therapeutic interventions, offering valuable insight into understanding and addressing neurodegenerative diseases [84]. Since EVs are derived from cells and contain a diverse range of proteins, nucleic acids, and lipids that can reflect the physiological and pathological states of donor cells, EVs are potential biomarkers for neurodegenerative diseases, such as Alzheimer's disease and Parkinson's disease [85–89]. Moreover, EV imaging enables scientists to identify specific EVs associated with disease progression. By using multiple imaging and analysis techniques, scientists can study the composition of EVs isolated from cerebrospinal fluid and blood to identify early indicators of neurodegenerative disorders, which is critical for early diagnosis and intervention [87, 90].

Moreover, in advancing therapeutic interventions for neurodegenerative diseases, EV imaging also plays an irreplaceable role. Recently, EVs have gained attention as potential carriers for targeted drug delivery to the central nervous system because of their ability to cross the blood–brain barrier [91, 92]. By loading therapeutic cargos and imaging agents into EVs simultaneously, researchers can track their biodistribution and optimize drug efficacy in the central nervous system using imaging techniques, which can help increase the precision of treatment [93, 94]. This imaging-assisted targeted therapy minimizes off-target effects and helps to improve the overall effectiveness of therapeutic interventions. Moreover, EV imaging can also be used to monitor treatment responses, noninvasively providing real-time feedback on the impact of therapy within the brain [95]. Like their application in cancer theronostics, the application of EV imaging in neurodegenerative diseases offers dual superiority in both biomarker discovery and therapeutic interventions, which will help us understand neurodegenerative diseases more deeply and improve clinical outcomes for each affected individual.

3.3 Tissue Injury

Like in the above applications, EVs can also be considered as biomarkers of tissue injury. Thus, the EV-based noninvasive imaging is a transformative technique for diagnosing tissue injury. After the injury, cells at the injury site undergoing stress or damage release EVs carrying a specific biomolecular profile to reflect the injury type and severity. All of these features, including the ability of EVs to traverse various biological barriers and the lipid bilayer of EVs, can protect cargos under harsh conditions, enhancing the accessibility of EVs for sampling [96]. These characteristics render EVs particularly attractive for minimally invasive diagnostic procedures, such as liquid biopsies, where the analysis of circulating EVs in bodily fluids can provide valuable information about the status of distant tissues [97, 98]. Researchers can obtain information of potential tissue injuries by analyzing individual EVs and their cargos via EV imaging modalities and other downstream analysis methods, allowing timely and auxiliary diagnosis [98]. This noninvasive approach holds promise for personalized medicine, offering a means to tailor regenerative strategies based on the specific molecular signatures detected through EV imaging.

Another key application of EV imaging in tissue injury is a visualization of EV-based therapies. In addition to their prominent stability, long circulation half-life and favorable safety profile, EVs possess unique homing features, which means that EVs prefer to target the cell type from which they are released [99, 100]. In addition, researchers can also engineer desired ligands on the surface of drug loaded EVs to target sites where the corresponding cell surface biomarkers are overexpressed [99]. Through imaging techniques, researchers can track the biodistribution of EVs, ensuring that therapeutic payloads reach the intended site of action. Moreover, it is possible to track EVs in vivo while indicating the location and severity of tissue injury [43], which allows for the integration of diagnosis and treatment via a single process. The combination of EV-based drug delivery and imaging technologies holds great promise for developing more precise and personalized therapeutic strategies for tissue regeneration.

4 Future Directions and Outlook

The future directions of EV imaging include advancing the investigation of cell-to-cell communication, early disease diagnosis, and precise therapeutic interventions. As technology continues to evolve, researchers should develop novel imaging contrast agents and imaging techniques to enhance the resolution and specificity of EV imaging both in vivo and in vitro. High-resolution imaging modalities, such as super-resolution microscopy and advanced imaging contrast agents, will visualize single EVs at the nanoscale, providing unprecedented details about their biogenesis, cargos, and interactions with recipient cells [38]. The integration of artificial intelligence and machine learning algorithms with EV imaging is also anticipated to play a remarkable role in analyzing large datasets generated from EV imaging studies, facilitating the identification of subtle patterns and correlations that may escape human observation.

Additionally, developing multiplexed EV imaging approaches is expected to be an important focus in future EV-related research. Multiplexed imaging makes it possible to simultaneously detect multiple EV-derived signals, including the size distribution, concentration, and protein and nucleic acid composition, providing a comprehensive and detailed molecular profile. This comprehensive information is very useful for more accurate disease diagnosis and monitoring, especially in complex conditions such as cancer and neurodegenerative diseases. A single-EV analysis technique based on a microfluid chamber and immune fluorescence imaging was designed for cancer EV analysis, which can analyze the EV population and perform robust multiplex protein biomarker measurements of individual EVs [30, 32]. The combination of fluorescent imaging and bioluminescent imaging used in EV therapy can both trace EV biodistribution and beneficial cargos in vivo [101, 102]. An increasing number of multiplexed EV imaging strategies, such as the above examples, will be designed to optimize EV therapy for enhanced precision, biodistribution, and therapeutic efficacy.

In conclusion, the future of EV imaging is characterized by the convergence of cutting-edge technologies, including high-resolution microscopy, artificial intelligence, multiplexed imaging, and noninvasive in vivo tracking. These advancements are poised to revolutionize our knowledge of the biological features of EVs and open new avenues for their diagnostic and therapeutic applications. As the field continues to evolve, interdisciplinary collaborations between researchers in biology, imaging sciences, and engineering will be crucial for realizing the full potential of EV imaging in healthcare and regenerative medicine.

References

1. Tkach M, Thery C. Communication by extracellular vesicles: where we are and where we need to go. Cell. 2016;164(6):1226–32. https://doi.org/10.1016/j.cell.2016.01.043.
2. van Niel G, D'Angelo G, Raposo G. Shedding light on the cell biology of extracellular vesicles. Nat Rev Mol Cell Biol. 2018;19(4):213–28. https://doi.org/10.1038/nrm.2017.125.
3. Andaloussi SEL, Mager I, Breakefield XO, Wood MJ. Extracellular vesicles: biology and emerging therapeutic opportunities. Nat Rev Drug Discov. 2013;12(5):347–57. https://doi.org/10.1038/nrd3978.
4. Kalluri R, LeBleu VS. The biology, function, and biomedical applications of exosomes. Science. 2020;367(6478):eaau6977. https://doi.org/10.1126/science.aau6977.
5. Zhang Y, Wu W, Pan X, Wang Y, Wu C, Lu L, et al. Extracellular vesicles as novel therapeutic targets and diagnosis markers. Extracell Vesicle. 2022;1:100017. https://doi.org/10.1016/j.vesic.2022.100017.
6. Kamerkar S, LeBleu VS, Sugimoto H, Yang S, Ruivo CF, Melo SA, et al. Exosomes facilitate therapeutic targeting of oncogenic KRAS in pancreatic cancer. Nature. 2017;546(7659):498–503. https://doi.org/10.1038/nature22341.
7. Aghajani Nargesi A, Lerman LO, Eirin A. Mesenchymal stem cell-derived extracellular vesicles for kidney repair: current status and looming challenges. Stem Cell Res Ther. 2017;8(1):273. https://doi.org/10.1186/s13287-017-0727-7.
8. Armstrong JPK, Stevens MM. Strategic design of extracellular vesicle drug delivery systems. Adv Drug Deliv Rev. 2018;130:12–6. https://doi.org/10.1016/j.addr.2018.06.017.
9. Todorova D, Simoncini S, Lacroix R, Sabatier F, Dignat-George F. Extracellular vesicles in angiogenesis. Circ Res. 2017;120(10):1658–73. https://doi.org/10.1161/CIRCRESAHA.117.309681.
10. Betzer O, Barnoy E, Sadan T, Elbaz I, Braverman C, Liu Z, et al. Advances in imaging strategies for in vivo tracking of exosomes. Wiley Interdiscip Rev: Nanomed Nanobiotechnol. 2020;12(2):e1594. https://doi.org/10.1002/wnan.1594.

11. Choi H, Mun JY. Structural analysis of Exosomes using different types of electron microscopy. Appl Microsc. 2017;47(3):171–5. https://doi.org/10.9729/AM.2017.47.3.171.
12. Zhang K, Chen S, Sun H, Wang L, Li H, Zhao J, et al. In vivo two-photon microscopy reveals the contribution of Sox9(+) cell to kidney regeneration in a mouse model with extracellular vesicle treatment. J Biol Chem. 2020;295(34):12203–13. https://doi.org/10.1074/jbc.RA120.012732.
13. Emelyanov A, Shtam T, Kamyshinsky R, Garaeva L, Verlov N, Miliukhina I, et al. Cryo-electron microscopy of extracellular vesicles from cerebrospinal fluid. PLoS One. 2020;15(1):e0227949. https://doi.org/10.1371/journal.pone.0227949.
14. Jafarzadeh N, Gholampour MA, Alivand M-R, Kavousi S, Arzi L, Rad F, et al. CML derived exosomes promote tumor favorable functional performance in T cells. BMC Cancer. 2021;21(1):1002. https://doi.org/10.1186/s12885-021-08734-3.
15. Jung MK, Mun JY. Sample preparation and imaging of exosomes by transmission electron microscopy. J Vis Exp. 2018;131:e56482. https://doi.org/10.3791/56482.
16. Sokolova V, Ludwig A-K, Hornung S, Rotan O, Horn PA, Epple M, et al. Characterisation of exosomes derived from human cells by nanoparticle tracking analysis and scanning electron microscopy. Colloids Surf B Biointerfaces. 2011;87(1):146–50. https://doi.org/10.1016/j.colsurfb.2011.05.013.
17. Mehdiani A, Maier A, Pinto A, Barth M, Akhyari P, Lichtenberg A. An innovative method for exosome quantification and size measurement. J Vis Exp. 2015;95:e50974. https://doi.org/10.3791/50974.
18. Zhou M, Weber SR, Zhao Y, Chen H, Sundstrom JM. Chapter 2—Methods for exosome isolation and characterization. In: Edelstein L, Smythies J, Quesenberry P, Noble D, editors. Exosomes. Cambridge: Academic Press; 2020. p. 23–38.
19. Dragovic RA, Gardiner C, Brooks AS, Tannetta DS, Ferguson DJP, Hole P, et al. Sizing and phenotyping of cellular vesicles using Nanoparticle Tracking Analysis. Nanomedicine. 2011;7(6):780–8. https://doi.org/10.1016/j.nano.2011.04.003.
20. Melo Sonia A, Sugimoto H, O'Connell Joyce T, Kato N, Villanueva A, Vidal A, et al. Cancer exosomes perform cell-independent microRNA biogenesis and promote tumorigenesis. Cancer Cell. 2014;26(5):707–21. https://doi.org/10.1016/j.ccell.2014.09.005.
21. Shin H, Jeong H, Park J, Hong S, Choi Y. Correlation between cancerous exosomes and protein markers based on Surface-Enhanced Raman Spectroscopy (SERS) and Principal Component Analysis (PCA). ACS Sensors. 2018;3(12):2637–43. https://doi.org/10.1021/acssensors.8b01047.
22. Sharma S, Rasool HI, Palanisamy V, Mathisen C, Schmidt M, Wong DT, et al. Structural-mechanical characterization of nanoparticle exosomes in human saliva, using correlative AFM, FESEM, and force spectroscopy. ACS Nano. 2010;4(4):1921–6. https://doi.org/10.1021/nn901824n.
23. Yurtsever A, Yoshida T, Badami Behjat A, Araki Y, Hanayama R, Fukuma T. Structural and mechanical characteristics of exosomes from osteosarcoma cells explored by 3D-atomic force microscopy. Nanoscale. 2021;13(13):6661–77. https://doi.org/10.1039/D0NR09178B.
24. Sharma S, LeClaire M, Gimzewski JK. Ascent of atomic force microscopy as a nanoanalytical tool for exosomes and other extracellular vesicles. Nanotechnology. 2018;29(13):132001. https://doi.org/10.1088/1361-6528/aaab06.
25. Park J, Hwang M, Choi B, Jeong H, Jung J-h, Kim HK, et al. Exosome classification by pattern analysis of surface-enhanced Raman spectroscopy data for lung cancer diagnosis. Anal Chem. 2017;89(12):6695–701. https://doi.org/10.1021/acs.analchem.7b00911.
26. Stremersch S, Marro M, Pinchasik B-E, Baatsen P, Hendrix A, De Smedt SC, et al. Identification of individual exosome-like vesicles by surface enhanced Raman spectroscopy. Small. 2016;12(24):3292–301. https://doi.org/10.1002/smll.201600393.
27. Li J, Li Y, Li P, Zhang Y, Du L, Wang Y, et al. Exosome detection via surface-enhanced Raman spectroscopy for cancer diagnosis. Acta Biomater. 2022;144:1–14. https://doi.org/10.1016/j.actbio.2022.03.036.

28. Russo M, Tirinato L, Scionti F, Coluccio ML, Perozziello G, Riillo C, et al. Raman spectroscopic stratification of multiple myeloma patients based on exosome profiling. ACS Omega. 2020;5(47):30436–43. https://doi.org/10.1021/acsomega.0c03813.
29. Guo S-C, Tao S-C, Dawn H. Microfluidics-based on-a-chip systems for isolating and analysing extracellular vesicles. J Extracell Vesicles. 2018;7(1):1508271. https://doi.org/10.1080/20013078.2018.1508271.
30. Lee K, Fraser K, Ghaddar B, Yang K, Kim E, Balaj L, et al. Multiplexed profiling of single extracellular vesicles. ACS Nano. 2018;12(1):494–503. https://doi.org/10.1021/acsnano.7b07060.
31. Bai J, Wei X, Zhang X, Wu C, Wang Z, Chen M, et al. Microfluidic strategies for the isolation and profiling of exosomes. TrAC, Trends Anal Chem. 2023;158:116834. https://doi.org/10.1016/j.trac.2022.116834.
32. Wang J, Ma P, Kim DH, Liu B-F, Demirci U. Towards microfluidic-based exosome isolation and detection for tumor therapy. Nano Today. 2021;37:101066. https://doi.org/10.1016/j.nantod.2020.101066.
33. Han Z, Peng C, Yi J, Zhang D, Xiang X, Peng X, et al. Highly efficient exosome purification from human plasma by tangential flow filtration based microfluidic chip. Sens Actuators B. 2021;333:129563. https://doi.org/10.1016/j.snb.2021.129563.
34. Lu Y, Tong Z, Wu Z, Jian X, Zhou L, Qiu S, et al. Multiple exosome RNA analysis methods for lung cancer diagnosis through integrated on-chip microfluidic system. Chin Chem Lett. 2022;33(6):3188–92. https://doi.org/10.1016/j.cclet.2021.12.045.
35. Tertel T, Giebel B. Amnis ImageStream—analysis of individual extracellular vesicles by imaging flow cytometry. Cytotherapy. 2020;22(5, Supplement):S55. https://doi.org/10.1016/j.jcyt.2020.03.075.
36. Clark RT. Imaging flow cytometry enhances particle detection sensitivity for extracellular vesicle analysis. Nat Methods. 2015;12(4):i–ii. https://doi.org/10.1038/nmeth.f.380.
37. Bebelman MP, Bun P, Huveneers S, van Niel G, Pegtel DM, Verweij FJ. Real-time imaging of multivesicular body–plasma membrane fusion to quantify exosome release from single cells. Nat Protoc. 2020;15(1):102–21. https://doi.org/10.1038/s41596-019-0245-4.
38. Verweij FJ, Balaj L, Boulanger CM, Carter DRF, Compeer EB, D'Angelo G, et al. The power of imaging to understand extracellular vesicle biology in vivo. Nat Methods. 2021;18(9):1013–26. https://doi.org/10.1038/s41592-021-01206-3.
39. Banerjee A, Lino M, Jesus C, Ribeiro Q, Abrunhosa A, Ferreira L. Imaging platforms to dissect the in vivo communication, biodistribution and controlled release of extracellular vesicles. J Control Release. 2023;360:549–63. https://doi.org/10.1016/j.jconrel.2023.06.039.
40. Li H, Huang H, Chen X, Chen S, Yu L, Wang C, et al. The delivery of hsa-miR-11401 by extracellular vesicles can relieve doxorubicin-induced mesenchymal stem cell apoptosis. Stem Cell Res Ther. 2021;12(1):77. https://doi.org/10.1186/s13287-021-02156-5.
41. Pan S, Pei L, Zhang A, Zhang Y, Zhang C, Huang M, et al. Passion fruit-like exosome-PMA/Au-BSA@Ce6 nanovehicles for real-time fluorescence imaging and enhanced targeted photodynamic therapy with deep penetration and superior retention behavior in tumor. Biomaterials. 2020;230:119606. https://doi.org/10.1016/j.biomaterials.2019.119606.
42. Chen C, Zong S, Wang Z, Lu J, Zhu D, Zhang Y, et al. Imaging and intracellular tracking of cancer-derived exosomes using single-molecule localization-based super-resolution microscope. ACS Appl Materi Interf. 2016;8(39):25825–33. https://doi.org/10.1021/acsami.6b09442.
43. Zhang K, Li R, Chen X, Yan H, Li H, Zhao X, et al. Renal endothelial cell-targeted extracellular vesicles protect the kidney from ischemic injury. Adv Sci. 2023;10(3):2204626. https://doi.org/10.1002/advs.202204626.
44. Salunkhe S, Dheeraj BM, Chitkara D, Mittal A. Surface functionalization of exosomes for target-specific delivery and in vivo imaging & tracking: strategies and significance. J Control Release. 2020;326:599–614. https://doi.org/10.1016/j.jconrel.2020.07.042.

45. Li R, Wang C, Zhou M, Liu Y, Chen S, Chai Z, et al. Heparan sulfate proteoglycan-mediated internalization of extracellular vesicles ameliorates liver fibrosis by targeting hepatic stellate cells. Extracell Vesicle. 2022;1:100018. https://doi.org/10.1016/j.vesic.2022.100018.
46. Chen C, Zong S, Wang Z, Lu J, Zhu D, Zhang Y, et al. Visualization and intracellular dynamic tracking of exosomes and exosomal miRNAs using single molecule localization microscopy. Nanoscale. 2018;10(11):5154–62. https://doi.org/10.1039/C7NR08800K.
47. Zhang K, Wang C, Wang R, Chen S, Li Z. Dual bioluminescence imaging of tumor progression and angiogenesis. J Vis Exp. 2019;150:e59763. https://doi.org/10.3791/59763.
48. Zhu L, Kalimuthu S, Gangadaran P, Oh JM, Lee HW, Baek SH, et al. Exosomes derived from natural killer cells exert therapeutic effect in melanoma. Theranostics. 2017;7(10):2732–45. https://doi.org/10.7150/thno.18752.
49. Zhang K, Li Z. Molecular imaging of therapeutic effect of Mesenchymal stem cell-derived exosomes for hindlimb ischemia treatment. Methods Mol Biol. 2019;2150:213–25. https://doi.org/10.1007/7651_2019_221.
50. Badr CE, Tannous BA. Bioluminescence imaging: progress and applications. Trends Biotechnol. 2011;29(12):624–33. https://doi.org/10.1016/j.tibtech.2011.06.010.
51. Ozawa T, Yoshimura H, Kim SB. Advances in fluorescence and bioluminescence imaging. Anal Chem. 2013;85(2):590–609. https://doi.org/10.1021/ac3031724.
52. Zhang K, Zhao X, Chen X, Wei Y, Du W, Wang Y, et al. Enhanced therapeutic effects of mesenchymal stem cell-derived exosomes with an injectable hydrogel for hindlimb ischemia treatment. ACS Appl Mater Interf. 2018;10(36):30081–91. https://doi.org/10.1021/acsami.8b08449.
53. Gong L, Weng Y, Zhou W, Zhang K, Li W, Jiang J, et al. In vivo CT imaging of gold nanoparticle-labeled exosomes in a myocardial infarction mouse model. Ann Transl Med. 2021;9(6):504. https://doi.org/10.21037/atm-21-981.
54. Betzer O, Perets N, Angel A, Motiei M, Sadan T, Yadid G, et al. In vivo neuroimaging of exosomes using gold nanoparticles. ACS Nano. 2017;11(11):10883–93. https://doi.org/10.1021/acsnano.7b04495.
55. Guo S, Betzer O, Perets N, Landau S, Offen D, Popovtzer R, et al. Extracellular vesicles tracking and quantification using CT and optical imaging in rats. Bio-Protocol. 2020;10(11):e3635. https://doi.org/10.21769/BioProtoc.3635.
56. Betzer O, Perets N, Barnoy E, Offen D, Popovtzer R. Labeling and tracking exosomes within the brain using gold nanoparticles. In: Nanoscale imaging, sensing, and actuation for biomedical applications XV: SPIE; 2018. p. 164–74.
57. Guo S, Betzer O, Perets N, Landau S, Offen D, Popovtzer R, et al. Extracellular vesicles tracking and quantification using CT and optical imaging in rats. Bio Protoc. 2020;10(11):e3635.
58. Liu T, Zhu Y, Zhao R, Wei X, Xin X. Visualization of exosomes from mesenchymal stem cells in vivo by magnetic resonance imaging. Magn Reson Imaging. 2020;68:75–82. https://doi.org/10.1016/j.mri.2020.02.001.
59. Busato A, Bonafede R, Bontempi P, Scambi I, Schiaffino L, Benati D, et al. Magnetic resonance imaging of ultrasmall superparamagnetic iron oxide-labeled exosomes from stem cells: a new method to obtain labeled exosomes. Int J Nanomed. 2016;11:2481–90. https://doi.org/10.2147/IJN.S104152.
60. Hu L, Wickline SA, Hood JL. Magnetic resonance imaging of melanoma exosomes in lymph nodes. Magn Reson Med. 2015;74(1):266–71. https://doi.org/10.1002/mrm.25376.
61. Busato A, Bonafede R, Bontempi P, Scambi I, Schiaffino L, Benati D, et al. Labeling and magnetic resonance imaging of exosomes isolated from adipose stem cells. Curr Protoc Cell Biol. 2017;75(1):3.44.1–3.15. https://doi.org/10.1002/cpcb.23.
62. Liu T, Li Z, Li X, Zhao R, Wei X, Wang Z, et al. In vivo visualization of murine melanoma cells B16-derived exosomes through magnetic resonance imaging. Biochim Biophys Acta Gen Subj. 2022;1866(2):130062. https://doi.org/10.1016/j.bbagen.2021.130062.

63. Almeida S, Santos L, Falcão A, Gomes C, Abrunhosa A. In vivo tracking of extracellular vesicles by nuclear imaging: advances in radiolabeling strategies. Int J Mol Sci. 2020;21(24):9443. https://doi.org/10.3390/ijms21249443.
64. Molavipordanjani S, Khodashenas S, Abedi SM, Moghadam MF, Mardanshahi A, Hosseinimehr SJ. 99mTc-radiolabeled HER2 targeted exosome for tumor imaging. Eur J Pharm Sci. 2020;148:105312. https://doi.org/10.1016/j.ejps.2020.105312.
65. Hwang DW, Choi H, Jang SC, Yoo MY, Park JY, Choi NE, et al. Noninvasive imaging of radiolabeled exosome-mimetic nanovesicle using 99mTc-HMPAO. Sci Rep. 2015;5(1):15636. https://doi.org/10.1038/srep15636.
66. González MI, Martín-Duque P, Desco M, Salinas B. Radioactive labeling of milk-derived exosomes with 99mTc and in vivo tracking by SPECT imaging. Nanomaterials. 2020;10(6):1062. https://doi.org/10.3390/nano10061062.
67. Khan AA, R TMdR. Radiolabelling of extracellular vesicles for PET and SPECT imaging. Nano. 2021;5(3):256–74. https://doi.org/10.7150/ntno.51676.
68. Zhu D, Lyu M, Huang Q, Suo M, Liu Y, Jiang W, et al. Stellate plasmonic exosomes for penetrative targeting tumor NIR-II thermo-radiotherapy. ACS Appl Mater Interf. 2020;12(33):36928–37. https://doi.org/10.1021/acsami.0c09969.
69. Attia ABE, Balasundaram G, Moothanchery M, Dinish US, Bi R, Ntziachristos V, et al. A review of clinical photoacoustic imaging: current and future trends. Photoacoustics. 2019;16:100144. https://doi.org/10.1016/j.pacs.2019.100144.
70. Han S, Lee D, Kim S, Kim HH, Jeong S, Kim J. Contrast agents for photoacoustic imaging: a review focusing on the wavelength range. Biosensors. 2022;12(8):594. https://doi.org/10.3390/bios12080594.
71. Piao YJ, Kim HS, Moon WK. Noninvasive photoacoustic imaging of dendritic cell stimulated with tumor cell-derived exosome. Mol Imaging Biol. 2020;22(3):612–22. https://doi.org/10.1007/s11307-019-01410-w.
72. Jang Y, Kim H, Yoon S, Lee H, Hwang J, Jung J, et al. Exosome-based photoacoustic imaging guided photodynamic and immunotherapy for the treatment of pancreatic cancer. J Controlled Release. 2021;330:293–304. https://doi.org/10.1016/j.jconrel.2020.12.039.
73. Ma N, Wu C, Meng Z. In vivo imaging and tracking of exosomes for theranostics. J Innov Opt Health Sci. 2021;14(06):2130005. https://doi.org/10.1142/S1793545821300056.
74. Sridharan B, Lim HG. Exosomes and ultrasound: the future of theranostic applications. Mater Today Bio. 2023;19:100556. https://doi.org/10.1016/j.mtbio.2023.100556.
75. Shi Y, Wang X, Zhang S, Yin H, Fan H, Tang Y, et al. Research progress in in vivo tracing technology for extracellular vesicles. Extracell Vesicles Circul Nucleic Acids. 2023;4(4):684–97.
76. Mohammadi M, Zargartalebi H, Salahandish R, Aburashed R, Wey Yong K, Sanati-Nezhad A. Emerging technologies and commercial products in exosome-based cancer diagnosis and prognosis. Biosens Bioelectr. 2021;183:113176. https://doi.org/10.1016/j.bios.2021.113176.
77. Zhao J, Xu L, Yang D, Tang H, Chen Y, Zhang X, et al. Exosome-driven liquid biopsy for breast cancer: recent advances in isolation, biomarker identification and detection. Extracell Vesicle. 2022;1:100006. https://doi.org/10.1016/j.vesic.2022.100006.
78. Wang S, Lin Y, Zhang Y, Qiu X, Pan Y, Yeung S-CJ, et al. Complex RNA world in small extracellular vesicles for liquid biopsy in cancer management. Extracell Vesicle. 2022;1:100015. https://doi.org/10.1016/j.vesic.2022.100015.
79. Li Q, Wang Y, Ling L, Qiao L, Chen H, Ding C, et al. Rapid and specific detection nanoplatform of serum exosomes for prostate cancer diagnosis. Microchimica Acta. 2021;188(8):283. https://doi.org/10.1007/s00604-021-04934-7.
80. Zhang S, Li D, Liu Y, Qin C, Tong L, Xu L. Multifunctional exosome-driven pancreatic cancer diagnostics and therapeutics. Extracell Vesicle. 2023;2:100022. https://doi.org/10.1016/j.vesic.2023.100022.

81. Qiao L, Hu S, Huang K, Su T, Li Z, Vandergriff A, et al. Tumor cell-derived exosomes home to their cells of origin and can be used as Trojan horses to deliver cancer drugs. Theranostics. 2020;10(8):3474–87. https://doi.org/10.7150/thno.39434.
82. Fang W, Li L, Lin Z, Zhang Y, Jing Z, Li Y, et al. Engineered IL-15/IL-15Rα-expressing cellular vesicles promote T cell anti-tumor immunity. Extracell Vesicle. 2023;2:100021. https://doi.org/10.1016/j.vesic.2022.100021.
83. Zhang J, Ji C, Zhang H, Shi H, Mao F, Qian H, et al. Engineered neutrophil-derived exosome-like vesicles for targeted cancer therapy. Sci Adv. 2022;8(2):eabj8207. https://doi.org/10.1126/sciadv.abj8207.
84. Sarko DK, McKinney CE. Exosomes: origins and therapeutic potential for neurodegenerative disease. Front Neurosci. 2017;11:82. https://doi.org/10.3389/fnins.2017.00082.
85. Ellison TJ, Stice SL, Yao Y. Therapeutic and diagnostic potential of extracellular vesicles in amyotrophic lateral sclerosis. Extracell Vesicle. 2023;2:100019. https://doi.org/10.1016/j.vesic.2022.100019.
86. Malm T, Loppi S, Kanninen KM. Exosomes in Alzheimer's disease. Neurochem Int. 2016;97:193–9. https://doi.org/10.1016/j.neuint.2016.04.011.
87. Van Giau V, An SSA. Emergence of exosomal miRNAs as a diagnostic biomarker for Alzheimer's disease. J Neurol Sci. 2016;360:141–52. https://doi.org/10.1016/j.jns.2015.12.005.
88. Wu X, Zheng T, Zhang B. Exosomes in Parkinson's disease. Neurosci Bull. 2017;33(3):331–8. https://doi.org/10.1007/s12264-016-0092-z.
89. Yuan L, Li J-Y. Exosomes in Parkinson's disease: current perspectives and future challenges. ACS Chem Neurosci. 2019;10(2):964–72. https://doi.org/10.1021/acschemneuro.8b00469.
90. Kandimalla R, Saeed M, Tyagi N, Gupta RC, Aqil F. Exosome-based approaches in the management of Alzheimer's disease. Neurosci Biobeh Rev. 2023;144:104974. https://doi.org/10.1016/j.neubiorev.2022.104974.
91. Haney MJ, Klyachko NL, Zhao Y, Gupta R, Plotnikova EG, He Z, et al. Exosomes as drug delivery vehicles for Parkinson's disease therapy. J Controlled Release. 2015;207:18–30. https://doi.org/10.1016/j.jconrel.2015.03.033.
92. Xu M, Feng T, Liu B, Qiu F, Xu Y, Zhao Y, et al. Engineered exosomes: desirable target-tracking characteristics for cerebrovascular and neurodegenerative disease therapies. Theranostics. 2021;11(18):8926–44. https://doi.org/10.7150/thno.62330.
93. Kojima R, Bojar D, Rizzi G, Hamri GC-E, El-Baba MD, Saxena P, et al. Designer exosomes produced by implanted cells intracerebrally deliver therapeutic cargo for Parkinson's disease treatment. Nat Commun. 2018;9(1):1305. https://doi.org/10.1038/s41467-018-03733-8.
94. Qu M, Lin Q, Huang L, Fu Y, Wang L, He S, et al. Dopamine-loaded blood exosomes targeted to brain for better treatment of Parkinson's disease. J Controlled Release. 2018;287:156–66. https://doi.org/10.1016/j.jconrel.2018.08.035.
95. Shi M, Liu C, Cook TJ, Bullock KM, Zhao Y, Ginghina C, et al. Plasma exosomal α-synuclein is likely CNS-derived and increased in Parkinson's disease. Acta Neuropathol. 2014;128(5):639–50. https://doi.org/10.1007/s00401-014-1314-y.
96. Vallabhajosyula P, Korutla L, Habertheuer A, Yu M, Rostami S, Yuan C-X, et al. Tissue-specific exosome biomarkers for noninvasively monitoring immunologic rejection of transplanted tissue. J Clini Invest. 2017;127(4):1375–91. https://doi.org/10.1172/JCI87993.
97. Sahoo S, Losordo DW. Exosomes and cardiac repair after myocardial infarction. Circ Res. 2014;114(2):333–44. https://doi.org/10.1161/CIRCRESAHA.114.300639.
98. Merchant ML, Rood IM, Deegens JKJ, Klein JB. Isolation and characterization of urinary extracellular vesicles: implications for biomarker discovery. Nat Rev Nephrol. 2017;13(12):731–49. https://doi.org/10.1038/nrneph.2017.148.
99. Tran PHL, Xiang D, Tran TTD, Yin W, Zhang Y, Kong L, et al. Exosomes and nanoengineering: a match made for precision therapeutics. Adv Mater. 2019;32:e1904040. https://doi.org/10.1002/adma.201904040.

100. Zipkin M. Exosome redux. Nat Biotechnol. 2019;37(12):1395–400. https://doi.org/10.1038/s41587-019-0326-5.
101. Lai CP, Kim EY, Badr CE, Weissleder R, Mempel TR, Tannous BA, et al. Visualization and tracking of tumour extracellular vesicle delivery and RNA translation using multiplexed reporters. Nat Commun. 2015;6:7029. https://doi.org/10.1038/ncomms8029.
102. Xia Y, Chen T, Chen G, Weng Y, Zeng L, Liao Y, et al. A nature-inspired colorimetric and fluorescent dual-modal biosensor for exosomes detection. Talanta. 2020;214:120851. https://doi.org/10.1016/j.talanta.2020.120851.

Chapter 16
Extracellular Vesicles: Main Challenges in Industrialization

Lu Lu and Xiaohu Ge

Abstract The sources and types of extracellular vesicles (EVs) are diverse, and primary and applied research in the past decade has shown tremendous potential and value in fields such as skincare, nutrition, and drug delivery. The novelty and heterogeneity of extracellular vesicles, as well as their early stage of development, present numerous challenges in the development of products. These challenges include scientific questions about extracellular vesicles, such as what they are, how they are produced, and their joint and individual characteristics. There are also challenges in the product development process, such as how to utilize the specific features of extracellular vesicles to develop what types of products, what quality control standards these products should have, and what production processes should be used. Compliance challenges also arise, such as obtaining legal status for extracellular vesicle products for development and addressing regulatory concerns with scientific and safety data. Market challenges exist, such as determining which extracellular vesicle products can win the market and users' favor. The road ahead is challenging but full of hope.

Keywords Extracellular vesicles · Challenges · Industrialization

1 Introduction

1.1 What Are Extracellular Vesicles?

Extracellular vesicles (EV) are highly conserved small membrane-surrounded structures derived from cells, present in extracellular space. It can be produced by nearly all kinds of organisms, from lower prokaryotes to higher eukaryotes. Generally, EV contains a variety of bioactive molecules, including lipids, proteins,

L. Lu · X. Ge (✉)
Tingo Exosomes Technology Co. Ltd, Tianjin, China
e-mail: gexiaohu@tingocell.com

© The Author(s), under exclusive license to Springer Nature Singapore Pte Ltd. 2024
Z. Li (ed.), *Extracellular Vesicle: Biology and Translational Application*,
https://doi.org/10.1007/978-981-97-5536-3_16

nuclear acids, and other macromolecules, depending on its species and cell types. According to their biological origin, EVs are mainly divided into exosomes, microvesicles, and apoptotic bodies. Among them, exosomes, with diameters of 30–150 nm, are secreted by fusion between cell membranes and multiple vesicle bodies (MVB), originating from the MVB membrane's inward budding [1–3]. Microvesicles are small membranous structures released by various types of cells, ranging between 0.1 and 1 μm in diameter. They are typically formed by budding and shedding from the plasma membrane [4–7]. Apoptotic bodies are large vesicles, can range in size from 1 to 5 micrometers, and are formed when the plasma membrane of a dying cell buds outward, enclosing portions of the cytoplasm, organelles, and nuclear fragments [8–10].

Meanwhile, Gram-negative bacteria and some Gram-positive bacteria can also release outer membrane vesicles (OMVs) through budding from the outer membrane of the bacterial cell envelope. In addition, plants can also produce extracellular vesicles, which are not entirely the same as the plant-derived nanoparticles in the present studies [11–14].

In general, EVs are believed to play a vital role in cell-to-cell communication by transporting their components, such as proteins, lipids, and nucleic acids, between cells as genetic information and signaling molecules, allowing cells to coordinate their activities and respond to external stimuli.

1.2 How Do EVs Work?

Generally, as studies recently show, EVs can work in at least four main ways [7]: first, various surface proteins and other molecules outside of EVs can interact with specific receptors on recipient cells, triggering signaling cascades and inducing particular cellular responses, such as free ligands. Second, EVs can directly transfer biomolecules, such as proteins, lipids, and nucleic acids, from the donor to the recipient cells by combining the vesicle membrane with the recipient cell membrane, allowing their contents to be delivered to the recipient cell without targeting specific cells. Third, EVs can be engulfed by the recipient cell by forming a membrane-bound vesicle, an endosome, which can then fuse with other intracellular compartments, such as lysosomes, releasing their contents into the cell without targeting specific cells. Fourth, in addition to triggering the signaling cascades in the recipient cells, some specific ligands carried on the EVs outside can lead to the internalization of EVs, allowing for targeted cargo delivery to particular cell types or tissues so that the released cargo can function directly in the recipient cells.

Recent studies have shown that EVs can be distributed differently under different administration routes and function in various modes. Studies suggest that (1) EVs can simultaneously function on donor cells, nearby cells, and other cells distant from the donor; (2) EVs have constituted a complex communication system between single-cell organisms, cells in multicellular organisms, and even species crossing species under particular conditions, such as parasitic relationship, symbiotic relationship, and food chain relationship in nature [15].

2 The Practical Value of EVs

As evidence is growing, EVs, such as exosomes, have gained increasing attention in various potential applications, including food, cosmetics, therapeutics, drug delivery, and diagnostics, due to their unique properties and ability to transfer biological information between cells, even across the species [7, 15–23].

EVs are produced by the donor cells; their components must fall within the scope of the donor's genetic or epigenetic background. It means that healthy and diseased cells can load their components separately into the EVs in humans or animals. Therefore, there is the probability of discovering the health status by detecting the components of EVs in the biofluid, such as blood and urine. ExosomeDx, now a biotech brand, pioneered exosome-based in vitro liquid biopsy diagnostics. They have successfully provided the LDT service based on liquid biopsy of urine-derived EVs to patients in prostate cancer testing for a few years [21].

All the EVs act as a whole communication system in a particular living biological system. Studies show that specific cohorts from an exact origin or having similar characteristics may play the same role as a functional intersection. It indicates that the natural or artificial payloads on or in EVs may serve as active ingredients in specific application scenarios. Furthermore, the difference in distribution mode in topical or systematic administration indicates that EVs may work as a vehicle for particular natural or artificial active components in various application scenarios under specific administration routines. The applications of food, cosmetics, medical devices, and therapeutics are classified artificially. EVs can be developed and used in these fields based on a similar underlying logic that EVs can be used in nature or after modification under all mentioned conditions.

Take milk exosomes as an example; the unmodified ones can directly function in moisturizing and anti-wrinkle by topical application and may promote digestive tract barrier by oral administration, while the modified ones can deliver cosmetic peptide crossing the skin barrier to improve the skin condition (data in-house) or deliver mRNA of immunogenetic peptide crossing the intestinal barrier to induce an immune response [22, 23]. Recently, more and more studies have confirmed the proof-of-concept of EV application. EVs from mesenchymal stem cells can promote wound healing and inhibit the inflammatory response. Bovine milk-derived exosomes loaded with drugs such as paclitaxel can efficiently kill the tumor cell. Exosomes conjugated with the monoclonal antibody anti-VEGFR also play an antiangiogenic role. As mentioned above, it may be the primary outlet of applications for EVs. So far, some therapeutics based on EVs from different cell lines have been in clinical trials, including unmodified stem cell-derived exosomes or engineered exosomes from 293 cells loaded with small molecules, proteins, and ASO. Meanwhile, Wanil Kim, ect. discovered that EVs derived from microorganisms, OMV, can also improve skin conditions by immune regulation. To some extent, the global enthusiasm for ongoing basic and applied research on EVs is a genuine reflection of the academic and industrial sectors' high expectations and aspirations for the broad applications of EVs, such as exosomes.

3 The Challenges EVs Faced in Industrialization

Although the academic community's enthusiasm for research on EVs, especially exosomes, remains high, and many efforts have been made to advance their practical applications, many challenges remain on the road to the industrialization and commercialization of EV-based products. Despite the promising applications of EVs, however, the entire scientific understanding of EVs is still on the way. Regardless of whether exosomes are used as raw materials for cosmetics, food, medical devices, or therapeutics, it is essential to understand their key features fully. A sufficient understanding can lead to adequate risk assessment in product lifecycle management, including but not limited to concept design, product design, product development, product manufacturing, product promotion, and sales. For manufacturers and regulatory agencies alike, the more understanding we have about EVs, the more advantageous it is, as it will be more controllable regarding product effectiveness and safety. In product lifecycle management, successfully developing a product that meets regulatory requirements is just the beginning. The value of EV products is demonstrated through successful sales, and the primary essential requirement to gain market recognition is to meet unmet market needs. The concept of starting with the end in mind and a thorough understanding of EVs are indispensable to achieving this final goal. Next, we will provide a preliminary overview of the potential challenges that the industrialization and commercialization of EV-based products may face [24].

3.1 The Challenges in Understanding EVs

All knowledge on EVs and their biogenesis, transport, cellular uptake, and functions is the foundation for practical applications of EVs. The great challenge for the academic and industrial community in practice may be the heterogeneity of EVs [7].

3.1.1 Heterogeneity Between Subtypes

Recent studies have shown that different subtypes of EVs have definitions based on their subcellular origins and biogenesis mechanisms. According to the description, all types of EVs are distributed in a wide range in sizes, and there is overlap between different subtypes. After that, they cannot be distinguished only based on a single characteristic, such as the size distribution, density distribution, or individual molecule marker. More attributes in a chain must be a proper strategy to identify EVs for specific usage.

3.1.2 Heterogeneity Between Sources

Different cells and single-cell organisms can release EVs into the same extracellular space, and single cells can simultaneously release different subtypes of EVs into the intercellular space. Therefore, EVs are distributed in a mixture under all biological conditions. It is not easy to obtain single-source EVs from such a heterogeneous mixture, such as bodily fluids, before knowing their unique characteristics or surface markers.

3.1.3 Heterogeneity in the Subtype and Among Individuals

The minimal unit of each EV cohort is a single nanoparticle. They may differ, at least in size and components, based on limited evidence because we have no efficient and reliable tools to explore the speculation mentioned above. A powerful tool for single EV research must be the first necessary driving force to explore the features further. Just as in the history of immune cell research, scientists initially discovered that immune cells could be divided into B and T cells. Further exploration based on differences in surface markers led to the identification of various subtypes of T cells. The study of EVs, their classification and functional activities will also inevitably undergo a gradual and in-depth process, similar to the history of immune cell research.

3.2 The Challenges on Research Tools of a Single EV

Heterogeneity is one of the fundamental aspects of our current understanding of extracellular vesicles. Therefore, to further advance our knowledge of EVs, it is inevitable that we will explore the direction of studying individual EVs. This is one of the necessary trends, as it contributes to a better understanding and utilization of EV heterogeneity.

Although there are currently existing tools or researchers that have developed some novel tools for analyzing single EVs, they still have their limitations, so they have not yet formed a driving force to advance the research of single EVs analysis rapidly. For example, transmission electron microscopy (TEM)-based single EV analysis, similar to other morphological analysis methods, is mainly used for qualitative analysis of EVs. However, TEM-based quantitative analysis is time-consuming, laborious, and inefficient with low throughput. Combined with more powerful artificial intelligence-based image analysis tools, it is possible to characterize multiple features such as morphology assessment, particle size distribution, and qualitative and quantitative analysis of markers. Nanoparticle analysis techniques based on Nanoparticle Tracking Analysis, light scattering, or Coulter's Principle can rapidly analyze size distribution and quantification. Still, they all face the problem of accuracy and precision, mainly when used for quality control. Researchers can detect only a few known protein markers, in addition to RNA, of individual nanoparticles. Their analysis accuracy may need to be more traceable, and particle counting precision needs to be improved. Instead, the omics information of the individual EV is the correct key to exploring the heterogeneity of EVs. Tao Luo et al. developed a single exosome mRNA analysis method based on high-throughput sequencing [25]. However, we still need more.

These shortcomings will gradually improve and eventually mature as relevant research continues to delve deeper and accumulate.

3.3 The Challenges in the Development Process of a Product Based on EVs

Generally, a product development process based on EVs has key steps, including requirement analysis, product design, marketing strategy plans, business analysis, product development, marketing, and commercialization. The main challenges of EV-based products are demand analysis, product design, and product development.

3.3.1 The Challenges in Product Design

The application of EVs is also characterized by significant heterogeneity. On the one hand, there is diversity in the mode of application, as EV-based products can be developed on their activity or nanomaterial properties with modification. On the other hand, there is diversity in the fields of application, as EVs can be used for cosmetics, food, and drugs or as vehicles for drug delivery carriers. Therefore, the primary task of requirement analysis is to determine what type of product to develop partially based on understanding and knowledge about specific sources of EVs. In the face of numerous choices, it is important to be pragmatic, understand trade-offs, and seize the right application areas.

Novel technologies may imply technological leadership, but also high risk in development and high cost in the early stage of application, which do not necessarily mean high technical barriers for a product. The obstacle lies in the high-tech that meets market demand, which constitutes the product's barriers. The developers of the EV-based product need to make every effort to establish a high degree of matching between product and market demand instead of being paranoid about technology. Even if EVs originate from the same source, such as milk EVs used for cosmetics or drug delivery, products must be designed with targeted specifications, formulation forms, purity, and content packaging to cater to different scenarios, further influencing the EVs' preparation process.

3.3.2 The Challenges in Product Development

The Challenges in the Application of Quality by Design(QbD)

QbD is a systematic, structured, science-based, and risk-based pharmaceutical development approach that incorporates quality and customer requirements into the design of products. It involves continuously identifying, evaluating, and understanding the critical sources of process variability and drug variability throughout the prescription/process development, pilot/ process scale-up, technology transfer, registration/validation batch production, commercial production, and lifecycle management processes. Controlling the production process ensures product quality in terms of safety and efficacy while exhibiting good process stability.

Starting with the end in mind as quality by design, we should first define the desired quality attributes of the EV product as a quality target product profile (QTPP), considering the needs of customers and consideration of the supervisory authority. Next, critical quality attributes (CQA) of EV products are delimited by identifying the critical product characteristics for ensuring safety and efficacy based on QTPP. Exploring the critical characteristics of the EV product is fundamental, followed by establishing quality standards. However, knowledge of EVs is blooming, but it still needs to be improved by learning more about EVs.

The Challenges in the Production of EVs

All EVs released into intercellular space originate from living or dying cells. Differ from cells, they stay in a relatively static state once released, while cells constantly change. However, the production of EVs is influenced by cells and is a relatively changing process, which can result in temporal heterogeneity of EVs. Temporal heterogeneity is a phenomenon encountered in the production of other recombinant biological products. However, the situation is even more complex for EVs due to the intricate attributes that make them complex subcellular structures of many biological molecules instead of one. EVs are generated by cells, the sources of which are diverse and complex, resulting in naturally varying EVs. Engineered cell lines, including eukaryotic and prokaryotic cells, can potentially minimize the heterogeneity of products within the population as other recombinant instead of eliminating them. Meanwhile, EVs derived from milk or plants still have a long way to go to achieve homogeneity of origination because of the heterogeneity from individuals of animal or plant populations and cells inside.

The Challenges in the Purification of EVs

This challenge differs in different application scenarios. For IVD, EVs serve as the vehicle of analytical substances, including protection against degradation; separating them from biofluid in the clinical laboratory may be unnecessary if the concentration is high enough in the EVs of the sample, and the sample has no interference, a specific test can be done directly against the substance in EVs; or if pretreatment, not aimed at EVs, is necessary, maximizing the recovery of EVs, instead of the purity is the critical point. However, enrichment may be required in some situations due to substances with low concentration of the analyzed in EVs, such as urine. The balance between substance concentration in EVs and EV recovery following pretreatment should be considered a priority. More recovery of EVs with the analyzed substance indicates less information loss. A pretreatment process that is low cost, automation, and high throughput is needed due to the requirement of work practice in the clinical laboratory. Most importantly, the clinical laboratory and doctors need valuable test items instead of methods.

In contrast, the purity of EVs is more important in industrial production, especially therapeutics. Stable large-scale production is the greatest challenge in the industrial segment, although capacity is currently in the hands of individual companies.

The available laboratory isolation methods for EVs include differential centrifugation (DC) and density-gradient centrifugation (DGC), besides size exclusion chromatography (SEC), and tangential flow systems (TFFs). The first two are unsuitable for industrial production as they are un-compatible with linear scaling-up of production capacity because there is no equipment with a larger scale. In addition, DC produces the EV product with low purity (around 40–60%) instead of DGC, making high purity (>98%) of EVs, based on UPLC-SEC (in-house data), while both methods are labor- and time-consumption and carry relatively high compliance risks in the manual operation process. An alternative method must be developed to remove the corresponding obstacles and stabilize and scale up the industrial development process.

SEC and TFF are part of traditional biopharma and vaccine industry processes and have been validated in recombination protein and virus production for decades. However, they are fresh for EV production, and they cannot achieve the goals of EV production separately. At the same time, the clinical trials of the Codiak company and our work confirm the availability of processes based on chromatography in addition to SEC. However, our in-house work indicates no universal process for all EVs. It is constant with the intrinsic heterogeneity of EVs, various sources with different contamination, and other EVs. The right way to solve the problems in the commercialization production of EVs as soon as possible is to integrate mature biopharma processes, such as chromatography and filtration, on a case-by-case basis at the present stage.

The Challenges in Virus Clearance

The production of EVs, like other biologics, typically involves using microbial or human/animal-derived cells, tissues, and body fluids as starting materials, which may carry the risk of viral contamination, in addition to accidental introduction during the production process. Regulations commonly require biologics to include at least one or multiple steps of the production process that can clear viruses by inactivating/removing them. So, when EVs are used in some ways, such as drug delivery systems administered in the IV routine, they should also be treated like other biologics. However, no specific regulations or laws currently target them. Standard virus elimination processes include using the Pasteurization method, dry heat method, organic solvent/decontaminant (S/D) treatment method, membrane filtration method, low pH incubation method, chromatography method, etc. However, they are not suitable for various reasons. Because viruses have a spatial size to EVs, membrane filtration and chromatography methods, including size exclusion, are unsuitable for removing viruses from EVs. EVs cannot withstand the membrane disruption caused by virus inactivation methods, including S/D treatment, nor low pH incubation. In addition, other heat-based virus inactivation methods can also cause significant damage to the structure and activity of biological molecules contained in EVs, such as 72 h of dry heat at 80 °C, 30 min of dry heat at 100 °C, and 10 h of wet heat at 65 ° C using the Pasteurization method. This is especially important for applications that require the biological activity of EVs themselves, such as the use of EVs derived from mesenchymal stem cells for tissue repair. So, EVs need a suitable virus clearance method under some conditions and it should explored in the present.

3.3.3 The Challenges in the Quality Control System

The quality control system is a structured framework implemented by organizations to ensure the consistency, reliability, and conformity of EV products. It encompasses various activities and methodologies, including establishing quality standards, quality planning, quality assurance, quality control, corrective actions, and continuous improvement to monitor and verify the quality of processes, materials, and end products, with the aim of meeting or exceeding the expectations of products based on EVs. The most significant challenges come mainly from establishing quality standards, which include determining the content of quality research, conducting methodological research, determining the scope and limits of quality standards, and developing and revising quality standards. These processes are closely related and support each other.

The Challenges in Determining the Content of Quality Research

The study of EV product quality is the foundation of establishing quality standards. Quality research content should be as comprehensive as possible, taking into account both general requirements and specific considerations. EVs, including exosomes, are derived from a complex population with significant interindividual differences and high heterogeneity within populations. The characteristics of the product, the preparation process used, and the stability of the research results should be fully considered in the development of the product to ensure that the content of quality research can fully reflect the product's characteristics and the changes in quality. Products generally need to consider their structural characteristics, physicochemical properties, characteristics of different formulations and product forms, application areas, interactions between various components in combination formulations, and the impact of excipients on the effectiveness and safety of the product. There is still limited understanding of the physical, chemical, and biological properties of EVs themselves and their relationship with efficiency, which, to some extent, has hindered the progress of related research.

The Challenges of Methodology Research

As mentioned above, EVs exhibit significant heterogeneity in size, cargo, and surface markers. Developing standardized methodologies to accurately characterize and quantify different subpopulations of EVs accurately is challenging. The research items and the experimental methods are typically chosen on the research objectives. Selection is generally based on literature, theories, or existing experiments. Conventional projects usually require consideration of the applicability of methods. The same project for different types of products may have additional permitted or required test methods. For example, detecting Chlamydia may involve chemical detection methods or quantitative fluorescent PCR methods. The specificity of the method is significant for identification purposes, such as morphological and biomarker identification. For inspection purposes, the focus is on specificity, sensitivity, and accuracy, such as purity and impurities in the product.

Today, the commonly used techniques to analyze EVs can generally be divided into two categories. The first category is bulk analysis techniques, such as liquid chromatography mass spectrometry, gas phase analysis, fluorescence quantitative PCR, and ELISA. These techniques are mature and widely used, but they can only differentiate heterogeneity between populations of EVs and cannot reflect heterogeneity within EV populations. The second category is single nanoparticle analysis techniques. Some of these techniques are mature, but have expensive equipment costs, making them inconvenient for quantitative detection, such as electron microscopy. Others, such as NTA, still need improvement in technical performance or product use. On the one hand, there is still room for improvement in the accuracy of the analysis. However, there is a lack of corresponding traceable standards for particle quantification, similar to flow cytometry, to ensure the accuracy of the analysis.

3.3.4 The Challenges of Logistics

Logistics is a part of enterprise supply chain activities that involve planning, implementing, and controlling efficient and cost-effective flow and storage of EV-based products from the source to the consumer to meet customer needs. Logistics centers on warehousing and transportation, and the process is heavily influenced by the stability of EV products.

Like most bioproducts, extracellular vesicles are generally considered relatively unstable. Laboratories typically use low-temperature and ultra-low-temperature storage for long-term preservation, which poses a significant challenge in developing EV products. The transportation and storage costs are high and inconvenient, hindering the widespread application of these products. In response to the limitations of low-temperature preservation, some researchers from academic institutions and companies have started to explore the development of freeze-drying processes for EVs, intending to achieve refrigerated or even room-temperature storage and transportation of EV products. These research findings indicate that EVs can be freeze-dried. However, more evaluation of the physicochemical and biological activities of freeze-dried EVs is still needed, and, more importantly, comprehensive stability studies are required to provide thorough confirmation. Even if the freeze-drying technique has been thoroughly proven to address EV product preservation and transportation issues effectively, it still faces other challenges, such as the high cost of industrial-scale freeze-drying, which also increases the difficulty of promoting EV products. In this case, possible further solutions may need to be considered from the following aspects: first, adopting other drying techniques with lower costs; second, preserving the product in liquid form and adding stabilizers to enhance the resistance of extracellular vesicles to factors such as temperature, thereby improving stability without using drying techniques; and third, strengthening the stability of EVs in liquid form through artificial modification. However, these strategies are primarily in the conceptual or early development stage and still have significant uncertainties.

3.3.5 The Challenges to the Safety of EVs

The complete characterization of EVs' contents and origination is a critical foundation to knowing their safety, and this is also significantly complex due to the heterogeneity and complexity of EVs, their originations, the biogenesis process of EVs, application scenarios, and the routine of administration. Due to the complexity, there may not be a complete unified conclusion indicating whether EVs are safe. The safety concerns of EV products in the present and future must be personalized, at least partially. Regarding the safety of EVs, two major aspects should be of significant concern based on the origin of the safety issue: endogenous and exogenous.

Firstly, the endogenous safety issues of EVs themselves come from their biological, physical, and chemical characteristics, which can only be gradually revealed as the research on them progresses. The natural variability of EVs, as an original and temporal heterogeneity, introduces uncertainties in their safety profile. These uncertainties include the loading of harmful or pathogenic payloads in EVs and the uncontrollable and unstable nature of their cargo and composition among different batches of EVs during production. In addition, unintended interaction between the non- or modified EVs and cells or tissues may cause improper effects or tissue damage. A safety assessment test is the most direct way to solve these problems. The biggest challenge is the enormous cost of conducting safety assessments, and it is not feasible for a single safety assessment test to address all safety concerns of regulatory agencies in different product sectors. Although the underlying technological logic may be similar, regulatory agencies for food, cosmetics, pharmaceuticals, and medical devices have independent and other compliance requirements regarding safety assessment items and methods.

Secondly, exogenous safety issues are generated during the entire process, from raw materials to the production and preparation of EVs. The biggest exogenous safety issue is the introduction of contamination, such as pathogenic bacteria and viruses, during the production or processing of raw materials.

The solution to this problem requires strict control over the source materials and the production processes. Taking EV products derived from bovine milk as an example, it is necessary to strictly manage the health of dairy cows, control the breeding environment, avoid the use of sick cows, and establish reasonable milk quality standards. Then strict management of the production process of milk-derived exosomes is required to ensure a clean and qualified production environment, prevent any introduction during the production process, and improve the quality control of milk EV products. Expensive sterile production processes can produce sterile EV products, or terminal sterilization processes can be used. However, using a sterilizing filter with a pore size of 0.22 μm can significantly decrease the number of EVs. Additionally, irradiation sterilization processes can partially break the chemical bonds of proteins, nucleic acids, and lipids in EVs. The choice of sterilization method must be carefully considered and selected based on a comprehensive evaluation. The great challenge in the future is that risks may become increasingly complex as the industrialization of EVs advances, and the EV sources will expand, which may bring about many unexpected situations.

3.3.6 The Challenges in the Field of Regulations

EVs have enormous potential for application development in various fields. Depending on their intended use and mode of action, EV products may fall within the regulatory framework of biologics, cosmetics, food, medical devices, and drugs. These fields are directly related to the fundamental health and safety of the general public, making them critical areas of regulation in countries around the world. The formulation and implementation of relevant laws and regulations aims to protect the legitimate rights and interests of the public, preventing health damage and significant economic losses caused by safety issues with related products. The development of EV products is inevitably subject to strict regulation by the relevant laws and regulations separately and faces several challenges in terms of legal rules. EVs are new biological materials with unique characteristics and complexities, and there is still a lot of uncertainty about their scientific and safety.

One of the main challenges is that these products need specific rules and guidelines. It is not possible to establish universal, targeted laws, regulations, and technical guidelines for several reasons: first, the significant differences in the physicochemical and biological properties caused by variations in the origin of EVs; second, the enormous differences in regulatory laws, regulations, and technical requirements at the underlying standards in different fields, such as food and drugs; third, the variations in regulatory logic and standards for the same field in other countries and regions; fourth, The drafting, formulation, and promulgation of laws and regulations is a relatively slow process that lags behind the development of technology and market demands. Taking milk exosomes as an example, based on their biological activity, milk exosomes can be used for skin care. However, in China, cosmetic ingredients are strictly limited to a detailed list of approved ingredients published by regulatory authorities, and milk exosomes are defined as one of the legally permissible ingredients for use rather than as a separate novel entity. On the other hand, in addition to specific-purpose ingredients, there are no specific restrictions on the use of cosmetic ingredients in the USA. Another challenge comes from the variability and heterogeneity of EVs and EV products. They can be derived from different origins, have varying compositions, exhibit distinct functions, and be used by various routines for other purposes. This variability makes it difficult to establish standardized testing methods and quality control criteria for regulatory purposes. Additionally, EV products' safety and efficacy assessment is a complex task. These products interact with various biological systems for different purposes and can have diverse effects. The regulatory agencies are most concerned about safety issues. Designing appropriate preclinical and clinical studies to evaluate their safety and efficacy is challenging, especially considering potential long-term and off-target effects. Lastly, the international harmonization of regulations for EV products is another challenge. Different countries and regions may have other regulatory requirements and approval processes, leading to delays and additional costs for developers who wish to market their products globally.

In general, the development of EV products requires a comprehensive and harmonized regulatory framework that addresses their unique characteristics and provides clear guidelines for safety and efficacy assessments.

3.3.7 The Challenges of the Market

The market challenge for EV products lies in customer awareness and understanding. Globally, strict regulations and high barriers to entry in the pharmaceutical industry result in a long development cycle and high investment requirements. As a result, many companies and individuals interested in extracellular vesicle products, particularly in beauty and cosmetic medicine, have entered the market with lower barriers. This has led to the proliferation of false concepts, market confusion, and a mix of good and bad products. This may raise concerns among regulatory agencies, leading to tighter regulations. Customers' understanding and perception of extracellular vesicle products may be intentionally or unintentionally misled or distorted. All stakeholders must address these antagonistic sentiments and provide accurate information.

The value and advantages of EVs must be effectively communicated as a novel and complex material. This requires active collaboration with medical professionals, industry associations, and specialized media. Furthermore, the fundamental point for a product to succeed in the market is its effectiveness, safety, and ability to meet market demands. A product must meet market needs to have value. The efficacy and safety can only be supported by sufficient technical accumulation, investment in funding, and time costs. Once extracellular vesicles transition from the laboratory stage to the market stage, the technological and production costs of the product become another essential basis for product promotion. Excessive costs will hinder rapid and large-scale promotion. Extracellular vesicles derived from bovine milk have a natural advantage compared to those derived from mammalian cells. Costs also directly impact the price of the final product, which further affects customer acceptance or market size. While ordinary consumers may not be susceptible to product compliance, the industry's healthy development must focus on product compliance. Only compliant products will have more robust vitality and longer life cycles.

4 Looking Forward to the Future

Despite facing numerous challenges on the way to industrialization, EVs have a promising future due to their broad scalability regarding raw materials and downstream applications. It is just a matter of time. The key to advancing the field lies in more profound research and understanding, mature transformation techniques, advanced analytical methods, efficient production processes, and improved regulatory systems. These shared goals will drive industrialization progress, and more attention and effective investment from the academy, industry, and market are needed.

References

1. Mateescu B, Kowal EJ, van Balkom BW, Bartel S, Bhattacharyya SN, Buzás EI, et al. Obstacles and opportunities in the functional analysis of extracellular vesicle RNA—an ISEV position paper. J Extracell Vesicles. 2017;6(1):1286095. https://doi.org/10.1080/20013078.2017.1286095.
2. Lener T, Gimona M, Aigner L, Börger V, Buzas E, Camussi G, et al. Applying extracellular vesicles based therapeutics in clinical trials—an ISEV position paper. J Extracell Vesicles. 2015;4:30087. https://doi.org/10.3402/jev.v4.30087.

3. Théry C, Witwer KW, Aikawa E, Alcaraz MJ, Anderson JD, Andriantsitohaina R, et al. Minimal information for studies of extracellular vesicles 2018 (MISEV2018): a position statement of the International Society for Extracellular Vesicles and update of the MISEV2014 guidelines. J Extracell Vesicles. 2018;7(1):1535750. https://doi.org/10.1080/20013078.2018.1535750.
4. Ratajczak MZ, Ratajczak J. Extracellular microvesicles/exosomes: discovery, disbelief, acceptance, and the future? Leukemia. 2020;34(12):3126–35. https://doi.org/10.1038/s41375-020-01041-z.
5. Clancy JW, Schmidtmann M, D'Souza-Schorey C. The ins and outs of microvesicles. FASEB Bioadv. 2021;3(6):399–406. https://doi.org/10.1096/fba.2020-00127.
6. Lei Q, Liu T, Gao F, Xie H, Sun L, Zhao A, et al. Microvesicles as potential biomarkers for the identification of senescence in human mesenchymal stem cells. Theranostics. 2017;7(10):2673–89. https://doi.org/10.7150/thno.18915.
7. Yáñez-Mó M, Siljander PR, Andreu Z, Zavec AB, Borràs FE, Buzas EI, et al. Biological properties of extracellular vesicles and their physiological functions. J Extracell Vesicles. 2015;4:27066. https://doi.org/10.3402/jev.v4.27066.
8. Santavanond JP, Rutter SF, Atkin-Smith GK, Poon IKH. Apoptotic bodies: mechanism of formation, isolation and functional relevance. Subcell Biochem. 2021;97:61–88. https://doi.org/10.1007/978-3-030-67171-6_4.
9. Ravichandran KS. Find-me and eat-me signals in apoptotic cell clearance: progress and conundrums. J Exp Med. 2010;207(9):1807–17. https://doi.org/10.1084/jem.20101157.
10. Nagata S, Segawa K. Sensing and clearance of apoptotic cells. Curr Opin Immunol. 2021;68:1–8. https://doi.org/10.1016/j.coi.2020.07.007.
11. Toyofuku M, Schild S, Kaparakis-Liaskos M, Eberl L. Composition and functions of bacterial membrane vesicles. Nat Rev Microbiol. 2023;21(7):415–30. https://doi.org/10.1038/s41579-023-00875-5.
12. Sartorio MG, Pardue EJ, Feldman MF, Haurat MF. Bacterial outer membrane vesicles: from discovery to applications. Annu Rev Microbiol. 2021;75:609–30. https://doi.org/10.1146/annurev-micro-052821-031444.
13. Kulp A, Kuehn MJ. Biological functions and biogenesis of secreted bacterial outer membrane vesicles. Annu Rev Microbiol. 2010;64:163–84. https://doi.org/10.1146/annurev.micro.091208.073413.
14. Schwechheimer C, Kuehn MJ. Outer-membrane vesicles from Gram-negative bacteria: biogenesis and functions. Nat Rev Microbiol. 2015;13(10):605–19. https://doi.org/10.1038/nrmicro3525.
15. Kalluri R, LeBleu VS. The biology, function, and biomedical applications of exosomes. Science. 2020;367(6478):eaau6977. https://doi.org/10.1126/science.aau6977.
16. György B, Szabó TG, Turiák L, Wright M, Herczeg P, Lédeczi Z, et al. Improved flow cytometric assessment reveals distinct microvesicle (cell-derived microparticle) signatures in joint diseases. PLoS One. 2012;7(11):e49726. https://doi.org/10.1371/journal.pone.0049726.
17. Yong T, Wei Z, Gan L, Yang X. Extracellular-vesicle-based drug delivery systems for enhanced antitumor therapies through modulating the cancer-immunity cycle. Adv Mater. 2022;34(52):e2201054. https://doi.org/10.1002/adma.202201054.
18. Herrmann IK, Wood MJA, Fuhrmann G. Extracellular vesicles as a next-generation drug delivery platform. Nat Nanotechnol. 2021;16(7):748–59. https://doi.org/10.1038/s41565-021-00931-2.
19. Elsharkasy OM, Nordin JZ, Hagey DW, de Jong OG, Schiffelers RM, Andaloussi SE, et al. Extracellular vesicles as drug delivery systems: why and how? Adv Drug Deliv Rev. 2020;159:332–43. https://doi.org/10.1016/j.addr.2020.04.004.
20. Liang Y, Duan L, Lu J, Xia J. Engineering exosomes for targeted drug delivery. Theranostics. 2021;11(7):3183–95. https://doi.org/10.7150/thno.52570.
21. Witwer KW, Buzás EI, Bemis LT, Bora A, Lässer C, Lötvall J, et al. Standardization of sample collection, isolation and analysis methods in extracellular vesicle research. J Extracell Vesicles. 2013:2. https://doi.org/10.3402/jev.v2i0.20360.

22. Zhang Q, Wang M, Han C, Wen Z, Meng X, Qi D, et al. Intraduodenal delivery of exosome-loaded SARS-CoV-2 RBD mRNA induces a neutralizing antibody response in mice. Vaccines. 2023;11(3):673. https://doi.org/10.3390/vaccines11030673.
23. Lu L, Bai W, Wang M, Han C, Du H, Wang N, et al. Novel roles of bovine milk-derived exosomes in skin antiaging. J Cosmet Dermatol. 2024;23(4):1374–85. https://doi.org/10.1111/jocd.16112.
24. Kimiz-Gebologlu I, Oncel SS. Exosomes: large-scale production, isolation, drug loading efficiency, and biodistribution and uptake. J Control Release. 2022;347:533–43. https://doi.org/10.1016/j.jconrel.2022.05.027.
25. Luo T, Chen SY, Qiu ZX, Miao YR, Ding Y, Pan XY, et al. Transcriptomic features in a single extracellular vesicle via single-cell RNA sequencing. Small Methods. 2022;6(11):e2200881. https://doi.org/10.1002/smtd.202200881.

SPRINGER NATURE

GPSR Compliance

The European Union's (EU) General Product Safety Regulation (GPSR) is a set of rules that requires consumer products to be safe and our obligations to ensure this.

If you have any concerns about our products, you can contact us on ProductSafety@springernature.com

In case Publisher is established outside the EU, the EU authorized representative is:

Springer Nature Customer Service Center GmbH
Europaplatz 3
69115 Heidelberg, Germany

The manufacturer's authorised representative in the EU is Springer Nature Customer Service Centre GmbH, Europaplatz 3, 69115 Heidelberg, Germany. If you have any concerns regarding our products, please contact ProductSafety@springernature.com

Printed and bound by CPI Group (UK) Ltd, Croydon, CR0 4YY
01/10/2026
02225910-0014